Advanced Series in Agricultural Sciences 5

Joe J. Hanan · Winfred D. Holley
Kenneth L. Goldsberry

Greenhouse Management

With 283 Figures

Springer-Verlag
Berlin Heidelberg New York 1978

Professor Dr. Joe J. Hanan
Professor Dr. Winfred D. Holley
Professor Dr. Kenneth L. Goldsberry

Colorado State University, Department of Horticulture
Fort Collins, CO 80523, USA

ISBN 3-540-08478-9 Springer-Verlag Berlin · Heidelberg · New York
ISBN 0-387-08478-9 Springer-Verlag New York · Heidelberg · Berlin

Library of Congress Cataloging in Publication Data. Hanan, Joe J., 1931–. Greenhouse management. (Advanced series in agricultural sciences; 5) 1. Greenhouse management. I. Holley, Winfred Davis, 1914– joint author. II. Goldsberry, Kenneth, L., 1932– joint author. III. Title. IV. Series. SB415.H34 631.5'4 77-21211.

Typesetting, printing and binding: Brühlsche Universitätsdruckerei, Lahn-Gießen.
2131/3130–543210

Preface

The change in greenhouse operation and technology in the last 20 years has been unprecedented. Photoperiodic control, mist propagation, greenhouse cooling, clean stock programs, CO_2 injection, to name a few, have all been inaugurated as regular greenhouse practices in this time. The introduction of new markets, new production centers, shifts in public attitudes, and the realization that greenhouse production is not simply growing crops, but the management of an enterprise in which people work, have combined to make this agricultural practice a challenging and rewarding vocation. The greenhouse grower, manager, and student who are training for this vocation have not had an up-to-date text book for many years. It has been our goal to bring both published and unpublished work together in this book, and to provide a bench mark from which we can continue to move forward.

It is not until a process of writing a text begins that one fully realizes how far we have come—and where we need to go. It is with some sadness that we realize that this book is not likely to remain long as an expression of the state-of-the-art. We do not expect it to be easy reading; for new terms, new technology, and new ways of doing things are not always easy. We have tried to develop technical subjects in such a way as to provide a basis to show the reason that we do some things in certain ways in greenhouse practice. We accept the blame if we have not been able to accomplish this objective to the satisfaction of the majority of our readers. We will undoubtedly be accused of being too shallow in some places, and too deep in others, and that is a cross one must bear with resignation.

American readers will find that we have attempted to use metric units as much as possible. In some instances, limitations have forced us to use English units so that the book may appear to be a hodge-podge of units, none making much sense. Perhaps this problem will help speed the adoption of the metric system in the United States greenhouse industry. To reduce the pain to an acceptable degree, we have tried to include a reasonably comprehensive set of conversion tables in the appendices.

When it comes to acknowledgments, we think the first to be thanked is an industry that has supported the three of us at Colorado State University in such a fashion that we were asked to write this book. In particular, the Colorado Flower Growers' Association has had a supportive relationship with the University that is probably unmatched in the world. Our jobs would not be here without their constant interest and help. With their innovative leadership to set the pattern, they have been joined in recent years by the Colorado Rose Committee and the

Colorado Bedding and Pot Plant Association. If nothing else, their often cogent, and sometimes embarrassing, questions serve to keep erudite professors' feet on the ground.

We would also like to acknowledge the following: the assistance and extensive use of unpublished work by Dr. John L. McKeever, Professor of Management, Colorado State University. We have also drawn freely from both published and unpublished work and thoughts of Dr. George Kress, Professor of Marketing at Colorado State University. We have long enjoyed an association with these two fine professors in mutual projects carried on with the assistance of the Colorado Flower Growers' Association. The latest Dutch marketing information was obtained with the help of Jan Hilverda, Jr., Aalsmeer, Holland. Dr. David E. Hartley, Departement of Horticulture, Colorado State University, provided most of the photographs in Chapter 10. Dr. A. M. Kofranek, University of California, Davis, also provided illustrative material for Chapter 10. Dr. Marlin N. Rogers, University of Missouri, Columbia, Dr. O. C. Taylor, University of California, Riverside, and Dr. H. E. Heggestad, U.S.D.A., Beltsville, Maryland, provided a number of pictures for Chapter 8.

Dr. J. J. Hesling, GCRI, England, and Dr. R. Baker, Department of Botany and Plant Pathology, Colorado State University, provided material for Chapter 9. Dr. Baker has contributed much—although he did not always realize it. It was with some trepidation that we included for the first time in a greenhouse management text a rather extensive chapter on disease and pest control. Arne Beisland, Agricultural College, Alnarp, Sweden, contributed with translations used in Chapter 2. Alvi O. Voigt, Pennsylvania State University, University Park, contributed both material and thoughts on Chapter 1.

Recognition is also given to the many individuals in the greenhouse manufacturing and construction business including: William Gunesch, Park Floral Company; Ray Doherty, Winandy Greenhouse Construction Company; Douglas McDougall and Harold Gray of Lord and Burnham Division; Donald E. McCrimmon, Nexus Corporation; Cliff Rough, Rough Brothers Greenhouses and Secretary for the National Greenhouse Manufacturers Association; Gary Schultz, Ikes-Braun Glasshouses; Robert Akins, Research Assistant and Dr. Jack Cermack of the Colorado State University Wind Engineering Program; and Acme Engineering Company, all of whom assisted in Chapters 3 and 4.

In the process of writing, we have gone through a number of typists. However, Judy Croissant, Department of Horticulture, Colorado State University, started it all with her help on the first two chapters, and Nancy Ballast and Becky Comfort, Department of Horticulture, got most of it dumped on their desks in the last-minute struggle to finish the text. We appreciate their competent typing, as well as the willingness of Dr. K. M. Brink, chairman, Department of Horticulture, to allow us to usurp secretarial time in a tight budget year. We also appreciated his support and encouragement, and salute him as one of the best investments this University ever made in terms of an administrative leader. Julia W. Hanan also contributed by undertaking the laborious process of assisting her husband in proofreading the galleys and final page proofs.

Finally, we dedicate this effort to those people, our friends in the industry and in the research institutions, with the help of whose efforts we have come this far—and upon whose efforts we will continue to move forward.

Fort Collins, January 1978 The Authors

Contents

Chapter 1 Introduction

1.1 The Industry

Diversity characterizes the greenhouse industry. It is diverse in size, in structures, in crops grown, in methods utilized, and in opportunities. This diversity is a strength, but perhaps its greatest weakness. Greenhouses are a means to overcome climatic adversity. They are, nevertheless, directly influenced by climate, by topography, by the technological status of the civilization about them, and by their history. As humans manage them, their appearance and operation reflect human attributes—for better or for worse. As with any human endeavor, a greenhouse cannot be considered separately from the social and political melieu in which it is operated and managed. Historically, our training schools and professional education methods have tended to emphasize greenhouse management, the growth and production of crops, to the exclusion of human factors and their interrelationships. Regardless of how it may appear in this book, there are several roads one may choose. The choices will depend upon the individual, his wants, his desires, the opportunities available to him. Certainly a greenhouse industry, as might conceivably be found in regions still using bullocks as prime movers (Fig. 1-1), will not have recognizable similarities to the same industry in a civilization capable of technology represented by Figure 1-2.

Fig. 1-1. Undeveloped technologies are not likely to support a viable greenhouse industry. Low income level, and lack of trained personnel with developed technology, prevents demand for greenhouse products and ability to fulfill any demand (Sebell, 1967)

Fig. 1-2. Modern greenhouses require high technology as typified by this example of a modern metropolis

1.1.1 Some General Characterizations

In spite of the diversity, there are some general conclusions that one can draw:

1. Greenhouses, as they are dealt with in this book, will usually be found in so-called "developed" countries. Advanced technology provides a demand through higher standards of living, and the means for satisfying a demand for ornamentals and foods.

2. As most industrialized countries are to be found in regions where climate prevents, or reduces, the choices of year-around production, these regions are likely to have the largest greenhouse concentrations.

3. Historically, greenhouse development has been aided when transportation to distant markets is prohibitive; when there is a requirement to cross jurisdictional boundaries; when political considerations restrict free commodity flow; and when the freshness and short life of the products in demand cannot be extended to maintain desirability when shipped.

4. Greenhouse production represents the most intensive of agricultural production methods. Structures used are costly, the required labor is high. On the other hand, with suitable management, the return is correspondingly high. Under certain conditions, greenhouse cultivation may be equal to highly developed, industrial manufacturing operations.

5. Energy requirements of greenhouses are high. They are gross consumers of energy derived from fossil fuels. Construction and operation (e.g. crops grown, etc.) are remarkably sensitive to raw material availability.

6. The majority of individual greenhouse operations can be considered as small businesses.

7. The industry as a whole represents a miniscule part of total worldwide agricultural operations—either when figured as a total of land under cultivation or total gross return. However, on a country-by-country basis, the industry can be a significant factor in a particular country's internal economy.

1.1.2 Size of the Industry

Table 1-1 is an estimate of the total world area in greenhouse production. These figures are very rough approximations as they do not include all countries such as Australia, China, etc. The industry is in a state of flux, and it is sometimes difficult to determine the dividing line between greenhouses and what are merely protec-

Table 1-1. Greenhouse area (ha) in the world

Country	Vegetables and fruit	Ornamentals	Total	Ref.
Holland	4680	2850	7530	Jacobs and Meyaard (1975)
Belgium	1700	400	2100	Jacobs and Meyaard (1975)
West Germany	1200	2300	3496	Jacobs and Meyaard (1975)
United Kingdom	1620	500	2120	Jacobs and Meyaard (1975)
Channel Islands	450	50	500	Jacobs and Meyaard (1975)
Ireland	180	20	200	Jacobs and Meyaard (1975)
Scandinavia	600	900	1500	Jacobs and Meyaard (1975)
France (mostly north)	400	400	800	Jacobs and Meyaard (1975)
Bulgaria	950	50	1000	Jacobs and Meyaard (1975)
Romania	?	?	+1200	Jacobs and Meyaard (1975)
Hungary	300	?	300	Jacobs and Meyaard (1975)
Poland	600	100	700	Jacobs and Meyaard (1975)
East Germany	?	?	1200	Jacobs and Meyaard (1975)
Czechoslovakia	?	?	750	Jacobs and Meyaard (1975)
Russia	?	?	4200	Jacobs and Meyaard (1975)
France (mostly south)	500	—	500	Jacobs and Meyaard (1975)
Spain	1800 (?)	—	1800 (?)	Jacobs and Meyaard (1975)
Canary Islands	400	25	425	Jacobs and Meyaard (1975)
Italy	5000	700	5700	Jacobs and Meyaard (1975)
Greece and Crete	1500	—	1500	Jacobs and Meyaard (1975)
Yugoslavia	120	—	120	Jacobs and Meyaard (1975)
Algeria	300	—	300	Jacobs and Meyaard (1975)
Morocco	60	—	60	Jacobs and Meyaard (1975)
Israel	50	120	170	Jacobs and Meyaard (1975)
United States	230	1851	2081	Dalrymple (1973)
United States	—	1964	1964	Fossum (1973)
Canada	?	?	299	Dalrymple (1973)
Turkey	?	?	2000	Dalrymple (1973)
Japan	349 (glass only)	243 (glass only)	592	Dalrymple (1973)
Total	22689	10509	43143	

Fig. 1-3. Inexpensive greenhouse construction in warm climate, largely for protection from rain

tive shelters. For example, if Japanese vegetable production in plastic tunnels were included, at least another 10000 ha would be added. It is likely that a large part of the acreage listed for Italy includes minimum structures, mostly for protection from rain. Similar types of minimum structures (Fig. 1-3) may be found in Israel, South and Central America, and parts of the United States. With good transportation, crops can be exported from suitable climatic regions to what would normally be considered a greenhouse production area. Examples are exports from Columbia, Costa Rica, Kenya, and Israel to Great Britain, West Germany, Scandinavian countries, and the United States. The viability of northern greenhouses, and the crops produced, can be directly influenced by import restrictions, available shipping space and transportation costs.

Note in Table 1-1, that total estimated land area in greenhouse production is less than 44000 ha. According to Krause (1976), field vegetables in Holland occupy nearly 41200 ha, compared to a total of 7530 ha for greenhouse production of all types. This comparison is not fair as the capital investment and return per unit area for greenhouses far exceeds that for field-grown crops. Still, greenhouses in Holland represent a much greater factor in that country's economy than in the United States where the total estimated land area in greenhouses is 2081 compared with 1.4 million ha in vegetables planted for processing and fresh market in 1970 (USDA, 1972a, b).

The preponderance of glasshouse production in vegetables in Europe reflects the climatic disadvantages of the region for fresh vegetable production; and, at least until after World War II, the numerous restrictions and jurisdictional boundaries across which products had to flow. The wide climatic variations in the United States, on the other hand, permit any vegetable to be produced outdoors the entire year. This, with an efficient and cheap transportation system, has resulted in a borderline economic situation for greenhouse vegetable production.

Table 1-2. Per capita flower and house plant consumption (Hylkema, 1976)

Switzerland	$ 32.20
Germany	28.15
Holland	23.15
Sweden	19.25
United States	9.77 (from Fossum, 1973)
France	9.75
Italy	8.75
United Kingdom	4.65

Dalrymple (1973) quoted Bailey as stating in 1891 that winter tomatoes could always find ready sale at prices of 40 to 80 cents per pound. Bailey's price range is generally higher than present prices for fresh tomatoes in U.S. supermarkets. There is an indication, with the more open trade policies of the European Common Market, of a decided shift from vegetable to ornamental production. A recent Grower magazine article (Anonymous, 1976) predicted 300 more Dutch flower growers in 1976.

An exception to competition from better climatic areas is production of crops where weight restrictions prevent long-distance shipping. This is the situation for pot plant and bedding crops, and particularly in the last two years, for foliage plants. As a rule, 800 km is about the maximum distance finished plants in soil can be shipped, although there are exceptions. The U.S. has rigid restrictions on plant imports with soil, so that foreign competition with potted crops is not likely to be important.

Within the last two years in the United States, there has been an unprecedented boom in foliage plant production. Alvi Voigt, in March, 1976, presented information showing a 676% increase in foliage wholesale value between 1970 and 1975. The increase between 1974 and 1975 was more than $ 73 million, with the possibility in 1976 of foliage plants exceeding the wholesale value of all other major ornamental crops combined—or about 53% of the market. Two reasons for this change may be the fact that the United States is approaching Europe in regard to conditions of urbanization and housing, and that the present generation has become more interested in the value of plants for recreational and aesthetic purposes. As Chapter 12 emphasizes, the opportunities for people in the industry are outstanding.

The effect of living standard and climate on flower consumption can be noted in Table 1-2. The countries with the highest per capita expenditure (Switzerland, Germany, Holland, and Sweden) are noted for their reasonably stable and expanding economies, as well as other attributes discussed in Chapter 12. The United States' ranking is relatively high, but still a very small part of personal expenditures as indicated by Voigt (1974; see Table 1-3). Personal expenditure in 1973 for greenhouse and related products was less than one-fourth of one percent of the total personal expenditure. The potential for an expanding market, paced by a vigorous industry, is outstanding. It contrasts favorably with greenhouse

Table 1-3. Personal consumption expenditures for the United States, 1973. (From Voigt, 1974)

Item	Billions $	%	Item	Billions $	%
Food and tobacco	178.676	22.2	Books and maps	3.790	7.2
Clothing, acessories, jewelry	81.274	10.1	Magazines, newspapers	5.028	9.6
Personal care	12.315	1.5	Nondurable toys-sports	7.677	14.7
Housing	116.367	14.4	Wheel goods, durable	7.019	13.4
Household operation	117.509	14.6	Radio-TV, records, etc	12.920	24.7
Medical care	62.726	7.8	Radio-TV repairs	1.666	3.2
Personal business	45.183	5.6	*Flowers, seeds, etc*	*1.964*	*3.8*
Transportation	109.228	13.6	Admissions to amusements	2.898	5.5
Recreation	*52.280*	*6.5*	Clubs and fraternal	1.349	2.6
Private education and research	13.225	1.6	Commercial participant	2.228	4.3
Religious and welfare	10.843	1.3	Pari-mutual net receipts	1.325	2.5
Foreign travel	5.595	0.7	Other	4.416	8.4
Total	805.221	99.9		52.280	99.9

vegetable production, where, once food needs are satisfied, there is little likelihood that significantly increased sales can be generated except with those individuals having a surplus discretionary income.

1.1.3 Structures and Capital

Chapter 3 deals with greenhouse construction. It is sufficient to emphasize the diversity that can be encountered as to size (Figs. 1-4 and 1-5) and construction materials (Figs. 1-6 through 1-9). Greenhouses have been constructed vertically, placed on top of buildings, buried in the ground, heated with solar radiation, propane, oil, gas, coal, etc., built of almost every material ready to the hand of man, and used for a variety of purposes ranging from cattle production to covering swimming pools. The average size business ranges between 4000–5000 m^2; or, assuming a total capital investment, exclusive of land, for a first-class, fiberglass-covered structure with fan-and-pad cooling and steam heat, of 50–60 dollars per m^2; a total capital investment of about $ 500000–600000 per ha. There are economies of size which is one reason accounting for the steady size increase of the individual businesses. Small hobby houses (Fig. 1-4) may have a per m^2 cost exceeding $ 250.00.

Initial investment will vary tremendously, depending upon crops grown, availability of materials and climatic conditions. The minimum structure in Figure 1-3, for example, will obviously cost much less than the one in Figure 1-6. Costs may range from less than $ 20.00 per m^2 for a single polyethylene film, wooden structure in the United States to nearly $ 100.00 per m^2 for a "first-class" production range. Economies can be achieved by using simple and cheap materials in the superstructure, eliminating raised benches and growing directly in the ground, or locating in a climatic area where costs for heating and cooling equipment are

Fig. 1-4. Greenhouse production range exceeding 150000 m^2, fiberglass covered in a climatic region requiring complete heating and cooling systems

Fig. 1-5. Small, fiberglass-covered hobby greenhouse

minimum. Chapter 3 provides some percentage figures that can be used to estimate costs. According to Krause (1976), 33% of the initial cost for Dutch greenhouses lies in the basic structure which, for a Venlo block, varies between $ 10 and $ 12 per m^2. Another 22% is usually required for heating, 17% for the building, 14% for equipment, 8% for land and 5% for tools and machinery. Land investment can vary, depending upon local conditions, and it is obvious that some land values will be high enough to preclude any investment for a greenhouse structure.

Fig. 1-6. End view of a double polyethylene, air inflated, ridge-and-furrow, quonset range

Fig. 1-7. Example of concrete superstructure with thin layer polyethylene for roof, sandwiched between layers of wire mesh, in a geographical location where wood is scarce and steel costly. Flat roof indicates that snow loads are nonexistent, and rainfall slight

To reiterate, any time a structure is raised to cover the land, investment per unit area far outweighs investment usually found for common agricultural activities.

Once an area is enclosed so that cultivation may proceed all year, costs of operation rise markedly. The structure reduces opportunities for large-scale mechanization, and the variability found in the greenhouse industry has so far

Fig. 1-8. Example of galvanized steel superstructure to be covered with fiberglass. Wooden gutters are supported on galvanized pipe posts

Fig. 1-9. An esoteric greenhouse design, probably covering a swimming pool. Note that cooling uses an evaporative cooler, forcing the air into the structure. This system is seldom feasible on a large scale. Greenhouses may assume many shapes and styles

precluded any special mechanization designs. As a result, labor input is the highest production cost. Again, information on production costs is hard to obtain since most individuals in the industry consider this proprietary information, and because uniform accounting procedures are seldom seen (see Chap. 11). For example, Krause (1976) provided some growing and production costs for Dutch greenhouses for various crops (Table 1-4). It is apparent that the dollar figures do not

Table 1-4. Costs of production in the Dutch greenhouse industry. (From Krause, 1976)

	Growing costs (per m^2) $	Total production costs (per m^2) $
Tomatoes (Jan to Aug) Cucumbers (mid Aug to Nov)	5.64–6.02[a]	13.16
Tomatoes (mid Feb to mid Aug) Two lettuce crops	4.14–4.51	11.28
Cucumbers (Jan to mid Jul) Tomatoes (mid Jul to mid Nov)	6.02	13.54
Roses	5.64	15.04
Year-around chrysanthemums	9.78	18.80
Freezias	7.82	12.03
Carnations	3.55	13.16
Anthuriums	6.58	15.84

[a] Figures converted from Dutch guilders to U.S. dollars; 1 guilder = 37.6 cents U.S.

include costs of depreciation, interest on borrowed money, administrative expenses, etc. When some of these items are included, costs per square meter will usually exceed $ 50.00, and, depending upon management, may approach $ 100.00 per m^2. It is not unusual to find labor costs varying between individual enterprises as much as 20–30%. Dalrymple (1973) showed estimated man-years of labor to vary between 3.5–9.6 per ha depending upon whether double cropping was practiced and the vegetable crop produced. Krause (1976) stated that five men per ha were required for rose production, 6.7 men for carnations, and four men per ha for year-around chrysanthemums.

Rising fuel cost in recent years has resulted in energy requirements becoming the second largest factor in greenhouse operation. Fuel and labor combined now often exceed 50% of the total production costs. A typical southwestern United States operation may have fuel costs ranging from $ 2.50 to $ 3.50 per m^2, depending upon the heating season's severity, and whether they are on natural gas. In the northeastern U.S., a grower may expect $ 10.00 per m^2 on oil.

None of the tables (Tables 1-5 and 1-6) in this chapter provide estimates for packaging, transportation, advertisement, and other selling costs. Table 1-6 does include a figure for grading and packaging carnations, but other selling costs for carnation producers may vary from 16 to 25%. Newcomers to the business often fail to consider the fact that a product must be sold, and this factor often determines profitability. An anonymous survey on greenhouse vegetable production in Colorado pointed out that small producers, selling locally, could usually maintain a small profit. If they were forced to package and transport their product any significant distance, however, competition of field-grown vegetables from other areas often made the operation unprofitable. When estimating costs for a new business, about 25% of the total estimated cost should be marketing costs, as well as sufficient operating capital of about $ 1.00–1.50 per m^2 per week to allow for the period between planting and when the product can be sold.

Table 1-5. Costs of production for a mixed crop operation, 22 500 m², approximately 68% in production. (From Bachman, 1975)

	Cost per m² [a] (producing area) $	Percentage
Labor	20.70	32.8
Supplies	7.92	12.5
Depreciation (building and equipment)	4.15	6.6
Utilities (fuel, water, power)	10.62	16.8
Taxes	1.47	2.3
Miscellaneous operating expenses (includes rental, trash removal, travel, telephone, etc.)	1.11	1.8
Insurance	0.31	0.5
Labor contract (75–76)	1.45	2.3
Interest on investment (9%)	6.60	10.4
Administrative cost	0.94	1.5
Down time (disease loss, etc.)	4.89	7.8
5% profit	3.01	4.8
Total	63.17	100.1

[a] Per m² of producing area per year.

Table 1-6. Estimated costs of production for a carnation producer of about 13000 m², 60% in production

	Cost per m² $	Percentage
Labor	25.04	33.5
Supplies (includes fertilizers, pesticides, plants, etc.)	7.84	10.5
Depreciation (includes building and equipment)	9.74	13.0
Utilities (includes fuel, power, and water)	9.46	12.6
Miscellaneous operating expense (includes rental, travel, telephone, etc.)	1.16	1.6
Insurance	1.34	1.8
Interest	7.71	10.3
Administrative costs	0.66	0.9
Grading	7.44	9.9
5% profit	3.56	4.8
Total	73.95	98.9

It is apparent that greenhouse production is a fuel intensive, labor intensive, and capital intensive enterprise. This does not prevent young people from eventually operating their own business growing plants. It does mean that good judgement and hard work will be required—together with some realistic, pragmatic decisions based upon the best information obtainable.

1.1.4 Questions to be Answered in Starting a Greenhouse Business

The upsurge in interest in starting, operating, or working in a greenhouse has been phenomenal. As with any goal, an individual must be able to formulate the questions to be answered in fulfilling a goal, and then obtain the answers. As with computers, a poor question only receives a poor answer, and could provide a worse solution than no question at all. As a summary to the text, and to provide a general review of the factors affecting greenhouse management, we list a few questions that one may ask before investing:

1. What is the marketing area in relationship to the crops grown? To whom is the product to be sold and what is the general standard of living of the consumers? What marketing systems are available? What will be the transportation costs? Who will be your competitors? What are possible costs for packaging and advertisement? How is the market likely to fluctuate with changing conditions? Chapter 12 provides information and an indication of the opportunities that may be available.

2. What are the climatic conditions—temperatures, available solar radiation, rainfall, humidity, wind velocities, and prevailing wind directions? Are there problems from local meteorological effects? From pollution? Chapters 2 and 4 provide additional information, and the answers to these questions will directly influence the type of structure and its initial costs (Chap. 3). The answers will also provide you with an indication of cultural procedures you may have to employ and limits to production.

3. What is the fuel and power availability? Can you use gas? Will you have to use oil? If coal is available, what restrictions on its use may increase initial investment and operating costs? Will you have to pay additional fees to bring energy supply to the greenhouse? Again, Chapter 4 provides some answers to these questions.

4. What is the water availability, and what is its quality? A good water source of high quality is cheap (Chap. 5). A poor water supply will limit growth, and may require additional investment for purification. Are there restrictions on well drilling, or on the size of the tap for domestic water?

5. What kind of soil is available? What are its drainage characteristics? Does it have a high water table (Chap. 4)? Is it shallow? Can you grow directly in the ground? Can you use it for your potting mixture? Does it have high fertility?

6. How much land do you need? How is its value likely to fluctuate? Is it in an area likely to urbanize? Is it likely to be surrounded by polluting industries?

7. Where are you going to obtain capital? What are your sources and how much will it cost you?

8. What are your managerial capabilities? Do you like people, and can you get them to work for you? Have you realistically assessed yourself? Chapter 11 discusses management and economics.

It may be that other questions have to be faced, or that some listed here will not be important. However, we feel that they cover the most important aspects of running a greenhouse, and every greenhouse operator will eventually be faced with a demand for information to answer the majority of them. It is not the purpose of this text to answer any of them completely, for the answers change with each condition and each individual. Our purpose is to provide a starting point.

1.2 Challenges to the Industry

The greenhouse industry is undergoing a rapid change. It is in a time of flux where problems of inflation, competition from new producing areas, fuel scarcities, and changes in social mores make it difficult to determine one's best alternative. We are at a period of time where the challenges have never been greater, and the opportunities for those willing to risk answering those challenges have never been more rewarding.

Barring unforeseen disaster, we can expect that the industry will be as vigorous as those who are its practitioners. The fact that costs will probably continue to increase, places higher demand on requirements for excellent management, with perhaps a touch of witchery for improving good choices. For the first time in recorded history, the world is seeing the limits of its resources. The shift in values and thought patterns could be shattering as we seek new energy sources to heat our greenhouses, or shift to more favorable regions for production. The importance of ornamental production to fulfill human cultural and aesthetic needs in an increasingly urbanized world, both to supply products and to provide a desirable working environment, will be the basis for its permanence. It is not prophetic to state that greenhouses 10 years from now will be considerably different from those we describe in this text.

It is probable that greenhouses will always be more diverse than other industries, and will, therefore, provide a greater range of opportunities for people to satisfy their personal needs. However, if we assume that labor costs continue to increase, then pressures for mechanization will continue to increase. Diversity in this instance becomes a curse, as engineering costs for design cannot be recovered when every establishment requires a different machine. There will be increased pressure to standardize greenhouse construction, and greater emphasis on crops capable of being mechanized. The machine and the crop must be developed simultaneously, making use of all controls and regulators (see Chap. 9) that will enhance mechanization. As with mechanization of other industries, capital costs will increase and labor will decrease. Whether the industry will be able to meet these two demands (lowered energy consumption and labor requirements) will depend upon its willingness to support activities toward these ends, to innovate, and to apply the innovations.

Another important challenge to the industry is innovation in marketing. We are seeing radical changes (Chap. 12) in methods. Unfortunately, there is no coordinated effort to promote new methods of marketing, but rather, the individuality of industry's practitioners, has permitted it to respond to changes that are occurring willy-nilly and not as the result of a concerted effort. If the industry desires to see prices increase, and, therefore, restrict itself to a highly limited market selling to those with a large discretionary income, then there is little hope that ornamentals will obtain a significant part of what discretionary income there may be. Flowers and plants must be made available to everybody, they must be priced in a range which the majority—in a possible time of income restriction—can afford, and they must be of a quality that competes favorably with other products. The pressure for greater production from an equal area at the same cost will not be less than it has been in the past. If the industry does not do this, then someone else will supply it to the detriment of an outstanding product, and to the misfortune of those of us who like to grow plants for their appearance as contrasted to plant production for meeting human physiological needs. To paraphrase Dickens: "It is the worst of times. It is the best of times. ..."

References

Anonymous: There'll probably by 300 more Dutch flower growers this year. The Grower **85(12)**, 621 (1976)

Bachman, L.: Production costs. Minn. State Florists Bull. Dec. 1, 1975

Dalrymple, D. G.: A global review of greenhouse food production. USDA, ERS Report **89**, 1973, 150 pp.

Fossum, T.: Trends in commercial floriculture. Crop production and distribution. A statistical compendium for the United States 1945–1970. The Soc. Amer. Florists Endowment, Washington, D.C., 1973

Hylkema, H. J.: Your flower market can grow! Can. Florist **71(7)** 9 (1976)

Jacobs, J. M., Meyaard, D.: Who'll survive? The Grower (Suppl.) **84(1)**, ix (1975)

Krause, W.: Mainstreet glasshouseville. The Grower (Suppl.) **85(4)**, 11 (1976)

Sebell, W. H., Jr.: Food and nutrition. New York: Time Inc., 1967, 200 pp.

USDA: Vegetables for processing. Revised estimates by states, 1964–1970. USDA, Statistical Bull. 494, 1972a, 49 pp.

USDA: Vegetables for fresh market. Revised estimates by seasonal groups and states, 1964–1970. USDA, Statistical Bull. 495, 1972b, 165 pp.

Voigt, A. O.: The fit of floriculture. In: Flower marketing information. Univ. Park, PA.: Penn. State Univ. 1974

Voigt, A. O.: Foliage phenomenon flabbergasting. In: Flower marketing information. Univ. Park, PA.: Penn State Univ. 1976

Chapter 2 Light

Greenhouses are converters of solar radiation. Their ability to do this efficiently depends upon location, structure, and arrangement. Greenhouses overcome climatic disadvantages so that solar energy can be put to economic use. A knowledge of basic principles is helpful in manipulating greenhouses and plant growth.

2.1 Principles

2.1.1 The Spectrum

The solar energy used by plants comprises a small part of the total spectrum shown in Figure 2-1. Regardless of whether radiation happens to be cosmic rays or a long radio wave, one may describe it and determine its behavior according to well-known principles. Radiation may be thought of as a series of waves, each wave containing a given amount of energy, depending upon its location in the spectrum. The radiation "quality", or its location in the spectrum may be described by measuring the number of waves passing through a point in one second (frequency), by counting the number of waves in a given distance (wave number), or by determining the length of the wave between two identical points (wavelength) (Fig. 2-2)—wavelength being the term applied in horticulture. A length of 500 nm denotes radiation that is seen by the human eye as green light (Fig. 2-1).

2.1.2 Energy

If wavelength is known, the amount of energy in a wavelength—or "packet"—can be calculated. A packet (quantum) of radiation with a length of 500 nm contains 4.0×10^{-12} ergs, or 57 kcal per mole. As wavelength decreases, energy per quantum increases, accounting for the penetration of X-rays, or the damage from nuclear radiation. It is this "quantum" energy that operates photosynthesis in green plants to produce food.

However, the energy absorbed by a plant may include several wavelengths, including visible energy (light), all streaming in on the plant at rates too high to measure individually. It is easier to use devices that measure all wavelengths at the same time. In practice, we often use the term "light", and describe the amount in "lux", "foot-candles" or "lumens". Technically, these terms refer to radiation as seen by the human eye. Plants do not respond to "light", they respond to "radiation" which may be at wavelengths that humans cannot "see" (Fig. 2-3). Most light

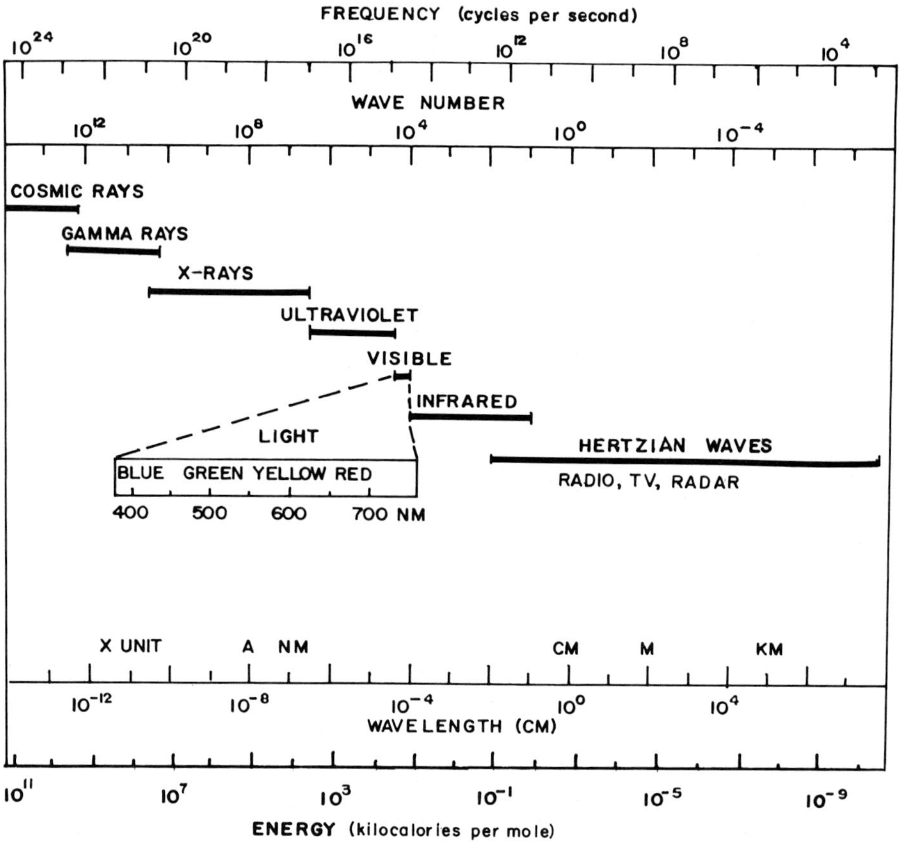

Fig. 2-1. Electromagnetic spectrum, showing position of visible radiation in respect to other parts of spectrum. (Adapted from Withrow and Withrow, 1956)

measuring instruments are adjusted to respond to radiation as we "see" it—not as plants "see" it. Radiation units will be "calories", "watts", "ergs", "Joules" or "Langleys". There is nothing wrong with using "light" terms as long as there is the realization of the difference. Watts and lux cannot be converted from one to the other unless conditions are carefully defined. Conversion factors and units are provided in the appendices, together with some definitions of important terminology.

2.2 Solar Energy

Sunlight is the primary radiation source. The amount received at sea level, on a flat surface is shown in Figure 2-4. The sensitivity of the average human eye is superimposed on the same graph. More than half of all solar radiation is at wavelengths longer than 700 nm, with a peak around 1000 nm. Much solar energy cannot be seen, but has an important effect on plant behavior. Wavelengths

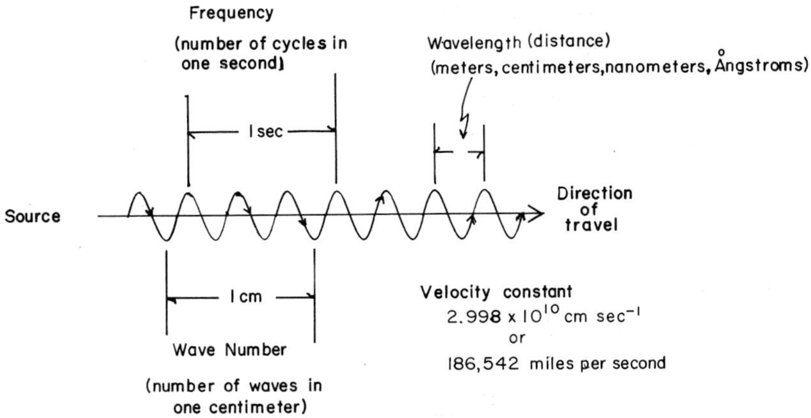

Fig. 2-2. Schematic representation of radiation, speed of which is constant. Common method of designating radiation important in plant growth is wavelength. Various methods of measuring radiation are related in following manner: Velocity of radiation = frequency X wavelength. Wave number = $\frac{1}{\text{wavelength}}$ energy = Planck's constant X frequency

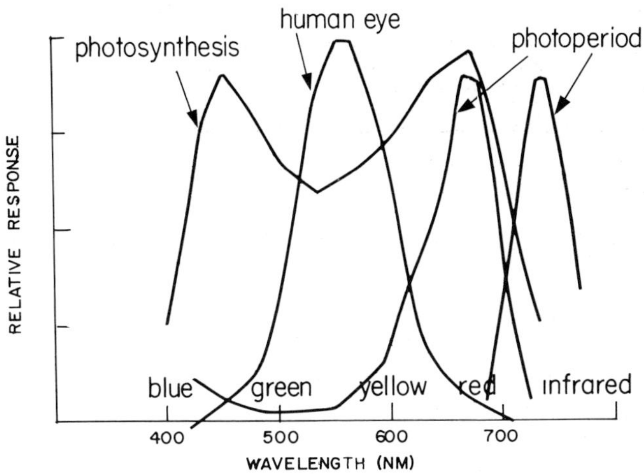

Fig. 2-3. Comparison of spectral response of plant processes with sensitivity of the human eye. (From Campbell et al., 1969)

shorter than 230 nm are filtered out by the atmosphere, and the peaks and valleys are caused by absorption of solar radiation by water vapor and CO_2 in the air.

The amount of energy actually reaching the earth's surface may be drastically modified by absorption, reflection, movement of the earth, topography, or clouds or dust. The earth's rotation alone accounts for much variation (Fig. 2-5), and local climate will further modify what is actually received. Quality and quantity of radiation is further modified by the greenhouse cover and superstructure.

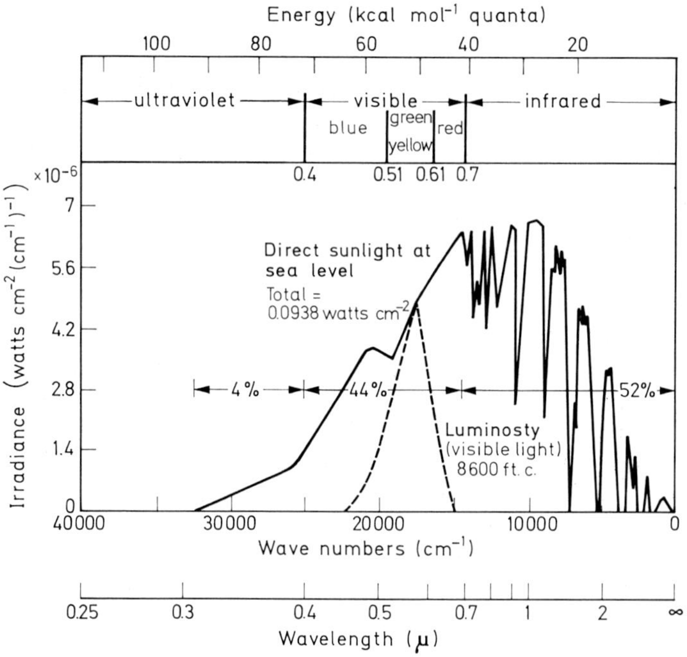

Fig. 2-4. Solar radiation on a clear day, reaching the horizontal earth's surface at noon, compared with sensitivity of an average human eye. Sudden dips in curve beyond 700 nm are absorption bands due to water and CO_2 in the atmosphere. Note that 52% of solar energy is at wavelengths longer than 700 nm (0.7 μ) and is not visible. (Adapted from Gates, 1962)

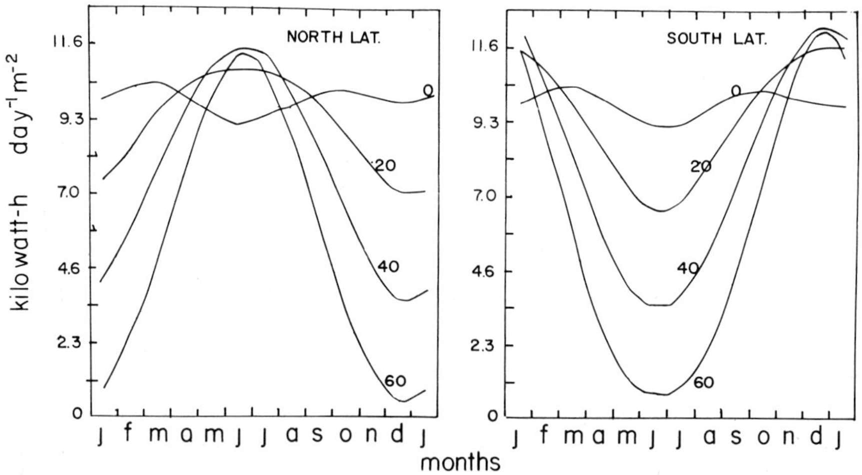

Fig. 2-5. Average solar radiation on a horizontal surface at sea level for different latitudes and times of the year. Variation results from tilt and movement of the earth about the sun (Albrecht, 1951)

2.2.1 Variation

As the result of the earth's spin and movement, solar radiation varies in a constant manner over the earth's surface (Fig. 2-5). This may be modified further by local climatic conditions. At latitudes much greater than 40° solar energy becomes the limiting factor in plant growth during winter. Other environmental factors must be adjusted accordingly for maximum growth.

While radiation generally decreases toward the earth's poles, there are variations due to prevailing winds, altitude, mountain ranges, or the presence of large water bodies. As an example, areas such as the Southwestern United States, regions of South America, Australia, and portions of Africa may be described as arid or semiarid with many days of clear sky. Provided there is sufficient water, adequate soil and temperature, these regions are highly productive. Hanan (1968) suggested that the number of clear days, or the total hours of sunshine, is a major factor when a location is evaluated for greenhouse use. For example, Figure 2-6 shows the average number of hours of sunshine for winter and summer in the United States.

Supplemental radiation from high intensity lights probably shows most benefit when the region has less than about 4.5 h of daily winter sunshine.

2.2.2 Daylength

A second important variation is the length of the light and dark periods. Depending upon latitude, daylength varies in a more uniform manner with season than does solar intensity (Fig. 2-7). Near the equator, the seasonal variation may be slight enough to neglect. At higher latitudes, however, the grower must adjust his cultural procedures if he happens to be flowering certain species outside their normal period.

2.2.3 Other Factors Affecting Solar Energy

A grower must take into account the local variations that can result from air pollution, man-made structures, and local topographic features. Figure 2-8 shows what might happen during winter around generating plants. Smog formation may reduce total radiation by as much as 30% (Schuck et al., 1970) in densely populated regions. Local climate may have cloud formation at certain periods of the day, or nearby structures or local land formations may shade the greenhouse in early morning or late evening hours. In low-lying regions, early morning periods can have frequent fog. Greenhouse location requires careful thought and close application to local meteorological records whenever possible. There may be a trade-off between disadvantages of low solar energy and the fact that the site has good soil, is close to the market, or simply is all that can be afforded.

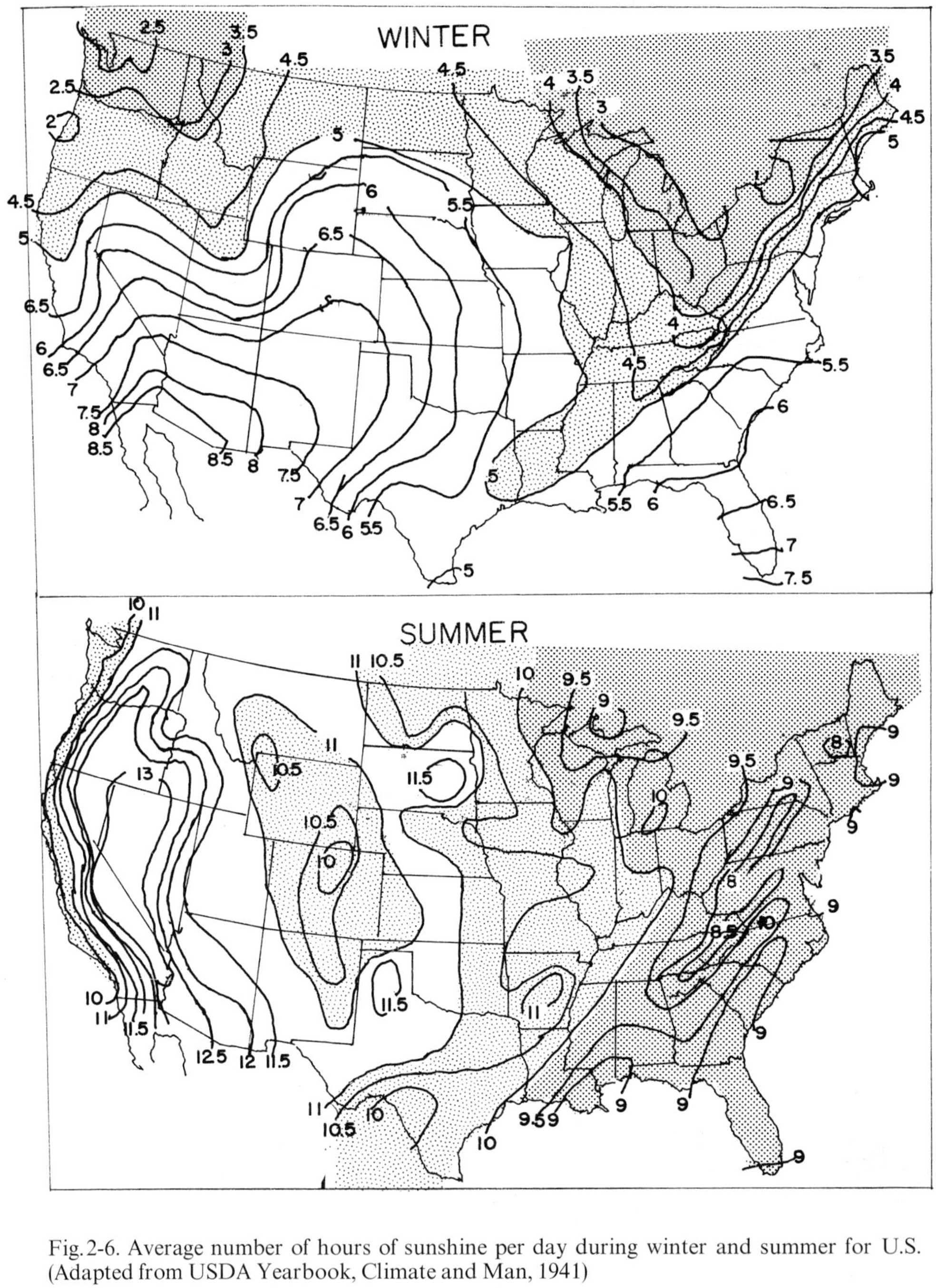

Fig. 2-6. Average number of hours of sunshine per day during winter and summer for U.S. (Adapted from USDA Yearbook, Climate and Man, 1941)

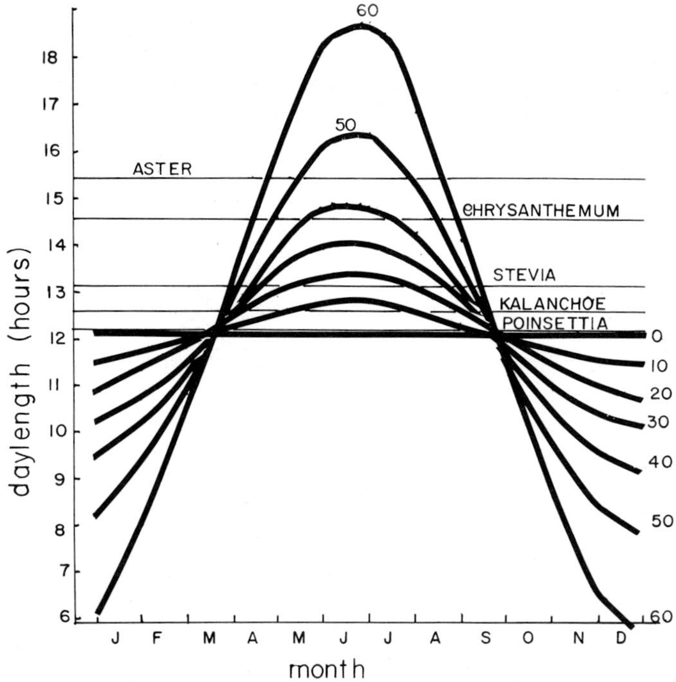

Fig. 2-7. Variation in daylength for different latitudes in northern hemisphere. Horizontal line for a species intersects the curve, showing critical daylength above or below which species will initiate flowers (List, 1966)

Fig. 2-8. Cloud formation during cold periods from an electrical generating plant. Plant growth in greenhouses under shadow will be directly affected

2.3 Supplemental Irradiation

Common usage refers to “lighting” plants in greenhouses with high intensity fixtures as “supplementary illumination”. In this chapter, we have converted energy units to watts throughout so that comparisons between sunlight and artificial light can be more easily compared.

2.3.1 Requirements

Solar radiation costs the grower nothing. Depending upon geographic location, however, there can be periods when natural radiant energy is deficient. The lower the radiant energy external to the greenhouse, the less efficient the greenhouse may be in transmitting usable energy. While a single leaf is light saturated at relatively low intensities (86–108 Wm^{-2}), full saturation of the entire crop may not be reached before approximately 500 $W m^{-2}$.

In Colorado, the average daily solar radiation during December is about 2300 W-h m^{-2} (Hanan, 1967), for an average of 161–333 W-h m^{-2} per hour. Canham et al. (1969), however, state that total daily radiation for Lea Valley, England, for three months of the year, to be below 350 W-h m^{-2}, and can fall as low as 87, or 11–44 W-h m^{-2} per hour. This is less than a third of the radiation expected for Colorado. Below 174 W-h m^{-2} per day, Canham et al. (1969) observed chrysanthemum flower initiation to be delayed. But, uniform flower initiation could be obtained if the light level during the first two weeks of short days was raised to 348 W-h m^{-2} per day. This could be obtained by supplementing natural daylight with artificial, equivalent to 75600 lux-h (7026 ft-c-h) or 6300 lux (21.1 $W m^{-2} h^{-1}$) at plant height for 12 h each day. This required 400 W mercury fluorescent lamps 1.1 m above a 0.9 m wide bed, spaced on 1.0 m centers.

At Lansing, Michigan, average daily radiant levels are about 872 W-h m^{-2} in December (Crabb, 1950), or approximately 109 W-h m^{-2} average per hour for 8 h. Carpenter (1974a, b) was able to achieve 162 $W m^{-2}$ on sunny days by lighting with 136 $W m^{-2}$. At supplemental intensities of 67.8 $W m^{-2}$, he was able to reach levels of 133 $W m^{-2}$. The highest intensities did not achieve a commensurate gain in rose growth.

According to the calculations of Good et al. (1974), supplemental lighting of about 43.2 W-h m^{-2} for 18 h daily will provide a positive benefit, but 28.8 W-h m^{-2} for 24 h daily will provide a higher return (Table 2-5). The amount of electrical energy that will be converted to useful work will vary with the source. For example, at a constant illumination of 10760 lux, absolute energy delivered can vary between 45.7 and 30.6 $W m^{-2}$ (Table 2-1).

Norton (1972) showed that lamps such as the high intensity sodium types provide significant savings although they are more expensive to install (Tables 2-2 and 2-4). However, Tables 2-1–2-4 show that the usable energy from various sources varies for an equal illumination level. We suggest that when the total daily radiation drops below 100 W-h m^{-2}, supplemental illumination may be useful. At 500 ppm CO_2, Aikin (1974) showed that individual leaf photosynthesis reached a

Table 2-1. Factors for converting foot-candle readings into absolute units (W m^{-2}) for common light sources (1 ft-c = 10.76 lux)

Light source	W m^{-2} ft-c^{-1}		
	300–800 nm (after Berneir, 1962)	400–700 nm (after Gaastra, 1959)	w m^{-2} 1000 ft-c (10760 lux)
Incandescent	—	0.0457	45.7
Incandescent (500 W)	0.0754	—	75.4
Fluorescent lamps			
White	0.0306	—	30.6
Warm white	0.0290	0.0303	30.3
De-Lux warm white	—	0.0342	34.2
Cool white	0.0328	0.0338	33.8
De-Lux cool white	—	0.0368	36.8
Daylight	—	0.0371	37.1
Gro-Lux	0.0966	—	96.6
Mercury (JH1-color imp)	0.0360	—	36.0
Sunlight	—	0.0432	43.2

Table 2-2. Comparison of lamps for supplementary irradiation (Norton, 1972)

	Efficiency (lumens per W)	Average life (h)
Incandescent	12–26	1000–3000
800 ma fluorescent	52–84	12000
1500 ma fluorescent	51–74	9000
Mercury (clear)	50–60	24000+
Phosphor mercury	50–60	24000+
Self-ballasted mercury	30	12000–14000
Metal halide	80–90	8000–10500
High pressure sodium	117	15000

Table 2-3. Energy output of selected lamps (Campbell et al., 1969)

Lamp	Input		Output			
	Lamp (w)	Total (w)	400–500 nm (w)	500–600 nm (w)	600–700 nm (w)	Total (w)
Incandescent	100	100	0.8	2.2	3.9	6.9
Fluorescent CW (40 w)	40	50	2.7	4.5	1.9	9.2
Fluorescent CW (1500 ma)	215	235	13.5	22.5	9.5	46.0
Fluorescent CW (1500 + 100 w)	315	335	14.3	24.7	13.4	52.9
Mercury phosphor	400	425	11.6	28.4	18.3	58.4
Metal halide	400	425	26.2	50.3	12.1	88.7
High pressure sodium	400	425	10.3	55.3	39.6	105

Table 2-4. Norton's (1972) comparison of potential plant growth lighting equipment. Illumination of 10 m^2 (100 ft^2) at 500 ft-c (5380 lux) from November through March. Data assume a 6-year depreciation. Costs based on new material, purchased in quantity and installed professionally in 1972

	High pressure sodium $	Metal halide $	Deluxe white mercury $	VHO fluorescent 2 lamps $
Annual energy cost 1 cent per lumen	8.53	8.37	8.01	8.37
Efficiency, maintained lumens	25024	14271	10015	9344
Cost of one fixture	149.93	92.28	66.98	33.00
Relative cost per 1 m^2, 12 h per day	9.10	11.90	12.80	13.00
Relative cost per 1 m^2, 24 h per day	11.20	14.60	17.00	18.10
Fixtures required per 10 m^2 per 500 ft-c	2	3.5	4.9	5.4
Total initial cost	224.18	149.07	115.95	82.22
Total annual cost, fixtures	16.25	23.78	25.90	34.24

maximum near 34000 lux—or about 138 W m^{-2}. The economics must be judged for each geographic area. One can assume that the higher the natural radiation level, the higher must be the supplement in order to achieve significant results.

2.3.2 Illumination Sources

Incandescent sources are seldom utilized for supplemental illumination. They are much too inefficient (Table 2-2). As noted in Figure 2-9, the output for incandescent sources steadily increase from about 350 nm, with most of the energy in the thermal wavelengths. Conversion of electricity to visible radiation is generally less than 5–7%.

Among the commonly used lamps are fluorescents and high intensity, mercury arc-discharge lamps which include the newer metal-halide and high pressure sodium lamps. Visible output varies with the particular type, with highest efficiency and wattage for the high pressure sodium. The latter has most of its energy concentrated in the red, whereas the metal halide types have a more evenly balanced spectral distribution between blue and red wavelengths (Table 2-3). Figures 2-10 and 2-11 show various spectral distributions for different fluorescents and the high pressure xenon. There is confusion due to the various trade names for each manufacturer's lamp, and the fact that manufacturers will publicize the most outstanding features of their particular product. To compare various lamps, they should be rated in lumens per watt (efficacy) and absolute energy between 400 and 700 nm. It may be desirable to use a source with lower output if installation expense is less, the complexity reduced and reliability improved.

Installations may cost the equivalent of new greenhouses. However, the advantage could result from better timing for the market. The grower is able to finish the crop at the proper time, whereas more production area may mean greater surplus production at the wrong time.

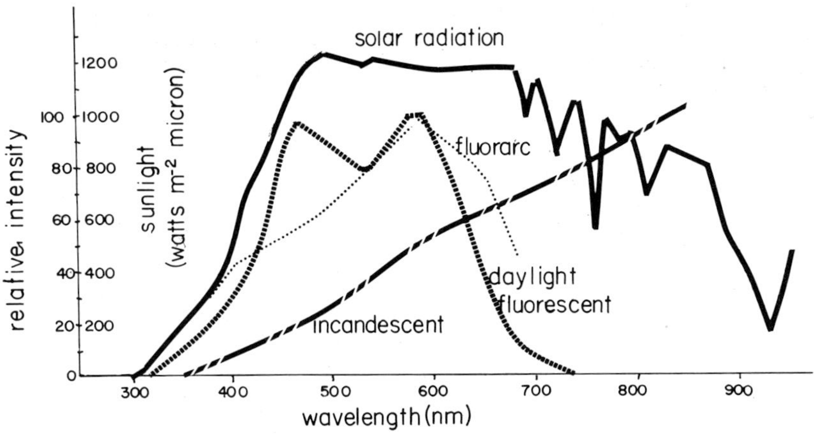

Fig. 2-9. Comparison of solar radiation spectrum at earth's surface according to Moon (1940) with various illumination sources (General Electric)

Table 2-5. Estimated benefits, costs and profitability for four alternative high intensity discharge lamp irradiation systems in production of 10000 rose plants (Good et al., 1974)

	12912 lux 1200 ft-c 18 h $	12912 lux 1200 ft-c 9 h $	8608 lux 800 ft-c 18 h $	8608 lux 800 ft-c 24 h $
Increased revenue				
25 cents per bloom	37500	15000	20000	30000
30 cents per bloom	45000	18000	24000	36000
Decreased costs due to fuel savings	2550	1275	1683	2244
Total benefits				
25 cents per bloom	40050	16275	21683	32244
30 cents per bloom	47550	19275	25683	38244
Increased costs				
Fixed				
Depreciation	15292	13579	10522	10522
Interest	7263	7263	5107	5107
Variable				
Water	250	100	140	200
Fertilizer	1500	600	840	1200
Electricity	11400	5400	7800	9000
Labor ($ 2.00/h)	4000	1600	2240	3200
Total costs	39705	28542	26649	29229
Net change				
25 cents per bloom	+ 345	−12267	−4966	+3015
30 cents per bloom	+7845	− 9267	− 966	+9015

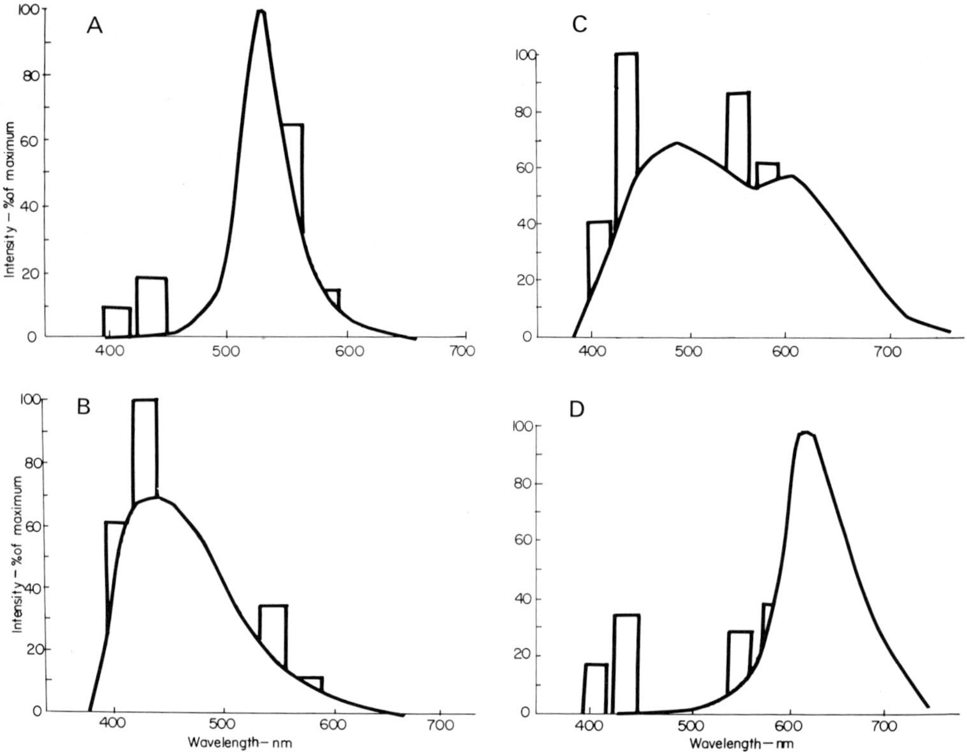

Fig. 2-10 A–D. Spectral energy distribution curves for various fluorescent lamps. (A) Green fluorescent lamp; (B) blue fluorescent lamp; (C) daylight fluorescent lamp; (D) red fluorescent lamp. (After Kleschnin, 1960)

2.3.3 Installation

Installations are as varied as lamp types. Fluorescents can be provided in various lengths as standard 40 W, high output, and very high output. These are often located in banks of two or more directly above the bench. The lamp fixture is likely to increase shading. An example of a metal halide installation is shown in Figure 2-12, and supplements radiation above several benches. A few commercial installations have tried a traveling lamp, moving at a constant speed from one end of the bench to the other (Smith et al., 1974).

The grower may be constrained by electrical availability. With airconditioning, maximum generating loads have shifted to the summer months, so that electrical companies can have a surplus capacity which they will sell at reduced rates. These problems can be important in deciding type of lamp and installation.

2.4 Effect of Radiation on Growth

The effect of radiation can be conveniently divided into three parts: (1) effect on flowering; (2) effect on photosynthesis; and (3) effect on plant temperature and

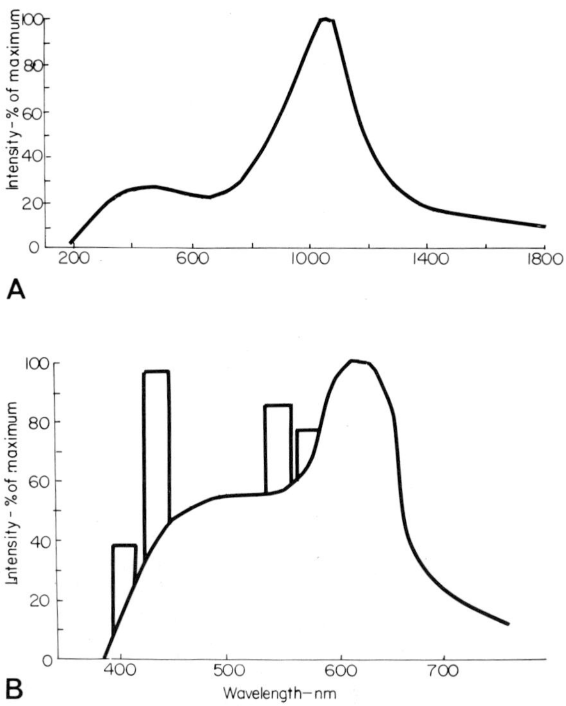

Fig. 2-11. (A) Spectral energy distribution for a xenon-arc lamp (Kleschnin, 1960).
(B) Spectral energy distribution for a warm white fluorescent lamp (Kleschnin, 1960)

Fig. 2-12. High-intensity, supplemental irradiation installation in commercial greenhouse

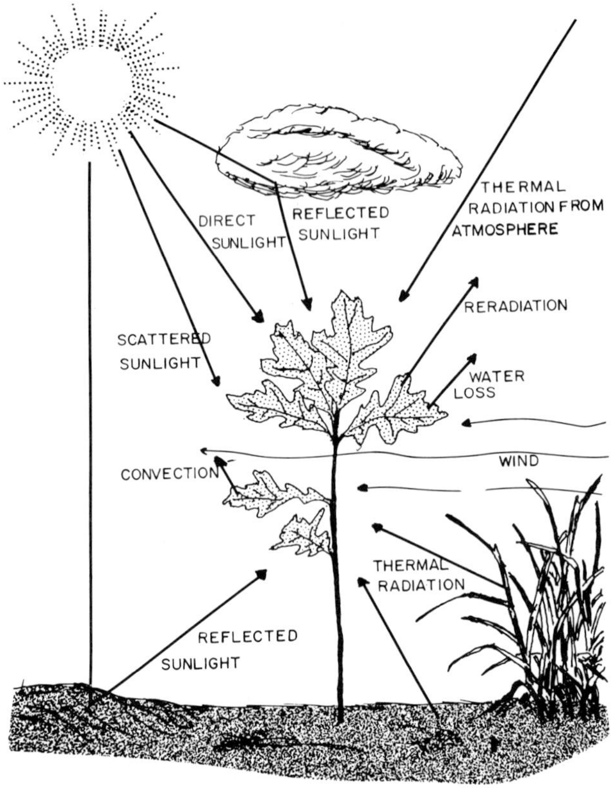

Fig. 2-13. Energy exchange by a whole green plant with its surroundings. (From Gates, 1962)

water loss. Photosynthesis and water loss are usually considered as being influenced by high light intensity. Flowering, however, can be determined by illumination levels less than 10.8 lux (1 ft-c or about 0.5–1 W m^{-2}).

2.4.1 Relationship to the Environment

Plants are closely coupled to their environment. Despite the many and useful tools that are available for us to use, the plant itself remains the best environmental indicator for the observant grower. Figure 2-13 shows how energy may be gained or lost by a single plant. If the greenhouse roof is cold, thermal energy will leave the plant with the effect of lowering the plant's temperature. If the evaporative pad is off in a dry climate, and the exhaust fans bring in dry air, water loss can increase with more solar energy being used to evaporate water. The plant temperature will be a summation of all incoming and outgoing energy exchanges. With these many options, we can appreciate why a particular plant reaction might be different from expected.

Greenhouse crops can be examined on different levels of sophistication. A microbiologist examines single cells, the physiologist observes certain reactions of a leaf, an anatomist may examine a particular organ, or a housewife may water a single plant. Much practical research has resulted from these studies, but they

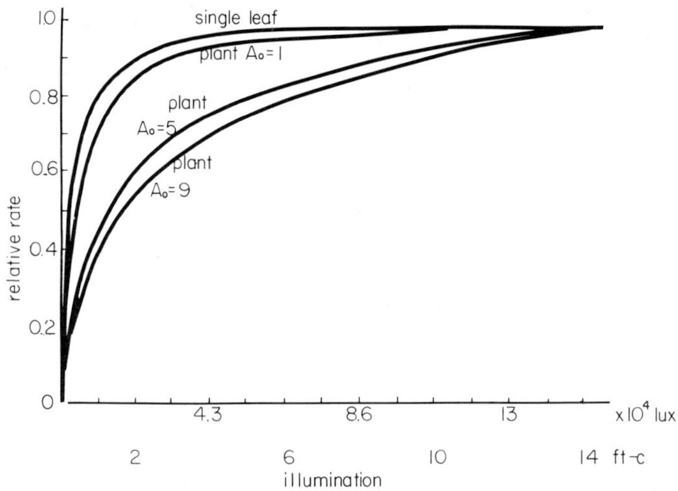

Fig. 2-14. Relative rate of gross photosynthesis as a function of illumination for a single leaf and for whole plants with Leaf Area Indices (LAI) of 1, 5 and 9. LAI is the ratio between total leaf area and the ground surface area. Values in ft-c are thousands. (After Davidson and Philip, 1958) (1000 ft-c = 43.2 W m^{-2}, approximately, using conversion factor obtained for sunlight by Gaastra, 1959; see Table 2-1)

often fail to explain the reactions of thousands of plants. The grower deals with populations, and his level of complexity is much higher. An acre of plants growing close together requires more light, more water, more CO_2, etc., and the crop response is often different than that of a single, isolated specimen. An example of the difference between single leaves and one or more plants growing closely together is to measure the change in photosynthesis with light as leaf area increases (Fig. 2-14). For a single leaf, photosynthesis increases rapidly with increasing light and reaches a maximum at a relatively low light intensity. As the thickness of the foliage increases, however, higher light intensities are required to achieve equal photosynthesis.

The most important item a grower must always keep in mind is that a single process like photosynthesis is influenced by several environmental factors. Use of all available solar energy may not be obtained because some other factor such as CO_2 level, water stress or nutrient deficiency may impose limits. It is the objective of a good grower to provide all factors in proper proportions so that his objectives (quality, yield, adaptation) are reached. In some cases, supply or "arrangement" of various environmental factors may be limited as a matter of economics. A common example is the use of salty irrigation water, which, while it reduces growth, is feasible, as better water would be prohibitively expensive.

2.4.2 Effect of High Intensity Radiation

2.4.2.1 Photosynthesis and Growth

Single leaves are light saturated at fairly low energy levels (Fig. 2-14). As intensity decreases, the conversion of light in photosynthesis becomes more efficient. Gaas-

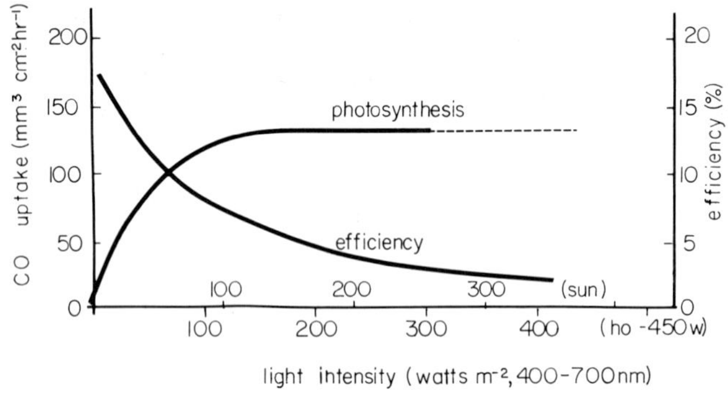

Fig. 2-15. Relationship between photosynthetic rate and efficiency of energy conversion as a function of radiant energy. (After Bonner, 1962)

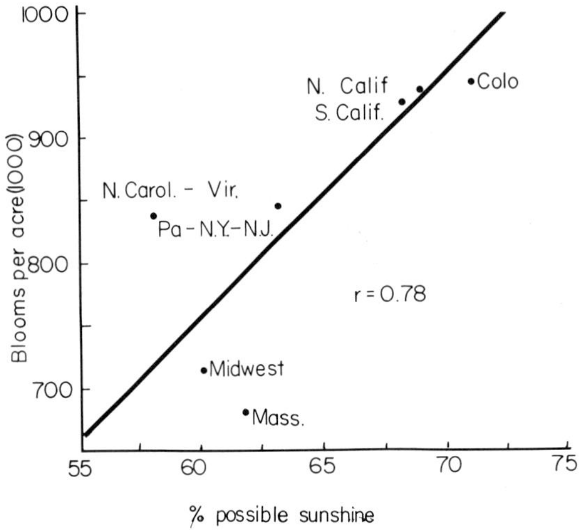

Fig. 2-16. Relationship between carnation production and percent of possible sunshine (Besemer, 1966) (one acre = 0.4 ha)

tra (1959) (Fig. 2-15) found that efficiencies approaching 20% can be achieved at levels ranging from one-tenth to two-tenths of full sunlight. In full sunlight, with enriched CO_2, efficiencies of 4% are common (Bonner, 1962). In the field, 2.0–3.5% has been achieved with rice and wheat. An average value for efficiency is about 2%.

Not all plants are equally efficient. Conversion of water and CO_2 follows different pathways, depending upon the species. Plants such as corn, sugarcane, chrysanthemum are called "C_4" plants as contrasted to "C_3" species such as Kalanchoe, Sedum and most succulents. C_4 plants are able to utilize light energy more efficiently. C_3 plants usually have higher respiration, and may actually release CO_2 in the light. The ability of C_4 plants to outproduce C_3 plants has led to efforts to breed the C_4 pathway into C_3 species (Loomis et al., 1971).

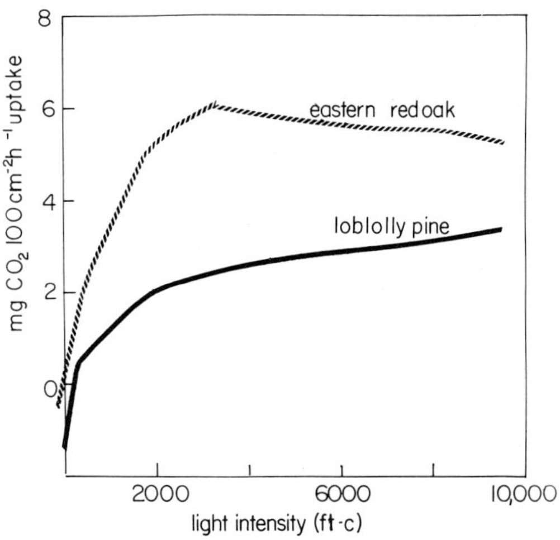

Fig. 2-17. Photosynthesis in tolerant (red oak) and intolerant (loblolly pine) species as a function of light intensity. Note that red oak reaches maximum CO_2 uptake at lower intensity than loblolly pine which continued to increase uptake as intensity increased. (After Kramer and Decker, 1944) (1000 ft-c = 10760 lux)

If food manufacture by plants becomes less efficient as light intensity increases, then why is yield usually higher with maximum sunlight (Fig. 2-16)? Higher light, as mentioned earlier, is required in order to bring all leaves in a bench to their maximum rate. There is competition between leaves in a dense canopy (Loomis et al., 1971). Upper leaves may be light saturated, but lower leaves will be shaded to different degrees. Quality of the radiation may be different inside the canopy. As the sun moves across the sky, different parts of the greenhouse bench will be irradiated at different times of the day, and the time when this occurs is often important. If it is assumed that no other factors are limiting, growth rates of the crop will usually be highest with maximum sunlight. Sunlight is the single most important factor limiting greenhouse production.

Not all species tolerate full sunlight. African violet, orchids, and many foliage plants will be damaged by full sunlight. It is sometimes desirable to reduce energy for purposes of improving quality or adaptation to conditions experienced in the consumer's hands. Figure 2-17 shows a typical example for photosynthesis in tolerant and intolerant species. Very many ornamental greenhouse plants show similar variations in response to radiant energy.

Light requirements vary with species, stage of growth and previous history. If two identical plants are placed in different environments for a period, and then both returned to the same condition for each, they will not respond the same. The problem of adaptation to the previous environments is indicated by Figure 2-18, which compares photosynthetic rates of different plants when their leaves have been preconditioned to different light levels. Plants preconditioned to high light will perform better under the same conditions. The same applies to plants conditioned to low energy levels. These responses may be expected if the environmental factor is a different watering treatment, fertilizer treatment or CO_2 concentration. Previous history, then, becomes very important when a problem is being diagnosed, and is very important with foliage plants that will be subjected to adverse conditions of dry atmosphere and low light in a home (Vlohos and Boodley, 1974).

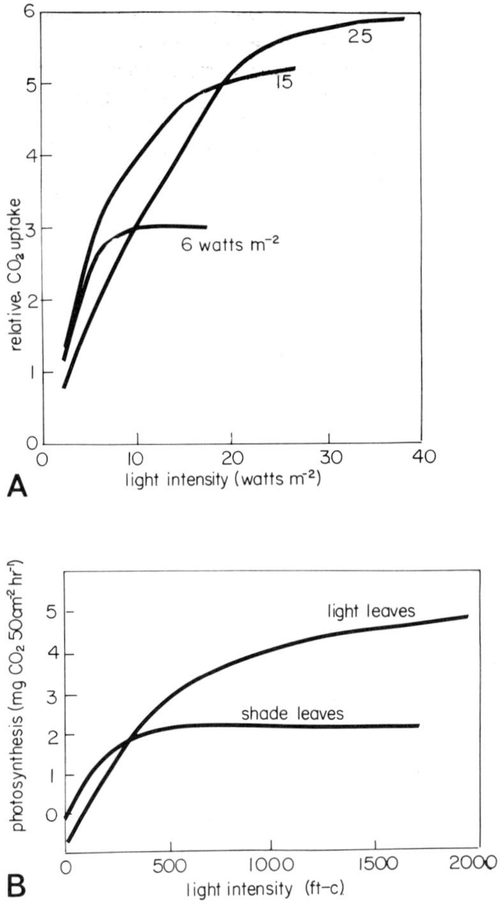

Fig. 2-18. (A) Light intensity curves for leaves adapted to three light levels of 6, 15, and 25 W m^{-2}. With adaptation to higher intensities, leaves showed a lower CO_2 uptake at low intensities and a higher saturation intensity. (After Wassink et al., 1956.) (B) Light intensity curves for leaves adapted to full light and to low light. (After Boysen-Jensen and Muller, 1929) (1 ft-c = 10.76 lux)

2.4.2.2 Plant Temperature and Water Loss

Since less than 5% of the total sunlight is employed in producing growth, the remaining energy must somehow be eliminated. There are several ways: (1) reflection; (2) transmission; (3) convection; (4) thermal reradiation as plant temperature rises; and (5) evaporation. The principal method is to change water in the plant to a gas which passes out through the stomates. Depending upon the circumstances, 70–90% of available sunlight is utilized by evaporating water. The importance is emphasized when, for some reason, the stomates close, preventing evaporation, and plant temperature rises (Fig. 2-19). As water is usually freely available in greenhouse culture, the rate of water loss will commonly have a direct relationship to the amount of sunlight. Water loss will follow a curve closely related to a curve for solar energy (Fig. 2-20). If the soil dries out, or if the root system is damaged, or if demand for water exceeds supply, the plant wilts. In any case, plant temperature rises. Mecklenberg et al. (1972) found temperatures on artificial turf could exceed surface temperatures on living grass by more than

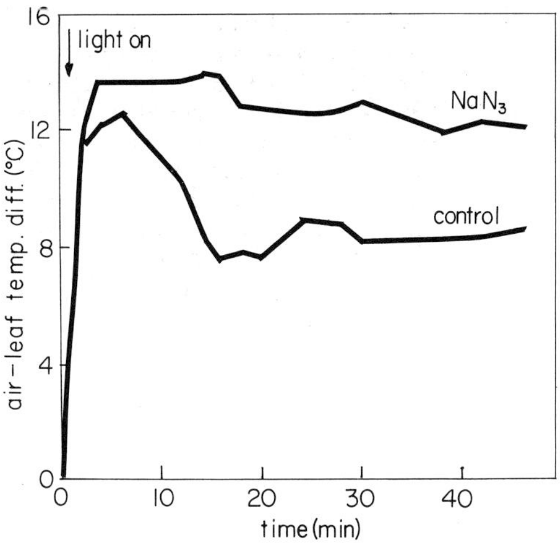

Fig. 2-19. Effect of stomatal inhibitor on leaf temperature of tomatoes. NaN_3 causes stomatal closure. When light is turned on, plant temperature is higher as less heat is utilized in evaporating water. (After Cook et al., 1964)

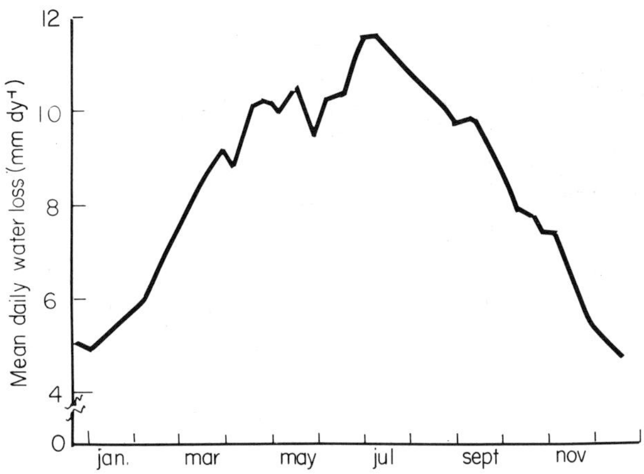

Fig. 2-20. Calculated water loss for carnations in glass-covered greenhouse for the year, assuming freely available water. (After Hanan, 1969)

60° F (33° C) under bright sunshine on football fields. The extremes could be reduced by heavy watering. Greenhouses with few plants, or an incomplete crop cover, are likely to be several degrees hotter even with maximum ventilation.

Most plants absorb radiation between wavelengths of 300–700 nm (Fig. 2-3), and again at wavelengths longer than 2000 nm. In the region where sunlight is most intense (Fig. 2-4), reflectance and transmission by green plants increases. By this adaptation, the problem of too much energy can be avoided. A good example of color's effect on temperature was shown by Hanan (1965) for red and white carnation flowers (Fig. 2-21). Under the same conditions, the greater energy absorption of red flowers will result in higher flower than air temperatures in full sunlight. White flowers will reflect more of the sunlight, and their temperatures will often be lower than the surrounding air temperature.

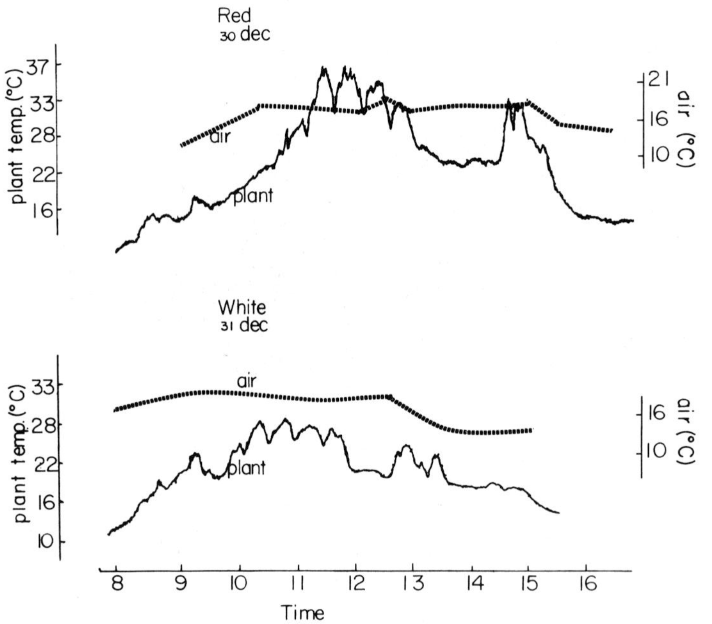

Fig. 2-21. Surface temperature of red and white carnation flowers in glass-covered greenhouse in December at Fort Collins, Colorado. No air movement. (After Hanan, 1965)

2.4.3 Low Intensity Radiation

2.4.3.1 Effect on Flowering

Growth control by changing the length of the dark and light periods is an important aspect of greenhouse culture. Beginning with the original research by Garner and Allard (1920), the opportunities for "photoperiodic" control have continuously expanded. Several examples of the effect of day and night length on flowering are shown in Figure 2-22. Photoperiodic response, however, is not restricted to flowering, but can include bulbing, tuber formation, elongation, leaf abscission, dormancy, bud break, seed germination and sex expression.

A condensed listing of plants sensitive to daylength is provided in Table 2-6. The response can be affected by temperature, and by other factors such as nutrition and sunlight. There are distinct varietal differences. That is, some varieties of a genus listed as normally sensitive to short days, may respond to different daylengths, change classification, or fail to react. Chemical growth regulators are

---

Fig. 2-22A–H. Several examples of photoperiodic effects on plants. (A) Aster, *left*: short days; *right*: elongation and eventual flowering under long days. (B) Petunia, *left*: long days; *right*: compact growth under short days. (C) Snapdragon, *left*: short days; *right*: long days. (D) Chrysanthemum, *left*: short days; *right*: long days. (E) Kalanchoe, *left*: long days; *right*: flowering under short days. (F) Ageratum, *left*: long days; *right*: short days. (G) Poinsettia, *left*: short days; *right*: long days. (H) Park oats, *left*: long days; *right*: short days ►

Fig. 2-22A–H

Table 2-6. Photoperiodic classification of plants. Condensed from Bickford and Dunn (1972)

Day-neutral	Short-day	Long-day
Cucumis sativus — cucumber	*Allium cepa* — onion[b]	*Callistephus chinensis* — China aster[a]
Fragaria chiloensis — Everbearing strawberry[b]	*Chrysanthemum morifolium* — chrysanthemum[b]	*Brassica rapa* — turnip
Gardenia jasminoides — Cape jasmine	*Glycine soja* — soybean[a]	*Antirrhinum majus* — snapdragon[a]
Gomphrena globosa — Globe-amaranth	*Chrysanthemum hortorum* — chrysanthemum[b]	*Petunia hybrida* — petunia[a]
Ilex aquifolium — English holly	*Cannabis sativa* — hemp	*Poa pratensis* — bluegrass[a]
Phaseolus vulgaris — Stringbean[b]	*Cosmos bipinnatus* — cosmos	*Cichorium endivia* — endive[a]
Zea mays — corn[b]	*Curcurbita* sp. — squash	*Dianthus barbatus* — sweet william[a]
Allium cepa — onion[a, b]	*Senecio cruentus* — cineraria	*Dianthus caryophyllus* — carnation[a]
Lathyrus odoratus — sweet pea[a]	*Solanum tuberosum* — potato[b]	*Iberis intermedia* — candy tuft[a]
Pisum sativum — garden pea[a]	*Zinnia* sp.	*Camellia japonica* — camellia[a]
Pelargonium hortorum — fish geranium[a]	*Lactuca sativa* — lettuce[a]	*Avena sativa* — oat
Viola tricolor — pansy[a]	*Chenopodium rubrum* — pigweed[b]	*Hibiscus syriacus* — althea
Fuchsia hybrida — fuchsia[a]	*Ipomoea batatas* — sweet potato	*Anethum graveolens* — dill
Capsicum fructescens — pepper[a]	*Kalanchoe blossfeldiana* — kalanchoe	*Phlox paniculata* — phlox[a]
Lycopersicum esculentum — tomato[b]	*Fragaria chiloensis* — strawberry[b]	*Oenothera parviflora biennis* — evening primrose[a]
Apium graveolens — celery[b]	*Xanthium pennsylvanicum* — cocklebur	*Delphinium cultorum* — larkspur[a]
Erysimum sp. — wallflower[a]		*Rudbeckia bicolor* — coneflower[a]
Hydrangea macrophylla — hydrangea[a]		*Beta vulgaris* — garden beet[a]
		Matthiola incana — stock
		Digitalis purpurea — foxglove[a]

[a] May be affected by temperature.
[b] May fit other categories, depending upon species or variety.

sometimes used to modify a plant response that usually requires a specific photoperiod. Some species have a "quantitative" response. That is, a given daylength is not an absolute requirement. Examples are carnation, cosmos, snapdragon and petunia. Other species are "qualitative" in that short or long days are an absolute requirement, such as kalanchoe, cocklebur, chrysanthemum and poinsettia. Still others require alternating periods of different light and dark. Bellflower *(Campanula)* flowers only when short days are followed by long days. Chrysanthemum and poinsettia will initiate flowers on short days, but if development is to continue, the days must be still shorter. Asters require long days for elongation; thereafter, the photoperiod is immaterial, but flowering will not occur before elongation.

Some care is required to distinguish between a true photoperiodic response, and a response due to the fact that a short day also reduces total photosynthesis.

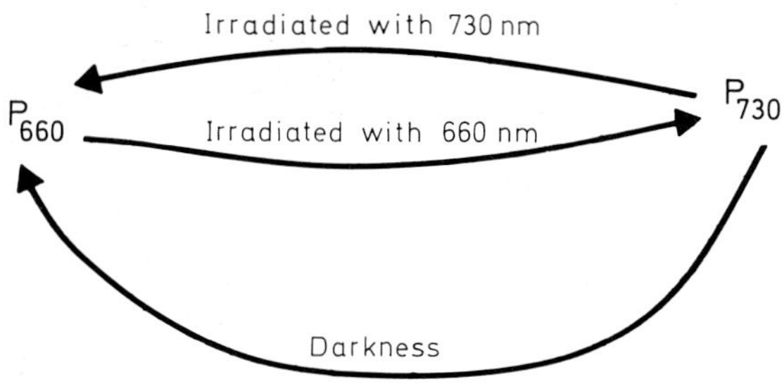

Fig. 2-23. Diagrammatic representation of phytochrome response to radiation quality and darkness. Shift to P_{660} in dark is temperature-dependent, and triggering of short day (long-night) response will depend upon time allowed for sufficient conversion to take place. Irradiation with incandescent light will shift P_{660} to P_{730}, inhibiting flowering in long-night plants

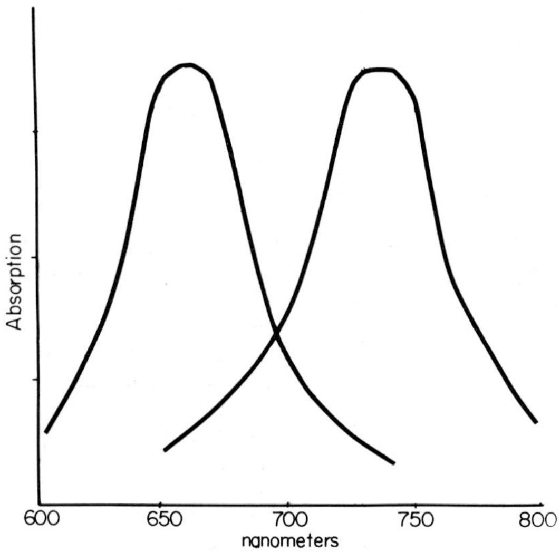

Fig. 2-24. Absorption of P_{660} and P_{730} as a function of wavelength. Greatest absorption occurs by P^{730} at 660 nm, while maximum absorption by P_{660} is at 730 nm or far-red light. (After Butler and Downs, 1960.) Reaction will occur at other wavelengths, but radiant energy must be more intense or for a longer time

Reduced growth may result simply from the fact that there is less time to manufacture food. Another problem, particularly in early research, was a high temperature effect from incandescent lighting under black cloth which gives an apparent response to lighting.

Species may have a temperature requirement that must be met before or after the photoperiodic requirement is fulfilled. Or, photoperiodism may be bypassed by subjecting the plant to a particular temperature regime. Sweet William, evening primrose, foxglove and beet require low temperature (vernalization) followed by daylength control in order to flower. According to Salisbury (1963), poinsettia is a short-day plant at high temperature, but a long-day responder at low temperatures. With some species of chrysanthemum and strawberry, the requirement changes with temperature.

2.4.3.2 Basic Process

The connection between plant environment that triggers the photoperiodic response is a pigment-containing system in the green tissue of plants. Figure 2-23 shows a simplified scheme of the reaction of the pigment, phytochrome, which responds to environment in one of three ways: (1) the wavelength of light irradiat-

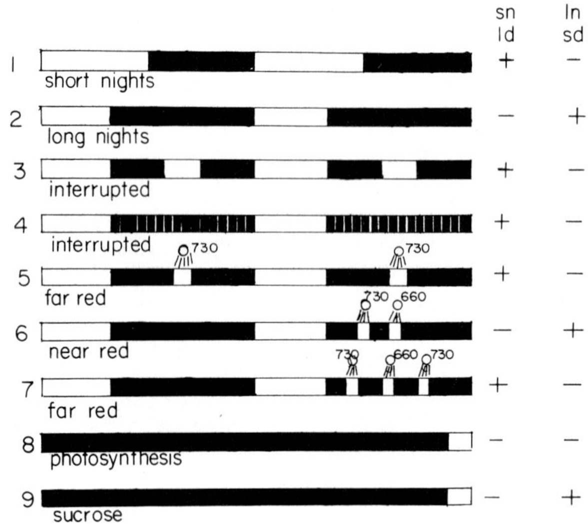

Fig. 2-25. Possible responses for long- or short-day plants under varying regimes of light and dark. *ln (sd)* long night; *sn(ld)* short night. A positive sign indicates initiation and development of flowers. Application of sucrose in *9* provides an energy base for the system to operate, normally provided by photosynthesis

ing the green tissue; (2) the length of the dark period; and (3), the temperature at which the dark conversion takes place.

Phytochrome can be in one of two configurations, P_{730} or P_{660}, the numbers referring to the wavelengths for which the configurations are most sensitive (Fig. 2-24). If green tissue is irradiated with light having a wavelength of 660 nm (near red), the phytochrome molecule will be shifted to P_{730}. P_{730} can be shifted to P_{660} by using a source with its greatest intensity at 730 nm (far red). With P_{730}, flowering in short-day plants will be prevented, or promoted in long-day plants. The reverse occurs with P_{660}. In darkness, P_{730} changes to P_{660}. The variation that can be expected with sensitive species is suggested in Figure 2-25.

In practice, the grower lengthens the dark period to induce flowering in chrysanthemums and kalanchoes under natural long days, and shortens the dark period in the winter to prevent flowering. It is the length of the dark period that determines response.

2.5 Daylength Control

Under natural conditions, the length of day and night varies over the earth's surface, according to season and latitude (Fig. 2-7). The change is fairly precise, although sometimes modified by local conditions of fog or persistant cloudiness. The latitude and species will determine whether shortening or lengthening of the dark period is required. If chrysanthemums are grown at latitudes less than 38°, artificial darkness will not be required as they will initiate and develop flowers regardless of season. To prevent flowering, nighttime is shortened by lighting with incandescent. Daylight duration can be obtained from Table 2-7, and with the requirements of the species in mind (Table 2-8), the decision can be made whether to lengthen or shorten the dark period.

Table 2-7. Duration of daylight (interval between sunrise and sunset) for different dates and latitudes. Note that civil twilight is not included. For photoperiodic purposes, one-half to one hour must be added to arrive at photoperiodically active light. For the southern hemisphere, enter a date about 6 months earlier or later as given in the table. Above 65° latitude, data is uncertain. (Abridged from List, 1966)

Day of month	Hours-minutes											
	Jan	Feb	Mar	Apr	May	Jun	Jul	Aug	Sept	Oct	Nov	Dec
Latitude 10° N.												
1	11–33	11–42	11–56	12–14	12–29	12–40	12–42	12–33	12–18	12–02	11–47	11–36
17	11–37	11–50	12–05	12–22	12–35	12–42	12–38	12–26	12–09	11–53	11–40	11–32
Latitude 20° N.												
1	10–57	11–16	11–45	12–20	12–52	13–16	13–19	13–02	12–32	11–57	11–25	11–00
17	11–05	11–32	12–03	12–38	13–07	13–20	13–12	12–48	12–13	11–40	11–10	10–56
Latitude 30° N.												
1	10–15	10–46	11–33	12–29	13–20	13–57	14–03	13–34	12–46	11–53	10–59	10–22
17	10–27	11–12	12–02	12–57	13–42	14–04	13–52	13–11	12–18	11–25	10–36	10–14
Latitude 40° N.												
1	09–23	10–10	11–18	12–39	13–54	14–49	14–58	14–16	13–05	11–47	10–29	09–33
17	09–42	10–47	12–00	13–20	14–27	15–00	14–42	13–41	12–24	11–06	09–55	09–20
Latitude 50° N.												
1	08–10	09–20	10–58	12–55	14–41	16–04	16–18	15–14	13–31	11–39	09–48	08–24
17	08–38	10–15	11–58	13–53	15–30	16–22	15–53	14–23	12–32	10–40	08–58	08–06
Latitude 60° N.												
1	06–03	08–00	10–28	13–17	15–58	18–17	18–43	16–51	14–10	11–28	08–43	06–28
17	06–51	09–23	11–55	14–45	17–18	18–50	17–57	15–30	12–44	10–02	07–24	05–54

Table 2-8. Some common lighting practices for responsive greenhouse crops, generally based on locations above 40° N latitude (1 ft-c = 10.76 lux, 1 W ft^{-2} = 10.76 W m^{-2})

Species	Long day Short day	Hours light (dark) required	Intensity		Remarks	Ref.
			ft-c	W ft^{-2}		
1. Aster (*Callistephus chinensis*)	LD	16	5–10	3–6	LD required for elongation, preceeding flowering, 4-h night break	Post (1950) Ball (1974)
2. Azalea (*Rhododendron*)	LD	16	25–40	8–10	LD promotes stem length and increases leaf size, supplemental irradiation	Bickford and Dunn (1972)
3. Carnation (*Dianthus caryophyllus*)	LD	24	5–10	3–6	Dawn-to-dusk at 6–7 expanded leaf pairs, for 2–4 weeks, promotes elongation	Ball (1974) Koon (1974)
4. Chrysanthemum (*Chrysanthemum morifolium*)	SD	(11)	5–20	3–12	SD required for flowering, at 60° F, cyclic lighting requires higher intensity to prevent flowering, dark period less than 7 h with 4-h night break	Post (1950) Ball (1974)
5. Gloxinia (*Sinningia speciosa*)	LD	16	10–20	5–10	Hastens flowering, daylight extension	Bickford and Dunn (1972)
6. Kalanchoe (*Kalanchoe blossfeldiana*)	SD	(11.5)	5–10	3–6	SD requirement for flowering, 4-h night break to prevent	Ball (1974) Post (1950)
7. Easter lily (*Lilium longiflorum*)	LD	16	15	5	Hastens flowering, 2 wks after emergence, 2–4 h night break	Wilkins et al. (1968a, b)
8. Petunia (*Petunia hybrida*)	LD	16	20	8–10	LD hastens elongation and flowering, variable with temperature, SD and high temperatures induce branching	Cathey (1964) Ball (1974)
9. Poinsettia (*Euphorbia pulcherrima*)	SD	(12)	1–5	2.5–5	SD to flower, LD plant at temperatures below 50° F	Bickford and Dunn (1972), Post (1950)
10. Fuchsia (*Fuchsia hybrida*)	LD	16	15	5	Daylight extension to enhance flowering	Sachs and Bretz (1962)
11. Snapdragon (*Antirrhinum majus*)	LD	16	5–10	3–6	Hastens flowering, night break 4 h at 5–10 leaf pairs	Langhans and Maginnes (1962)

12. Stephanotis *(Stephanotis floribunda)*	LD	14	5–10	2.5–5.0	LD to flower, 4 h night break, night temperatures of 65° F	Post (1950) van der Veen and Meijer (1959)
13. Strawberry *(Fragaria sp.)*	LD	16	5–15	2–9	LD to promote runner formation and flowering, prevent dormancy at night temperatures of 43° F or lower, 8 h light period (SD) to initiate flowers at any temperature, LD thereafter	Went (1957) van der Veen and Meijer (1959)
14. Calceolaria *(Calceolaria herbeohybrida)*	LD	16	5–10	0.5–3	Earlier flowering, 8 h daylight extension or 5 h night break, temperature must be below 50° F to initiate flowers	Bickford and Dunn (1972)
15. Cineraria *(Sinningia cruentus)*	LD	16	5–10	0.5–3	Respond similarly to Calceolaria, incandescent may produce softer growth	Bickford and Dunn (1972)
16. Centaurea *(Centaurea sp.)*	LD	16	5–10	0.5–3	Requires LD to flower, day-neutral at low temperatures	Post (1950) Bickford and Dunn (1972)
17. Camellia *(Camellia japonica)*	LD	16	10–20	5–10	LD promotes flowering at high temperatures, minimum night temperature of 60° F required for flower initiation	Post (1950) Bickford and Dunn (1972)
18. Begonia *(Begonia sp.)*	SD	(12)	5–15	1–5	Small flowered variety, Lorraine, flowers on SD, large flowered Elatior, goes dormant, tuberous rooted types produce tubers on SD, semituberous or fiberous rooted similar	Bickford and Dunn (1972) van der Veen and Meijer (1959) Post (1950)
19. Christmas cactus *(Zygocactus truncatus)*	SD	(15)	—	—	Flowers on SD, 4-wk period	van der Veen and Meijer (1959)
20. Tall marigold *(Tagetes erecta)*	SD	(15)	—	—	Hastens flowering on SD	Ball (1974)

2.5.1 Lengthening the Dark Period

The most common method is to cover the crop with an opaque material that reduces light intensity below 22 lux (2 ft-c). The methods are numerous, depending upon area to be covered, bench arrangement and greenhouse construction (Fig. 2-26). The most common material, a black, sateen cloth, is pulled over the plants between 1600 and 1830, and removed between 0700 and 0900. The times may vary, depending upon the local conditions. Black cloth is expensive; for example, if two raised benches, 120 ft (40 m) long, are to be shaded, a cloth about 20 × 135 ft (7 × 45 m) would be required. At 1970 prices, the cost for cloth would amount to over $ 200.00. A cheaper substitute is 4–6 mil black plastic. The cost would be less than $ 50.00.

Black cloth is permeable to water vapor, however, reducing the possibility that moisture may condense on the vegetation. Whatever the shading material, there is danger of excessive heat under the shade if pulled on clear summer days. Late pulling can be practiced, fans can be used to pull air under the cloth, and special provisions are sometimes necessary to dry out the plants to prevent condensation. Labor requirements for pulling shade cloth over large areas can be expensive, and mechanical methods are usually preferred.

Fig. 2-26. Example of black cloth shading across a greenhouse. Fans pull air under the cloth to prevent excessive temperatures

The amount of light allowed under the cloth depends upon the species. Poinsettias are sensitive to intensities less than 10 lux (1 ft-c), and turning on the greenhouse lights at night to check temperatures, or outside street lights, have been known to delay flowering. Tears, leaks between cloth junctions, and worn spots on the cloth should be avoided. Occasional 24-h skips in shading are not disastrous, but the crop can be delayed. It is not good to skip over the weekends.

2.5.2 Shortening the Dark Period

Incandescent lights to break the dark period into short cycles are most common (Figs. 2-27 and 2-28). Lights are installed above the benches. According to Ball (1974), one 60-W bulb every 4 ft (1.2 m), 60 ins (1.5 m) above the soil is sufficient for a single 4 ft (1.2 m) bed. For two beds, a single row of 100-W bulbs every 6 ft (1.9 m); or for three beds, a 150-W bulb every 1.9 m, is adequate. For a 20 ft house (6 m), a single row of 300-W reflector bulbs (Fig. 2-28) every 10 ft (3 m) will be sufficient. Low intensity lighting of this type requires about 16 watts per m^2 (1.5 W per ft^2). A minimum of 22 lux (2 ft-c) is required, although 54 lux (5 ft-c) at plant height is preferred.

Commonly, the night is broken into two short periods by lighting for 4 h between 2200 and 0200 h. The length of the darkness should be shorter than 7 h. Assuming 2 h per night for two benches, having above them 21 100-W light bulbs, the power consumption would be 4.2 kW-h. At 3 cents per kW-h, the cost would be about 13 cents per day to light the benches. The cost can be reduced by two-thirds by breaking the night into a series of shorter periods, or "cyclic" lighting. As noted in Figure 2-25, a series of short lighting periods will affect photoperiodic

Fig. 2-27. Lighting installation in Florida to prevent flowering

Fig. 2-28. 200-W bulb with internal reflector commercially employed for photoperiodic control

Fig. 2-29. Traveling fluorescent lamps to provide photoperiodic control on chrysanthemums. System operates similarly to cyclic lighting, as each plant is irradiated for a short period as contrasted to continuous irradiation in middle of night. System must ensure sufficiently high intensity, and lights must travel from one end of house to the other to provide proper frequency

Table 2-9. Carrying capacity of wire and distance from supply to load. (See Appendix for conversion to metric). (Abridged from USA and USAF, Electricians Manual, 1957)

Load (A)	Minimum wire size (AWG)[a]	Wire size (AWG)[a] Distance one way from supply to load (ft)							
		50	100	150	200	250	300	350	400
120 V, single phase									
15	14	14	10	8	6	6	4	4	4
25	12	10	8	6	4	4	2	2	2
35	12	8	6	4	2	2	1	1	0
45	10	8	4	2	2	1	0	0	2/0
55	8	6	4	2	1	0	2/0	2/0	3/0
65	8	6	4	2	0	2/0	2/0	3/0	4/0
75	6	6	2	1	0	2/0	3/0	4/0	4/0
85	6	4	2	1	2/0	3/0	3/0	4/0	
95	6	4	2	0	2/0	3/0	4/0		
220 V, three phase									
15	14	14	14	12	10	8	8	8	6
25	12	12	10	8	8	6	6	6	4
35	12	12	10	8	6	6	4	4	4
45	10	10	8	6	6	4	4	2	2
55	8	8	8	6	4	4	2	2	2
65	8	8	6	4	4	2	2	2	1
75	6	6	6	4	2	2	2	1	0
85	6	6	6	4	2	2	1	0	0
95	6	6	6	4	2	1	1	0	2/0
125	4	4	4	2	1	0	2/0	2/0	3/0

[a] AWG = American wire gage, B & S. A gage of 1 = 289 mils = 0.289 inches = 7.3 mm. 1 mil = 0.001 inch = 0.025 mm. As gage increases, wire diameter decreases.

response similarly to one long lighting cycle. The result is to reduce the total time power will be required. However, the system complicates controls. As worked out by Cathey et al. (1961 a, b), the area to be lit is divided into five sections so that each receives 6 min of light every 30 min (20% of the time). There is an intensity versus time relationship so that if the time of actual lighting is reduced, the intensity of that light must be increased. At the level suggested, 110 lux (10 ft-c) is necessary. The lighting can be reduced to 3 s every minute for 4 h on some chrysanthemums species, but intensity should be increased to 220 lux (20 ft-c). Response often varies between varieties, so that the grower should always test response to cyclic lighting on a small scale before full conversion.

There are other systems that have been employed as shown in Figure 2-29. While the cost in bulbs and wiring is reduced, there is an increased investment in the mechanical system to move the fluorescent lamps down the greenhouse aisles.

High voltage bulbs are usually cheaper for wiring installations as wire size decreases with increasing voltage (Table 2-9). Operating an incandescent above or below its rated voltage will change its lifetime as well as its light output. A 120-V lamp operated at 125 V will produce 16% more light, but its life span will be

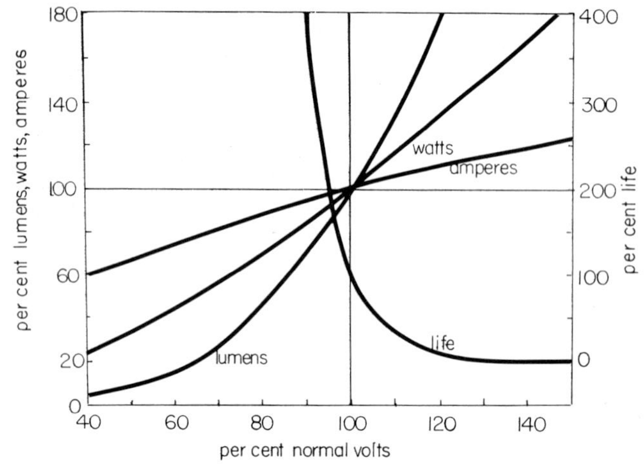

Fig. 2-30. Diagrammatic relationship between voltage, amperage, and wattage of incandescent lamps and life and output

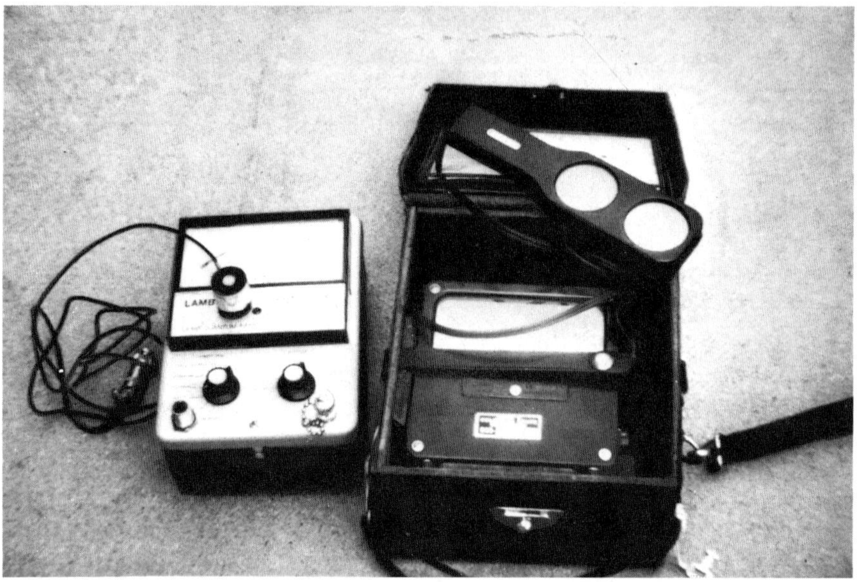

Fig. 2-31. Two types of photometric sensors. *Left:* "Lambda meter", evaluates in lux and is cosine corrected. Sensors are interchangeable to measure photosynthetically active radiation in "Einsteins" or total radiation in gm-cal cm^{-2}. *Right:* "Illuminometer", with output in foot-candles, noncosine corrected, but color-corrected for human eye sensitivity. Latter usable only for direct radiation, perpendicular to receiving surface, and is most common type employed in the U.S.

reduced by 38%. The same lamp at 115 V will produce 13% less light, consume 6% less power, and last 62% longer (Fig. 2-30). To avoid duplicating heavy wiring, or increasing the load factor (which may determine base electrical cost), growers sometimes utilize exhaust fan capacity for the lighting system.

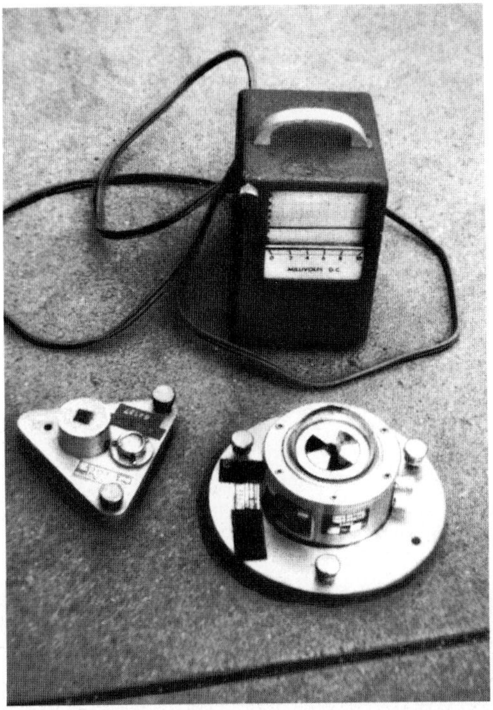

Fig. 2-32. Two types of radiometric sensors for measuring radiation between 3 and 3000 nm, together with an inexpensive recording device for measuring output from the silicon cell *(left)*. The glass-covered sensor *(right)* is an Eppley pyranometer, standard U.S. Weather Bureau radiometer for measuring short-wave solar radiation

2.6 Measurement of Radiation

Light measurement may be difficult or easy, depending upon accuracy required and how the measurement is made. Very few sensors respond equally well to all wavelengths. Costs for measuring devices may range from a few dollars to several thousand. But, some simple device pays dividends to the grower in terms of safety and information: A photometric device will ensure that the lighting system is adequate to prevent delayed and nonuniform flowering. Other devices will permit the grower to monitor radiation through the greenhouse cover, allowing timely practices to be started before sunlight is drastically reduced due to dirt or weathering. Continuous measurements will allow him to program watering frequency, improve crop timing, and set temperatures in accordance with available sunlight.

2.6.1 Photometric Sensors

Photometric sensors are devices that measure energy in the visible spectrum. They may be fitted with filters so the sensor responds similarly to the human eye. Two examples are provided in Figure 2-31. Even camera light meters can be used if their limitations are understood. Simple homemade systems can be put together with little cost. Units are in foot-candles, lux or lumens. Much of the radiation in sunlight or artificial source may be excluded by photometric devices and so considerable caution needs to be used in converting units.

2.6.2 Radiometric Sensors

These devices are employed to measure radiant energy over the entire spectrum in some absolute unit as watts, ergs, calories or langleys. Figure 2-32 shows two types, the Eppley being rather expensive and is the U.S. Weather Bureau's standard apparatus for measuring solar radiation between wavelengths of 300–3000 nm. The silicon cell, if suitably calibrated for sunlight, can give reasonable information for an approximate cost of about $ 250.00.

References

Aikin, J. W.: Photosynthesis in roses. I. Effect of light intensity. Colo. Flower Growers' Assoc. Bul. **289**, 1–3 (1974)

Albrecht, F.: Intensität und Spektralverteilung der Globalstrahlung bei klarem Himmel. Arch. Met. Geophys. Bioklim. **3**, 220–243 (1951)

Ball, V.: The ball red book. West Chicago: Geo. J. Ball, 1974, 502 pp.

Bernier, C. J.: 1962. Measurement techniques for the radiant energy requirements of growing plants. Reprint No. 48. Nat. Tech. Conf. Illum. Eng. Soc., Dallas, Texas, 1962

Besemer, S. T.: An economic analysis of the carnation industry in the United States. M.S. Thesis, Colo. State Univ., Ft. Collins, 1966, 124 pp.

Bickford, E. D., Dunn, S.: Lighting for plant growth. Kent, Ohio: Kent State Univ., 1972, 221 pp.

Bonner, J.: The upper limit of crop yield. Science **137**, 11–15 (1962)

Boysen-Jensen, P.: Die Stoffproduktion der Pflanzen. Jena: Gustav Fischer, 1932

Boysen-Jensen, P., Muller, D.: Die maximale Ausbeute und der tägliche Verlauf der Kohlensäureassimilation. Jahrb. Wiss. Botan. **70**, 493–502 (1929)

Businger, J. A.: The glasshouse (greenhouse) climate. In: Physics of the plant environment, eds. van Wijk. Amsterdam: North Holland, 1966, 382 pp.

Butler, W. L., Downs, R. J.: Light and plant development. Sci. Am. **230** (6), 56–63 (1960)

Campbell, L. E., Klueter, H. H., Cathey, H. M., Krizek, D. T., Bailey, W. A.: Light sources used in horticulture in the USA. Symp. Electricity and Artificial Light in Horticulture, Littlehampton, England. March 17, 1969. Paper No. 4/5, 1969, 18 pp.

Canham, A. E., Cockshull, K. E., Hughes, A. P.: Supplementary illumination of Chrysanthemums. Manuscript, Electrical Res. Sta., Univ. of Reading, Shinfield, 1969, 2 pp.

Carpenter, W. J.: Rose lighting, comparing HID lamps and intensities. Roses, Inc. Bull. May, 1974, 56–63 (1974a)

Carpenter, W. J.: High intensity supplemental lighting of roses with lucalox and multivapor lamps. Roses, Inc. Bull. Dec., 1972, 39–43 (1974b)

Cathey, H. M.: Prescriptions for annual plants. Light and chemicals. Florists' Rev. **133** (3458), 64–66; **134** (3450), 23–25 (1964)

Cathey, H. M., Bailey, W. A., Borthwick, H. A.: Cyclic lighting—to reduce cost of timing chrysanthemum flowering. Florists' Rev. **129** (3330), 71–22, 72–75, 94–95 (1961b)

Cathey, H. M., Borthwick, H. A.: Cyclic lighting for controlling flowering of Chrysanthemums. Proc. Am. Soc. Hort. Sci. **78**, 545–552 (1961a)

Chang, J.: Climate and agriculture. Chicago: Aldine, 1968, 304 pp.

Cook, G. D., Dixon, J. L., Leopold, A. C.: Transpiration: its effects on plant leaf temperature. Science **144**, 546–547 (1964)
Crabb, G. A., Jr.: Solar radiation investigations in Michigan. Tech. Bull. 222, Mich. State Coll., East Lansing, 1950
Davidson, J. L., Philip, J. R.: Light and pasture growth. UNESCO Arid Zone Res. 25, 1958
Forsythe, W. E.: Smithsonian physical tables. Misc. Collection 120. Smithsonian Inst., Washington, 1964, 877 pp.
Gaastra, P.: Photosynthesis of crop plants as influenced by light, carbon dioxide, temperature, and stomatal diffusion resistance. Med. Landbou. Wageningen. **59**, 1–68 (1959)
Garner, W. W., Allard, H. A.: Effect of relative length of day and night and other factors of the environment on growth and reproduction in plants. J. Agr. Res. **18**, 553–606 (1920)
Gates, D. M.: Energy exchange in the biosphere. New York: Harper and Row, 1962, 151 pp.
Gates, D. M.: A. Heat, radiant and sensible, Chapter I. Radiant energy its receipt and disposal. Meteorol. Monographs **6**, 1–26 (1965a)
Gates, D. M.: Heat transfer in plants. Sci. Am. **213**, 76–84 (1965b)
Gates, D. M., Keegan, H. J., Schleter, J. C., Weidner, V. R.: Spectral properties of plants. Appl. Optics **4**, 11–20 (1965)
Good, D., Goodrich, D., Langhans, R.: Economic analysis of the use of high intensity discharge lamps in rose production. Roses Inc. Bull. 53–56 (1974)
Hanan, J. J.: Preliminary measurements of flower temperature. Colo. Flower Growers' Assoc. Bull. **188**, 1–4 (1965)
Hanan, J. J.: Colorado solar energy. Colo. Flower Growers' Assoc. Bull. **210**, 6 (1967)
Hanan, J. J.: Advantages of clear-day climatic regions for greenhouse production. Colo. Flower Growers' Assoc. Bull. **217**, 1–3 (1968)
Hanan, J. J.: Water loss and stress in carnations grown under glass. Colo. Flower Growers' Assoc. Bull. **233**, 1–3 (1969)
Hanan, J. J.: Some observations on radiation in greenhouses. Colo. Flower Growers' Assoc. Bull. **239**, 1–4 (1970)
Kleschnin, A. F.: Die Pflanze und das Licht. Berlin: Akademie, 1960, 619 pp.
Klueter, H. H., Cathey, H. M., Downs, R. J., Bailey, W. A.: Artificial lighting in greenhouses. Ann. Meeting. Am. Soc. Agr. Eng., Ft. Collins, Colo. Paper No. 64–433, 1964, 6 pp.
Koon, G.: Timing carnations in Colorado with lights. Colo. Flower Growers' Assoc. Bull. **284**, 3–4 (1974)
Kramer, P. J., Decker, J. P.: Relation between light intensity and rate of photosynthesis of loblolly pine and certain hardwoods. Plant Physiol. **19**, 350–358 (1944)
Landsberg, H. E.: Solar radiation at the earth's surface. Solar Energy **5**, 95–98 (1961)
Langhans, R. W., Maginnes, E. A.: Temperature and light. In: Snapdragons, a manual of culture, insects and diseases and economics of snapdragons, ed., Langhans, R. W. Ithaca, N.Y.: N.Y. State Flower Growers' Assoc., 1962
List, R. J.: Smithsonian meteorological tables. 6th ed. Smithsonian Inst., Washington. Smiths Misc. Collection 114, 1966, 527 pp.
Loomis, R. S., Williams, W. A., Hall, A. E.: Agricultural productivity. Ann. Rev. Plant Physiol. **22**, 431–468 (1971)
Mecklenberg, R. A., Rintelmann, W. F., Schumaier, D. R., Den Brink, C. V., Flores, L.: Effect of plants on microclimate and noise reduction in the urban environment. HortScience **7**, 37–39 (1972)
Moon, P.: Proposed standard solar radiation curves for engineering use. J. Franklin Inst. **230**, 583–617 (1940)
Norton, R. A.: Cost comparisons and comparison of lamps for supplementary greenhouse lighting. Paper presented to Western Region, Am. Soc. Hort. Sci., Eugene, Oregon, 1971, 2 pp.
Norton, R. A.: Information on supplementary illumination of greenhouse crops. Paper presented at the Western Region, Am. Soc. Hort. Sci., Eugene, Oregon, June, 1972, 2 pp.
Post, K.: Florist crop production and marketing. New York: Orange-Judd, 1950, 891 pp.
Sachs, R. M., Bretz, C. F.: The effect of daylength, temperature and gibberellic acid upon flowering of Fuchsia hybrida. Proc. Am. Soc. Hort. Sci. **80**, 581–588 (1962)
Salisbury, F. B.: The flowering process. Elmsford, N.Y.: Pergamon, 1963

Schuck, E. A., Altshuller, A. P., Barth, D. S., Horgan, G. B.: Relationship of hydrocarbons to oxidants in ambient atmosphere. J. Air Pollution Control Assoc. **20**, 297–302 (1970)
Smith, C. V., Kingham, H. G.: A contribution to glasshouse design. Agr. Meteor. **8**, 447–468 (1971)
Smith, W. T., Downs, R. J., Bonaminio, V. P.: An economic system of supplying supplemental light in greenhouses. Am. Soc. Agr. Eng., Paper No. 74–4042. Summer Meeting, Okla. State Univ., Stillwater, 1974, 10 pp.
USA-USAF: Electrical wiring. U.S. Army U.S. Air Force Tech. Man. TM 5-760/TO 00-105A-10, 1957, 91 pp.
U.S.D.A.: Climate and Man. USDA Yearbook, House Doc. No. 77, Washington, D.C., 1941, 1248 pp.
Veen, R. van der, Meijer, G.: Light and Plant Growth. Eindhoven: Philips, 1959, 161 pp.
Vlohos, J., Boodley, J. W.: Acclimatization of Brassica actinophylla and Fiscus nitida to interior environmental conditions. N.Y. State Flower Industries Bull. **50**, 2–5, 7 (1974)
Wassink, E. C., Richardson, S. D., Pieters, G. A.: Photosynthetic adaptation to light intensity in leaves of Acer pseudoplantanus. Acta Botan. Neerl. **5**, 247–256 (1956)
Went, F. W.: The experimental control of plant growth. Waltham, Mass: Chronica, 1957
Wilkins, H. F., Waters, W. E., Widmer, R. E.: Influence of temperature and photoperiod on growth and flowering of Easter lilies (Lilium longiflorum Thunb., 'Georgia,' 'Ace,' and 'Nellie White'). Proc. Am. Soc. Hort. Sci. **93**, 640–649 (1968a)
Wilkins, H. F., Waters, W. E., Widmer, R. E.: The influence of carbon dioxide, photoperiod and temperature on growth, and flowering of Easter lilies. (Lilium longiflorum Thunb. 'Ace' and 'Nellie White'.) Proc. Am. Soc. Hort. Sci. **93**, 650–655 (1968b)
Withrow, R. B., Withrow, A. P.: Generation, control and measurement of visible, and near-visible radiant energy. In Radiation biology, Vol. III, ed. A. Hollaender, pp. 125–258. New York: McGraw-Hill 1956

Chapter 3 Greenhouse Construction

3.1 Historical

Historians do not document the exact inception of growing plants under cover, but Lemmon (1962) recorded the writings of Plato, who, in the 4th Century BC, indicated in his *Phaedon* that plants were grown under protection.

Lemmon also noted that Sir Joseph Banks mentioned the forcing of dessert fruit in Roman times under thin sheets of mica called "Muscovy glass" *(lapis specularis)*. No elaborate structures were used; in most instances plant materials were grown in pits and covered with sheets of mica. Heat was obtained from decomposing manures and hot air flues.

One of the first references of glass use was in 1385 in the Bois de Duc in France, where they grew flowers in glass pavilions facing south (Lemmon, 1962). Taft (1926) included information on the glass house constructed in Apothecaries Garden, Chelsea, England, which had tall side walls of glass, but an opaque roof. It was not until approximately 1700 that glass roofs were constructed and for the next 100 years, few improvements were made. Greenhouse construction started in the United States following the Civil War. Census reports show that there was only one commercial greenhouse prior to 1800; three before 1820, 178 in 1860, in 1890, 4659 facilities covering 38823247 ft^2, and in 1910 there were 105 million ft^2 (Wright, 1917). Fossum (1973) extracted data from the 1970 census and recorded 6487 establishments representing 213939000 ft^2 in floriculture production in the United States.

One of the first greenhouse builders in the United States was Frederic A. Lord, who constructed his houses, in Buffalo, New York, in 1855. In 1870, Mr. Lord moved to Irvington, New York, and in 1872 entered into partnership with W.A. Burnham under the name of Lord and Burnham. The Lord and Burnham company was incorporated in 1883 (Taft, 1926).

In 1888 the firms of Hutchings and T.W. Weathered and Sons, both of New York City, added construction departments to their greenhouse heating businesses. A great deal of competition was displayed among all of the greenhouse companies until the 1900s. Around that era Lord and Burnham Company merged with some of the competitors and also expanded into Canada. Greenhouse construction flourished until the prosperity level diminished during the late 1920s.

Many greenhouse manufacturers contributed to the industry in the early 1900s. The Foley Greenhouse Manufacturing Company constructed several institutional facilities and supplied parts for commercial houses across the nation. The A. Dietsch Company was instrumental in helping to develop the greenhouse vegetable programs in the eastern United States with their greenhouses and construction crews. Both the Foley and Dietsch companies closed their doors in the mid 1950s. The John C. Moninger Company (Moninger, 1913) supplied complete greenhouses throughout North America. Their catalog, a hardback book, contained more than 350 pages of parts, prices and designs. The American Greenhouse Manufacturing Company was very active in the east and was identified by the use of a scissor-type truss in their houses. After standing for more than 50 years, some of these structures collapsed in the snow storm of 1975. There was some type of merger with American and Moninger, but American started failing around 1929. Two men, Mike Winandy, construction superintendent, and P. L. McKee, sales manager, both with the John C. Moninger Company decided to build their own businesses. Mr. McKee transformed the staggering American Greenhouse Company into the now active National Greenhouse Manufacturing Company and Mike Winandy started the Winandy Greenhouse Construction Company, which is still active.

In 1914 the Ickes-Braun Mill Company was actively engaged in making sash for cold frames. Within four to five years they were manufacturing pipe greenhouse structures and attaching their sash products as part of the covering. In the early thirties the Schultz family became involved with the greenhouse portion of the business. Since that time the name has changed to Ickes-Braun Greenhouses and more recently Ickes-Braun Glasshouses.

The Metropolitan Greenhouse Manufacturing Company was very active until shortly after World War II when it closed its doors. During WW II several greenhouse companies turned their interests to manufacturing items for our national endeavors. The Burnham Corporation turned to constructing floating bridges, hand grenades, aircraft carrier decks and other supplies (*Our 100 Years*, 1956). Since WW II few greenhouse manufacturing companies have emerged. In 1946 Rough Brothers started supplying parts and constructing facilities in the east. The Nexus Corporation began operations in 1967 and has been constructing in the Central and Northwest United States. Double A Truss Manufacturing Company has been active on the West Coast.

The main commercial greenhouse manufacturers in the United States are:

Lord and Burnham Div.
Burnham Corporation
Irvington, New York 10533

National Greenhouse Company
P. O. Box 100
Pana, Illinois 62557

Nexus Corporation
2250 19th Street
Denver, Colorado 80202

Ickes-Braun Glasshouses Div.
Roper Corporation
P. O. Box 147
Deerfield, Illinois 60015

Rough Brothers, Inc.
P. O. Box 16010
Cincinnati, Ohio 45217

Double A Truss Manufacturing Company
320 Wetmore
Manteca, California 95336

3.2 Planning a Greenhouse

It is assumed the greenhouse operator has already evaluated the market potential that would be related to a new greenhouse business or expansion of present facilities. In either case, it is important that plans for all phases of greenhouse construction be developed and a tentative budget formulated. Such a procedure is part of good management. The plans will be required by the local building authority and complete plans, budget and projected income solicited by all potential financial lending agencies.

3.2.1 Basic Considerations

3.2.1.1 Location

"The location of a range of glass for commercial purposes, where the elements of expense and profit are to have first consideration, is of great importance." These words by Liberty Hyde Bailey (1900) are just as important today as they were then. Bailey went on to point out that the desirability of a location is also based on the adaptability and value of the land, cost of fuel delivered, ample and inexpensive water and the proximity to market.

3.2.1.2 Utilities

One of the major considerations for a greenhouse complex is utilities. Compromises in marketing potential and geographical location may have to be considered when utilities, especially fuel, are sought. The energy crisis of the 1970s curtailed greenhouse construction and prudent planning will be required to meet the utility needs of the operator in the future.

3.2.1.3 Climate

Another factor involved in location planning is weather conditions. What is the yearly available solar energy? How much moisture falls, summer and winter? What are the maximum and minimum temperatures and their duration? What are the hail and wind belts? Is air pollution a potential problem? Information on all of the foregoing questions allows the greenhouse operator to determine the degree to which he can maintain near optimum environmental conditions for plant growth. In some areas of the world, the availability of light is paramount (Lawrence, 1963; Smith, 1967), but where solar energy is abundant, other climatological factors take precedence. Information on climatic conditions can be obtained from government Climatological Reports, local weather stations and, if one has enough interest, much can be learned through communications with some of the "old-timers" in a community.

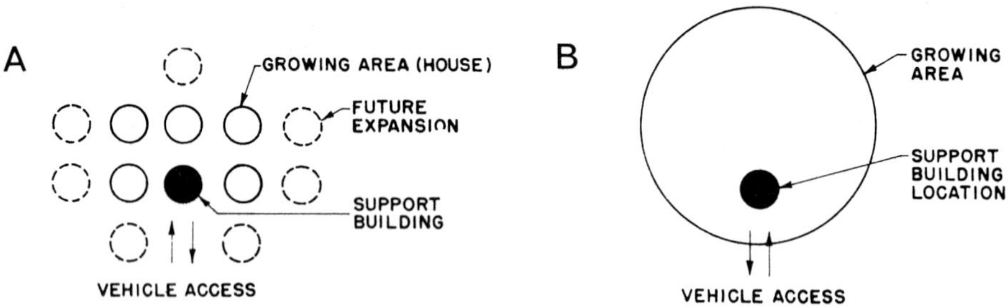

Fig. 3-1A and B. Concepts for designing a greenhouse complex (Ross and Aldrich, 1976). (A) Individual houses and direct house shipping. (B) Large enclosed area and central shipping

3.2.1.4 Topography

Another important consideration is the topography and soil type of the proposed building site. A building site of uniform grade, with a slight slope is most desirable. If ground beds are to be considered using the existing soil, little or no grading should be accomplished unless there is assurance the remaining soil will be reasonably consistent in texture, structure and depth. Water drainage in the subsoil horizons is also very important. Many greenhouses have been built on land where poor drainage exists and there are continuous nutrient, disease, and water problems.

3.2.1.5 Basic Layout

One of the merits of being able to build a new greenhouse facility is the opportunity to provide adequate growing, work, shipping, and storage space. Not all greenhouse complexes need the same type of facilities, but all need to be planned and laid out so expansion can occur easily and material handling systems employed.

The impact of the bedding plant business in the United States has contributed more to the planning of industry related facilities, than all the recommendations from extension personnel or professors for the past 50 years. Growers switching from other floricultural crops to bedding plants have found they need to modify their present facilities to meet the needs of handling a different crop.

A diagram, Figure 3-1, developed by Ross and Aldrich (1976) presents a general concept that is adaptable to any type of greenhouse operation.

Based on the "operational analysis" conducted by Ross et al. (1974), greenhouse facility "layouts" should eliminate, if possible: poor arrangement of working, storage and growing areas, thus eliminating backtracking, distances, and obstructions. One must consider material handling and mechanization.

3.2.2 Support Facilities

Support facilities in some instances are considered a luxury, but are necessary in order to obtain the greatest return from the production area and are required for

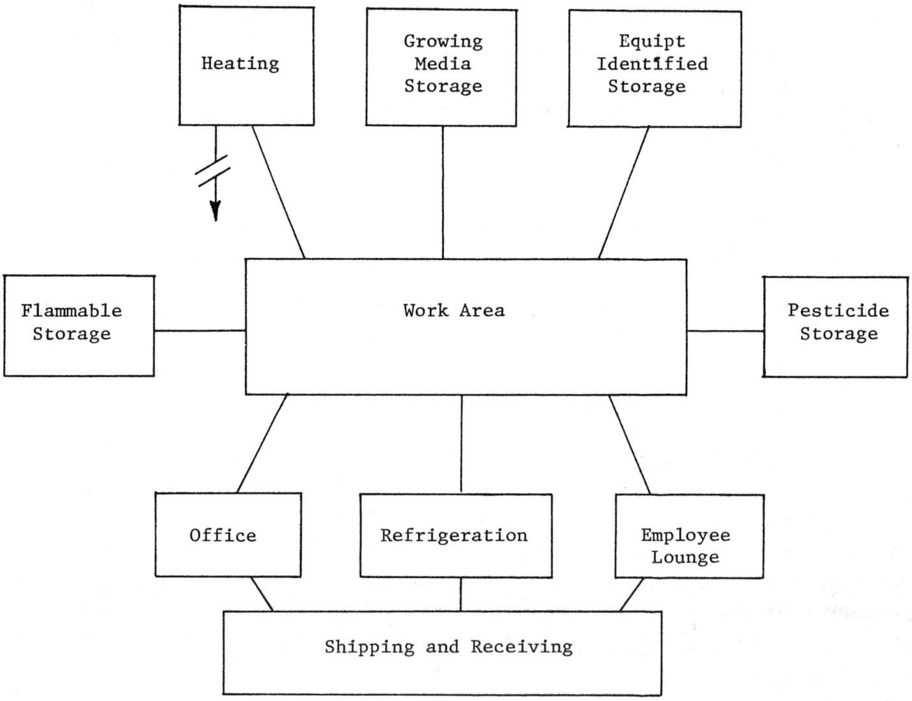

Fig. 3-2. The interrelationships of the support facilities that should be considered when planning a greenhouse complex

efficient management. Figure 3-2 provides a chart of the major components of the support facilities.

Every greenhouse operation can use refrigeration facilities in many ways including bulb rooting, cold induction, cut flower storage, and for holding seedlings and pot plants. Government agencies are also demanding that flammable and pesticide storage facilities be created. Such facilities are required to be separate from other support or production areas. A third facility that is becoming a necessity is an employee lounge and lunch room. A better employee-employer relationship is going to be thrust on the greenhouse industry in the near future.

The utilization of a shipping or work area depends on the type of operation and the crop grown. Some greenhouse operators prefer to ship directly from the production area and also do much of their potting or transplanting there (Ross et al., 1974).

The office area has become an important part of the greenhouse complex, especially for those operations that have environmental control panels and communication centers. The greenhouse of the future will not only be monitored for temperature, but some programs will utilize television cameras to monitor plant conditions, employee production and shipping and handling activities, all from the office.

3.2.3 Production Facilities

3.2.3.1 Design for Use

1. Type of Crop. The type of crop considered should be one of the primary criterion used to determine a suitable greenhouse design. Bedding plant facilities do not require as much "head room" for growing as would be required in a rose or carnation range. On the other hand, the type of automatic or mechanical requirements may dictate a similar design for growing all floriculture crops. The bedding or pot plant grower may move all containers with a forklift, tractor, or overhead tram. The carnation grower often leaves plants in production two to three years and constructs elevated walks. In some areas of the world it is not unusual to see pot plants on two or three shelf levels in the greenhouse. Greenhouses used for vegetable production often incorporate tractors, planters, and harvesters into their program and the structures must be designed to meet these needs—head room in a plant growing facility must be considered.

If a grower anticipates definite crops, that would be compatible to a specific greenhouse design, then he would probably decrease the cost of his initial investment and no doubt lower his operating costs for years to come by purchasing a specific structure.

Coupled with "design for use" is the need to consider economics. The economics of heating and cooling efficiency, construction costs and durability. One greenhouse construction contractor claims "the higher the eave the easier it is to maintain a uniform inside temperature". Temperature is considered later, but the additional building materials for such a design not only contribute to a higher initial cost, but also increased fuel requirements, due to a greater heat loss and enlarged air volume.

2. Bench Type. Several bedding and pot plant growers have eliminated benches in their production areas. One grower places his poinsettia crop on a concrete floor and follows it with bedding plants. Another covers the ground with gravel and has asphalt walks ... the bench has been completely eliminated. Many other wholesale bedding plant growers have also eliminated benches. Plant production at ground level could prove more costly for the grower due to temperature differentials which will be discussed in Chapter 4.

In recent years, many cut flower growers have started planting directly in ground beds. Wood is usually installed around the perimeter to retain the growing media, for attachment of water systems and to act as a kick board to decrease the possibility of contamination from the walk areas (Fig. 3-3).

Where subsoil conditions prevent good drainage, perforated tile and gravel can be installed in the bottom of a two or three foot trench, dug with a backhoe. The topsoil or modified mix is then placed on top of the gravel and the perimeter boards constructed.

Ground beds are often difficult to pasteurize or fumigate. Some growers inject steam through the tile, if it is clay or a material that will withstand the temperature. Ground benches are basically an elevated ground bed (Fig. 3-3). A box or bench with side walls is constructed on piers. Many growers of cut flowers and

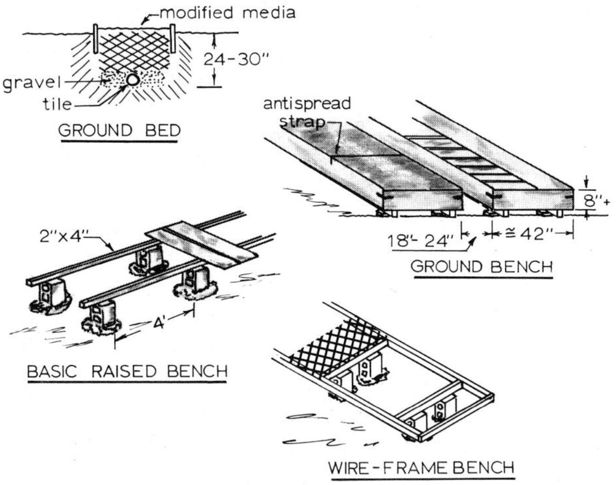

Fig. 3-3. Basic construction designs for greenhouse benches. Note concrete blocks are placed on small piles of concrete to facilitate leveling and promote rigidity

vine crops prefer this type of low growing area because potential disease, water, and growing media problems are easier to control.

Elevated benches have been a tradition in design and only in the past decade have growers realized certain types are prohibited economically. If short species, or single-crop types of cut flowers are grown in raised benches, then sound bench legs, bottoms, and sides will be necessary to withstand the maximum load. The monolithic concrete bench has proven satisfactory in many greenhouses. Some are 50 years old and those in the Colorado State University Research Greenhouses have been installed 30 years. Some of the redwood benches are also 30 years old, but are deteriorating.

Many greenhouse operators have designed their own benches and they have proven valuable. Walker and Duncan (1973b) described six types of raised benches that can be constructed by the greenhouse operator (Table 3-1). Almost any combination of material can be and is used.

Table 3-1. Basic materials that can be used for greenhouse benches. (Adapted from Walker and Duncan, 1973b)

Bottom	Legs	Sides
Corrugated transite	Pipe, concrete, aluminum	Flat transite
Flat transite	Pipe, concrete, aluminum	Flat transite
Wood Redwood Cedar Treated fir, etc.	Concrete block	Wood, none
Welded wire	Wood, concrete block	None
Expanded metal	Concrete block	None
Snow fence slat	Concrete block	None
Concrete	Concrete	Concrete

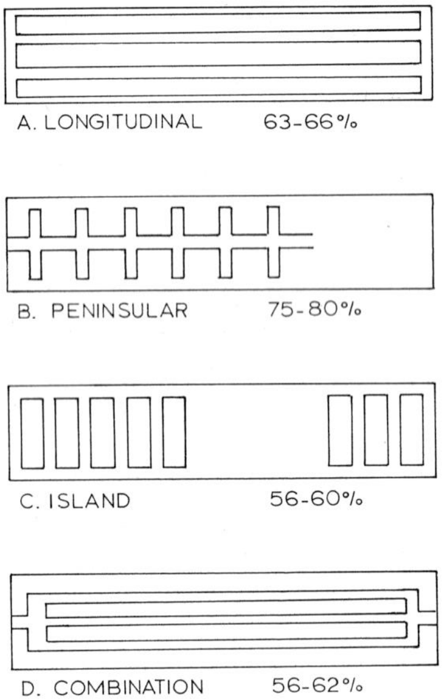

Fig. 3-4 A–D. Bench arrangements that can be adapted to any type of growing or marketing facility. The approximate percent of ground area covered by each bench design is also presented

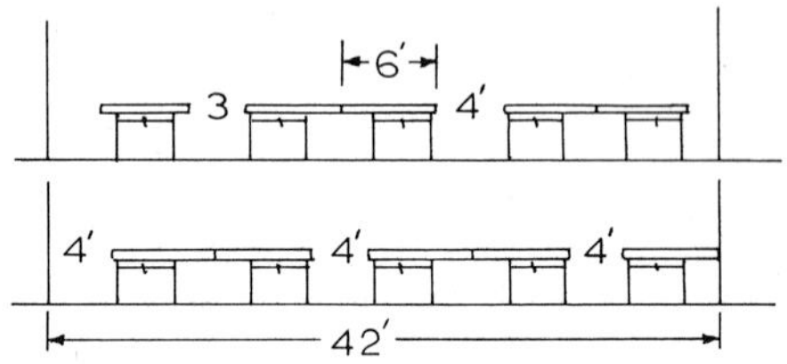

Fig. 3-5. Space-saving movable benches that can provide 86–90% growing area

Bench arrangement and utilization can be the controlling factor on a profit and loss statement. A good manager will design benches to fill the greatest ground area possible. Many greenhouse facilities are limited to two or three bench arrangements or combinations thereof (Fig. 3-4). Greenhouse structures have been designed with a definite correlation between the number of longitudinal benches that will fit into a facility, and the width of the structure. Most longitudinal and island benches have been traditionally designed 42 in wide although some may be 60 in. The 42 in bench is more convenient for reaching plant materials in the center of the bench. Bedding and pot plants are often grown on peninsular benches which can be 60–72 in wide and still convenient. There is also a height of bench above ground versus width correlation for ease of handling plant materials.

The combination and tiered benches have been reintroduced in the past few years especially in the retail floricultural outlets for display purposes. Due to lower light requirements of foliage plants, the tiered bench can provide an excellent gross return for the ground area covered.

Benches tomorrow—Many growers and industry-related personnel have been considering bench designs for decades. The ideal situation would be to have 100% of the ground area covered and still be able to meet the cultural requirements of a given crop. Nursery Supply, a Denver-based pot plant organization, has eliminated benches for pots and converted to suspended pipe holders. The truss design also acts as a manifold for attaching watering tubes.

The Lieder and Son Corporation of Chicago has installed the new Simtrac "aisle eliminator bench system." The system, Figure 3-5, provides approximately 20% more growing area than the standard longitudinal bench design. Individual benches can be cranked back and forth to make an aisle between any bench or in combinations. Aisle width must be considered as the era of greenhouse mechanization approaches and there will be new ideas for bench designs to meet the needs of every operation.

3.2.3.2 Mechanization

Ross et al. (1974) stated that: "Material handling and mechanization was found to be a major area for engineering research", in the bedding plant industry. Every effort should be made to decrease overhead and the greenhouse operator must include as much mechanization as is economically possible into his plans, when designing any type of greenhouse facility. Most growers place mechanization as one of their top priority needs (Voigt, 1973).

Obtaining one new piece of equipment that can reduce overhead is not necessarily being mechanized, but it is a start. For the most part being mechanized means developing labor-saving systems that will reduce labor input and the number of times materials are handled. Bartok (1974) listed eight factors that should be considered when developing labor-saving systems. It was pointed out, however, once a system is selected and installed, employees should be trained to utilize its capabilities. Factors listed by Bartok are:

1. The size and layout of existing greenhouses
2. Future expansion plans
3. The use of standard equipment and its cost per unit moved
4. Other uses for the equipment during the year
5. The amount of growing space that will be lost
6. The amount of labor that will be saved
7. Distance the material has to be moved
8. Maintenance

Mechanization can work bilaterally—take the components of a crop to the worker or take the employee to the crop. In Japan, a suspended rail system is being used to water, feed, and spray pot plants spaced on the ground plus transport disbudders and "order pullers". The research station in Wageningen, Holland (*Approach to Mechanization*, 1975) is conducting studies with totally movable pallets for pot plants that cover an entire greenhouse. They have developed a machine that will "pot" plants and place them on a wire grid pallet. The pallet is suspended in a nutrient solution for a period of time. After growth is achieved, the pallet, mounted on rollers, is pushed to the marketing area.

Plans for black cloth "shading" systems for photoperiod control should be included in all new greenhouse structures. Many systems are automated and cover thousands of square feet in minutes. Most black cloth systems can be modified for future use, especially during periods of probable energy shortages. Unpublished research by Tristan (1975) integrated a netting material on a Simtrac Total Shade System (Trade Name) to control the summer light intensity. The same system can be employed to develop a smaller volume for greenhouse cooling purposes by installing a clear plastic film.

Equipment for mechanization systems is not necessarily greenhouse oriented. Many growers use ingenuity and adapt equipment from other industries to their system. Bartok (1974) listed 51 pieces of equipment, their potential use and sources, that can be considered in greenhouse production. The value of such equipment is only realized if facilities are predesigned or redesigned for their incorporation.

3.2.4 Greenhouse Orientation

The orientation of a proposed greenhouse structure should be included in the "layout" plans. By combining the climatological data available, one should be able to locate growing facilities for minimum effects of adverse winds and maximum available solar energy.

3.2.4.1 Wind Effects

Most geographical areas have at least one prevailing wind and many have two; one during summer months and another during the winter. Greenhouses with double top ventilators are adaptable to such conditions because a leeward vent can always be opened if needed.

The direction of the prevailing wind does not always indicate the direction of the higher velocity winds. On days when exhaust fans are operating and high winds keep the louvers closed, grèenhouse temperatures can climb. Many greenhouse exhaust fans are the nonloading types and cannot function against winds approaching 20 mph.

Damaging winds are generally associated with storms and there is no way to predict how they will affect the greenhouse structure. Only sound craftsmanship will sustain the majority of storms.

Wind can be modified by constructing fences of varying heights or growing trees and shrubs for windbreaks. Information that is available indicates a solid windbreak, which creates turbulence, is much less effective than one which allows a small amount of wind to pass through; windbreaks open in the lower portion allow the wind to push through the openings and shoot upward on the leeward side (Woodruff, 1954).

Greenhouse operators should consult wind experts before constructing a windbreak as some designs could cause turbulence over a greenhouse range and property damage.

3.2.4.2 Shading Effects

Shading by surrounding terrain, buildings, and plant materials can affect plant growth and perhaps cost the grower money. The subject is well covered by Walker and Duncan (1971) and is, therefore, presented in part in this text.

> "When planning the construction, give primary consideration to obtaining maximum sunlight exposure during those 'short' days of midwinter when the sun is lowest in the sky. Maximum sun altitude (angle of sun above earth's horizon) occurs at noon and varies from a high on June 21 to a low on December 21 in the northern hemisphere. Solar altitudes for selected latitudes (distance measured in degrees north and south from the equator) and times of day are given in Table 3-2. The latitude of 40° would correspond to Denver, Colorado; 38°

Table 3-2. Solar altitudes for selected latitudes and time (Walker and Duncan, 1971)

Latitude	Date	Solar altitudes at specified times, degrees (angle β of Fig. 3-6)		
		8:00 a.m. and 4:00 p.m.	10:00 a.m. and 2:00 p.m.	12:00 (noon)
42°	Dec 21	4°	19°	25°
	Mar 21 and Sept 21	22	40	48
	June 21	37	59	72
38°	Dec 21	7°	22°	29°
	Mar 21 and Sept 21	23	43	52
	June 21	37	60	76
34°	Dec 21	9°	26°	33°
	Mar 21 and Sept 21	24	46	56
	June 21	37	62	80
30°	Dec 21	11°	29°	37°
	Mar 21 and Sept 21	26	49	60
	June 21	37	63	84

ANGLE β	RATIO L/H
10	5.67
20	2.75
30	1.73
40	1.19
50	0.84
60	0.58
70	0.36

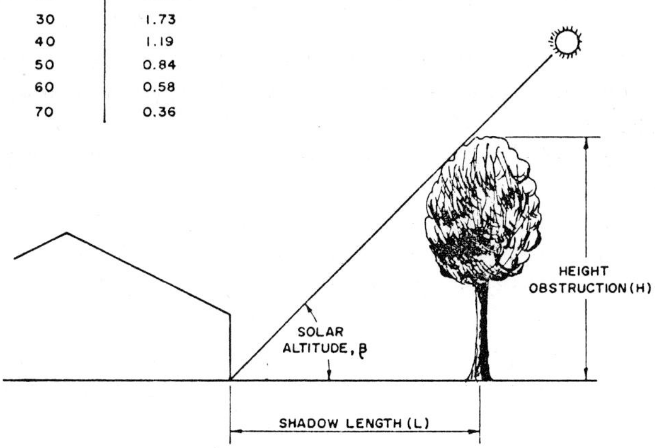

Fig. 3-6. Ratio of shadow length and obstruction height for selected solar altitudes (Walker and Duncan, 1971)

to Lexington, Kentucky; 34° to Little Rock, Arkansas; and 30° to New Orleans, Louisiana. At solar noon, the sun is located due south. This means that the building site should preferably have an open southern exposure. If the land slopes, it should ideally slope to the south.

"Do not build near large trees, buildings, or other obstructions which will shade the building. Figure 3-6 gives the ratio of shadow length and height obstructions for selected solar altitudes. To determine how far away an obstruction must be to prevent a shadow on the greenhouse, multiply these ratios by the obstruction height. As a general rule, no objects taller than 10 feet should be within 27 feet of the greenhouse in either the east, west, or south direction. Even objects this tall will cast long shadows in the early morning when the sun is particularly low in the sky. It is desirable to control the land immediately south, east, and west of greenhouses to prevent tall structures from being erected in future years."

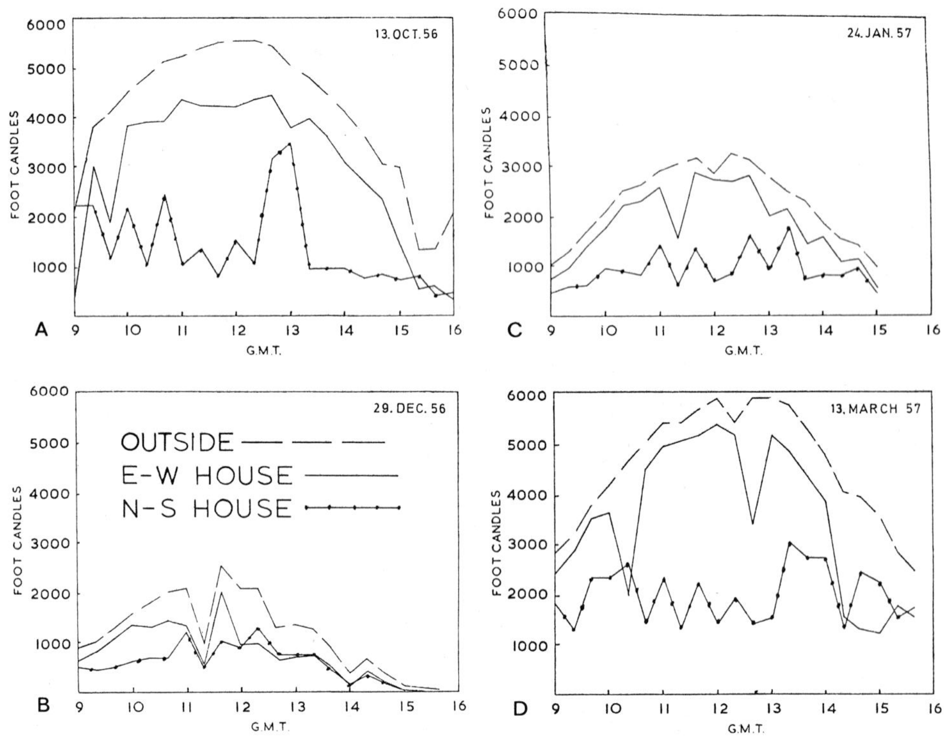

Fig. 3-7A–D. Average daily variation of insolation received in glasshouses with different orientations in southern England (Lawrence, 1963)

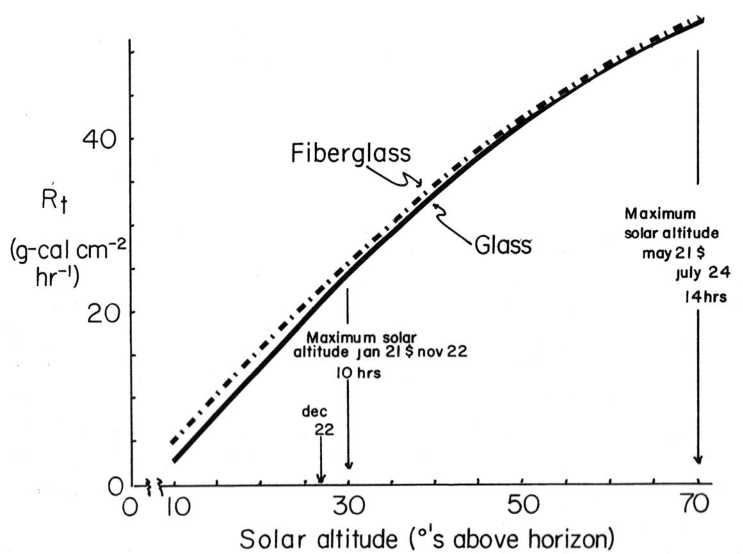

Fig. 3-8. Relationship between mean total shortwave radiation inside a N–S FRP and an E–W glass covered greenhouse as a function of solar energy (Hanan, 1970)

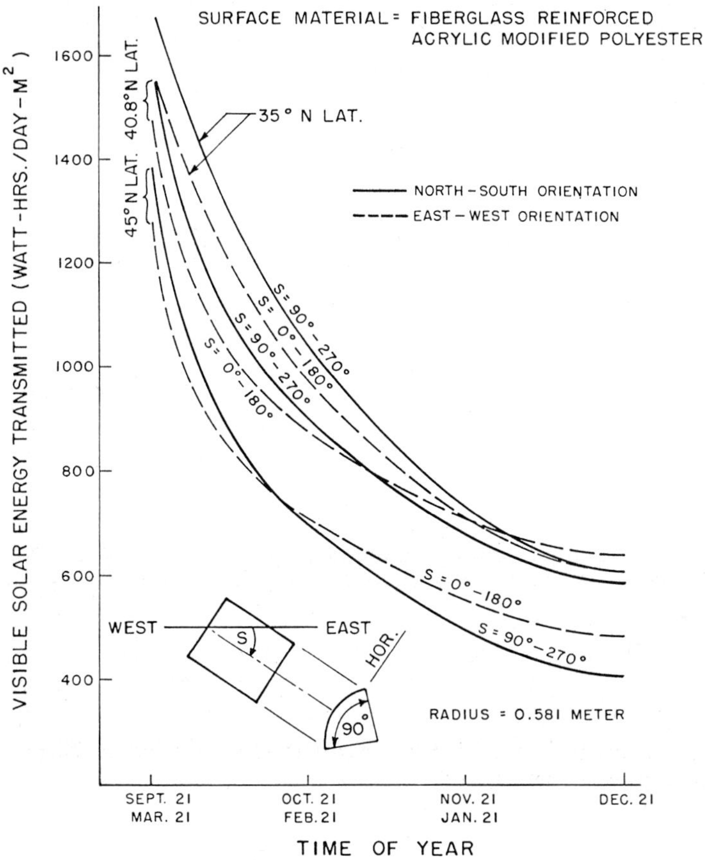

Fig. 3-9. The solar transmission characteristics of FRP panels oriented in different positions (Aldrich et al., 1966)

3.2.4.3 Solar Energy Effects

Through the years there has been much concern over the proper orientation of greenhouse structures in relation to the compass rose. Lawrence (1963) demonstrated that an E–W oriented glasshouse received more solar radiation than did N–S structures (Fig. 3-7) in Bayfordbury, England. Harnett (1974) reported that studies by L.G. Morris, at the National Institute of Agricultural Engineering, verified the work by Lawrence and showed further that insolation received in an E–W multispan house was greater than a N–S oriented glasshouse. Manbeck and Aldrich (1967) found that an east-west oriented house is preferable in northern latitudes above 40–45° in the winter, but at other times of the year and at lower latitudes the N–S orientation is preferred.

Hanan (1970) compared the total shortwave radiation received in a N–S oriented FRP covered house to an E–W glasshouse and found that little difference occurred (Fig. 3-8). He concluded that as far as solar transmittance is concerned, at latitudes approximately 42° N, there was little advantage of an E–W oriented house over a N–S house whether glass or FRP was used.

Aldrich et al. (1966) studied the transmission characteristics of FRP using different orientations (Fig. 3-9). They found that time of year, location and orientation all affected the energy transmitted through FRP.

3.3 Standards for Construction

Few architects or engineers have a background in greenhouse construction and design. Consequently, many cities, counties, and states have based design requirements on the Uniform Building Code standards. In short, greenhouses in most areas of the world must conform to the same standards of design and construction as those used for homes and other small buildings.

During the past quarter century, greenhouses have been redesigned by growers, amateur engineers, and architects and plant enthusiasts. Configurations have been numerous and superstructure materials unbelievable. Oil well sucker rod, electrical conduit, plastic pipe, rolled metal and fiberglass reinforced plastic have been used for greenhouse framework, along with the standard materials of steel pipe, angle iron, wood, and aluminum.

Little, if any, attention has been given to the structural strengths of construction materials, especially by neophytes in the greenhouse business. Some greenhouse manufacturers have also overlooked structural needs in order to provide a more competitive product.

It is apparent that the geographical location has a definite bearing on structural needs and possibly greenhouse design. Structures in Wisconsin must consider ice and snow loads, those in Colorado are concerned with wind factors, while those in southern California are primarily interested in keeping rain and potential frost off plants. Fire codes will also vary from city to city and state to state.

Greenhouse manufacturers could conceivably manufacture greenhouse designs for 10–15 different total load conditions, but they cannot afford to customize each job and, therefore, retain only two or three designs, based on load factor needs.

3.3.1 Building Codes

3.3.1.1 Uniform Building Code

Most city and state governments have used the Uniform Building Code (1970) as a guide for construction requirements. Some building departments have modified the code, providing leniency; others have adopted it as a minimum standard. An example is the live load requirements per square foot on a greenhouse roof; the city and county of Denver, Colorado, requires a rating of 30 lbs per ft^2. Some of the surrounding counties will accept a 50% lower figure.

3.3.1.2 ANSI

In July, 1972, the American National Standards Institute, Inc. accepted a committee report revising the requirements for minimum design loads in buildings and structures and have forwarded it to the National Bureau of Standards (NBS) for consideration and adoption. Previous adoptions by NBS, 1924, 1945, and 1955, have been the basis for the Uniform Building Code. Several building code agencies have already accepted the report and are rapidly modifying their building codes. It is anticipated the American Standard Building Code Requirements for Minimum Design Loads in Buildings and Other Structures, A 58.1-1972, commonly known as ANSI A 58.1-1972, will be accepted shortly. It may however, be revised or withdrawn at any time.

1. Design Loads. Structures will withstand a certain weight applied to them either vertically or horizontally before they will collapse. The weight imposed, commonly referred to as load includes: (ANSI A 58.1-1972, 1.2)

Dead load—Weight of all permanent construction, heaters, water pipe, all fixed service equipment, when carried by the structural members.

Live load—Weights superimposed by use; (in greenhouse would include hanging baskets, shelves, attached vine crops, etc.) but not wind, snow, earthquake or dead load.

Wind load—Loads caused by wind blowing from any horizontal direction. They are determined by the correlation of mean recurrance interval and the fastest mile speed for 25, 50, or 100 years.

All permanent structures require wind loads to be at least calculated from the 50-year recurrence mean unless "structures have no human occupants or, where there is negligible risk to human life, a 25-year mean recurrence interval may be used" (ANSI, 1972) (Fig. 3-10). Wind loads are based on the height of a building above ground (greenhouses will fall in the "less than 30 feet" category) versus the surrounding terrain such as in cities or hilly areas, suburban or rolling ground and open country. Local codes may dictate which wind map is to be used.

2. Wind Load Calculations. The intent of this section is to present the basic components of wind load calculations. Complete information and exact situations can be obtained from ANSI A 58.1-1972.

The effective velocity pressures can be calculated from the map by converting the wind speed using the formula: (ANSI A 58.1-1972, 6.3.4)

a) $q_{30} = 0.00256\ V_{30}$, where:

q_{30} = basic wind pressure in pound-force per ft^2

V_{30} = basic wind speed in miles per hour. Thus:

$q_{30} = 0.00256 \times 70^2$

12.25 = lbs per ft^2, a force created by a 70-mph wind on a ft^2 of side wall.

Other factors such as wind gusts and height of facility can be included in the final calculations, but are not in this example. The final wind load factor *(W)*, for a greenhouse, is calculated by the formula:

b) $W = q_{30} \times q_p$, where:

q_p = external pressure acting at local positions.

The external pressure coefficients have been extracted from tables (ANSI, 1972) and shown as positive or negative (suction) pressures acting on three com-

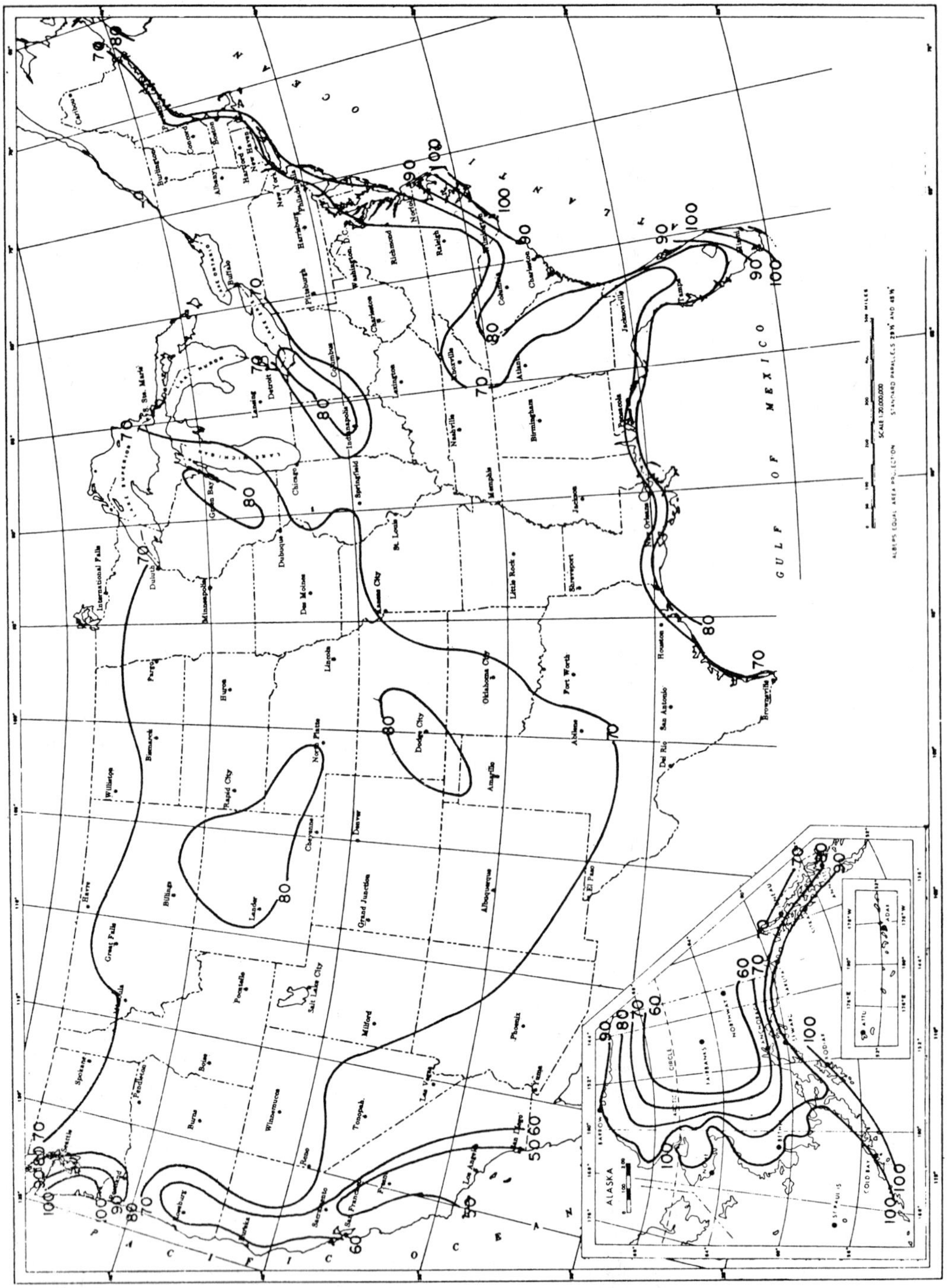

Fig. 3-10. Basic wind speed in miles per hour. Annual extreme fastest-mile speed 30 ft above ground, 25-year mean recurrence interval (American National Standard A 58.1-1972)

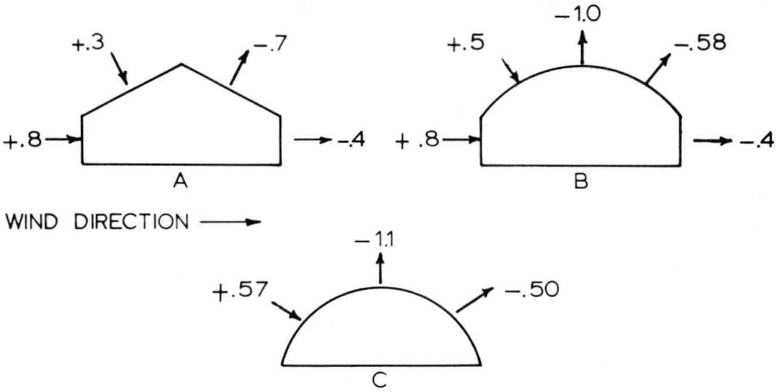

Fig. 3-11. Wind pressure coefficients for various roof designs. (Adapted from ANSI A 58.1-1972)

mon greenhouse structures (Fig. 3-11). Using the type C greenhouse and the vertical pressure coefficient (q_p) equals -1.1 therefore:

$W = 12.25(-1.1)$

$W = -13.79$ lbs per ft^2 of suction created by a 70-mph wind across the top of a quonset type greenhouse roof.

3. Snow Loads. Greenhouse-type structures should also be designed for snow load conditions. Snow load designs are based on climatological data collected by the U.S. Weather Bureau and depicted on the snow load map (Fig. 3-12) as isolines of ground snow load in pound-force per square foot. Once again, because a greenhouse has "negligible use to human life" a 25-year mean recurrence map is used. The minimum snow loads for free standing, ridge and furrow, pitched or curved roofs are determined by multiplying the ground snow loads given in Figure 3-12 by the coefficient for snow load (C_s). The basic snow load (ANSI, 1972, 7.2.1) coefficient (C_s) has been set at 0.8 and is decreased or increased according to the following conditions:

1. Decreased load when a roof slope exceeds 30° and snow slides off.
2. Decreased load when a roof has a clear exposure in wind swept areas.
3. Increased load when a non-uniform accumulation occurs on pitched or curved roofs.
4. Increased load when valleys are formed by multiple roofs.

Other conditions of variable snow loads are also described in ANSI A 58.1-1972, 7.2.1, but the above criteria will generally meet the needs of the greenhouse industry. ANSI (1972, 7.2.1) recommends that a reduction in the snow load coefficient should not reduce the snow load factor less than 12 lb ft^{-2}. Walker and Duncan (1973a) recommend nothing less than 10–12 lb ft^{-2}.

There are other load factors included in ANSI A 58.1-1972 but they will not be considered here. It should be noted however, that there are combinations of loads that must be considered in greenhouse construction. They include:

Dead load (D) + live load (L) + wind load (W) + snow load (S) + additional loads (A). The additional loads can be those described in ANSI A 58.1-1972, 4.1, i.e. rain, ice, earth, and hydrostatic pressure, horizontal components and others.

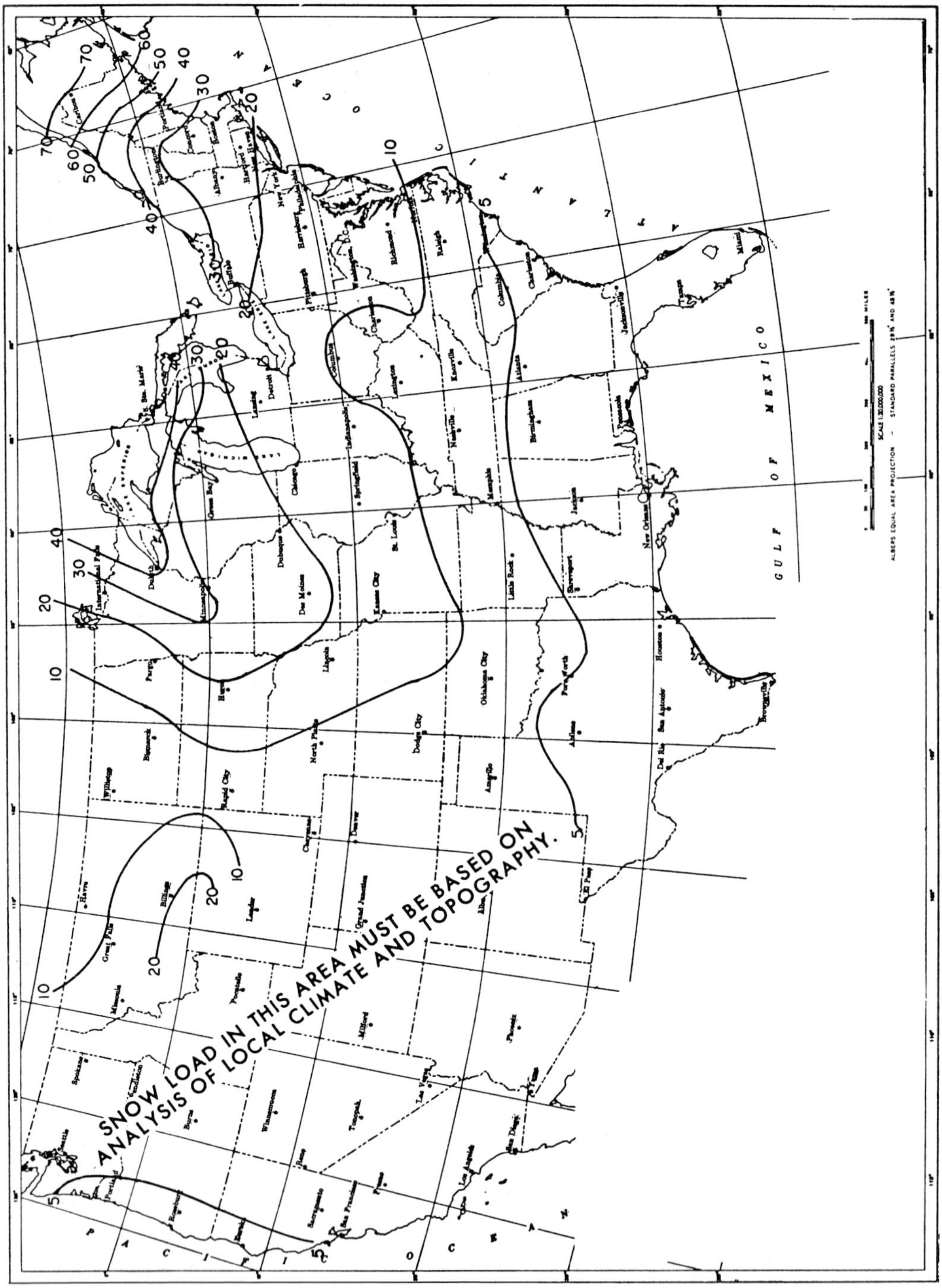

Fig. 3-12. Snow load in pound-force per square foot on the ground, 25-year mean recurrence interval (American National Standard A 58.1-1972)

3.3.1.3 National Greenhouse Manufacturers' Association (NGMA)

The NGMA was established in 1959 to develop standards for greenhouse design and construction. It is apparent that greenhouse structures do not fit completely the criteria set forth in the Uniform Building Code or the more recent ANSI A 58.1-1972.

On November 12, 1975, NGMA adopted a standard related to design conditions. The adopted standard is to be presented to the different coding bodies throughout the United States. The complete standard is herewith presented to acquaint the reader with the basis for design criteria as established by greenhouse engineers (Rough, 1976).

NGMA Standard

Design Conditions

Design members to carry the following loads:

1. Dead load.
2. Live load minimum 15 lbs per ft^2 on horizontally projected area. Live load may be reduced to minimum 10 lbs in area below 10 lb line on map (Fig. 3-14).
3. Wind load minimum 20 lbs per ft^2 on vertically projected area below 30 ft.

Wind loads must be increased to a maximum of 25 lbs in areas indicated on the map (Fig. 3-13). Wind loads may be reduced to a minimum of 15 lbs in areas indicated on the map. Horizontal wind pressures may be reduced one-third of value listed above in height zones of 10 ft or less.

In designing for the above loads, the loads may be considered to act in any of the following combinations:

1. Dead load plus live load.
2. Dead load plus wind load plus $\frac{1}{2}$ live load.
3. In addition to above, roof bars, purlins and rafters shall be capable of carrying a minimum 100 lb concentrated load at the center of any span.

The above criteria apply only to greenhouses heated to a minimum of 50° F.

Maximum allowable deflection shall be minimum 1/120 of clear span.

Steel members shall be designed in accordance with current American Institute of Steel Construction specifications. Aluminum members shall be designed in accordance with current Aluminium Association specifications for aluminum structures.

Based on ANSI A 58.1-1972, the minimum load design adopted, No. 3 above, is based on a 90 mph wind. It should be noted however that the live load, No. 2 above, includes a 10 lb ft^{-2} snow load and 5 lb ft^{-2} for weight superimposed by use. Snow load cirteria were also adopted by NGMA, Figure 3-14.

Other standards for greenhouse construction have been adopted by NGMA including heating and ventilation criteria. A complete set of standards may be obtained by writing the National Greenhouse Manufacturers' Association, P.O. Box 128, Pleasantville, New York 10570.

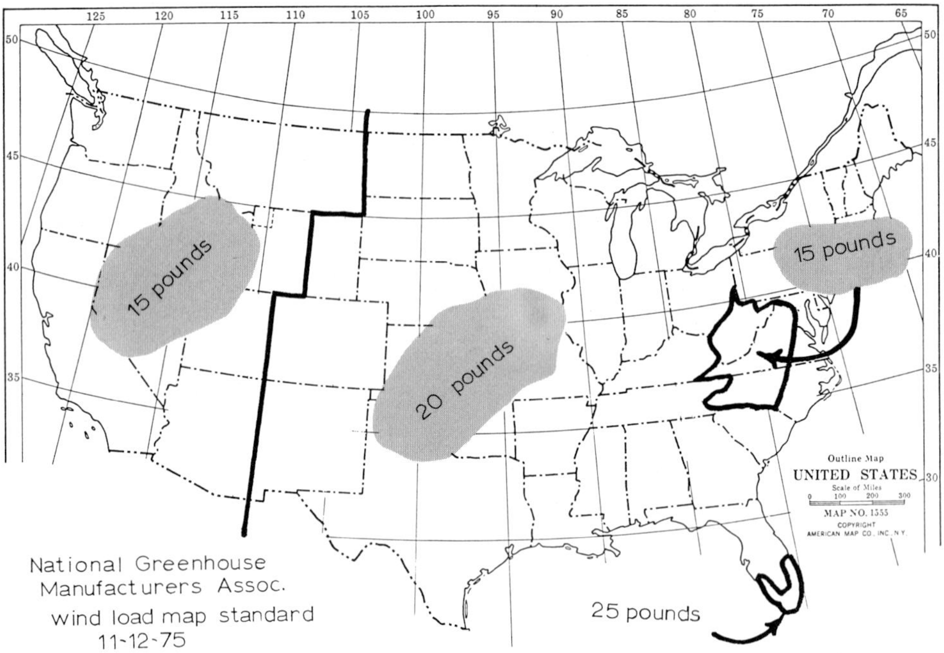

Fig. 3-13. Wind load map adopted by the National Greenhouse Manufacturers' Association, November 12, 1975

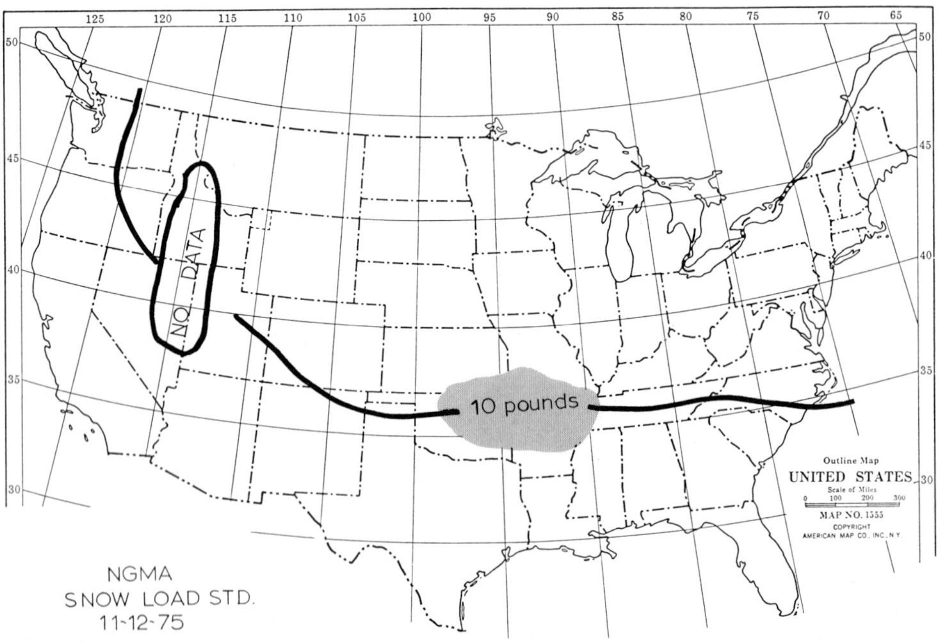

Fig. 3-14. Snow load map adopted by the National Greenhouse Manufacturers' Association, November 12, 1975

3.4 Structural Materials and Methods

3.4.1 Materials

Bailey (1900), Wright (1917), Taft (1926), and Post (1950) all presented instructions on the erection of wood or wood-pipe structures. The superstructure and framework of smaller houses were usually made entirely of cypress wood. Several of these houses are standing today, some have been remodeled, others should be torn down for safety reasons. The larger greenhouse styles of the early 1900s were patterned after European types, and were often constructed with a pipe or angle iron superstructure and wooden sash bars for setting glass (Fig. 3-15). One such single house, 120 ft wide and 600 ft long, was constructed by Moninger (1913).

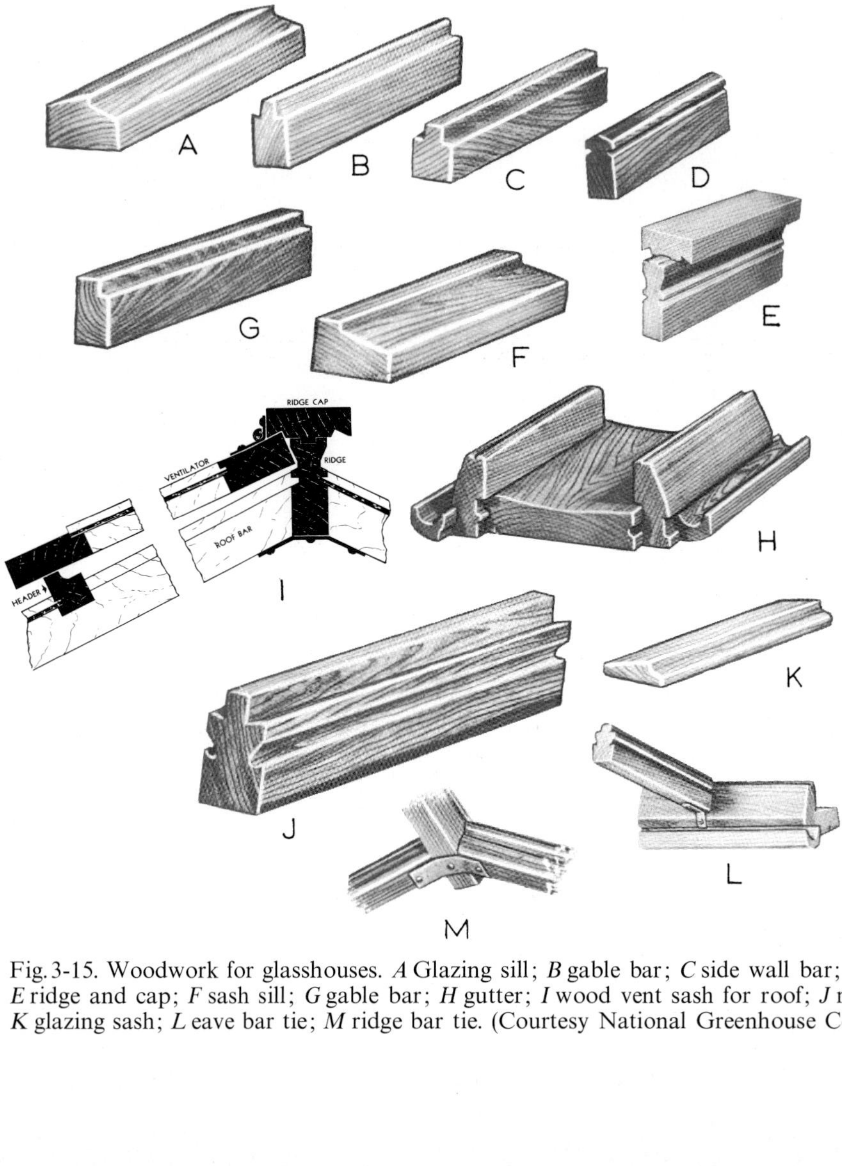

Fig. 3-15. Woodwork for glasshouses. *A* Glazing sill; *B* gable bar; *C* side wall bar; *D* ridge; *E* ridge and cap; *F* sash sill; *G* gable bar; *H* gutter; *I* wood vent sash for roof; *J* roof bar; *K* glazing sash; *L* eave bar tie; *M* ridge bar tie. (Courtesy National Greenhouse Company)

Allowable Loads on Beams (Approximate)

Example: 1,200 lbs, 12' span
2x8 adequate for strength
2x8 adequate for 1/240 deflection
2x10 needed for 1/360 deflection
2x6 adequate for shear
Select 2x8 or 2x10 depending on deflection requirements

Lumber dried below 19%
Construction Douglas Fir
No 1 Southern Pine, or equiv.

Total Uniform Load, Lbs

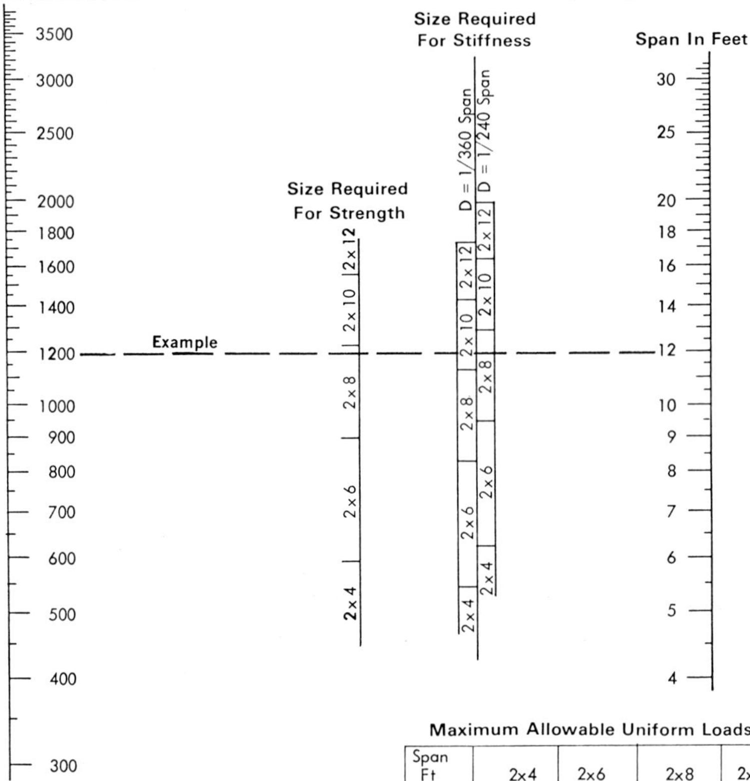

Beams

Use a straight-edge to line up the total load with the span. Note where the straight-edge crosses the strength and stiffness lines.

1/360 stiffness: plaster ceilings, home floors, some building codes.

1/240 stiffness: farm buildings, except lintels over large windows.

Select member size with adequate strength and stiffness.

Check in shear table

Maximum Allowable Uniform Loads In Shear

Span Ft	2x4	2x6	2x8	2x10	2x12
3	1180				
4	1111	1855			
6	1049	1688	2462		
8	1020	1614	2310	3080	
10		1575	2228	2933	3700
12		1548	2176	2845	3559
14			2140	2785	3464
16			2145	2741	3398
18				2708	3344
20				2682	3307

Fig. 3-16. Nomograph of approximate allowable loads on wood beams. (Adapted from a chart by Weyerhauser Co.: Structures and Environment Handbook, 1972)

Allowable Loads On Columns (Approximate)

Lumber dried below 19%
No 1 Southern Pine, Douglas Fir or equiv.

Examples:
1. 14,000 lbs, 8' length: 4x6 needed
2. 24,000 lbs, 5' length: 4x6 needed (load exceeds max for 4x4, List B)
3. 3,000 lbs, 20' length: 6x6 needed (length exceeds max for 4" thick columns, List D).

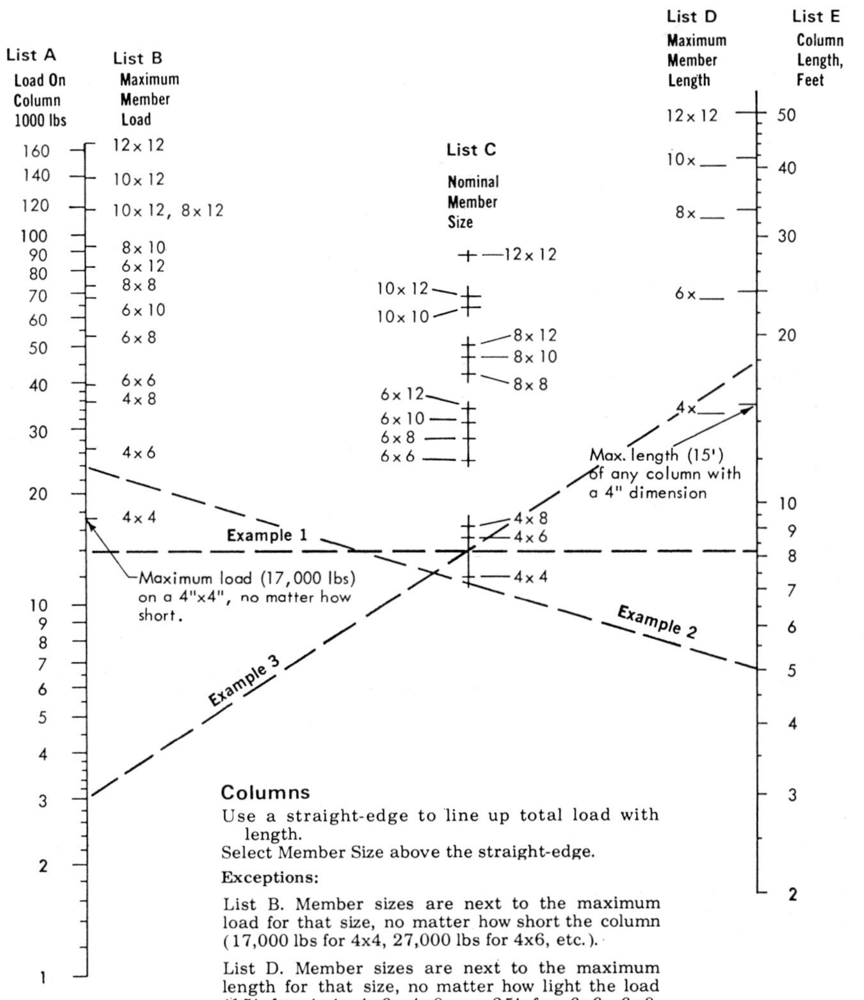

Columns

Use a straight-edge to line up total load with length.
Select Member Size above the straight-edge.

Exceptions:

List B. Member sizes are next to the maximum load for that size, no matter how short the column (17,000 lbs for 4x4, 27,000 lbs for 4x6, etc.).

List D. Member sizes are next to the maximum length for that size, no matter how light the load (15' for 4x4, 4x6, 4x8. . .; 25' for 6x6, 6x8, 6x10. . .).

Fig. 3-17. Nomograph of approximate allowable loads on wood columns (SEH, 1972)

3.4.1.1 Wood

Wood still plays a part in greenhouse construction, especially for the "do it yourself" grower and some large commercial greenhouse operators who have looked closely at the economics of construction.

Cypress and redwood are seldom, if ever, used in the superstructure of commercial greenhouses because of the economic and strength factors. Depending on the geographical location, the most common construction woods are Douglas fir, hemlock, spruce, pine, and some cedar. Trusses, columns, beams or purlins made of these materials have different load bearing capabilities and in most instances, if maintained, will remain sound for many years. The following nomographs (Figs. 3-16 and 3-17) on beam and column loads are presented to acquaint the potential builder with some strength characteristics of wood (SEH, 1972). One should always be aware of the engineering requirements in any type of construction around the greenhouse. When considering wood for superstructure, acquire the help of a knowledgeable person to assist in planning and design.

3.4.1.2 Steel

The increased desire for more light energy in greenhouses and to cover more ground area with one "clear span" unit, led to the development of the steel structures. Manufacturers of the early 1900s were using steel pipe, angle or channel for the greenhouse superstructures, to which wooden bars were adapted.

Today, high tensile strength pipe and formed or tubular steel comprise the basic components of many greenhouse superstructures. The major components are hot-dipped galvanized while parts such as bolts and screws are zinc plated for ease of installation.

Corrosion on steel components can usually be controlled by proper maintenance. Care should be taken not to allow fertilizer, or other corrosive materials to come in contact with structural parts for long periods of time.

Some greenhouse operators prefer to use black pipe or wrought iron components painted with rust inhibitors. In many instances, such components will last a lifetime.

3.4.1.3 Aluminum

Aluminum components for greenhouse construction were introduced in the early 1950s. Aluminum is light in weight, easy to handle and is not adversely affected by most greenhouse conditions. Most aluminum components are extruded and more than 25 extrusions are used in the greenhouse industry (Fig. 3-18). Aluminum extrusions are made by heating the aluminum alloys to 800° F, when the metal becomes plastic in nature, a hydraulic ram forces it through a hole cut in a die (Aluminum, 1968).

Corrosion of aluminum can occur. It is associated with the flow of electric current between various anodic and cathodic regions. There are several types of corrosive attacks, but the most common form is pitting (Binger et al., 1967).

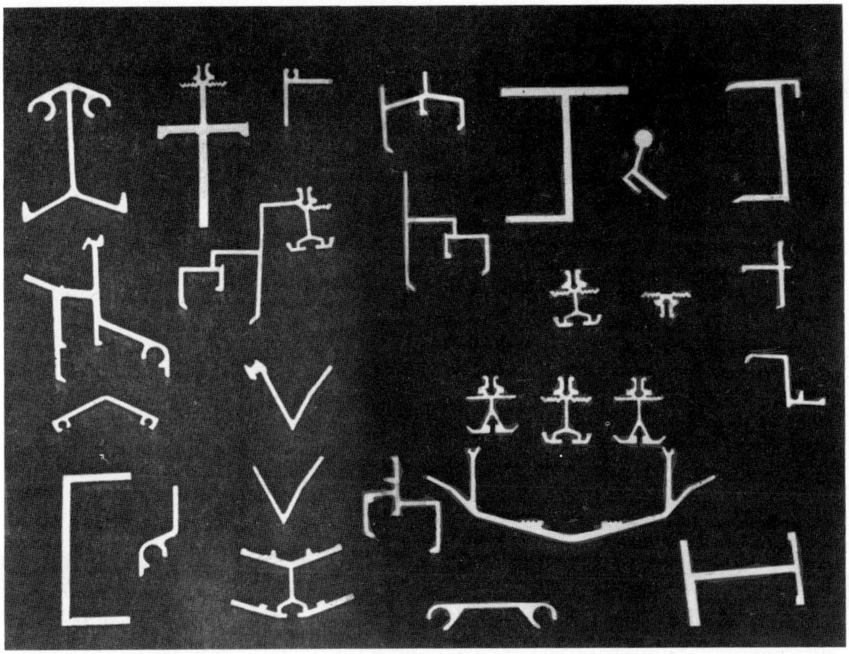

Fig. 3-18. Aluminum extrusions used in greenhouse construction. (From Rough Brothers' Greenhouse Catalog, 1976)

Aluminum parts placed in soil or concrete are often attacked especially where fertilizer materials are involved. A second type of corrosion, "fretting", often occurs. It is caused by two metals rubbing together, a black by-product is usually created and can be controlled with proper lubrication. Silicone or "Teflon" base lubricants are often used to lubricate moving aluminum parts.

All of the materials used in greenhouse construction have different load bearing capabilities (Table 3-3). These capabilities can be decreased or increased by the influence of many factors. It is advisable for one to be aware of load factors before he hangs pots and other items from every inch of available truss or purlin.

3.4.2 Erection

The ease of erection lies with the proper alignment of columns. Various contractors and manufacturers approach the initial phases of construction differently. Some use trusses on every column (Fig. 3-19A), others build the gable ends in conjunction with the side walls and place trusses in the center (Fig. 3-19B). In either case, proper placement of all columns is very important. Most construction companies drill column holes, pour concrete and then place the columns in the concrete, checking for height and alignment before the concrete sets. The base of exterior columns require some modification (Fig. 3-20) for ease of continued con-

Table 3-3. Approximate live load factors of purlin materials used in greenhouse structures. Load is based on a 12-ft span[a]

Cross section	Material	lbs/lineal ft
L	Angle 2–1/2 × 2–1/2 × 3/16	
	Aluminum	24
	Steel	40
[	Channel 2 × 4 × 5/32	
	Aluminum	133
	Steel	200
I	I-Beam (L & B) 2–1/8 × 3 × 5/32	
	Aluminum	61
	Steel	116
O	Pipe 1–1/4 O.D.	
	STD Steel	25
	High tensile	60
□	Tubular Steel 2 in square	
	Aluminum	25
	Steel	75
⎍	Formed Steel	30
▮	Wood 2 × 4 in	
	No. 1 fir	25
▮	Wood 2 × 6 in	
	No. 1 fir	52

[a] Data obtained from greenhouse engineers who use each particular configuration.

struction. The installation of concrete foundations, except in special instances, is no longer considered economically feasible.

Every component added to the greenhouse superstructure increases rigidity (Fig. 3-21). Anti-sway bracing (Fig. 3-22A) and gable bracing are vital parts of greenhouse construction. The knee braces (Fig. 3-22B) are important additions when extra long columns are required or construction occurs in geographical areas of potentially high velocity winds.

The addition of the covering requires some planning and coordination. Depending on the time of year or geographical location, one must consider the need for heating or cooling the moment the structure is covered. Thus, it is desirable to have most of the heating and ventilating equipment operational before the house is completely enclosed. The prevailing wind should also be considered when applying covers, especially glass or FRP panels. Most construction companies install the roof first, starting with the windward side of the greenhouse. Secondly, the gable ends and side walls on the windward side are enclosed, followed by the leeward additions. Such a system of construction prevents potential covering destruction in case of high winds and a "ballooning" effect.

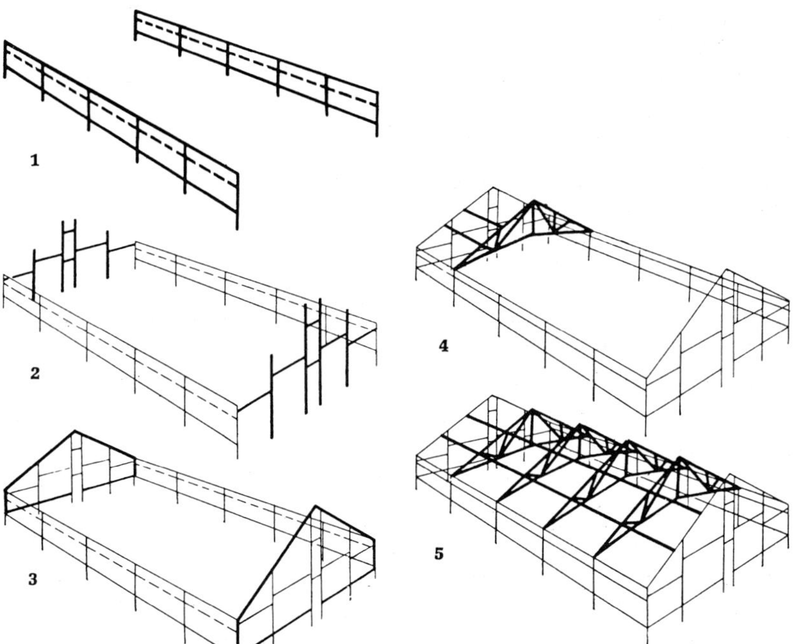

B

Fig. 3-19. (A) Erecting trusses on every column. (Courtesy Ickes-Braun Glasshouses.) (B) Erection schedule of greenhouse structures with center trusses only. (Courtesy Lord and Burnham)

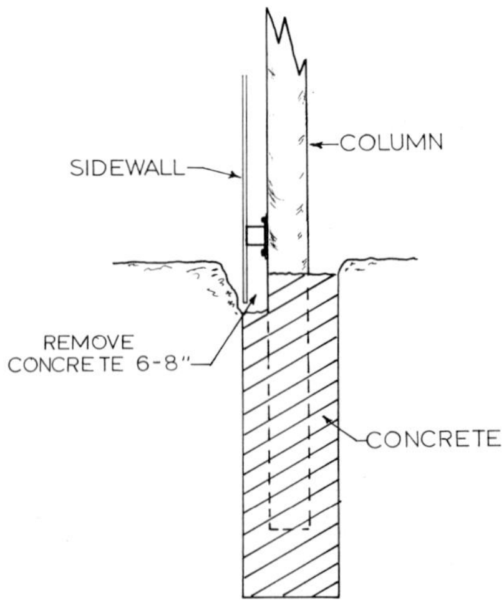

Fig. 3-20. Cut to be made in setting concrete on exterior sides of columns, so side walls can extend neatly into the ground

3.4.3 Glazing

3.4.3.1 Glass

The advent of bar caps (Fig. 3-23) eliminated maintenance on many greenhouses. Bar caps have been on the wood bars of the Colorado State University greenhouses for 29 years. Presently the bedding putty has hardened and cracked. Reglazing should have taken place at least six years ago.

The aluminum bar, introduced in the 1950s, has been almost maintenance free. Some greenhouse operators have, however, questioned the problems of heat transfer through aluminum parts, especially glazing bars. Edward Owen Engineering Ltd., of Hampshire, England, calculated the heat loss radiated from a "traditional" aluminum bar extrusion with the web outside the greenhouse

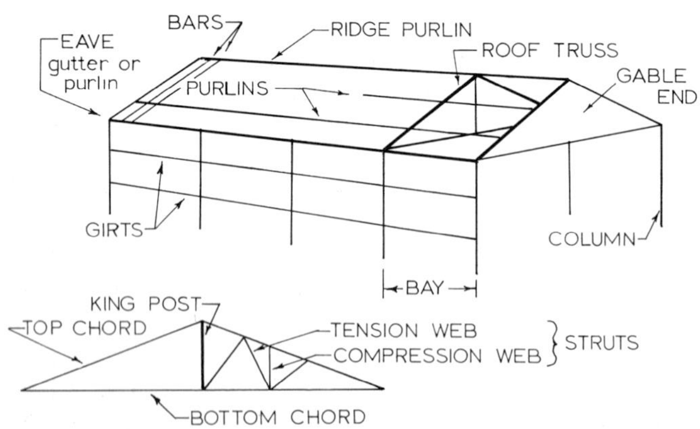

Fig. 3-21. Component parts of greenhouse superstructure

Fig. 3-22. (A) Anti-sway bracing in roof and side wall sections. (Courtesy Nexus Corporation.) (B) Knee braces on ridge and furrow structure. (Courtesy Ickes-Braun Glasshouses)

(Fig. 3-23 A). Based on a three bay, 9.1 m span house, 45.7 m long, it was estimated that every meter length of the glazing bar had an exposed surface of approximately 1.735 cm^2 acting as a fin radiator. The total house has 473.2 m^2 of exposed surface which would have a heat loss of 179000 BTU h^{-1}, if there was a 15° C temperature differential. Edward Owen Ltd. designed a glazing bar, (Fig. 3-23 B), with the web inside the greenhouse and a plastic bar cap on the outside, thus there is negligible heat loss by conduction. The longevity of the plastic bar cap is not known. In areas of high solar energy, photo degradation could be very rapid.

When a greenhouse roof is glazed, the panes are lapped approximately 1 cm. Engineers have found that a 6.2 k h^{-1} wind across the greenhouse roof causes 2.5 airchanges per hour inside houses glazed with 61 × 61 cm overlapped glass (Owen, 1975). The house described above could have a heat loss of 250000 BTUs h^{-1}, due to overlapped glass.

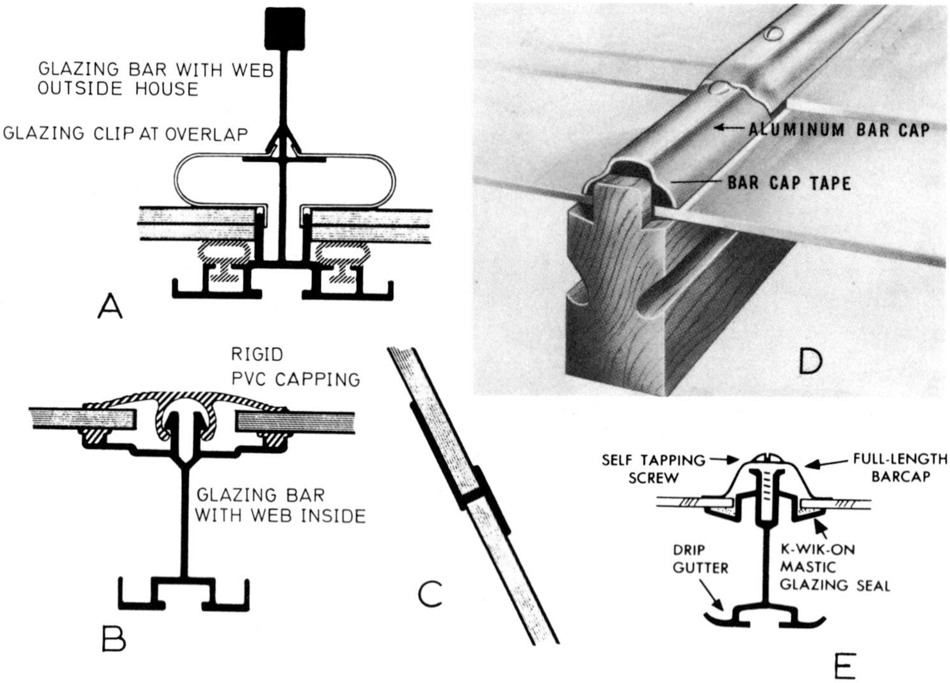

Fig. 3-23A–E. Bars for glazing with glass. (A) Aluminum bar with structural web outside; (B) aluminum bar with web inside; (C) plastic "H" butting strip for roof glass; (D) traditional cap for wood bars; (E) aluminum bar and cap. (Courtesy Lord and Burnham, National Greenhouses and Edward Owen Greenhouses)

3.4.3.2 FRP Panels

The era of fiberglass reinforced plastic panels (FRP) as greenhouse coverings has contributed to many design modifications. The combinations of aluminum extrusions and FRP panels have aided greenhouse construction, leading to reduced costs. Many greenhouse designs have been modified to adapt FRP panels (Fig. 3-24).

Attachment of FRP panels to the greenhouse structure has been a continuous evolution. Most construction companies have developed every "short cut" possible and obtain new ideas with every job. The use of self-tapping screws (Fig. 3-25) has been one of the biggest assets to rapid construction.

Many fiberglass manufacturers recommend the use of a mastic between the laps of FRP panels. The use of such a material eliminates dirt accumulation between laps but increases the cost of installation. It is also a questionable practice where gas fired unit heaters are used, as a degree of air exchange is desirable to prevent plant pollution problems. Further discussion on this subject is in Chapter 8.

The storage of FRP panels prior to installation also merits some consideration. If panels are stored where moisture can seep between them and have direct

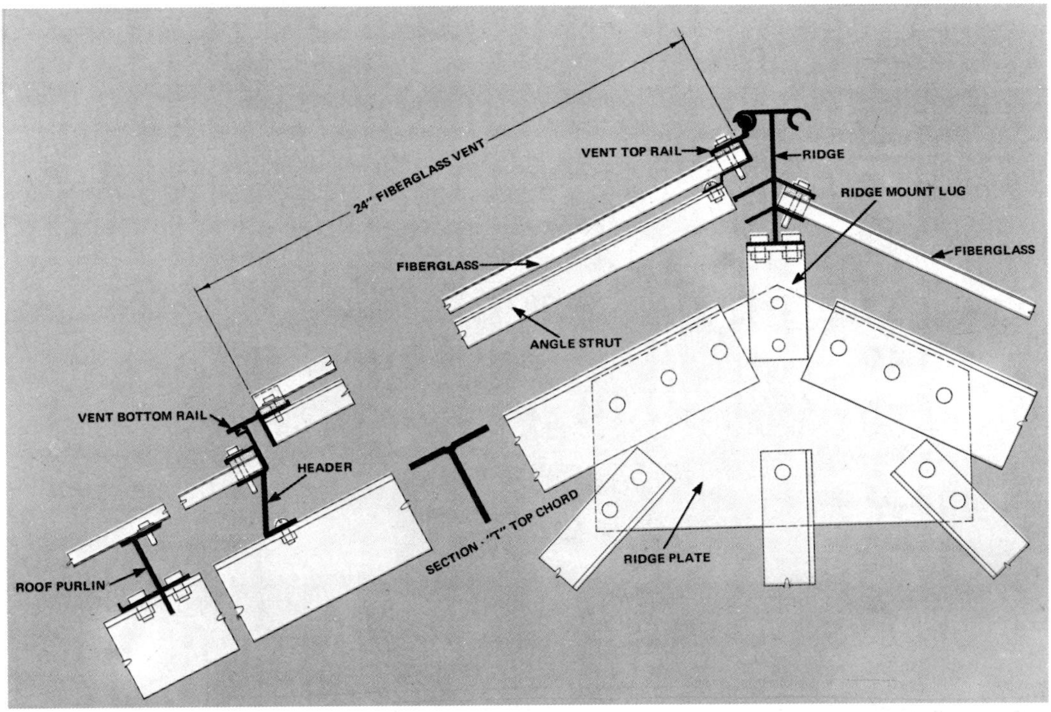

Fig. 3-24. Structural details of fiberglass panel installation on a ridge with ventilator. (Courtesy National Greenhouses)

exposure to solar radiation, damage will occur. The combination of moisture, high temperatures, and perhaps organic chemicals, interact and irreversible milky white areas form in the panels. FRP panels should always be stored in a dry area where no direct sunlight can reach them.

3.4.3.3 Plastic Films

Many methods of covering greenhouse structures with film plastics have been developed in the past five years. Methods of attachment have been given the greatest consideration. The simple methods using wood (Fig. 3-26) have given away to extruded aluminum parts. Most of the plastic film holders are designed to hold 4–12 mil-thick film.

When installing film on a structure, attach it to the gable ends first, then pull it to each side for final attachment.

3.5 Greenhouse Coverings

One of the greatest revolutions in greenhouse history occurred in the two decades following World War II, and might be attributed to the technology gained there-

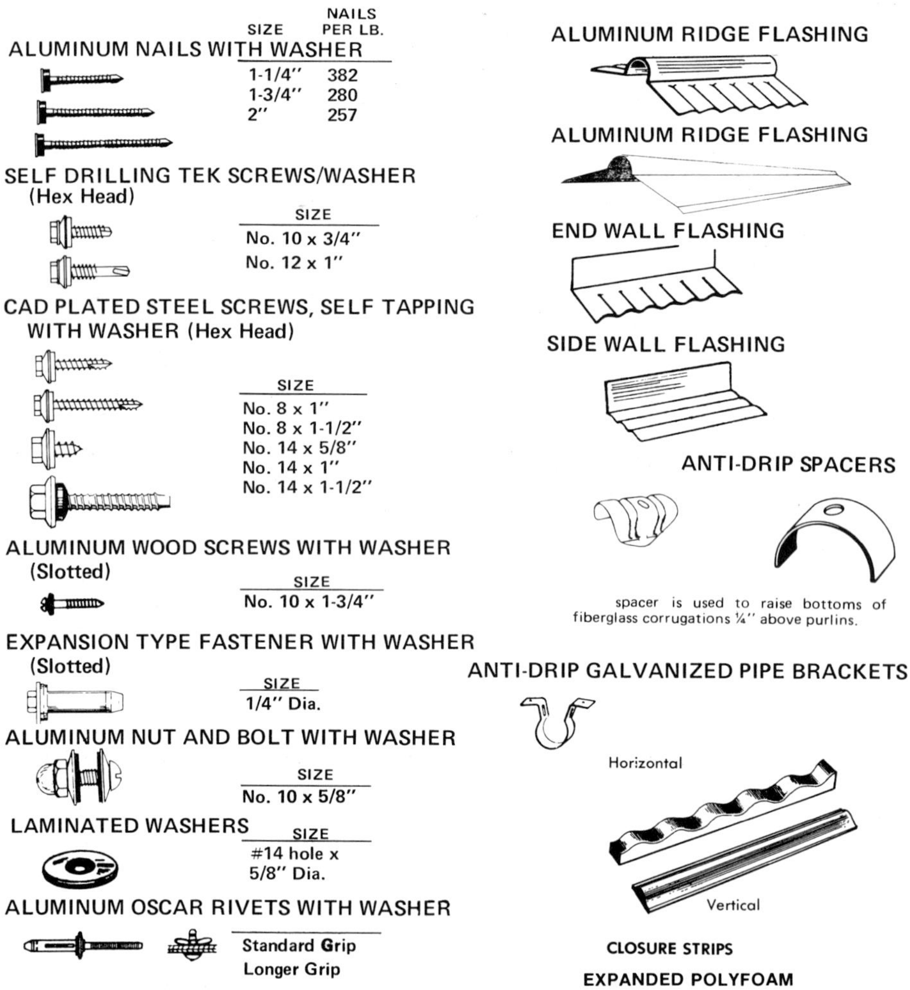

Fig. 3-25. Accessories used to attach FRP panels to superstructure. (Courtesy A. H. Humment Seed Company)

in. Greenhouse researchers throughout the world were considering the potential uses of plastics as a substitute for glass as a covering. Plastics were considered to be more economical, easier to install, and they did provide good light transmission. A historical overview is timely at this point, so that the merits and faults of the various coverings can be evaluated.

3.5.1 Glass

During the past six centuries, glass has been the main transparent medium used to provide natural light in protected environments for plant growth. Through the

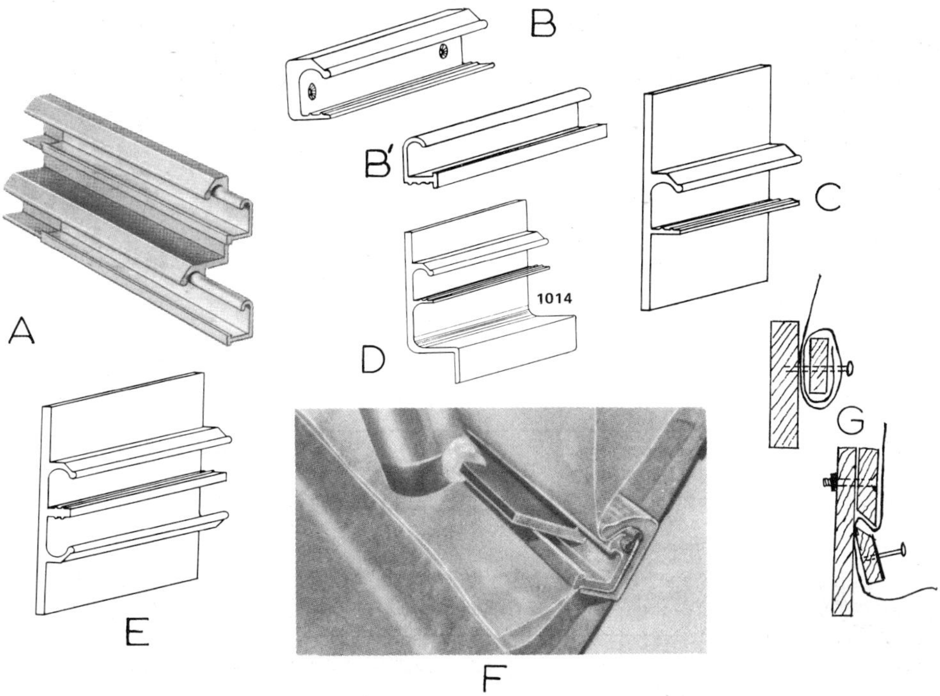

Fig. 3-26 A–G. Accessories for attaching plastic film coverings to superstructures. (A) Aluminum extrusion for attaching two layers of film for an inflated system; (B) simple extrusion with B′ insert; (C) aluminum extrusion that can be bent for attaching film to the gable ends of quonset type structures; (D) baseboard extrusion; (E) eave (roof-sidewall) extrusion; (F) special tool for inserting and removing inserts; and (G) wooden attachments. (Courtesy National Greenhouses, A. H. Hummert and Stuppy Greenhouse Supply)

early seventeenth century, arguments, and theories about covered garden buildings were common. The first greenhouse building material to come under close scrutiny was glass. For several hundred years, only two types of glass had been used. One was broad glass which was made by dipping a metal cylinder in molten glass, then stripping off the glass and ironing it out. Broad glass tended to be uneven in thickness and was usually streaked. The second type, crown glass, was made by spinning a circle at the end of a glass blower's pipe. It had a greenish cast and was favored by gardeners. By 1883, sheet glass was being produced. It was said of the garden house: "one of the greatest improvements made in their construction since the substitution of roofs of glass for those of opaque materials" (Lemmon, 1962).

Sheet glass, for greenhouse use, has been classified in many ways. A grade of first, second, third, etc., indicated the degree of perfection. The lower the number, the less the impurities. Glass can be obtained as single or double strength. "Single strength" weighs approximately 16 oz per ft^2 and the "double strength" weighs 24–26 oz. The double strength is recommended for greenhouse construction because of its resistance to breakage. In the early 1900s, greenhouse glass could be obtained in sizes ranging from 7 × 9 in to 26 × 26 in and ranged in thickness from 1/16 to 1/4 in. The standard "accepted" size was 16 × 24. Starting in the late 1940s, most U.S. greenhouses were covered with a 20 × 20 in double strength glass. Some European greenhouse manufacturers have gone to panes 26 × 65 in. In recent years, several "Dutch type" greenhouses have been built in the U.S. using the larger glass (Nicholson, 1887; Bailey, 1900; Taft, 1926; Lemmon, 1962).

Glass used in the construction of plant covers today is still largely soda-lime-silica glass. For many years the Federal Government has maintained specifications for all types of glass. The following requirements for greenhouse glass are taken from the lastest specifications (Fed. Spec. DD-G-004516, 1966).

Designation:	Double strength, having a minimum and maximum thickness of 0.115 and 0.134 in respectively.
Cut size:	The length and width cut size tolerance is 1/32 in.
Quality:	Glass may contain defects of any size or intensity, but shall contain no stones which may cause spontaneous breakage.
Sheet:	Greenhouse quality, intended for use in greenhouse glazing or similar applications where quality is unimportant.

The definition of sheet glass in the Federal specification is: "Transparent, flat glass having glossy, fire-finished, apparently plane and smooth surfaces, but having a characteristic waviness of surface".

3.5.2 Plastics

There are many plastic resins used in industry and most of them have been classified into two categories, thermosetting and thermoplastic. Thermosetting resins are those materials that undergo a chemical polymerization reaction or "cure" upon initial heating. Reheating does not reverse the process or change the physical condition. Thermoplastic materials when reheated, merely change physical condition and become soft or hard when cooled (Sonneborn, 1964). The first thermoplastic, cellulose nitrate, given the trade name "celluloid" was discovered in 1868. Because of the highly flammable characteristics of cellulose nitrate, it was not marketed and, in the 1930s, cellulose acetate was introduced. The earliest known thermosetting plastic, a phenol formaldehyde resin called "Bakelite," (Trade Name) was developed in 1909 (Dresser, 1957). Undoubtedly, the greenhouse industry could not survive without the aid of plastics.

3.5.2.1 Film Materials

The first flexible film, a thermoplastic, was developed by British chemists Faucett and Gibson in 1933. The first ton of film called polyethylene was produced in England in 1938 with production starting in the U.S. in 1943 (Dresser, 1957). The earliest known use of polyethylene film for plant protection and growth in the United States was suggested by Emmert (1954).

Another film commercially identified as Flex-O-Glass (Trade Name), a vinyl acrylic, was developed by Harold Warp in 1924. This film has excellent light transmission and fair weathering characteristics but has never been promoted for large greenhouse coverings (Warp, 1971). It is available through lumber and hardware stores throughout the United States.

The polyvinyl chloride (PVC) films have been used to some degree for greenhouse coverings in the United States. Two major factors that have limited PVC use are the lack of wide widths and an electrostatic condition that causes dust to collect on the outer surface.

Film plastic researchers in Japan have overcome major problems with PVC films including dust free materials. Greenhouses used for vegetable and fruit production comprise 95% of the plastic covers and only 5% are under glass. Flowers and ornamentals are still basically grown under glass (36%) and only 6% of the total area is under plastic. Japan is the worlds largest user of vinyl film (Kondo, 1975). The greenhouse industry should reconsider use of PVC films, if they can be produced economically.

From 1957 to 1960, Colorado State University conducted research on several film materials for greenhouse coverings. One of the film products evaluated was "Tedlar" (Dupont Registered Trademark). This polyvinyl flouride film, 2.0 mil-thick and 3 ft wide had to be stretched and held in place with tack strips. The stresses created by the wind and cold temperatures caused the "Tedlar" and other films to contract and tear, starting at the nail holes.

All of the film materials were difficult to install and proved impractical as greenhouse coverings. They included Mylar (Dupont Registered Trademark) (polyester), Kodapak (a cellulose), and polystyrene.

The most common film plastic is polyethylene, it has become a household item used for sandwich, clothes, and trash bags as well as curtains, tablecloths, gloves, and paint dropcloths. Polyethylenes used for household purposes do not make good greenhouse coverings because of their UV instability when exposed to solar radiation. Ultraviolet light causes film degradation. Both antioxidants and ultraviolet absorbers must be added to keep the material in a stabilized form. The benzophenes, substituted benzotriazoles, and acrylonitriles and related compounds are used as stabilizers (Keveren, 1973).

1. Film Standards. When new materials are developed, it is imperative that methods of classification and testing be designed. In 1960, the Society of the Plastics Industry, Inc. requested the assistance of the Department of Commerce in establishing a commercial standard for polyethylene sheeting. The standards were developed by film producers, distributors, and users in cooperation with the National Bureau of Standards, published in 1961, revised in 1969. The "Standard for Polyethylene Sheeting" (construction, industrial, and agricultural applications) is classed as National Bureau of Standards Voluntary Product Standard PS 17–69. It established: "(1) dimensional requirements for standard sizes and types of various products, (2) technical requirements for the product, and (3) methods of testing, grading, and marking these products".

Requirements described in Standard PS 17–69 that would be of interest to the greenhouse operator are materials content, package weight, and thickness. The sheeting should consist of basically ethylene copolymers with lesser amounts of other polymers. It may contain additives such as pigments or stabilizers. The

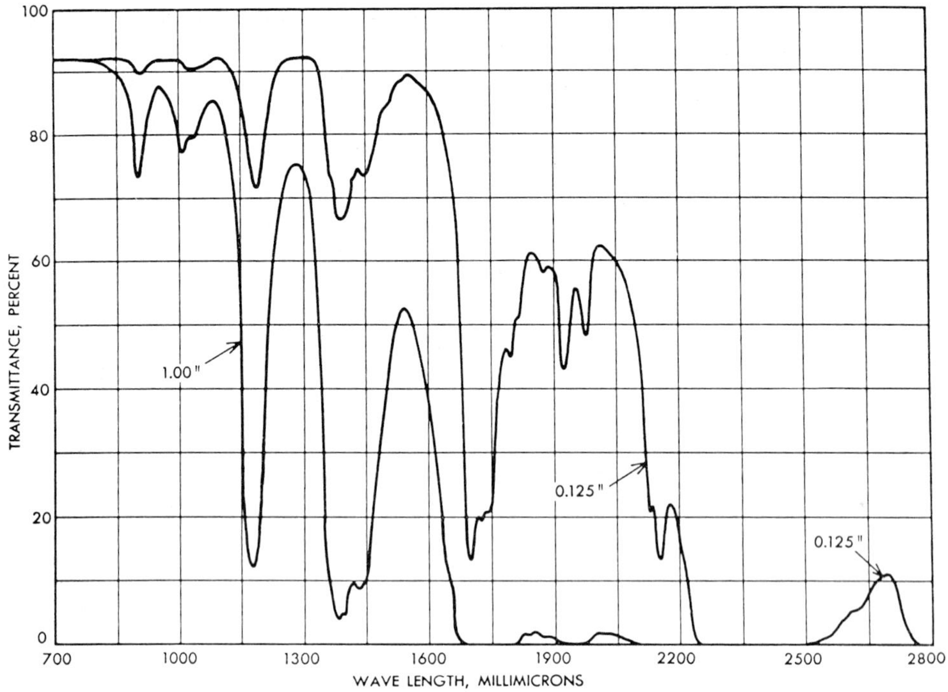

Fig. 3-27. The effects of Plexiglass thickness on the transmittance of infrared wavelengths (Plexiglass, 1964)

minimum thickness should not be less than 80% of the nominal thickness when determined by the method described in paragraph 4.6.1.1 of the Standard.

When large bubbles or irregular areas form in the polyethylene film used in "air-inflated" structures, the product standards are not being met. The actual weight of the package or roll can be determined and should agree with the manufacturer's specifications. Any irregularities in product specifications should be reported to the Weights and Measures division within the State Department of Agriculture.

2. *Rigid Plastics.* The late 1950s brought new technology to the greenhouse covering industry. Rigid plastic coverings were being considered because they provided adequate light transmission, facilitated rapid construction, and depending on the material, decreased building costs. The following three products compose the major materials generally considered for enclosing growing facilities for plant production.

1. *Acrylic Sheets.* Acrylic monomers were reported as early as 1843, but it was not until 1901 that Dr. Rohm of Germany reported the development of acrylic materials and in 1927 directed the first commercial production of the resin (Frederick, 1939; Leyson, 1944; BPI, 1967; Or-chem, 1967). Plexiglass and Lucite (Registered Trade Names) are two of the better-known flat acrylic sheets. They can be obtained with different thicknesses, colors, and wavelength transmittancy charac-

Table 3-4. Transmittance values for colorless, white translucent (W) and transparent gray and bronze (Plexiglass acrylic sheets, 1967)

Color number (white)	Solar energy transmittance $^3/_{16}$ in thick	Light transmittance $^1/_8$ in thick	$^3/_{16}$ in thick	$^1/_4$ in thick
Colorless	85%	92%	92%	92%
W-2075	73%	82%	77%	72%
W-2067	66%	71%	61%	53%
W-2159	62%	64%	52%	43%
W-2254	59%	60%	48%	40%
W-2447	52%	53%	42%	36%
W-7138	37%	42%	32%	26%
W-7328	27%	32%	23%	17%
W-7420	18%	23%	15%	11%
W-7414	11%	17%	11%	7%
W-7508	8%	8%	5%	4%

Color number	Transmittance (all thicknesses) Solar energy	Visible light
Colorless	85%	92%
Gray No. 2538	27%	16%
Gray No. 2537	41%	33%
Gray No. 2064	44%	27%
Gray No. 2074	32%	14%
Gray No. 2094	55%	45%
Gray No. 2514[a]	62%	59%
Gray No. 2515[a]	74%	76%
Bronze No. 2370	20%	10%
Bronze No. 2412	35%	27%
Bronze No. 2404	56%	49%
Bronze No. 2539[a]	62%	61%
Bronze No. 2540[a]	75%	75%

[a] Non-standard colors.

teristics. The near infrared transmittance of Plexiglass is dependent on thickness (Fig. 3-27), thus a "greenhouse heat effect" can be controlled to a degree. Colorless acrylic is highly transparent, but the transmission characteristics can be modified by color content (Table 3-4) as well as thickness. Even though acrylic sheets are highly weather resistant, approximately 10 times more resistant to impact than greenhouse glass and considered to transmit 90–95% of available solar energy, they are cost prohibitive for commerical greenhouse construction.

2. *Polyvinyl Chloride (PVC).* Rigid PVC panel production began in the United States around 1959, but a light-stable product for greenhouse glazing could not be manufactured. The Japanese produced a panel for export in the late 1950s, but it proved to be light-sensitive and turns "yellow" within 18 to 24 months and after 10 years, it turns black and becomes brittle. A higher-quality PVC was manufactured in Israel in the late 1960s and has remained relatively

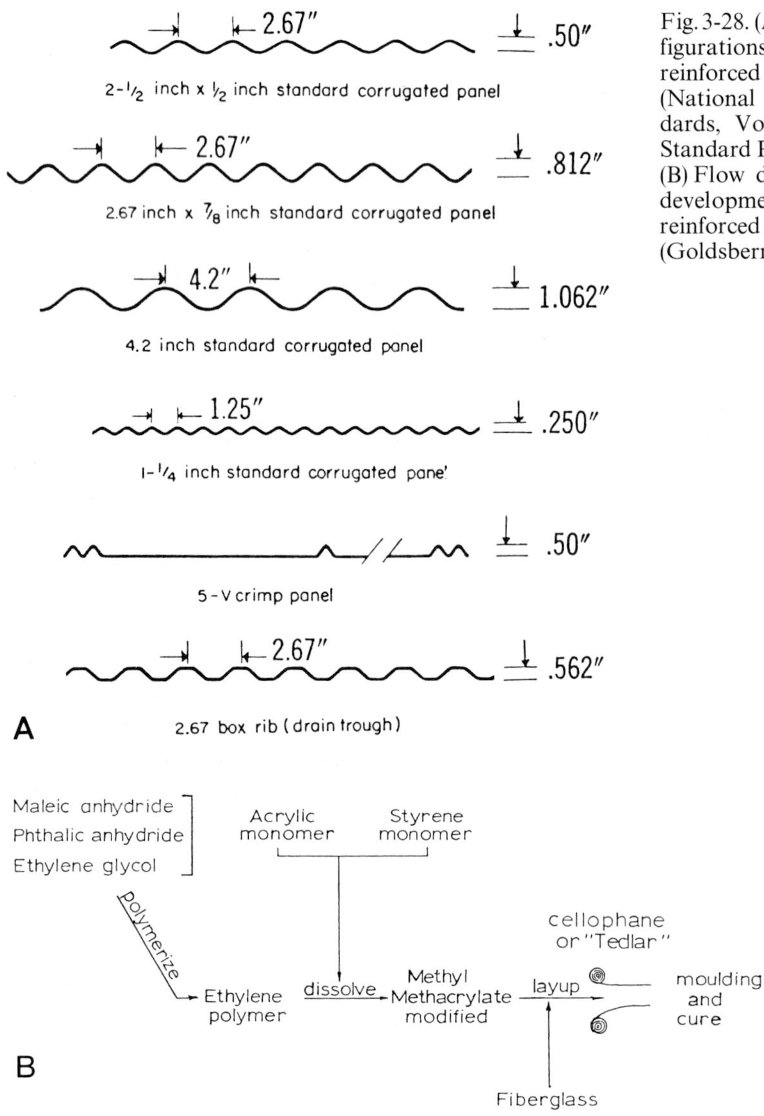

Fig. 3-28. (A) Common configurations of fiberglass reinforced plastic panels (National Bureau of Standards, Voluntary Product Standard PS 53–72). (B) Flow diagram showing development of fiberglass reinforced plastic panels (Goldsberry, 1969)

clear for approximately five years but became extremely brittle. To date, it appears that there is no rigid PVC material adaptable to greenhouse use.

3. Fiberglass Reinforced Plastic Panels (FRP). The first known FRP panels in the United States were produced in approximately 1947. These translucent thermosetting panels were used mainly as skylights in corrugated metal buildings. The panels were composed of polyester resin, catalyst, filler, and glass mat (Shand, 1955). After approximately two years exposure, the panel started eroding, exposing the glass fibers.

By the mid 1960s the FRP panel had been improved with the addition of an acrylic monomer. Figure 3-28 shows a generalized flow diagram of materials for the present day production of glass reinforced plastic panels (FYGM, 1920; O'Keefe, 1953; BPI, 1967). New formulations have improved the FRP panel providing increased weatherability, transmission stability, and longevity.

One of the earliest evaluations of FRP as a cover for plant growth was by Holley in 1956 (Holley and Baker, 1963). In 1958, Carpenter (1961) built a greenhouse structure of wood and FRP for plant growth evaluations. In the summer of 1959, Briggs (1961) compared the growth of carnations under glass, several plastic films and FRP. From 1958 to 1960, White (1960) evaluated the growth response of several plant species grown under clear FRP. Work at Colorado State University in 1966 (Holley et al.) indicated that the frost type fiberglass yielded greater growth than clear FRP, PVC panels or glass. Most research in the United States to date, indicates that plants grown under translucent FRP panels are equal to or superior to those grown under glass.

Harnett (1972) reported on research at Efford EHS where five identical houses were "clad" with four plastics and the fifth with glass. They concluded that FRP was not a suitable cover under United Kingdom solar energy conditions and should be disregarded in future designs.

Inferences are often made regarding the superior growth of plants achieved under polyethylene compared to other plastics. Two investigations (Dallyn and Sheldrake, 1968; Schales et al., 1968) indicated no significant differences in the yield or quality of tomatoes grown under four types of plastic, including FRP.

FRP panel standards. The Society of the Plastics Industry, Inc., in August, 1955, requested the cooperation of the U.S. Department of Commerce Commodity Standard Division to assist in the development of a Commerical Standard for polyester fiberglass reinforced building panels. A committee of manufacturers, users, and interested parties combined efforts and the first Commercial Standard (CS 214-57) was published in 1957. It was amended in 1961 and modified again in 1971 as a voluntary product standard. On January 1, 1972, the U.S. Department of Commerce, National Bureau of Standards Voluntary Product Standard, PS 53-72, Glass-Fiber Reinforced Polyester Structural Plastic Panels, became effective.

A second standard, that can be considered as minimum conditions, is found in the *Federal Specifications* on Plastic Panels, Corrugated, Translucent Glazing L-P-500C, February 5, 1974. These specifications correlate well with the Voluntary Products Standard.

Some of the factors that are important to FRP users are herewith presented and comments made. In depth information should be obtained directly from the Voluntary Product Standard, PS 53–72. Reference is often made to the American Society for Testing and Materials (ASTM) procedures, which are not described in full.

Materials — The components should consist of polyester resins, glass fibers, and can have minor amounts of other resins, stabilizers, etc. There is no method of testing for material content.

Classification — Plastic Panels have been classified as: Type I—General Purpose, Type II—Fire Retardant (limited flammability).

Length and Width —	The tolerance on nominal lengths and widths is $\pm 1/4$ in.
Weights —	The weight is based on ounces per square foot of panel. Only three weights (grades) are described in the product standard, Table 3-5. It should be noted that the tolerance is 10% of the weight.
Configuration —	Many configurations are available (Fig. 3-28). The product standard indicates that the configuration is directly related to the transverse load strength and panel weight.
Color —	The panels shall have uniform color when visually examined from a distance of 10 ft. Minor differences in intensity of color caused by non-uniform distribution of glass fiber shall not be cause for rejection.
Light Transmission —	The nominal light transmission factor shall be agreed upon between purchaser and seller. The tolerance shall be $\pm 5\%$ of the nominal factor when measured by the method described in ASTM D 1494-60 (1969). ASTM evaluated the transmission characteristics of FRP panels in a specially designed box using fluorescent bulbs as a light source.
Appearance —	The panels shall be free from foreign inclusions, cracks, crazing, die lines, pinholes, striations (slight channels or ridges) as determined by visual inspection.
Flammability —	The rate of burning is determined in accordance to ASTM D 635-68. Type I panels shall have a flame spread of less than 2.0 in per min and Type II, less than 0.35 in per min. It should be pointed out that this rate is to take place under special controlled conditions and does not correlate with the flammability potential under actual use.

Table 3-5. Fiberglass reinforced plastic panel weights and tolerances (National Bureau of Standards Voluntary Product Standard PS 53-72)

Weight of panel oz/ft^2	Tolerance oz/ft^2
8	± 0.8
6	± 0.6
5	± 0.5

Several factors related to the voluntary standards should be discussed briefly. First, there is no standard for a 4 oz panel. If the 10% tolerance was applied to this panel, the minimum weight would be 3.6 oz ft^{-2}, not 4.0 oz. It is also possible that a 4.5 oz ft^{-2} panel can be sold to one customer and a 5.5 oz to someone else, both being classed as a 5 oz panel and one will outwear the other. Lastly, there is no apparent requirement in telling the buyer how to properly store an FRP panel shipment.

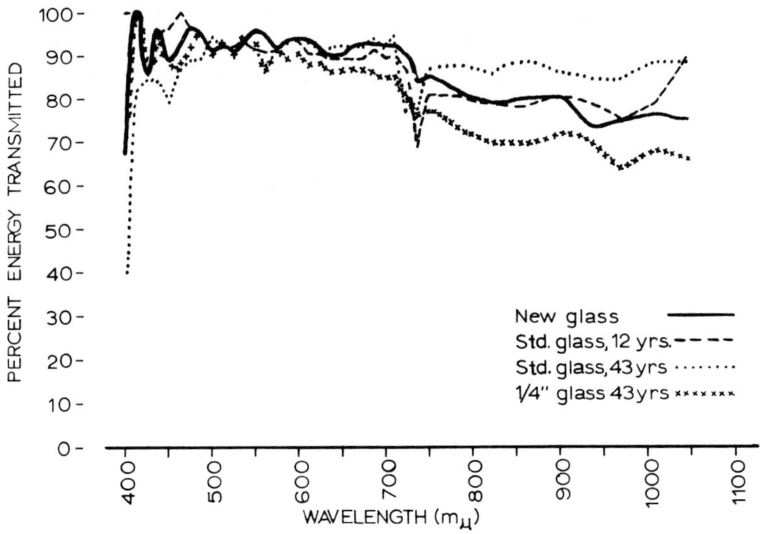

Fig. 3-29. Percent of unobstructed solar energy transmitted by panels of greenhouse glass (Goldsberry, 1967 b)

3.5.3 Cover Transmission Qualities

An early account of heat transmission through glass and its effects on plant growth was demonstrated when the Kew Gardens palm house was constructed in approximately 1833 (Leuchars, 1854; Lemmon, 1962). The intense light passing through the glass was assumed to cause a "burning" on the palm leaves and a pale yellowish-green tinted glass was recommended and was manufactured to decrease heat transmission. The theory may have been correct, but the tinted glass was later replaced with colorless glass.

With the advent of plastic covers, a number of studies on their radiation transmission properties compared to glass have been conducted. Trickett and Goulden (1958) found that the transmission characteristics of glass, comparable to that used on greenhouses, was 20% less than for glass the thickness of film plastics. Goldsberry (1967 b) found that the light transmission characteristics of greenhouse glass were basically unaffected by age and thickness, Figure 3-29. The etching created by air pollution or "scum" accumulated over a period of time contributes only to minor changes.

Williams (1968) evaluated the light and infrared transmittance of several film and rigid plastics in Hawaii. After one-year exposure, he found that the visible light transmittance of the vinyl films decreased 6–16% based on thickness, the rigid PVC from 1 to 14%, depending on color, no reduction for the Mylar (Trade Name), and only 1% loss through the FRP panel. The infrared transmission capability was severely reduced in the thicker PVC films, colored PVC panels, and Mylar after one year.

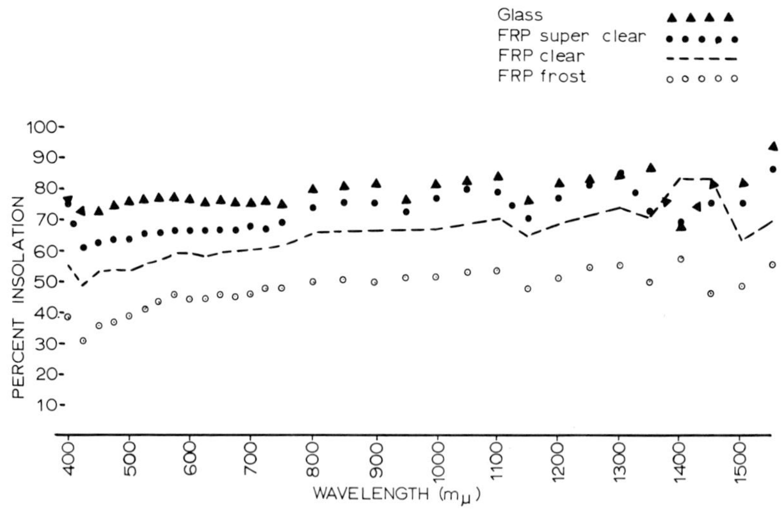

Fig. 3-30. Spectral distribution of energy transmitted through four greenhouse covers (Goldsberry, 1968a)

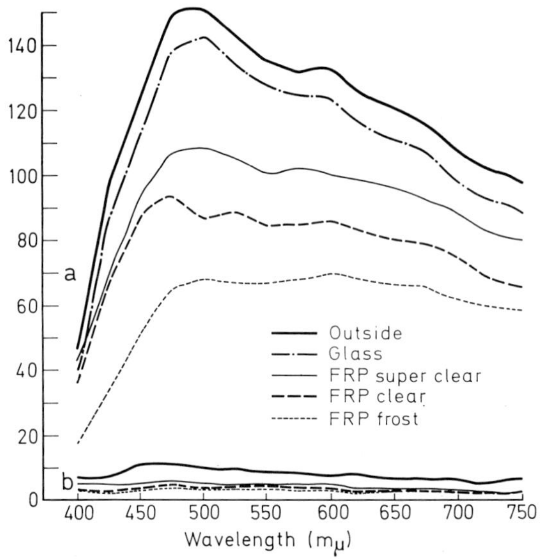

Fig. 3-31. Spectral intensity under four covers: *a* May 13, 1968, an unobscured day; *b* May 15, 1968, a day of total overcast (Goldsberry, 1969)

Excellent plant production has been obtained in structures covered with glass and FRP panels (Fig. 3-30) on frost, clear and superclear at Colorado State University (Goldsberry, 1968a, 1969). The transmission of insolation through these covers on an unobscured day and a day of total overcast were observed. It was apparent that during periods of total overcast, there were no significant differences between the spectrums transmitted by the four coverings (Fig. 3-31).

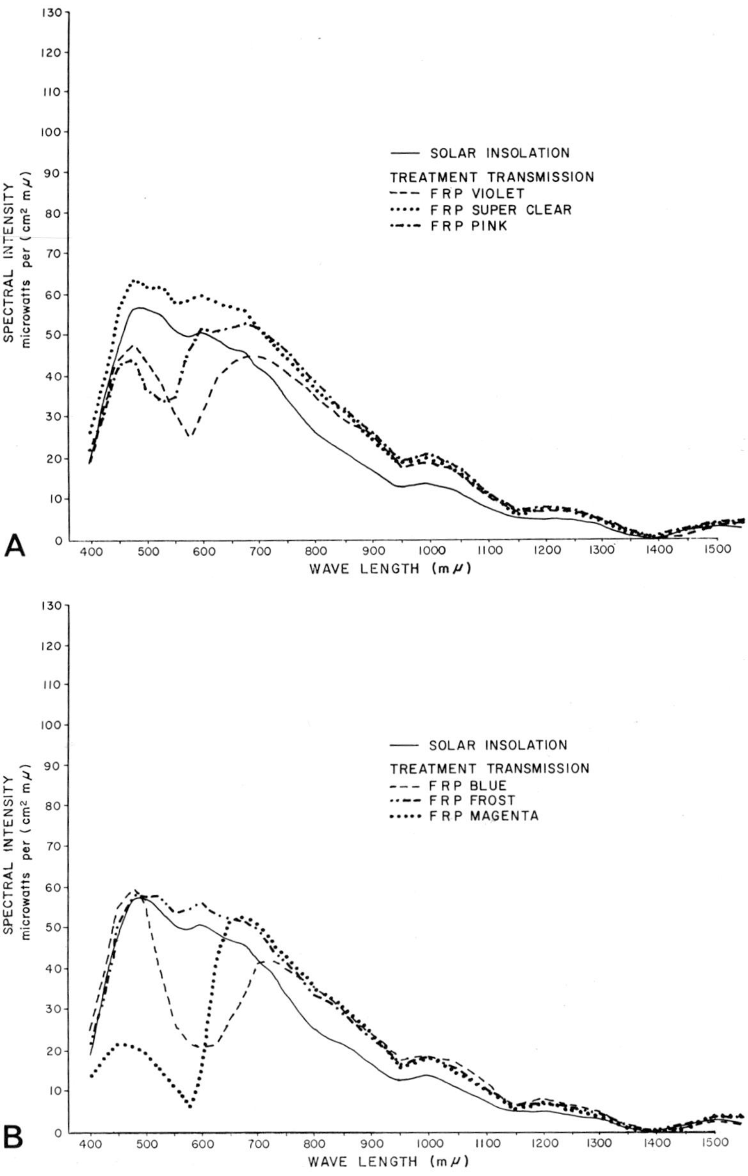

Fig. 3-32A and B. Mean spectral curves of unobscured light received under tinted FRP compartment coverings between November 20, 1968, and January 9, 1969. (A) Violet, superclear, and pink. (B) Blue, frost, and magenta (Goldsberry, 1969)

The energy transmitted through a covering is directly related to its cleanliness. In 1973, a grower noticed the accumulation of cement dust on his greenhouse roof, from a nearby concrete batch plant. After cleaning a 100 ft^2 section of roof on a glass and FRP covered house, pyranometers were placed under the cleaned and dirty sections and compared. There was 22.5% less energy received through

Table 3-6. The mean surface temperatures and BTU quantities obtained in six plant growth compartments with FRP panel covers

Cover	Surface temperature °F			Heat transmission BTUs
	Foliage	Black	White	
Violet	71.93	102.20	71.20	351.64
S. clear	74.93	107.00	75.80	401.18
Pink	74.47	99.00	75.53	341.59
Blue	74.07	102.93	74.27	305.08
Frost	75.40	106.20	75.07	389.94
Magenta	72.53	100.13	74.60	317.66
Incoming insolation		83.2[b]	50.46[b]	386.88
Q=HSD[a]	1.41	1.75	0.76	11.00

[a] Tukey's honestly significant difference.
[b] Not considered statistically.

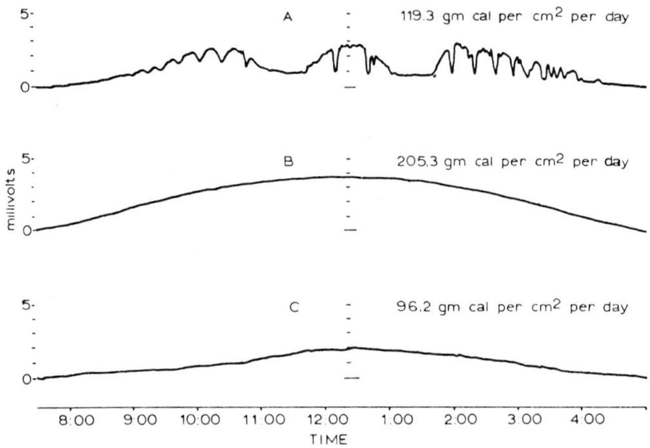

Fig. 3-33. Solar energy curves received under greenhouse coverings of *A* sash bars and glass and *C* FRP clear compared to *B* incoming energy (January 21, 1968) (Goldsberry, 1970b)

the dirty glass and only 4.8% less through the dirty fiberglass. The cement evidently stuck more to the glass than to the fiberglass.

The transmission characteristics of tinted fiberglass reinforced plastic panel covers provide definite differences in available solar (Fig. 3-32) and thermal (Table 3-6) radiation (Goldsberry, 1969, 1971b). The transmission characteristics and plant responses obtained in a clear FRP covered chamber were compared to those in five chambers covered with tinted panels of violet, pink, magenta, blue, and frosted white. The final results indicated that nonpigmented coverings transmitted the most energy and contributed to the best growth. It is possible, however, to utilize some types of pigmented covers to aid in controlling the growth of

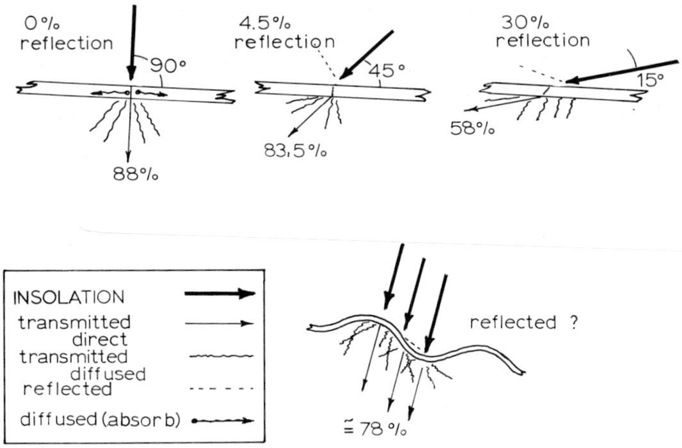

Fig. 3-34. Insolation striking glass *(top)* or corrugated FRP panel will be reflected, absorbed, or transmitted. (Adapted from Taft, 1926; Walker and Slack, 1970; Duncan and Walker, 1973)

some plants or provide more optimum conditions for certain phases of plant growth.

Walker and Slack (1970) found that some greenhouse covers will permit thermal radiation transmittance whereas others block its passage. It should be noted that those blocking thermal radiation tend to help decrease heat loss in the winter (Table 3-7).

McLaughlin and Sheldrake (1973) measured the effects of the greenhouse superstructure on the amount of direct light reaching a greenhouse bench (Table 3-8). Their data substantiated the results obtained when measuring solar energy transmitted through an FRP panel and glass covered house at CSU, which indicated that 58.1% of the available solar energy reached plants grown in the glasshouse and only 49.2% in the FRP covered house (Goldsberry, 1970b). The sash bars, trusses, and pipes in the glass house, Figure 3-33, and the diffusion characteristics of the FRP contributed to lower energy levels than were outside the greenhouse.

3.5.4 Optical Properties of Covers

Insolation falling on greenhouse covers is transmitted, reflected, and absorbed. The cover, color, configuration, thickness, and cleanliness determine the degree to which energy will be directed. Covers that allow a high percentage of light energy to pass through, are classed as transparent. A translucent cover causes a diffusing condition, breaking up the rays. The transparent cover allows both direct and diffuse light energy to pass through. Glass and clear plastics transmit a high percent of direct energy, creating shadows on and in plant canopies (Fig. 3-34). The "hammered" glass and translucent plastics transmit a lower amount of direct

Table 3-7. Percent solar and thermal radiation transmittance of certain materials (Walker and Slack, 1970)

Film type	Solar transmittance[a]		Thermal transmittance
	Single glazing	Two layer glazing	Single glazing
Polyethylene, clear	93	88.0	
Polyethylene, commercial clear	76 (89)	(81)	70.8
Polyethylene, UV	74 (88)		
Glass	86 (90)	75	4.4
Polyvinyl, clear	86 (91)	(84)	12.0
Polyvinyl, haze	(89)	(82)	
Mylar (polyester)	86 (90)	80	16.2
Rigid, fiberglass	18 (78)[b]	(64)	1.0

[a] Data without parentheses is direct transmittance, data with parentheses is total transmittance.
[b] Some newer fiberglass materials have 85–95% total transmittance.

and total light and more diffused rays. The "frost" colored FRP panels have been recommended in geographical areas of high light because of the superior responses of plant growth obtained in past research (Holley et al., 1966; Goldsberry, 1968a).

Trickett and Goulden (1958) evaluated the ultraviolet transmission characteristics of covers and found that in no case does the UV cutoff point lie within the wavelength range of photosynthetic activity (400–720 nm). An example of the absorption of some light energy being converted to heat as it passes through a cover was demonstrated in research with tinted panels (Goldsberry, 1969). During the course of the investigation, the presence of condensate was noted on the inside surfaces of the superclear and frost FRP coverings and its absence on the tinted panels. The inside surface temperatures of the panels were measured with an infrared thermometer, and it was found that the superclear and frost panels were 4–14° F cooler than the tinted covers. The elimination of condensate was attributed to the absorption of light energy by pigment particles thereby increasing the temperature of the surrounding plastic and glass media, thus evaporating the condensate. The colorless and frosted type FRP panels remained cool due to the absence of pigments.

3.5.5 Physical Properties of Covers

The physical properties of the greenhouse cover must be adequate to large geographical areas. Duncan and Walker (1972) conducted studies to determine the physical properties of several greenhouse coverings. The complete section of their talk on the subject is reprinted.

"Covering materials must be physically resistant and durable to naturally occurring weather elements. Forces such as wind, rain, hail, and foreign objects frequently act on the greenhouse covering. The material which can best withstand these forces is most satisfactory for greenhouse covering.

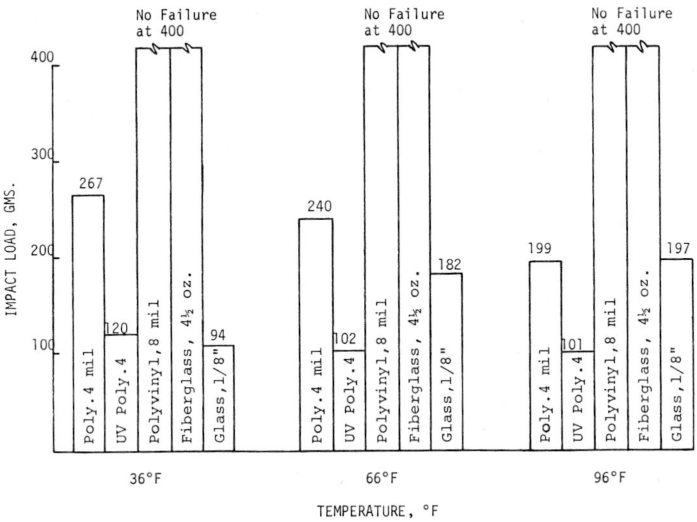

Fig. 3-35. Impact load in g for 50% failure of greenhouse glazing materials as influenced by temperature (Duncan and Walker, 1972)

Table 3-8. Light readings at bench level (foot candles) (McLaughlin and Sheldrake, 1973)

Outdoor reading	10000
Under newly cleaned glass	8800
Under new polyethylene (2 layers)	8300
Under one year old poly (2 layers)	7200
Under small sash bar (glass)	3200
Under large truss support (glass)	1400
In 2 × 6 in rafter shade (poly house)	5800

"Manufacturers generally publish certain strength data for their covering materials. Often this does not include all methods of possible loading nor is the information presented consistently among manufacturers. Walker and Slack (1970) tested and reported in detail the strength properties of several materials. Two methods of loading were used: Impact test and lateral load test. These tests were patterned after ASTM D 1709-62T and 1502-60, respectively, and the results for relative impact strength of selected covering materials at three different temperatures are shown by Figure 3-35.

"The results indicate that, with the exception of glass an increase in temperature results in a decrease in impact strength. For the 36° F condition, all materials evaluated proved to have greater impact strength than regular glass. At the higher test temperature, glass was stronger than 4 mil UV-resistant polyethylene. No damage occurred with the maximum impact load of 0.88 pound to the 8 mil polyvinyl and $4^1/_2$ oz glass. A study of the effect of plain polyethylene film thickness indicates a linear relationship between thickness of the material and impact strength. The exact relationship between these tests and the severity of hail on the materials is not known, but relative strengths should indicate the order of resistance.

"The steady and continuous load that might be applied by snow or other similar possibilities is illustrated by the lateral load test. The results of the load tests at three different temperatures for a 24-in wide by 24-in long panel are listed in Figure 3-36. The plastic film

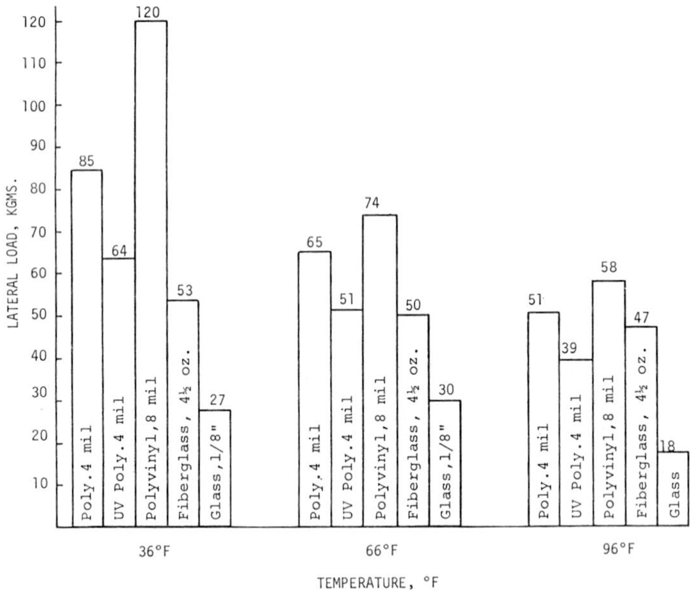

Fig. 3-36. Maximum load in kg ag 6 in deflection or failure for 24 × 24 in laterally loaded glazing materials as influenced by temperature (Duncan and Walker, 1972)

materials showed a decrease in load as the temperature increased. As would be expected, the thicker materials carried more load per unit of deflection than the thinner ones. However, if the values in Figure 3-36 are divided by the mil thickness of the film, the regular polyethylene film has the largest load-carrying capacity per unit of actual deflection, being approximately twice as strong as the polyvinyl film and about 50% stronger than UV polyethylene. The rigid materials did not appear to be affected by the temperature. All of the materials proved stronger than glass. It should be recognized that these tests were conducted with new material and *the load-carrying capacity of the plastic film would decrease with deterioration because of weathering*."

3.5.6 Weatherability of Covers

3.5.6.1 Film Plastics

In the early 1950s, the only greenhouse cover that had withstood the "tests of time" was glass. New plastic products were being "peddled" as covers, but little was known of their longevity. Most of the evaluations were made by the manufacturer, in some instances in cooperation with universities (Emmert, 1955).

In 1962, the National Agricultural Plastics Association Testing Committee initiated a program to evaluate 20 different plastic films at six test sites—California, Mississippi, New York, Ontario (Canada), Oregon, and Virginia (Massey, 1965). Variations in performance existed between sites and products, but general conclusions were made (Table 3-9).

A second test was completed in 1965. It was noted that the weatherable life of 6 mil polyethylene films had increased 66% over the 1962 evaluations, a credit to expanded technology (Moore, 1969).

Table 3-9. Average life of film plastics exposed on a test rack in six geographical locations for three years (Massey, 1965)

	Thickness	Duration (months)
UV-inhibited polyethylene	2 mil	11.0
	4 mil	14.5
	6 mil	19.5
Vinyls	4 mil	18.5
	6 mil	26.0

Fig. 3-37. Degradation of a 6 mil vinyl film after 20 months service on a Colorado State University quonset structure

All plastic covers are susceptible to photo degradation. Both polyethylene and vinyl films are affected by ultraviolet light, if inhibitors, absorbers and stabilizers are not incorporated. Unstabilized polyethylene film will become brittle and tear like paper when exposed to solar radiation. Even polyethylene with inhibitors will gradually disintegrate when the film is stressed over supports and heat builds up in the area of contact. Higher temperatures enhance degradation.

Vinyl films have a color change during degradation (polyethylene remains clear to cloudy) and small brown areas at points of contact gradually enlarge and eventually decompose or tear (Keveren, 1973). Figure 3-37 shows how vinyl film decomposed on a Colorado State University structure.

Painting of the greenhouse superstructure with white or aluminum paint decreases the possibility of rapid film degradation. It has also been suggested that the outside of the film be painted in areas of contact (Keveren, 1973). One should be sure a nontoxic paint is used.

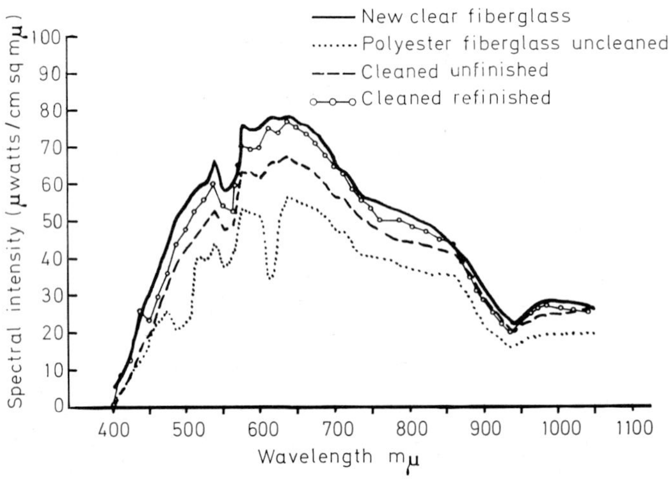

Fig. 3-38. Spectral distribution of insolation transmitted through weathered (66 months) FRP panel when cleaned and refinished (Goldsberry, 1967a)

3.5.6.2 Rigid Plastics

Approximately 24–36 months after rigid FRP panels were being installed on commercial greenhouses, growers in many parts of the country became aware of panel discoloration and "blooming" of the fibers. Goldsberry (1967a) evaluated the transmission characteristics of a $5^1/_2$-year-old panel that was removed from a greenhouse roof. The total heat transfer of the uncleaned section of panel was 26.8% less than unobstructed solar radiation and 20% less than through new acrylated fiberglass. The spectral distribution and intensity obtained through uncleaned, washed and refinished sections of the panel were compared to new acrylated material. Transmission characteristics were greatly improved by washing a section of the panel with a tri-sodium phosphate solution and then removing the remaining loose fibers with coarse steel wool followed by a rinse with water (Fig. 3-38). The refinished portion was comparable to the new panel.

Some older FRP panels were observed to have excessively reduced light transmission characteristics. Such panels appeared to have a fungus growing on the surface. Durrell and Goldsberry (1970) verified the premise and identified the fungus as *Aureobasidium pullalans*. They noted further that FRP panels, after a period of time, tended to have surface cracks which formed a polyagonal areolar pattern. The cracks, the result of shrinkage, were filled with cells and hyphal strands of the fungus *A. pullalans*, (Fig. 3-39). *A. pullalans* is a common fungus found on bark of trees and painted surfaces and is known to attack old books. It is readily disseminated by wind and water and when attached to FRP panels, has been found to decrease the light transmission characteristics by 50%.

The observations of Durrell and Goldsberry led to an evaluation of the surface characteristics of recently manufactured and refinished FRP panels (Goldsberry, 1971a). New panels were found to contain bubbles; exposed glass fibers and numerous types of cracking (Fig. 3-39 B–D). All of the deformities affect the weatherability and longevity characteristics of the panels.

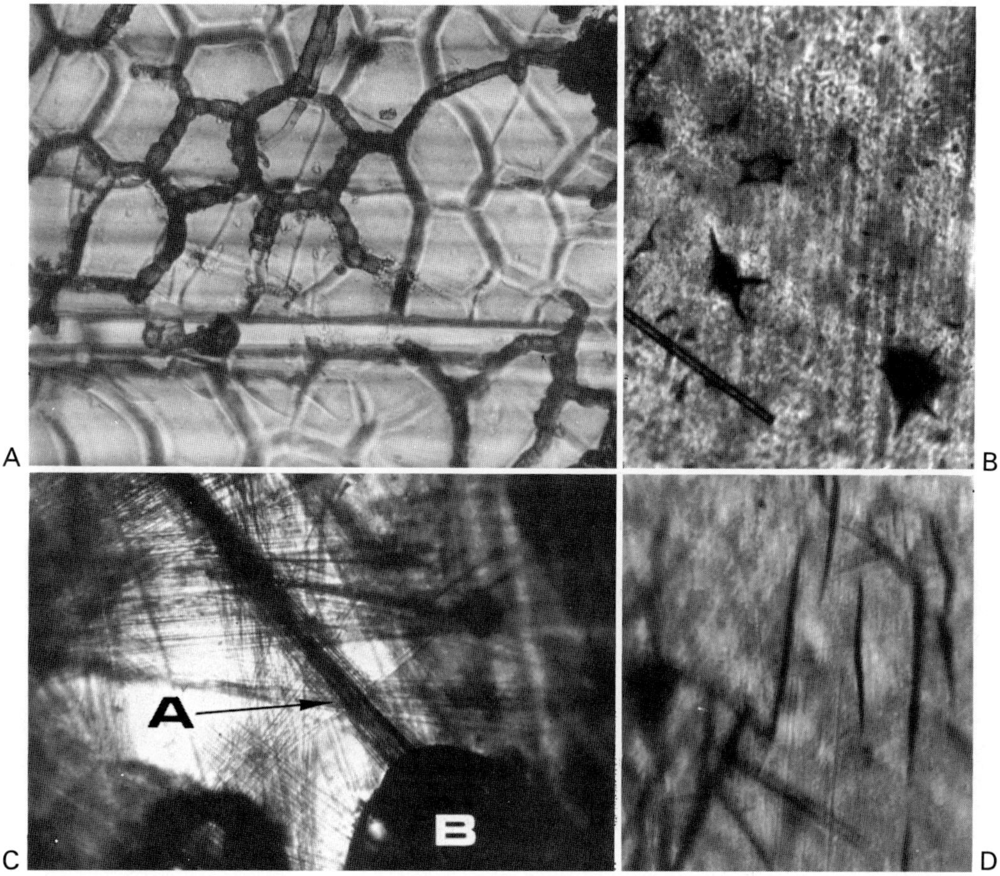

Fig. 3-39. (A) Hyphae of *A. pullalans* growing in polygonal surface cracks of a weathered FRP panel. (B) Bubbles and adjacent cracking found in a new FRP panel. (C) *Arrow A* points to a glass strand stained with dye from an exposed air bubble *(B)* in a new panel. (D) Parallel cracking in FRP panels (Goldsberry, 1971a)

The National Agricultural Plastics Association (NAPA) initiated an FRP panel evaluation in 1968 to determine the weatherability of numerous FRP panel products. Fifteen specimens of 5- and 6-oz acrylic modified greenhouse grade panels were placed in 7 locations in North America. Hartman (1973b) reported substantial differences in both appearance and post exposure transmissivity of various sample groups. The results tended to confirm previous reports that weatherability varied with grade and possible differentiation occured among FRP panel brands.

The longevity of an FRP panel is directly related to its "weatherability" providing it is not physically harmed. Four factors contribute to the weatherability of FRP panels. Hartman (1968) described three of the factors and the fourth is based on several years of observation at Colorado State University.

1. Photo degradation: Polyester resins, being organic in nature are susceptable to ultraviolet radiation, the same as human skin. UV screening agents in locations can protect skin and thus prevent rapid color changes from red to brown. Organic UV screening and absorbing agents protect polyester panels and limit the degree of discoloration to hues of yellow.

2. Oxidation: All organic and inorganic materials are affected by the presence of oxygen in that they are decomposed. The sliced apple on the kitchen counter, through oxidation, changes color and gradually decomposes. Plastic panels and films are constantly being oxidized, resulting in glass fiber bloom, brittleness and/or disintegration.

3. Surface erosion: The elements of wind, moisture, and soil particles all contribute to the longevity of any type of surface. When the polyester resin surface of an FRP panel is removed, the glass fibers tend to appear. Cracks, and holes become more numerous and dirt actually enters the panel contributing to an off color and decreased light transmission.

4. Thermo degradation: The fourth factor which is often overlooked in greenhouse cover degradation processes is the effect of temperature. Temperature in the gable of a greenhouse will reach 110 to 120° F during periods of high insolation. Thermal expansion is of sufficient magnitude to be considered in greenhouse coverings, thus the temperature differentials confronting FRP panels contribute greatly to degradation. Any change in the linear dimension of a solid, (width, height, length) is known as linear expansion. The coefficient of linear expansion of a solid is the change in length per unit length per degree change in temperature (Shortley and Williams, 1955). It is hypothesized that the components of FRP panels all have different coefficients of expansion and temperature changes influence the degree of photo degradation, oxidation, and erosion.

3.5.6.3 FRP Surface Durability

Shortly after FRP panels were manufactured for greenhouse use, manufacturers became aware of the need to improve the surface durability. A "resin rich" acrylic modified coating on the exposed surface side of panels was included in the manufacturing process by one company. Another organization provided a pebbled resin enriched surface which was to improve light transmission and help maintain a cleaner panel. The later type panel has only been moderately successful when compared to the "resin rich" product.

Several FRP panel "refinishers" have been marketed for restoring the surface. The application of refinishers to a well washed and smooth panel does improve the transmission characteristics (Goldsberry, 1967a), but after a period of time, usually 6 to 8 months, they start flaking off.

A second evaluation of refinishers was conducted by Goldsberry and Homan (1975) on three-year old FRP panels. The experiment was designed to evaluate trisodium phosphate (1 lb/gal water); Colgate-Palmolive car cleaner ($^1/_2$ oz/gal water) and Spic and Span ($^1/_2$ cup/gal water) and water as cleaners before refinishing, plus four refinishers. The results indicated that the washing treatments did not contribute more to the bonding of one refinisher than another. Flaking of the refinishers in this experiment was comparable to the results obtained in earlier tests. It is assumed that surface cracking continues and the differences in elasticity of the panel and refinisher causes the refinishers to loosen.

The longevity of the FRP panel has been greatly enhanced with the introduction of the polyvinyl flouride film (PVF) by the DuPont corporation. The PVF film, Tedlar, is laminated to the polyester panel at the time of its manufacture by replacing the cellophane on top of the resin layup and allowing it to chemically bond as it cures (Wilson, 1968). "Tedlar" is transparent to visible light and, therefore, has a UV screening agent added to protect the resin at the film-resin interface (Wilson, 1971). High quality acrylic modified panels and good production procedures must be used to successfully laminate "Tedlar" to the polyester sheet. "Tedlar" laminated to low quality, nonacrylic modified sheets has not prevented yellowing in the panels. It is assumed that the yellowing in such instances is due, in part, to thermo degradation.

Some greenhouse growers have experienced problems with delamination of "Tedlar" film. In most instances the problem is associated with the panel manufacturing process. DuPont has developed specifications regarding the lamination of "Tedlar" to FRP panels. A portion of technical bulletin TD-26 is included for information purposes:

> "This specification covers the performance requirements of fiberglass reinforced plastic panels prefinished with "Tedlar" PVF film. The panels shall be either standard styrene polyester, acrylic modified polyester, fire retardant polyester, or epoxy. Glass content shall be approximately 25%.
>
> "The film shall be integrally bonded to one or both sides of the panel during the cure cycle of the resin and the laminate shall be capable of withstanding the following:
>
> Non-peelable[1]—after 30 min immersion in a boiling 5% calgonite[2] solution.
>
> Non-peelable—after 24 h immersion in boiling water.
>
> No visible surface erosion, fiber exposure delamination, crazing, or blistering—5000 h Atlas Weather-Ometer.
>
> Fabricated panels shall be installed in accordance with panel manufacturers' instructions."

Only time will determine the true value of a "Tedlar" coated panel on commercial greenhouse installations. The economics related to the initial cost versus the depreciation and replacement time will be the main factors.

It is imperative that greenhouse grades of plastic materials be used as covers. Such grades are not available through building supply or hardware stores.

3.6 Greenhouse Designs

A desire to obtain the greatest amount of insolation possible, cover the maximum ground area for the least cost and have a structurally sound facility, have been the basic criteria for the development of several types of greenhouses in the past century.

---

[1] Albisetti, C.J.: Technical Consulting Services, E. I. DuPont De Nemours and Co. Wilmington, Delaware 19898. Personal communication, December (1976). (An unpublished change in the specification.)

[2] Made with the commercially available Calgon water softner, Consumer Products Co. Pittsburgh, Pennsylvania 15230.

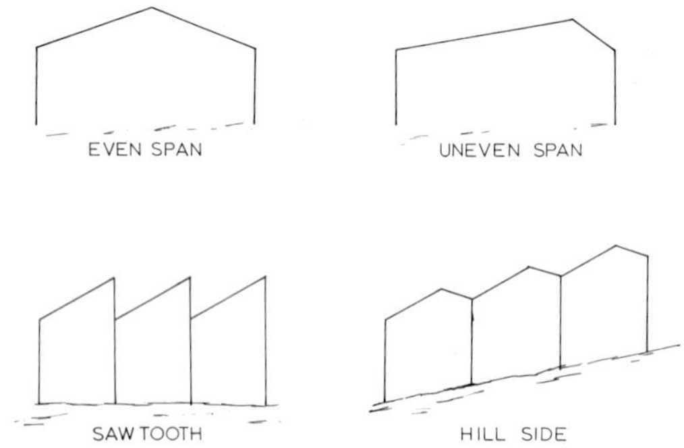

Fig. 3-40. Greenhouse styles still in use around the world

The glass covered structures have been limited in style. Many greenhouse structures classed as "lean-to or shed roof", even and uneven span, and "hillside" houses are still found throughout the world (Fig. 3-40). Some of these types are economically impractical and would not meet today's requirements of controlled environment. The sawtooth styles, however, have been instrumental in the development of the floriculture industry in Central America (Holley, 1969) as well as other countries and are being considered in the U.S.

Until recent years, the pitch of the glasshouse roof has been of paramount importance. Based on physical phenomena related to the interception of the sun's "rays", (Table 3-10) designers found that a roof pitch of 35° was most satisfactory and less than 26° deterred snow removal and increased the possibility of "drip" due to condensation on the underside of the glass (Wright, 1917).

The twentieth century has provided the greenhouse industry with numerous roof pitches, shapes and designs. Researchers in Europe and other northern latitude countries have been concerned with greenhouse designs because of minimum light energy conditions.

Lawrence (1963) discussed the merits of four basic greenhouse shapes. The structure providing the maximum light transmission is the curved or semicircular roof. Since it is almost impossible to use glass on such a structure, the British designed a house with sloping side walls, which provides more light transmission

Table 3-10. Percent of solar radiation lost when intercepting a glass greenhouse covering at different angles (Wright, 1917)

Angle of sun ray (°)	Loss by reflection (%)
60	2.7
50	3.4
40	5.7
30	11.2
20	22.2
15	30.0
10	41.2

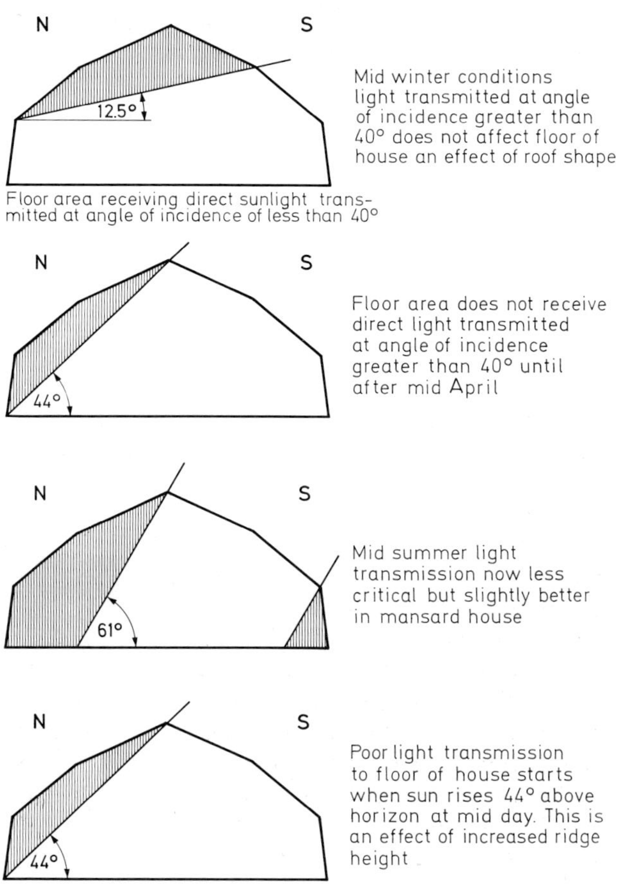

Fig. 3-41. Mansard house. Transmission of direct sunlight at angle of incidence less than 40° latitude 52° (Smith, 1967)

efficiency than the conventional peak design. Thus, the Mansard design, which allows solar energy to be transmitted at an angle of incidence never greater than 40°, is recommended for the British greenhouse growers (Fig. 3-41).

The introduction of plastic coverings and modern cooling techniques have allowed several modifications in greenhouse design in the past twenty years.

3.6.1 Free Standing Structures

Most greenhouses manufactured in the United States are patterned after the standard peak or arc type structures (Fig. 3-42).

3.6.1.1 Peak Roof

Structures are generally designed to be covered with glass or FRP panels. Some growers are covering "peak" structures with double layer film materials for air inflated purposes; in many instances, as an interum structure until they are financially able to install permanent coverings.

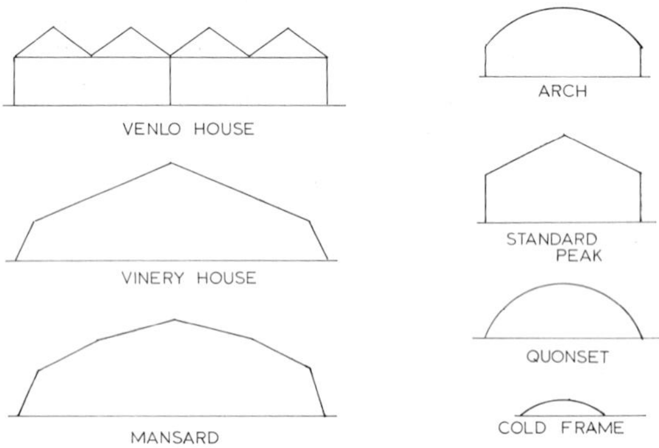

Fig. 3-42. *Right:* basic greenhouse structures common in the United States. *Left:* structures used in European area

One of the latest additions to the free standing design is the adaptation of industrial warehouse type framework to greenhouse needs (Butler Style, 1976). A 125 × 500′, 4 oz FRP covered house, with a 24′ ridge and 10′ eave was completed in 25 working days. The extremely low pitched roof was built in Waco, Texas, where snow is practically unknown. Such a structure would be questionable further north in some geographical locations.

3.6.1.2 Arch or Curvilinear

Roof structures, as described by Taft (1926), have become the leading design of the 1970s. They have been developed not because of the light transmission considerations, but due to economic factors; they can be constructed for approximately 25% less cost than a peak roof structure. The curved or arched roof is also easily adaptable to both rigid and film coverings.

The arch type structure has been modified into several designs. Hardly a year goes by, when one of the greenhouse manufacturers or greenhouse operators introduces a "new and better" structure.

The quonset style structure can be designed low to the ground and used as temporary cold frames or placed on columns of any height to meet the crop requirements. The value of the arch or curved roof greenhouse is outstanding.

3.6.2 Air-Inflated System

The greatest boon to the United States greenhouse industry in the last half century has been the air-inflated system of greenhouse construction. Dr. William Roberts (1968) an Agricultural Engineer at Rutgers University was instrumental in introducing the concept.

Air-inflated structures have a double layer plastic film cover that is kept relatively rigid and in place pneumatically. The double film, perferably 6 mil, is pulled snugly over the superstructure and sealfastened to the structure at all outer edges.

Inflation is accomplished by using a small, approximately 1/70 hp, squirrel-cage blower to pump air between the layers. It is recommended (Roberts, 1969) that the makeup air be taken from outside the greenhouse, because it is usually dryer and helps prevent condensate between the layers.

A maximum static pressure between the two film layers should range from 0.2 (6 mm) to 0.5 (12 mm) inches of water, depending on external wind pressures. A pressure of 0.2 in is sufficient in 20–25 mph winds. In January, 1971, a Colorado State University Monsanto 602 (Trade Name) covered, air-inflated quonset (20 × 50′) withstood winds of 30–70 mph over a 24-h period that contained a peak gust of 110 mph. The damper on the blower was completely open during this period of time.

The static pressure between the film layers can be measured with a simple manometer as described by Sheldrake (1971): "One end of a piece of 0.25 in (6 mm) clear plastic tubing can be pushed through a small hole in the inside film layer. The tube remaining in the greenhouse should be bent to form a 10–12 in (25–30 cm) U-tube and attached to the greenhouse superstructure with a small

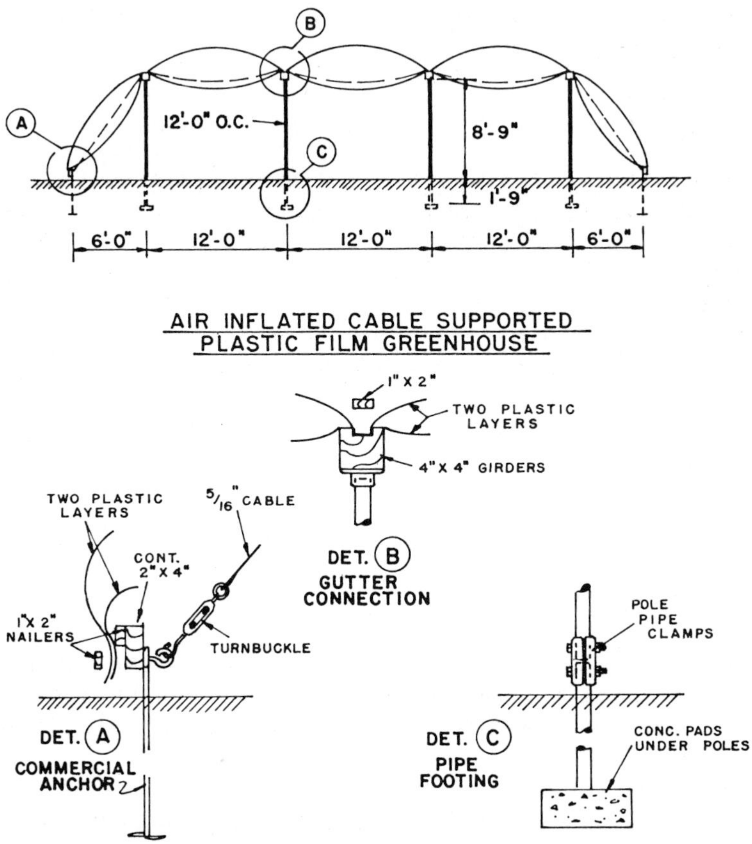

Fig. 3-43. Air-inflated cable supported greenhouse (Roberts, 1971)

ruler positioned behind it. After 6–8 in (15–20 cm) of water is poured into the U tube, the water-column pressure can be measured by observing the difference between the level of the water on both sides of the U. A difference of 0.25 in (6 mm) is adequate. (0.2 in water column is approximately one pound per ft^2)."

A recent innovation of the air-inflated system involves a flat roof superstructure of steel, wood, or cable. Scientific Farms of San Antonio, Texas, has a channel iron-frame air-inflated structure that is used for tomato production. Roberts (1971) described research conducted on an air-inflated cable supported greenhouse. The structure (Fig. 3-43) has potential in many areas of the world and could be easily modified to a cloth type house for summer use.

The air inflated system has been adapted to both peaked roof (Sheldrake, 1971) and arch greenhouse superstructures which in turn have been developed as facilities ranging from animal shelters to swimming pools.

Air supported structures have been considered (Roberts, 1971) and constructed in Ohio for vegetable production. It appears that such facilities are not feasible without more research.

3.6.3 Ridge and Furrow Configurations

The basic configuration for commercial greenhouse structures in the United States is the ridge and furrow house. They are least expensive to build, conserve ground area and require less fuel for heating than the free standing types. Such structures are generally used for production of a single plant species such as carnations, roses, pot mums or crops that can tolerate comparable environmental conditions.

The greenhouse of the 70s combines the free standing peak or arch structures to form a ridge and furrow facility that can be readily covered with greenhouse grade plastic materials and even include the air inflated system (Fig. 3-44).

The only limitations on the area covered by a ridge and furrow structure is the distance between the fan and cooling pad area. This should be limited to approximately 200 ft, but will be discussed in more detail in the chapter on temperature.

There are some disadvantages to the ridge and furrow structures, namely reduced light due to shadows from the eaves and other superstructure, and the possibility of snow accumulation over the eaves (Taft, 1926). The plastic covers on arched ridge and furrow houses are less likely to present light problems than standard glasshouses, especially in "high light" areas. Snow load over the eaves is still a potential problem. Several greenhouses are underdesigned for weather conditions that only exist once in 25 years (Fig. 3-44B). Others are underdesigned for any snow load, especially at the eaves. Ridge and furrow houses must be designed and constructed to withstand all the load factors or they could fall like dominoes.

One of the first growers to employ the air-inflated system on an arch structure was Art Van Wingerden of Granville, Illinois (Ball, 1972). He designed and constructed six acres of 12 ft wide ridge and furrow houses. The houses are of the low profile arch, thus decreasing the wind load factor.

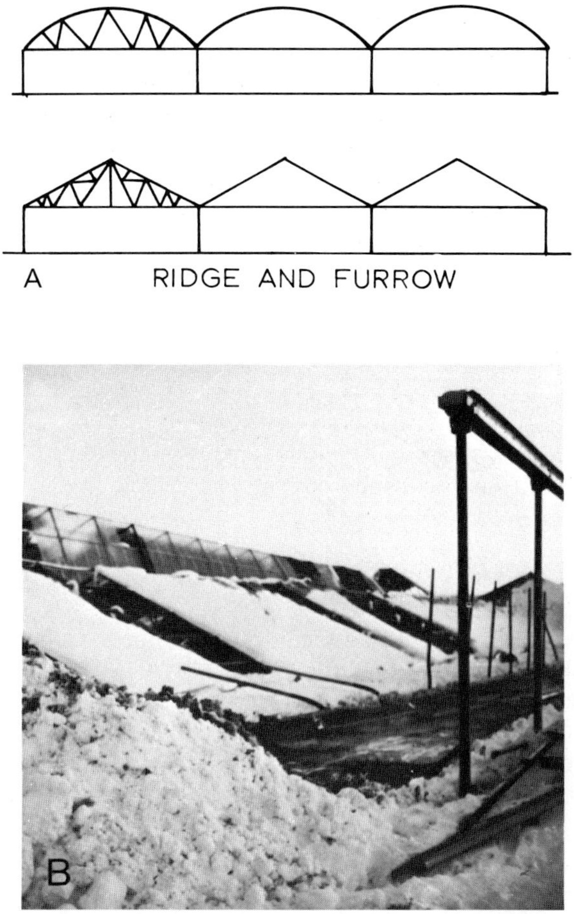

Fig. 3-44. (A) Ridge and furrow configurations of greenhouse structures. (B) Peak ridge and furrow house under construction when wind and snow storm struck

3.7 Perils to the Greenhouse Structure

3.7.1 Maintenance

The greenhouse like other buildings is susceptible to several destructive forces and conditions. Two of the major perils are the lack of preventive and corrective maintenance. One of the strong points needed by a greenhouse manager is the ability to see potential problem areas in facilities, equipment and grounds. Many greenhouse operators think that keeping records is too laborious, especially on such factors as structural paint schedules, fan belt replacement, addition of boiler compound, fertilizer injector maintenance, fumigation equipment inspection, greasing and oiling bearings of motors, and cleaning steam traps. A general trend in the greenhouse has been, "do not bother with it until it quits running or falls down".

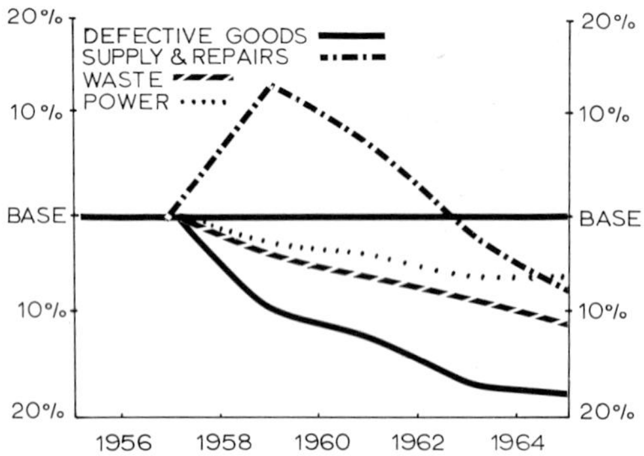

Fig. 3-45. The effects of a preventive maintenance program on some of the factors related to textile manufacturing over an eight-year period (Henderson, 1965)

3.7.1.1 *Corrective Maintenance*

Should be conducted when the need is discovered. This type of maintenance includes repairing hoses, fixing vent closures, cleaning cooling pad drains, changing fan belts and solving other minor problems.

3.7.1.2 *Preventative Maintenance*

Means planning ahead and checking equipment and facilities that appear to be functioning well, on a regularly scheduled program (Suber, 1976).

In a talk to university physical plant administrators (Henderson, 1965), a textile manufacturer told of the preventive maintenance program initiated by their company. They first tried to determine the most important areas where such a maintenance program would be beneficial. They started with each piece of production equipment and completely rennovated it where it stood. Renovation was scheduled again depending on the kind of equipment, every 12–18 months. Between times, every 12–24 weeks, the most critical and sensitive parts were inspected, repaired or adjusted. Experience showed a lack of lubrication and often the use of improper lubricants as a primary problem. The company also extended the program to auxiliary equipment, utilities, buildings and rolling stock. The total value of the improvements (Table 3-11) was impressive. Improvement did not happen overnight, but was a gradual process (Fig. 3-45). Such a program for a greenhouse operator can be equally rewarding as it is directly related to profits.

Table 3-11. The total improvement made from a preventive maintenance program, 1956–1964 (Henderson, 1965)

Production	+14%
Direct labor	−7
Supply and repairs	−8
Defective goods	−18
Waste	−11
Power	−6
Maintenance labor	−14

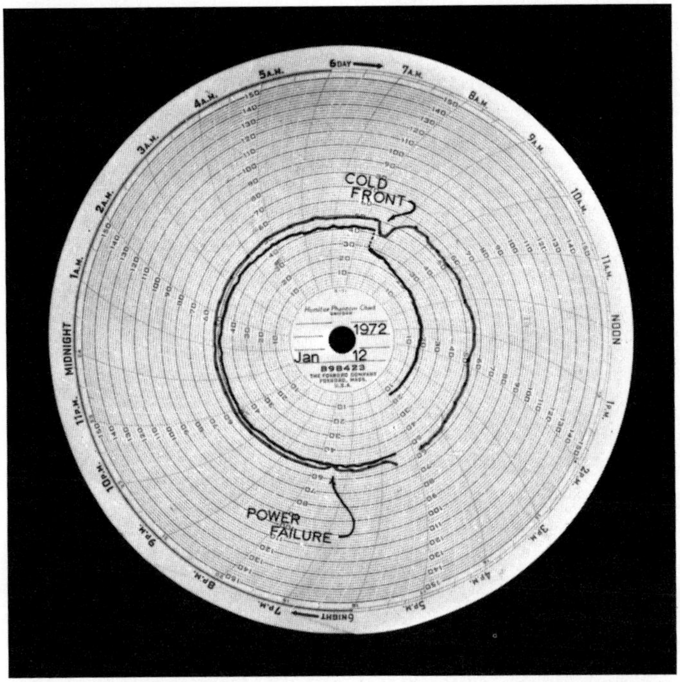

Fig. 3-46. *Outer line:* temperature in a rose house with the roof off and the air inflated film plastic put on at 8:30 a.m. *Inner line:* outside air temperature (Goldsberry, 1973)

3.7.2 Wind

One of the most destructive forces around a greenhouse is wind. The wind invariably starts blowing when film or rigid plastics are being installed and many thousands of square feet have been lost in minutes. Walker and Duncan (1973a) describe the effect of opening a door on the windward side of a greenhouse during high winds. The positive pressure created inside the greenhouse, plus the lift effect on the leeward side created a force far greater than the "full potential wind force." Several greenhouses have been subjected to similar conditions during periods of construction and cover materials have been destroyed.

On January 11, 1972, at Fort Collins, winds, that had reached velocities of 30 mph for seven of the first eleven days of the month, turned to near tornado conditions. At 9:30 p.m., following a power failure with outside temperatures of 40–50° F (Fig. 3–46), the wind velocity reached 110 mph and completely removed the FRP covering on the south side of an east-west oriented, Colorado State University research rose house. All other facilities, including an air inflated house were unharmed. The following morning, when the wind stopped, as a cold front reached the area, an emergency air inflated roof was installed and the rose research project was saved (Goldsberry, 1973). January of 1976 brought devastating gales to England and destroyed many plastic tunnels and greenhouses. One structure was completely removed, leaving no debris and only exposed lettuce plants (Norman, 1976). The superstructure of many of the film "clad" tunnels was made of 3/4 in tubing. The Lee Valley Experiment station is now recommending nothing less than 1 in tubing. These examples perhaps represent only once in a 100 year winds, but proper design and precautions should eliminate some of the potential destruction.

3.7.3 Hail

One of the most feared perils to a glasshouse operator is hail. Many greenhouse operations have been totally destroyed within minutes by hail. Some of the older glasshouse operators still drape 1 in mesh hail screen over their houses each spring and remove it in the fall.

FRP covered greenhouses are often left "starred" by hail. Many growers feel that starring reduces light and is deterimental to the panel. Simulated hail studies reported by Hartman (1973a) indicated that starring was basically a "cosmetic" effect and did not affect the light transmission characteristics of panels. Hail occurs in different sizes, degrees of hardness, and in different shapes. In the late 1960s, hail larger than baseballs put several holes in a FRP greenhouse roof. On another occasion, ping pong sized hail damaged the roof. More research is needed to determine the effects of starring, especially on "Tedlar" laminated panels.

3.7.4 Fire

Because of the flammability characteristics of FRP panels, greenhouse fires appear to have become more prevalent. Greenhouse fires have not been uncommon, even before the advent of FRP panels.

Glasshouse fires have started in service buildings, cooling pads, and with black shade cloth. Heat and smoke are the damaging factors in a glasshouse fire. Intense heat immediately rises to the ridge and cracks the glass. If smoke is heavy, it spreads throughout the house and blackens any remaining glass and superstructure.

Several fires have been started by "pig-tail" type light sockets used for photoperiod purposes. They have shorted out and ignited black shade cloth. Oil fired CO_2 generators have also been known to "shoot" a glowing piece of soot onto black shade cloth, causing it to ignite (Wetzel, 1976).

Film covered greenhouse fires are usually contained more easily because the heat immediately melts the film and allows the products of combustion to escape. Unless the cooling pad material, black shade cloth, wooden benches or some other flammable material "feeds" the fire, it will not spread readily.

Fiberglass covered greenhouse fires have been of some concern to horticulture industries since the late 1960s. In February, 1968, approximately 80000 ft^2 of FRP covered greenhouse burned in 12 min in the Denver, Colorado, area. The fire was started by welders working in the cooling pad area; the cooling pads were

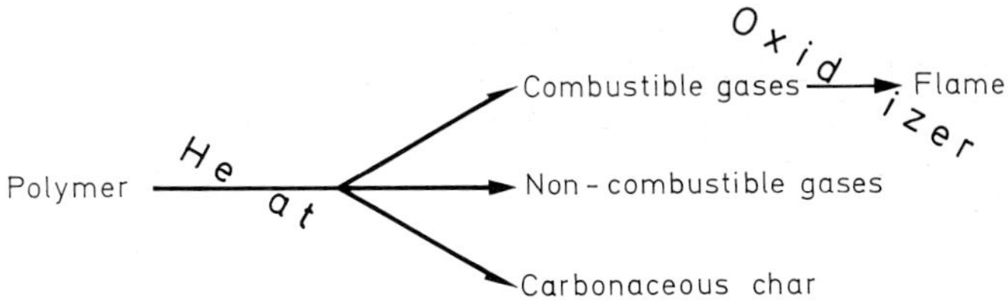

Fig. 3-47. Pyrolytic decomposition of many polyester plastics (Buck, 1951)

ignited and resulted in complete loss of the facility. A second fire occurred on July 4th of the same year. Children playing near a FRP covered greenhouse threw a lighted sparkler on the roof. Numerous FRP greenhouse fires across the nation, plus the Denver area losses, led to a cooperative study between Colorado State University, Florists' Mutual Insurance Company of Edwardsville, Illinois, Colorado Flower Growers' Association, several FRP manufacturers and fire detection companies. A program was developed to determine the characteristics of FRP greenhouse fires and how they might be prevented, detected, and controlled.

A graphic description of the decomposition of a polymer, such as the acrylic modified polyester used in glass reinforced panels, provides background for understanding some of the aspects of greenhouse fires (Fig. 3-47). It is further described by Buck (1951):

"In the combustion of any polymer, decomposition first occurs as the result of heat supplied from a source. At the same time, any combustible gases from this decomposition are heated to their ignition temperatures and if an adequate oxidizing agent is present, these gases ignite. This burning provides additional heat to propagate further decomposition and ignition. The simple requirements for a flame then are heat, a material which will provide combustible gases, and an oxidizing agent."

The Colorado State University greenhouse flammability studies were initiated in October, 1969, when a $20 \times 40'$ pipe greenhouse structure was constructed, oriented E–W, for use in a series of controlled fires (Fig. 3-48). Five phases in the study were planned and FRP, PVC, and polyethylene were used as covers. A standard evaporative cooling pad system (west end), sprinkling systems, ionization detectors, and exhaust fans (east end) were involved in different phases.

Observations and data collected from the series of controlled fires have led to the following conclusions:

It is apparent exhaust fans contribute to the rapid spread of combustible by-products and fire. The first step in controlling a greenhouse fire is to de-energize the exhaust fan system if possible.

Even though ionization detectors energized the alarm seconds after the first two controlled fires were started, in large volume greenhouses they proved ineffective unless the units were spaced 20 ft apart in the ridge of each greenhouse. Such an installation would be too costly.

The PVC and fire rated FRP panels installed as barriers against flame spread were of no value. They might help if there is no wind or operating exhaust fans.

Sprinklers with fine droplets were of no value because the slightest breeze in the ridge area dissipated the water. Sprinklers with very large droplets can possi-

Fig. 3-48. Controlled fire in FRP covered greenhouse, Colorado State University

bly prevent flame spread, providing they are energized early enough and are able to keep the FRP roof panels at temperatures below 800° F.

The fire rated FRP panels did slow the flame spread when there were minimum combustible products feeding the fire, and no wind or exhaust fans operating (Fig. 3-49). When temperatures are high enough to cause the volatilization of combustible by-products from the rated panels, they will burn slowly and, if detected early enough, can be extinguished.

3.7.4.1 Causes of Greenhouse Fires

Based on reports from fire departments, horticultural supply salesmen and insurance companies, the following factors have contributed to greenhouse fires:

Storage: Many growers, especially those producing miscellaneous crops, have numerous storage areas. For some it is the rear lot and can be considered a "junk yard". For others it is a shed or the boiler room where fertilizers and liquid combustibles are kept out of the weather. In any case, most storage areas are prime locations for rapid movement of fire, if there happens to be a source of ignition.

Electrical: Many areas of the nation allow greenhouse owners to do their own electrical work. Improper wire size, overloaded circuits, and faulty equipment can lead to overheating and a source of ignition. The licensed electrician has also been known to improperly splice and wrap wires resulting in an electrical short in the circuit box.

Heating: Gas fired unit heaters are often taken for granted while the boiler is considered infallible by many growers. Low water "cut offs" and

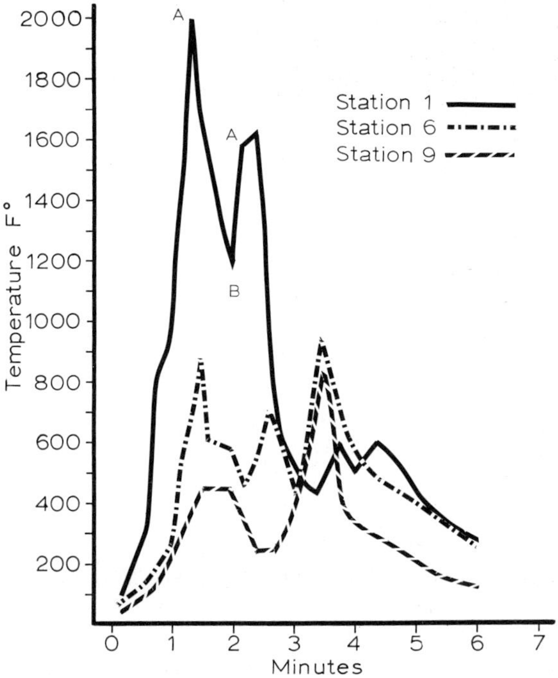

Fig. 3-49. Temperatures recorded in the ridges of a greenhouse covered with fire rated and standard FRP panels. *Station 1:* above area where fire was started; *Station 6:* midway in house; *Station 9:* above exhaust. Exhaust fan was turned off at Points *A* and on at *B*

other safety equipment have failed. Bird nests, dirt, or spider webs have been found in the burner sections of gas fired unit heaters and new units have been installed without orifices.

Inventors: For some unknown reason, the greenhouse operator is a genius at designing substitute equipment, temporary electrical hookups, modifications for the boiler, or plans for facility expansion. Some geniuses have heaped fuel on fires, others have started them.

People: Several greenhouse fires have been attributed to carelessness of employees, owners, children, and the general public. Welding, grass fires, and 4th of July sparklers have all supplied the source of ignition.

Facilities: Many of the present greenhouses do not meet structural load, stress, and design requirements of the building or fire codes. Gas unit heaters attached to the superstructure have caused structural stress, warpage, and breakage leading to gas pipe leaks and electrical shorts. The ignition of black shade cloth has been a major contributor to greenhouse fires; in some instances, there has been a lack of maintenance, faulty wiring and exploding light bulbs to ignite the cloth.

Plumbing: The improper use of materials and installation of gas pipe has been a contributing factor in greenhouse fires. The codes may be adhered to, but the workmanship can be faulty.

3.7.4.2 *Ways to Prevent or Contain Greenhouse Fires*

Management not only includes growing and business, but also awareness. Every greenhouse owner, operator and/or manager should heed the following points:

Cleanup:	If an owner will establish some housekeeping policies and not only maintain them himself, but encourage his employees to develop a sense of pride, a better working environment will be established and fire potential decreased. Include in the cleanup program weeds, lumber, trash, combustible liquid, and old equipment, both inside and outside the greenhouse.
Construction:	Do not cut corners when constructing new greenhouses, remodeling or installing temporary equipment. Use nonflammable materials in areas of high fire risk. Fire rated panels, corrugated sheet metal, and other non or low "flame spread" materials can be effectively used to prevent or contain fires in cooling pad areas.
Design:	Design facilities for fire control by connecting greenhouse sections with nonflammable materials. Place all combustible supplies in a building designed for them; do not store them in the greenhouse or boiler room. Consult your insurance agent about your proposed plans, large, or small, in new or old greenhouses before construction starts—your insurance rates could be reduced. Design with the safety of the employees as top priority.
Heating equipment:	Before each heating season starts, the greenhouse operator should check every unit heater. Make sure they are free of foreign material, have burner settings that provide a blue flame and check the motor. Many unit heater motors are not designed to operate 24 h a day, and when they are used on a "tube" system they can overheat and wear out bearings. If you heat with boilers, have them inspected and always go with the inspector while he inspects your equipment. Help locate potential problems, and do not try to cover them up.
Electrical:	Many greenhouse operators do their own electrical work. If you are not acquainted with amperage, wire size, volts, circuit loads and the related equipment and tools, hire an electrician that knows what he is doing. Make sure no wires are connected to the steel greenhouse frame without using insulators (wood is not necessarily a good insulator). Weather proof electrical materials should be used in areas of high moisture. Have the main electrical power supply switch box installed in a convenient and visible location in the service area.
Shade cloth:	Consider the use of polyethylene shade "cloth" or an equivalent material. Standard black cloth is highly

flammable and should not be too near the electrical system used for lighting. Install the lighting system to meet the codes; it will cost money, but save dollars in the long run.

Smoking areas: A visitor, employee, or even the owner can flip a match or cigarette into the cooling pad, fuel, or storage areas. Designate areas for smoking and enforce their use.

Life safety: Every employer must consider the lives of his employees. Conduct fire drills at least four times a year and assign specific duties to key employees. When an alarm sounds, have key employees turn off the gas and electricity, escort employees and locate the fire and man fire extinguishers. Have a fire alarm system that is energized automatically upon fire detection or that can be manually sounded. The alarm should run on both AC and DC current.

Employee education: Each employee should be acquainted with every aspect of fire prevention, detection, and control. Conduct periodic seminars to discuss methods of preventing all types of fires around a greenhouse. Burn a scrap of cooling pad, an old glass bar or a few square inches of FRP panel for observation. Let the employees become acquainted with the smell of the by-products of each material so they may detect fires early and sound an alarm.

Maintenance: Preventive maintenance is one of the best investments available to greenhouse operators—yet one of the least applied programs. "Check" lists for the care of motors, temporary or permanent electrical "hook ups", heating equipment, compressors, pumps, and the like should be standard measures for fire prevention.

Fire department coordination: Most fire departments do not check greenhouses. Acquaint the personnel at the closest fire house with your facility and employee education and life safety programs. Ask them to assist you and make recommendations.

Do not take your greenhouse for granted, keep the flame out!

3.7.5 Snow

The worst snow in 100 years hit Michigan and Ohio in December, 1974. More than 20 in fell along Lake Erie causing greenhouse damage and 15 acres of greenhouses were lost in the Toledo and Cleveland, Ohio, areas. Most collapsed houses were ridge and furrow, covered with plastic or glass.

Several greenhouses have been caught in snow storms prior to being completed (Fig. 3-44 B). In December, 1972, a FRP covered greenhouse under construction in the Denver, Colorado, area was hit by hard winds preceeding a storm

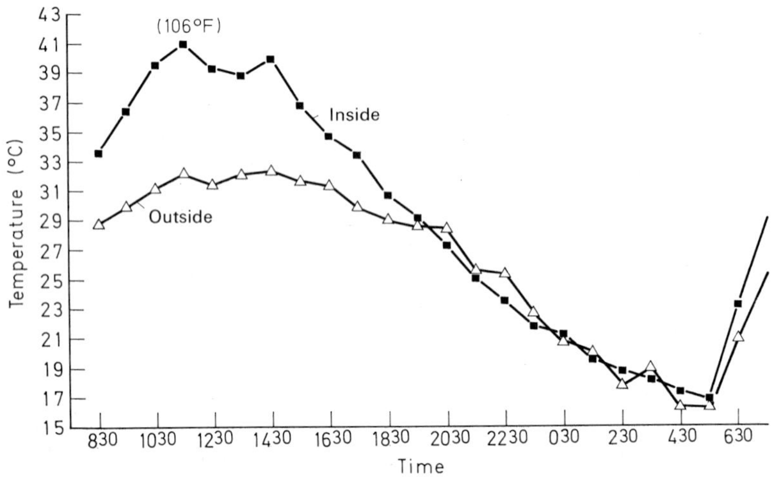

Fig. 3-50. Air temperatures reached in a closed double poly house on July 8, 1976 (Goldsberry and Sherry, 1976, unpublished)

front and the roof sections were lifted out of alignment. Immediately following the wind, a six inch snow fell and the majority of the new structure collapsed. On another occasion, snow fell first, then the wind started and created drifts five feet high in one end of the gutters of a ridge and furrow house and it collapsed.

The discussion previously in the greenhouse design section indicated that snow load factors are not included in many greenhouse designs. Snow loads during construction and heavy snows when the greenhouse is operational, are potential problems. The increased price paid for a structure with more load bearing capacity will often be saved in time and effort over a lower priced facility that will collapse in light winds and minimal snow fall.

3.7.6 Summer Temperatures

Many greenhouses, especially bedding plant facilities, are closed during the summer months and intense temperatures can be reached, especially in the roof area. On a day of intermediate cloudiness, a small 15 × 40′ air inflated greenhouse was closed and ventilating equipment turned off from 0830 to 0630 the following day (Fig. 3-50). Conditions in larger greenhouses can be worse and more destructive. High temperatures contribute to the degradation of plastics, cause wood to dry out and split, paint to blister or peel, and in some instances, grease to be forced out of bearings by expansion. Aluminum and steel structural parts will expand and contract causing glass to crack or bolts to loosen. Good managers will include such facilities in their preventive maintenance programs and at least keep doors and ventilators open and a single exhaust fan operational.

3.8 Construction Costs

The intent of this section is not to provide the reader with a series of prices related to greenhouse construction, because it would be impossible and inaccurate. Many factors enter into the cost of developing any greenhouse facility—land, water and other utilities, structure, contents, labor, and numerous incidentals.

Almost every publication of most trade magazines has an advertisement for a greenhouse ranging in price from 87 cents to \$ 1.20 or even \$ 2.11 per ft^2. It is doubtful a "turn key" facility can be purchased for these prices when one considers all of the factors actually involved.

In the early 1970s, a grower could purchase 15000 ft^2 of FRP covered greenhouses for \$ 1.12 per ft^2. What did the purchase include? All but the cooling and heating equipment, electrical wiring, and concrete for setting columns. These additions were to be paid by the purchaser along with the erection costs. Added to these additional costs is the fact the purchaser paid for unloading all materials, which were prepaid.

A realistic cost is also not forthcoming when a grower pays an "X" price for a complete facility in 1972 and in 1976 requests bids on a duplication of the same structure; the lowest bid in 1976 being 10% lower than the 1972 price.

Halsey (1975) suggests that greenhouse construction costs be based on an "adjusted present value concept" which allows the owner to view the change in true value over a period of years. He used a discounted cash flow analysis as a tool to evaluate the true value of five structural options: (1) Dutch-type glass, (2) Aluminum-Tedlar FRP, (3) Aluminum-Standard FRP, (4) Aluminum-double poly, and (5) Wood-Tedlar FRP. It was assumed that each structure would have approximately the same costs for interior needs such as cooling, heating, and cultural equipment. Since many growers expect a 10% minimum return on their investment, the costs were discounted at that level.

"Table 3-12 shows years 0 through 30, followed by a column entitled, 'Discount Coefficient.' The discount coefficients calculated for the respective years at 10% are from discount tables. To convert future actual costs to present value equivalents, each actual cost is discounted by multiplying it by its corresponding discount coefficient. For example, using Option 1, the cost for year 1 is (0.9091) (\$ 100). This procedure is continued to the 30th year with the products summed, including year 0. The result gives the 'present value' of \$ 95,586." Halsey's analysis of the five options, Table 3-12, revealed that the inflatable "double poly" house (4) was the "least-cost" alternative in present real values and would, therefore, be the best investment for a grower. It should be noted, however, that the aluminum-Tedlar FRP house (2) is almost as equally desirable as option 4.

The analysis showed the Dutch house as the least expensive alternative in actual (future) costs, but when discounted at 10% it is the most expensive in present values.

Approximately 18–35% of the cost of greenhouse construction is involved in erection (Bartok, 1971; Rough, 1976), depending on the type and design. Labor for construction presents the largest variable in computing costs. It will vary from month to month and often depends on the location and travel time of the con-

Table 3-12. Expected costs and present values for five alternative greenhouses of 1 acre[a]. Discounted at 10% (Halsey, 1975)

Year	Discount coefficient	Dutch type glass Option I		Aluminum-tedlar Option II		Aluminum-fiberglass Option III		Aluminum-doublepoly Option IV		Wood-tedlar Option V	
		Actual cost $	Present value $	Actual cost $	Present value $	Actual cost $	Present value $	Actual cost $	Present value $	Actual cost $	Present value $
0	1.0000	92 783	92783	61 420	61 420	55 757	55 757	47196	47196	46609	46609
1	0.9091	100[b]	91	300	273	100	91	100	91	300	273
2	0.8264	100	83	300	248	100	83	5763	4763	300	248
3	0.7513	100	75	300	225	100	75	100	75	300	225
4	0.6830	100	68	300	205	100	68	5763	3936	300	205
5	0.6209	100	62	300	186	13504	8385	100	62	300	186
6	0.5645	100	56	300	169	100	56	5763	3253	300	169
7	0.5132	100	51	300	154	100	51	100	51	300	254
8	0.4665	100	47	300	140	100	47	5763	2688	300	140
9	0.4241	100	42	300	127	100	42	100	32	300	127
10	0.3855	3585[b]	1382	19466	7504	13504	5206	5763	2222	49523	19091
11	0.3505	100	35	300	105	100	35	100	35	300	105
12	0.3186	100	32	300	96	100	32	5763	1836	300	96
13	0.2897	100	29	300	87	100	29	100	29	300	87
14	0.2633	100	26	300	79	100	26	5763	1517	300	79
15	0.2394	10	24	300	72	13504	3233	100	24	300	72
16	0.2176	100	22	300	65	100	22	5763	1254	300	65
17	0.1978	100	20	300	59	100	20	100	20	300	59
18	0.1799	100	18	300	54	100	18	5763	1037	300	54
19	0.1635	100	16	300	49	100	16	100	16	300	49
20	0.1486	3585	533	19466	2893	13504	2007	5763	856	49523	7359
21	0.1351	100	14	300	41	100	14	100	14	300	41
22	0.1228	100	12	300	37	100	12	5763	708	300	37
23	0.1117	100	11	300	34	100	11	100	11	300	32
24	0.1015	100	10	300	30	100	10	5763	585	300	30

Table 3-12 (continued)

Year	Discount coefficient	Dutch type glass Option I		Aluminum-tedlar Option II		Aluminum-fiberglass Option III		Aluminum-doublepoly Option IV		Wood-tedlar Option V	
		Actual cost $	Present value $	Actual cost $	Present value $	Actual cost $	Present value $	Actual cost $	Present value $	Actual cost $	Present value $
25	0.0923	100	9	300	28	13504	1246	100	9	300	28
26	0.0839	100	8	300	25	100	8	5763	484	300	25
27	0.0763	100	8	300	23	100	8	100	8	300	23
28	0.0693	100	7	300	21	100	7	5763	399	300	21
29	0.0630	100	6	300	19	100	6	100	6	300	19
30	0.0573	100	6	300	17	100	6	100	6	300	17
Total		102753	95586	108586	74485	125777	76627	129478	73233	154055	75727

[a] 1 acre = 43560 ft = .4047 ha.
[b] Annual maintenance or recovering costs.

struction crews. Many growers feel they can purchase a greenhouse structure and use their own personnel for erection at a lower cost than contracting the work. In some instances this is true, if the knowledge of erection, proper tools and time are available.

When greenhouse construction prices are compared, one must consider all facets. Moninger (1913) expressed his company's approach in words that are of equal value today: "We will furnish you with the same houses at the same price and any house we make, on the same price basis ... You will be treated to the same square deal we gave them, you will like our way of doing business and you will get your money's worth. We were not the highest in price on the Bassett and Washburn job and we were not the lowest. Just a fair honest price for fair honest goods."

References

Aldrich, R. A., White, J. W., Manbeck, H. B.: Transmission of solar energy through rigid plastic greenhouses. In: Proc. 7th Natl. Agr. Plastic Conf. Cottes, D. J. (ed). Univ. Kentucky. October 31–November 2, 1966, pp. 79–91

American National Standard Building Code Requirements for minimum design loads in buildings and other structures, A 58.1-1972. New York: American National Standards Institute, Inc. 1972

Anonymous: Aluminum, how it's made and where it's used. Aluminum Association of America, 1968

Anonymous: Approach to mechanization brings crop to the worker. The Grower. Sept. 6, 1975

Anonymous: December snow storm causes damage for BPI'ers. BPI News Bedding Plant Incorporated, 4479 Seneca Drive, Okemos, Missouri 48864, January 1975

Anonymous: "Butler-Style" framing systems entering greenhouse industry. Growers Talks **39** (9), 8–10 (1976)

Axlund, D. S., Besemer, S. T.: Greenhouse insulation experiment. San Diego Gas and Electric Co. Publication 1974, 22 pp.

Bailey, L. H.: Encyclopedia of American horticulture. London: MacMillan 1900

Ball, V.: Changing trends in growing structures. Growers' Talks **36** (8), 1972, 24 pp.

Banpo Polyester Industry, Ltd. B. P. L. 37, 6-chome, Sumie-Naka, Sumiyoshi Ku. Osaka, Japan: Bull. No. 25, 1967

Bartok, J. W., Jr.: Comparative costs of greenhouse construction. Conn. Greenhouse Newslett. No. 43, Univ. Connecticut, Nov. 1971

Bartok, J. W., Jr.: Labor-saving devices for commercial greenhouse operations. Paper 74-4518 presented at the American Society of Agr. Engineers Meeting. Chicago, Illinois, Dec. 1974

Binger, W. W., Hollingsworth, E. H., Sprowls, D. O.: Resistance to corrosion and stress corrosion. In: Aluminum Properties, physical metallurgy, and phase diagrams. VanHorn, R. (ed). Metals Park, Ohio: Am. Soc. Metals, 1967, Vol. I

Briggs, R. A.: The effects of glass and fiberglass on carnation growth. Unpubl. Masters thesis. Fort Collins, Colorado, Colorado State Univ. 1961

Buck, G. S., Jr.: Fire-resistant textiles. In: Kirk-Othmer encyclopedia of chemical technology. New York: Wiley, 1951, Vol. VI, p. 543

Carpenter, W. J.: Reinforced plastics: results of heat and light studies considered. Florists Rev. **3310**, 53–54 (1961)

Dallyn, S., Sheldrake, R., Jr.: Comparison of polyethylene and new "clear-type" fiberglass in tomato production. Natl. Agr. Plastics Conf. Proc. **8**, 27–31 (1968)

Dresser,T.O.J.: Reinhold plastics application series. I. Polyethylene. New York: Reinhold 1957
Duncan,G.A., Walker,J.N.: Review of greenhouse coverings. Paper 72-406. Am. Soc. Agr. Engineers. Annual Meeting, Hot Springs, Arkansas, 1972
Durrell,L.W., Goldsberry,K.L.: Growth of Aureobasidium pullulans on plastic greenhouse roofing. Mycopathol. Mycol. Appoicata. **42**(3–4), 193–196 (1970)
Edward Owen glasshouses. Supplement to The Grower. Nov. 15, 1975
Emmert,E.M.: University of Kentucky builds a greenhouse covered with polyethylene. Agr. Newslett. **2**(5), 92–93 (1954)
Emmert,E.M.: Low cost plastic greenhouses. Progress Report No.28. Agr. Exptl. Station. Univ. Kentucky, 1955
Fifty years of glass making. Pittsburgh,Pa.: MacBeth-Evans Glass Company, 1920
Fossum,M.T.: Trends in commercial floriculture crop production and distribution—a statistical compendum for the United States, 1945–1970. Soc. Am. Florists Endowment, July, 1973
Frederick,D.S.: Acrylic resins. Modern Plastics **17**(2), 22–24 (1939)
Goldsberry,K.L.: Light transmission of refinished fiberglass. Colo. Flower Grower's Assoc. Bull. 202, February, 1967a
Goldsberry,K.L.: Spectral transmission of greenhouse glass. Colo. Flower Grower's Assoc. Bull. 208, August, 1967b
Goldsberry,K.L.: Transmissivity of greenhouse coverings as related to carnation growth. In: Proc. 8th Natl. Agr. Plastics Conf. Univ. of Calif. San Diego, California, 1968a
Goldsberry,K.L.: Spectral transmission of some plastic greenhouse coverings. Colo. Flower Grower's Assoc. Bull. 213, January, 1968b
Goldsberry,K.L.: Plant growth response to the transmissivity of glass reinforced thermosetting plastic panels. PhD thesis. Iowa State Univ., Ames, Iowa, 1969
Goldsberry,K.L.: Flammability Control of FRP greenhouse coverings. Progress Report I. CFGA Bull. 242, July, 1970a
Goldsberry,K.L.: The availability of light energy for plants grown in greenhouses. Colo. Flower Grower's Assoc. Bull. 235 (1970b)
Goldsberry,K.L.: Evaluating performance of FRP panels for greenhouse coverings. In: Proc. 10th Natl. Agr. Plastics Conf. Univ. Illinois. Courter,J.W. (ed) 1971a
Goldsberry,K.L.: Plant responses to light quality created by tinted FRP. In: Proc. 10th Natl. Agr. Plastics Conf. Univ. Illinois. Courter,J.W. (ed) 1971b
Goldsberry,K.L.: Air-Inflated System for an Emergency Roof. Colo. Flower Grower's Assoc. Bull. 276, May 1973, pp.3 and 4
Goldsberry,K.L., Homan,T.: Refinishing FRP greenhouse panels. Colo. Flower Grower's Assoc. Bull. 297, March, 1975
Gunesch,W.: Pers. Commun. Park Floral Company, 1632 Broadway, Englewood, CO 80202, July, 1976
Halsey,L.A.: Time is money for greenhouse operators. Florida Flower Grower. **12**(1), (1975)
Hanan,J.J.: Some observations on radiation in greenhouses. Colo. Flower Grower's Assoc. Bull. 239, April, 1970, pp.1–4
Harnett,R.F.: Rigid plastic still not a serious challenge to glass. The Grower. October 14, 1972
Harnett,R.F.: The advantages of east–west multispans. The Grower. July 13, 1974
Hartman,H.: Fiberglass-reinforced panels in greenhouse construction—present developments and future prospects. 23rd Ann. Tech. Conf. Reinforced Plastics/Composites 355 Lexington Ave, New York Inst.: Soc. Plastics Ind. Inc., 1968
Hartman,H.: The effects of hail-produced "starring" on fiberglass-reinforced plastic greenhouse panels. Fiberglass Reinforced Plastics Cover, Reinforced Plastics/Composites Inst. 355 Lexington Ave, New York, 10017, 1973a
Hartman,H.: National agricultural plastics testing program. Phase III. In: Proc. 11th Natl. Agr. Plastics Conf. Simpson, Ill.: Natl. Agr. Plastics Assoc., 1973b, pp.210–220
Henderson,R.E.: Preventive maintenance. In: Proc. 52nd Annual Meeting, Assoc. of Physical Plant Administrators of Universities, May, 1965. North Carolina State University

Holley, W. D.: Frontiers in floriculture. Colo. Flower Grower's Assoc. Bull. 229. June 1969
Holley, W. D., Baker, R.: Carnation production. Dubuque, Iowa: Brown 1963
Holley, W. D., Goldsberry, K. L., Schroeder, M. L.: Progress report on greenhouse coverings. Colo. Flower Grower's Assoc. Bull. 189, January, 1966
Keveren, R. I.: Plastics in horticultural structures. Rubber Plastics Res. Ass. G.B. Shawbury, Shrewsbury SY4 4NR, England, 1973
Kondo, T.: Plastics in agriculture in Japan. Plasticulture December, 1975
Lawrence, W. J. C.: Science and the glasshouse, 3rd. London: Oliver and Boyd 1963
Lemmon, K.: The covered garden. London: Museum Press Ltd. 1962
Leuchars, R. B.: A practical treatise on hot-houses. Boston, Mass.: Jewett & Co. 1854
Leyson, B. W.: Plastics in the world of tomorrow. New York: E. P. Dutton, 1944
Manbeck, H. B., Aldrich, R. A.: Analytical determination of direct visible solar energy transmitted by rigid plastic greenhouses. Trans. of ASAE, **10**, 564–567 (1967)
Massey, P. H., Jr.: The national agricultural plastics testing program. Phase I. In: Proc. 6th Natl. Agr. Plastics Conf. Oebker, N. F. (ed). Tucson, Arizona: Univ. Arizona 1965
McLaughlin, M., Sheldrake, R.: Light, greenhouse coverings, and plant growth. In: Proc. 11th Natl. Agr. Plastics Conf. Simpson, Il.: Natl. Agr. Plastics Assoc., 1973, pp. 70–74
Moninger, J. C.: Everything for the greenhouse. Chicago, Il.: Fred Klein 1913
Moore, E. L.: Report of the national agricultural plastics testing program. Phase II. In: Proc. 9th Natl. Agr. Plastics Conf. Goldsberry, K. L. (ed). Colorado State Univ. 1969
NIAE's inflatable plastic house light loss same as EW venlo. The Grower. December 6, 1975, 1027 pp.
Nicholson, G.: Dictionary of gardening, Div. III. London, England: L. Upcott 1887
Norman, B.: The house which vanished. The Grower. 85 (7), February 14, 1976
O'Keefe, P.: Glass reinforced plastics. Materials Methods **37**, 119–134 (1953)
Or-chem Topics No. 23. Philadelphia, Pa.: Rohm and Haas 1967
Our one hundred years. DesPlains, Il.: Burnham Corp., 1956
Plexiglas—design, fabrication, and molding data. Bull. PL—53F. Philadelphia, Pa.: Rohm and Haas Co., May, 1964
Plexiglas acrylic sheets. Publication PL 693a. Stanford, Conn.: Commercial Plastics and Supply Corp., October, 1967
Post, K.: Florist crop production and marketing. New York: Orange Judd, 1950
Roberts, W. J.: Double covering a film greenhouse using air to separate the layers. In: Proc. 8th Natl. Agr. Plastics Conf. Hall, B. R. (ed). Univ. of Calif. San Diego, Ca.: February, 1968, pp. 38–44
Roberts, W. J.: Developments in double covering film houses using air inflation. In: Proc. 9th Natl. Agr. Plastics Conf. Goldsberry, K. L. (ed). Colo. State Univ. October, 1969, pp. 173–181
Roberts, W. J.: Air-inflated and air-supported greenhouse structures. In: Proc. 10th Natl. Agr. Plastics Conf. Courter, J. W. (ed). Univ. of Ill. November, 1971, pp. 103–109
Roberts, W. J., Mears, D. R.: Double covering a film greenhouse using air to separate film layers. Transaction's Am. Society Agr. Engineers. **12** (1), 32, 33, and 38 (1969)
Ross, D. S., Aldrich, R. A.: Production operation analysis. In: Bedding plants, 2nd Mastalerz, J. W. (ed.). Univ. Park, Penn.: Pennsylvania State Univ 1976
Ross, D. S., Aldrich, R. A., Persson, S. S. P.: An engineering analysis of bedding plant production operations. Paper 74-4515 presented at the Am. Soc. Agr. Engineers Meeting. Chicago, Il.: December, 1974
Schales, F. D., McNeil, M., Smeal, P. L.: A study of environment and plant responses in greenhouses covered with different plastics. Natl. Agr. Plastics Conf. Proc. **8**, 23–37 (1968)
Shand, E. B.: Glass engineering handbook. Corning, New York: Corning Glass Works, 1955
Sheldrake, R., Jr.: Air makes the difference. Am. Veg. Growers, January, 1971
Shortley, G., Williams, D.: Elements of physics, 2nd ed. New York: Prentice Hall, 1955
Smith, F. G.: We need to adopt new thinking on light transmissions. The Grower. November 4, pp. 846–850 (1967)
Sonneborn, R. H.: Fiberglass-reinforced plastics. New York: Reinhold, 1964
Specification "Tedlar" PVF film laminated to fiberglass reinforced plastics. Technical Information Bull. TD-26. E. I. DuPont DeNemours and Co, December (1965)

Structures and environment handbook. Midwest Plan Service. Agricultural Engineering Bldg., Iowa State Univ., Ames, Iowa, 50010, 1972

Structures and environment handbook. Midwest Plan Service. Agricultural Engineering Bldg., Iowa State Univ., Ames, Iowa 50010, 1975

Taft, L. R.: Greenhouse construction. New York: Orange Judd Co., 1926

Trickett, E. S., Goulden, J. D. S.: The radiation transmission and heat conserving properties of glass and some plastic films. J. Agr. Eng. Res. **3**, 281–285 (1958)

U.S. General Services Admin., Fed. Supply Service. Glass, plate, sheet, figured (corrugated and float for glazing, mirrors and other uses). Interim Fed. Specification DD-G-00451b. August 22, 1966

Voigt, A.: Bedding plant problems. BPI News. Bedding Plants, Inc. East Lansing, Michigan, November, 1973, pp. 3–4

Walker, J. N., Duncan, G. A.: Greenhouse location and orientation. Univ. Kentucky. Misc. Publication 397, 1971

Walker, J. N., Duncan, G. A.: Greenhouse structure. Univ. Kentucky Agr. Engineering. Extension Service Bull. AEN-12, 1973a

Walker, J. N., Duncan, G. A.: Greenhouse benches. Univ. Kentucky Agr. Engineering. Extension Service Bull. AEN-13, 1973b

Walker, J. N., Slack, D. C.: "Properties of greenhouse covering materials". Transactions of ASAE, **13**(5), 682–684 (1970)

Warp, H.: Historical development of plastics for agriculture. In: Proc. 10th Natl. Agr. Plastics Conf. Courter, J. W. (ed). Univ. of Illinois 1971

White, D. G.: Rigid fiberglass for greenhouse plants. Oklahoma State Univ. Exp. Stat. Proc. Ser. P-349. April, 1960

Williams, D. D. F.: Performance of greenhouse plastics in Kula, 1966–1967. Tech. Prog. Rpt. 166, Hawaii Agr. Exp. Stat. Univ. of Hawaii (1968)

Wilson, C. L., Loomis, W. E.: Botany. 3rd ed. New York: Holt Rinehart and Winston, 1962

Wilson, D. D.: Fiberglass surfaced with "Tedlar" PVF film for greenhouse cover. Proc. 8th Natl. Agr. Plastics Conf. Hall, B. (ed). Univ. of California, San Diego, California, 1968, pp. 32–37

Wilson, D. D.: Greenhouse coverings of fiberglass surfaced with "Tedlar" PVF film. Proc. 10th Natl. Agr. Plastics Conf. 1968, pp. 160–167

Woodruff, N. P.: Shelterbelt and surface barrier effects on wind velocities, evaporation, house heating, and snow drifting. Tech. Bull. **77**, Kansas Agr. Exp. Stat. (1954)

Wright, W. J.: Greenhouses, their construction and equipment. New York: Orange Judd Co. 1917

Chapter 4 Temperature

4.1 Introduction

It is instructive to read early texts, Laurie and Kiplinger (1948), and Post (1950), and compare the changes and greater detail of crop temperature requirements in recent publications (see Table 4-1). Temperature is an environmental factor which the greenhouse grower can manipulate easily, yet requirements for precision are great. The greenhouse manager is faced with a variety of conflicting requirements which must be carefully thought out and planned to avoid difficulty; yet he must be ever ready to change temperature to offset conditions over which there may be no control. Thus, temperature manipulation fulfills a variety of needs, such as: (1) timing the finished crop to coincide with periods for maximum return (Christmas, Easter, etc.); (2) fitting the crop to market requirements (keeping life, acclimatization, etc.); and (3) overcoming adverse factors such as low light, low CO_2, disease, or inadequate fertilization.

4.1.1 Definitions

Temperature refers to the relative "hotness" or "coldness" of an object. It is a means of relating "intensity" of heat between two or more objects. Temperature and heat are two different things. A cubic meter of air at the same temperature as a cubic meter of water does not contain the same amount of heat. Air has a lower capacity to transfer heat to an object at a lower temperature as compared to water. Temperature refers to the ability to transfer heat to another colder region and is expressed by some arbitrary scale such as degrees Celsius (° C), degrees Fahrenheit (° F), or degrees Kelvin (° K). Heat may be expressed in terms of calories (cal), British Thermal Units (BTU), or Joules (J) (see Appendix A, Table 10). Temperature is "intensity", whereas heat can be thought of as capacity to do work, or energy. The distinction between the two (temperature and heat) is important as the two interact to influence each other.

A second term that is frequently used in the literature is "optimum" temperature. What is "optimum" is often in the eye of the beholder and varies accordingly. As will be noted, the "best" temperature at which a plant can be grown changes with a number of factors:

1. Species, cultivar, and clone.
2. Available radiant energy (light) which is influenced by the species, planting arrangement, density, greenhouse structure and its cover, and the climatic conditions of the location.

3. Water availability which is influenced by the physical and chemical properties of the root medium, location, irrigation system, cultural practices, and season.
4. Carbon dioxide concentration and presence of pollutants.
5. Previous history of the plant.
6. Stage of growth of the plant, age, and the particular plant structure or process which happens to be most important at a particular period in time.
7. Nutrition of the crop, fertilization practices, and water quality.
8. The marketing period for which the crop is being grown, the kind of product which is to be produced, and the method of packaging and handling the product is to be given.

As Went emphasized in 1957b, it is not sufficient to think in terms of two or three factors with temperature interactions, we must think "multi-dimensionally". Even sophisticated computer programs cannot provide all answers. But, an observant, conscientious and intelligent grower, with sufficient experience, is capable of maintaining near "optimum" temperatures.

4.1.2 Climatic Factors of Importance

We have indicated that greenhouses are means to overcome climatic adversity. That does not say that factors external to the greenhouse are unimportant. As noted in Chapter 2, climatic light factors determine greenhouse productivity and cultural practices. Likewise, temperature influences heating and cooling systems as discussed later in this chapter, and plays a considerable role in initial capital costs. An example is provided in Figure 4-1 which shows seasonal temperature variations at different latitudes. Obviously, structures located in the vicinity of Puerto Rico need to be minimum as compared to Sioux Falls, U.S.A. where buildings will be required to withstand snow loads with substantial heating and cooling equipment. However, while no heating may be required at locations near the equator, high humidity may prevent effective evaporative cooling and production of certain crops. Latitude, therefore, is a main factor in influencing greenhouse temperature control.

A second factor is altitude. For every 120 m increase in elevation, flowering of unprotected crops is delayed an average of four days. The effect on agronomic crop production is schematically shown in Figure 4-2 for two different latitudes. A practical example of altitude and latitude interactions in the floral industry is the burgeoning production in such areas as Columbia and Kenya, where fortuitous combinations of high altitudes close to the equator permit nearly optimum temperatures for specific crops outdoors the year around. Again, greenhouse structures are needed merely to prevent rain from damaging blossoms.

A third factor is the proximity of large bodies of water combined with prevailing onshore winds that tend to ameliorate climatic extremes. It is not happenstance that the major horticultural production areas of the world are located in regions where the maritime influence (e.g., California, Mediterranean) reduces daily and seasonal temperature flucuations (Fig. 4-3), increases the length of the growing season, and moves back the dates of spring and fall frosts. Again, the effect on greenhouse construction and temperature control is to reduce the need

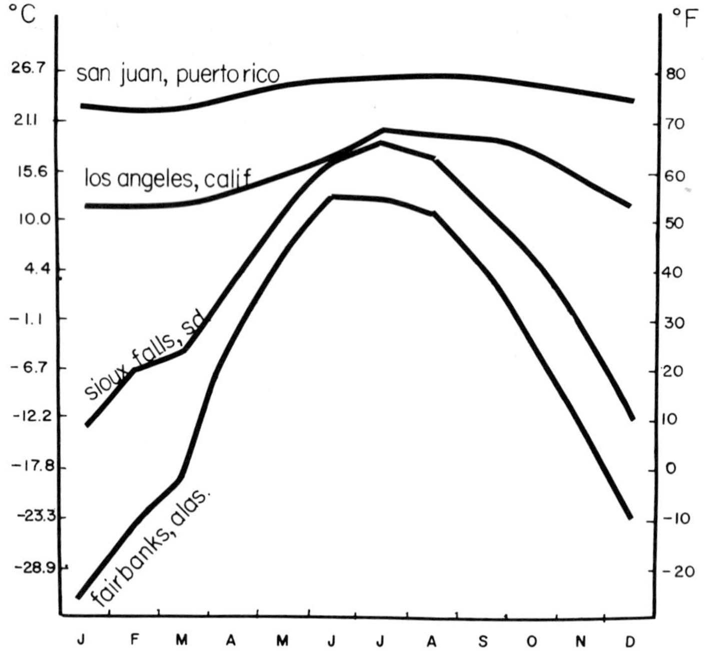

Fig. 4-1. Mean monthly temperatures for various locations in the world (Went, 1957a). Note that latitude is a primary indication of mean temperature variation to be expected at a particular location. Seasonal temperature changes increase the higher the latitude, or the further one moves north or south of the equatorial regions typified by Puerto Rico

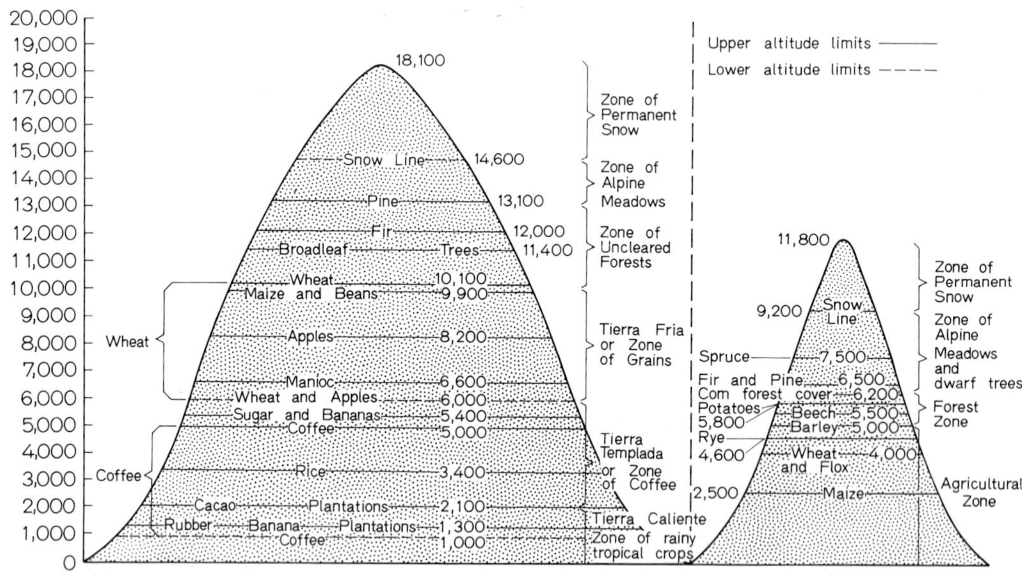

Fig. 4-2. Diagrammatic illustration of the effect of altitude on temperature and unprotected crop production at two different latitudes. The peak at left represents an 18800 ft mountain at the equator (0° lat), the 11800 peak on the right would be similar to a latitude of 40–45°, north or south (Wilsie, 1962)

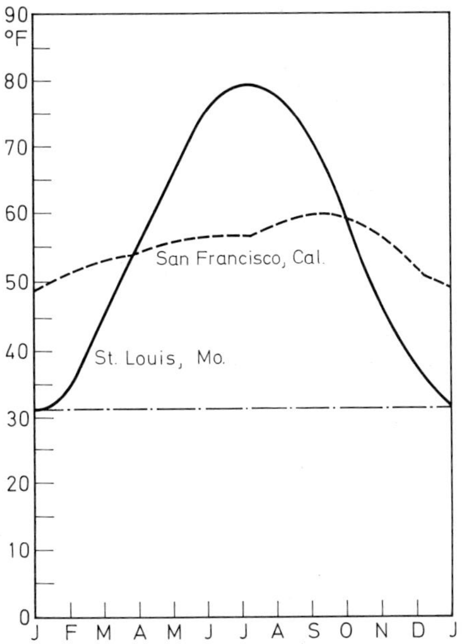

Fig. 4-3. Mean daily temperatures at San Francisco, California, and St. Louis, Missouri. This illustrates differences in seasonal temperature extremes between a maritime and continental region, approximately on the same latitude (Wilsie, 1962)

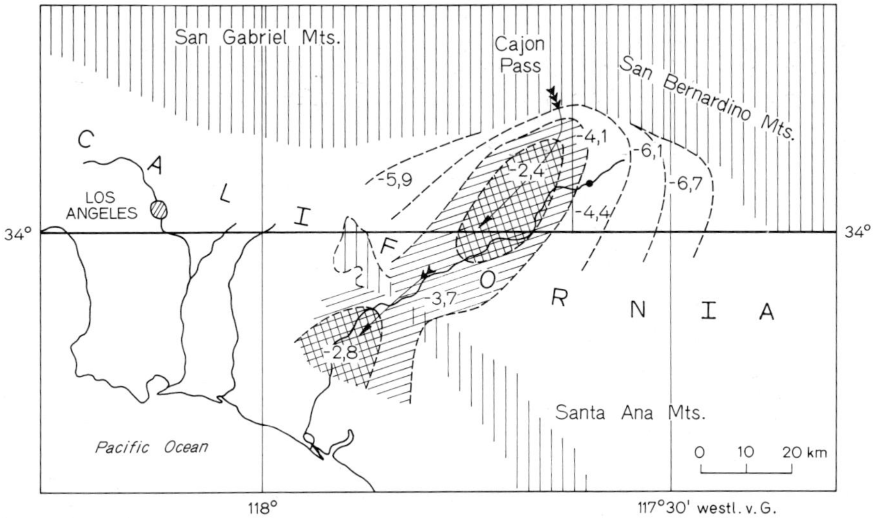

Fig. 4-4. Effect of local topography on minimum temperatures during the night. Note that the minimum temperatures are higher in the path of the main airflow from the Cajon Pass, and progressively decrease as one moves out of the main air path (Young, 1921)

for extensive structures and elaborate heating and cooling installations. Continental climates are characterized by temperature extremes between seasons and between day and night.

Other factors are conditions of local topography which influence wind movement, air drainage, and sometimes the duration of sunlight. The air movements up and down high mountain valleys reduce temperature extremes (Fig. 4-4), permitting apple and peach production at relatively high altitudes in Colorado. Geiger (1965) in his outstanding text on climate near the ground, thoroughly discussed factors of windbreaks, slopes, cities, etc., and the importance they have in relation to local temperatures. These same factors influence greenhouse temperature control, and siting of a greenhouse in regard to air drainage is nearly as important as siting of an orchard. Locating a greenhouse in a frost pocket will increase fuel costs, and poorly air-drained sites can drastically increase costs of disease control.

There is not enough space to discuss all aspects that a careful operator should consider in siting greenhouses. It is sufficient to emphasize that some work in obtaining information of area and local climatic conditions can play an important part in success of the enterprise, and the ease of temperature manipulation. Latitude, altitude, maritime or continental influences, local topography, wind direction and velocities, humidity and pollution factors should all be thoughtfully considered whether one is building a new greenhouse or planning to purchase an existing establishment.

4.2 Temperature and Growth

4.2.1 Factors Determining Plant Temperature

In Chapter 2, with the aid of Figure 2-13, the energy exchanges between a single plant and its surroundings were discussed. Heat (energy) can be reflected, transmitted, absorbed, conducted, reradiated, or used to evaporate water. The latter generally requires over 500 calories per gram of water evaporated. Obviously, if radiant energy is absorbed, plant temperature must rise until an equilibrium is reached at some higher value. The new temperature equilibrium is determined by:

1. The fact that plants will reradiate thermal energy to their surroundings—the amount depending upon the temperature differential between plant and the surrounding area (greenhouse roof, heat pipes, etc.).

2. The removal of energy by conduction and convection processes. Or, the flow of cooler air over warm leaves, an example of which is provided in Figure 4-5. As air movement increases, more heat is conducted away and plant temperature comes closer to the air temperature. If air temperature is higher than plant temperature, heat can flow in the opposite direction, tending to raise plant temperature.

3. The amount of energy used in photochemical processes such as photosynthesis. Generally, as pointed out in Chapter 2, this is so small as to be commonly ignored in assessing energy balances of growing crops.

4. The evaporative rate of the plant. As a rule, between 70 to more than 90% of absorbed solar radiation is utilized to evaporate water.

5. The final equilibrium temperature is influenced by the capacity of the plant to store heat—or thermal heat capacity of the tissue and its mass. Therefore, temperatures of thin leaves will vary faster and farther than thick leaves, or large

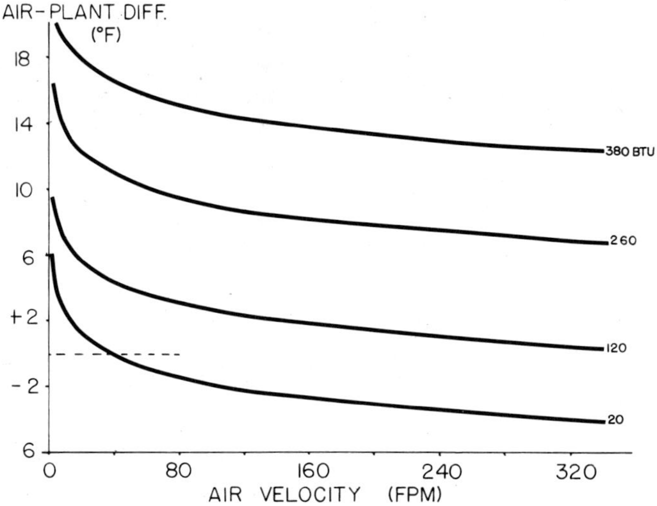

Fig. 4-5. Effect of air velocity at different solar radiation intensities on differences between flower and air temperature of carnations (Hanan, 1969). The vertical axis is the deviation of flower temperature from the air temperature at different air velocities and radiation intensities of 380, 260, 120, and 20 BTU h^{-1} ft^{-2}. Note that at low intensities, flower temperature dropped below air temperature at wind speeds above 40 ft min^{-1} Air temperature maintained at 70° F. The relationship changes slightly as air temperature changes

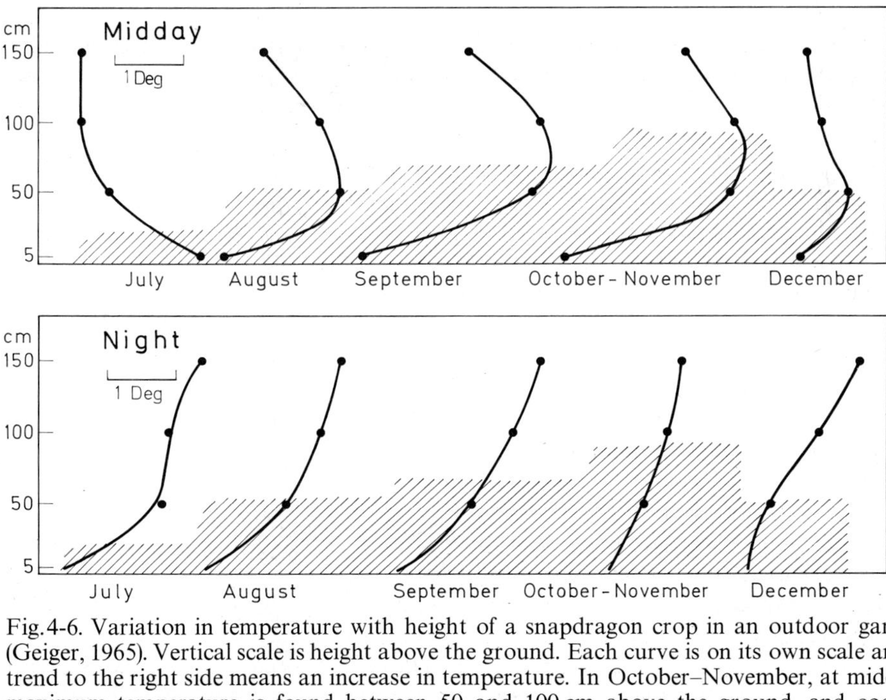

Fig. 4-6. Variation in temperature with height of a snapdragon crop in an outdoor garden (Geiger, 1965). Vertical scale is height above the ground. Each curve is on its own scale and a trend to the right side means an increase in temperature. In October–November, at midday, maximum temperature is found between 50 and 100 cm above the ground, and coolest temperature at zero height

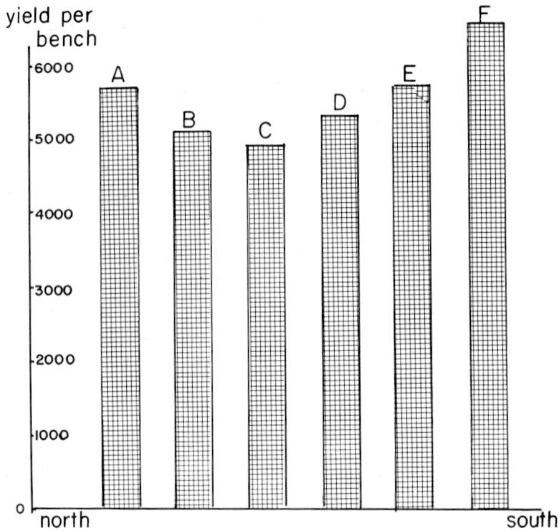

Fig. 4-7. Variation in yield of roses across a fiberglass-covered, east-west oriented, steam-heated greenhouse for one year. *Bench A* was on the north, *Bench F* the south (Hanan, unpublished). Largely due to differences in light

masses such as flower buds or storage organs, when there is a change in the environment.

The final effect is to eliminate possibility of controlling individual plants, and particularly whole crops, at any single temperature—particularly under rapidly changing conditions of sunlight, air movement and air temperature. As the crop grows, and changes shape, and as the season progresses with variations in the sun's position in the sky, one will find the crop temperature varies accordingly. An example is provided in Figure 4-6. During early stages of crop growth, maximum light interception is on the soil surface, so the temperature will be the highest there. At night, reradiation from the soil surface cools the ground so it is the coolest. As the crop grows, maximum interception of solar energy moves to some point above the soil surface so that maximum temperatures will often be found in the same area as maximum light interception. Much research has been conducted on light interception by agronomic crops, but very little on greenhouse crops. A short study by Hall and Hanan (1976), for a one-year-old carnation crop, showed light energy to be nearly zero at 0 and 50 cm above the soil surface in the center of a 105 cm wide bench. With a crop 125 cm tall, total energy received was only 31% of that outside, 12.5 cm in from the vertical outside surfaces of the crop, 90 cm above the ground. Is there any wonder that thick, heavy crops will loose their leaves on the lower parts of their stems when there is practically no energy for them to utilize in food manufacture?

Not only will temperature vary with distances into the crop canopy, but wind velocity, humidity, and CO_2 concentrations will also vary. In this single most important area, we have practically no information by which we might improve greenhouse productivity and understand why crops behave as they do. As suggested by Hanan and Heins (1975) for carnations, the effect of varying planting density on production is largely during the first period of growth. Once the crop has filled a space to which it is confined (bench), productivity ceases to be a function of the number of plants per unit area and is a function of vegetative surface area exposed to radiant energy.

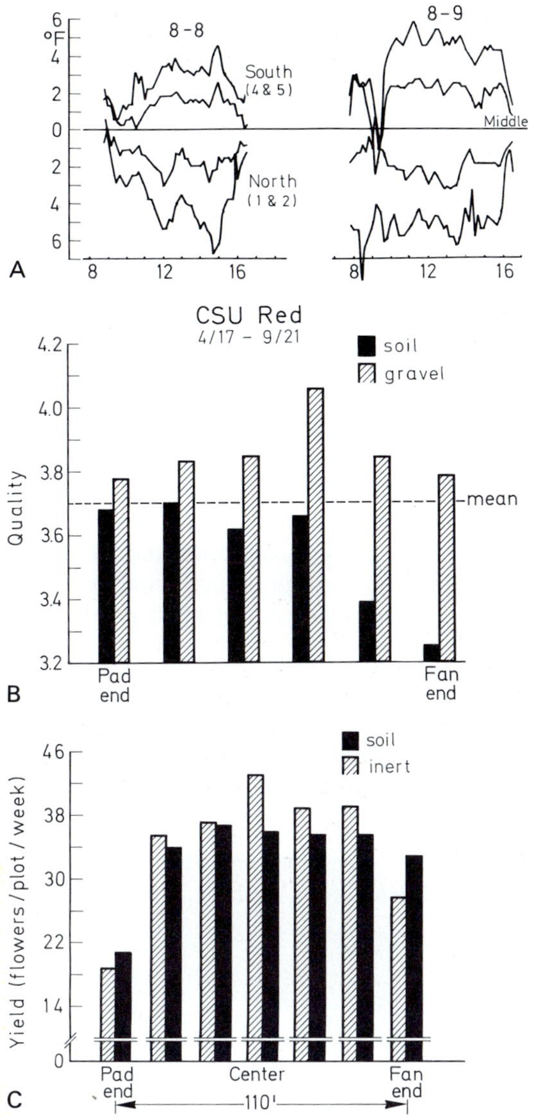

Fig. 4-8 A–C. Effect of temperature variations in a 106-ft greenhouse equipped with evaporative fan and pad cooling, during the cooling season on production and flower quality of carnations grown in two root media (Hanan et al., 1973). (A) Air temperature in an aspirated shelter at the center of the house was taken as zero, and temperatures recorded similarly at other locations are plotted as the differences from the middle for August 8 and 9. The lower curves represent temperature toward the pad end, the upper curves the temperatures toward the exhaust fan end of the greenhouse. (B) Average mean grade of carnations as a function of bench position, maximum possible grade = 5 for all highest grade, or 2 for all lowest grade. (C) Average number of cut flowers per plot per week for the period April 17 through September 21 as a function of bench position

Even if air temperatures are controlled uniformly throughout the greenhouse space, plant response remains a partial function of position. Hanan (Fig. 4-7) showed significant variation in rose yield between benches in the center of the house and sides, with the south bench in an east-west oriented greenhouse providing the highest yields. The reasons for these results are due largely to light variations in the house and amount of sunlight intercepted. Usually, air temperatures in greenhouses vary from one location to another, depending upon the type of heating system, its arrangement, wind direction and velocity, greenhouse orientation and type of structure and crop arrangement. A very good example of the effects of such air temperature variations on production is provided in Figure 4-8.

It is not feasible to install sufficient cooling equipment to eliminate temperature rise from evaporative pads to exhaust fans in a commercial range. Thus, consistent air temperature variations during the cooling periods may be expected. Other sections of this chapter provide additional information on common temperature variations.

The final result is to preclude any single temperature being maintained for any significant period in a commercial greenhouse. Thus, when one speaks of "optimum air temperatures" for a particular crop, he is essentially saying that an air temperature "X", as measured at some location in the greenhouse, under the usual prevailing conditions, at the time the growth was observed, provided the greatest average return—whatever that return may be in terms of yield, quality, time to flower, etc. The actual plant temperature might have been something else and the results may not be applicable elsewhere under different climatic conditions, greenhouse cover, heating system, etc. Over a sufficient period, empirically gained information and practical experience combine to provide reasonable estimates of "optimum temperatures" for growth under prevailing conditions. Published optimum temperatures, despite obvious improvements in the last 30 years, remain rather nebulous. A good grower becomes aware of his greenhouse, and by careful planning, learns to place cultivars responding best to slightly higher temperatures in the warmest locations, others in cooler locations. Or, in pot plant production, temperature differentials may be utilized by moving the crop to various areas in order to hasten or delay maturity, or hold for sale.

4.2.2 Effects

Important in the ultimate return from a crop are the basic physiological processes common to all plants such as: photosynthesis, respiration, translocation, ion uptake, transpiration, pigment formation, reproduction, bulbing, elongation, and

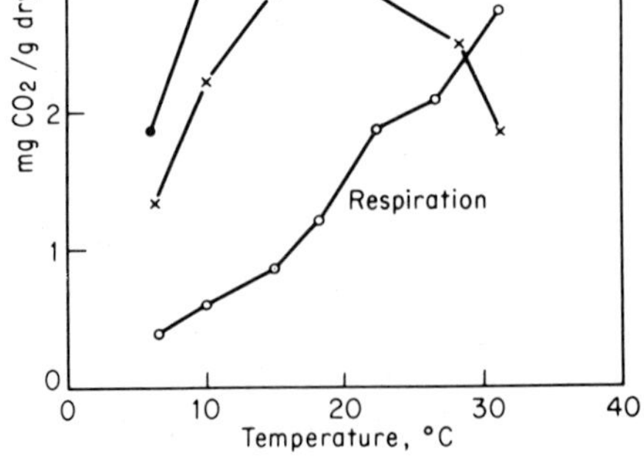

Fig. 4-9. Effect of temperature on respiration, apparent net and true photosynthesis (Stålfelt, 1937)

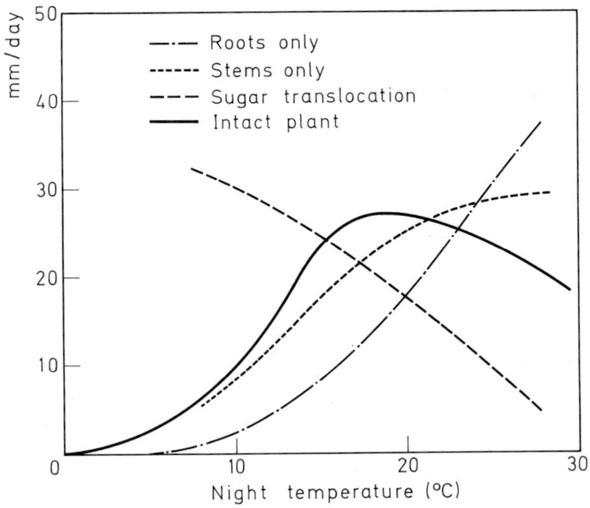

Fig. 4-10. Effect of night temperature on elongation of different plant parts (Went, 1956)

Fig. 4-11. Variation in flower size of CSU Red Sim carnations as the result of differences in the day temperature regime (Hanan, 1959)

many others. Temperature influences all of these processes in various ways and to different degrees. The classical example of this is Stålfelt's relationships between temperature and the rate of respiration compared with true photosynthesis (Fig. 4-9). Over a limited range at which plant growth will occur, chemical processes will commonly be doubled for every 10° C rise in temperature. Respiration shows a continuous rise as temperature increases, and so also does photosynthesis (Fig. 4-9). But, photosynthesis can also be limited by available energy and CO_2 level. As respiration is the utilization of materials provided by photosynthesis to build the plant, a high enough temperature is possible for many plants where

Fig. 4-12. Effect of four day temperatures on carnation growth when produced at two different times of the year. The best day temperature for maximum elongation shifts with the season, or the amount of radiant energy available (Hanan, 1959)

there is a net decrease in the sugar production. The practical effect is weak and spindly growth from too high temperatures in the greenhouse—especially during low light periods. This is a fairly simple example of a temperature effect on plants, and the importance of interactions between two or more basic processes. Temperature affects physical processes such as transpiration. A rise in leaf temperature can increase vapor pressure inside the leaf so that, if the outside vapor pressure

Fig. 4-13 A–C. Various effects of low temperatures on growth of some commercial greenhouse crops. (A) Red "flushing" in the cultivar White Pikes Peak carnation. (B) "Bullheads" or "cabbage" head roses resulting from low temperatures in combination with rapid fluctuations. (C) "By-passing" in spray chrysanthemums caused by low temperatures at the time of flower initiation and development

remains the same, transpiration will be faster. If the soil is cold, water and ion uptake and root growth will be reduced, perhaps to a point where all top growth ceases regardless of the air temperature.

The effect of temperature on processes such as elongation will vary with the plant part about which one is concerned. That is, a 20° night temperature which perhaps causes a 15 mm per day elongation of a root may not result in the same elongation rate for stems. When the total plant is concerned, maximum elongation may occur at some intermediate temperature as suggested in Figure 4-10.

Eventually, the average temperatures sustained by a crop results in a product which is a compromise. Figures 4-11 and 4-12 provide two examples. Figure 4-12 emphasizes the importance of climatic variations external to the greenhouse on response. But, despite our conclusion that precise and accurate plant temperatures are exceedingly difficult to maintain, and that final results are compromises; we should not assume that temperature manipulation can be sloppy. As discussed later in this chapter, carelessness in temperature measurement, improper location of sensing devices and related equipment, inadequate maintenance, and failure to monitor and plan suitable temperatures often results in serious monetary loss to the operator (Fig. 4-13).

4.2.3 Optimum Temperatures

We defined "optimum temperature" earlier. It is, essentially, the best compromise resulting in the desired response. A grower, therefore, should always define his objectives first; determine how he can achieve those objectives second, and third, assess the capability of his equipment and facilities. Table 4-1 surveys a part of the existing English literature concerning optimum temperatures. A number of conclusions can be drawn from a study of this table:

1. Most species require diurnal temperature fluctuations. Species that are indigenous to tropical areas (most foliage plants for example) will usually grow better at constant day-night temperatures (see Fig. 4-1 for the reason).
2. The best temperature usually varies with season and location (total available solar energy).
3. Optimum temperature varies with age (Figs. 4-14 and 4-15).
4. Optimum temperature varies with the stage of growth.
5. Optimum temperature varies with the particular process such as rooting, germination, floral initiation, bulbing, etc. (Figs. 4-15 and 4-16).
6. Optimum temperature varies with grower objectives.

What Table 4-1 does not show is the important differences that can exist between varieties in a particular species. The literature is relatively silent on this subject, although publications provided by various companies will often suggest optimum temperatures for their plant materials. Word-of-mouth and accumulated practical experience by individual growers are generally most common.

The temperatures provided in Table 4-1 were originally published in Fahrenheit degrees. We have converted them to Celsius and rounded them to the nearest whole degree. Use of Table 13, Appendix A to convert them back to °F may result in slightly different values from those originally published. Also, one degree Celsius equals 1.8° F, so that, in effect, precision of the stated temperature requirement is reduced.

Table 4-1. Some suggested "optimum" temperatures for greenhouse crops (°C). (For conversion to Fahrenheit, see Appendix A, Table 13)

Plant	Temperature		Remarks	Ref.
	Day	Night		
Aster	—	10	Night temperature, 3–5° C higher in day	Ball (1975)
Azalea	—	10	Temperature for stock plants	Larson (1975)
	— 17–27	15	Rooting medium temperature and air temperature	Larson (1975)
	—	13 (13–18)	13 to maintain vegetative growth, will grow adequately over range, days 5–8° C higher	Love (1975)
	—	16–18	Most cultivars require 18 or above for flower initiation. Some will initiate down to 16. Interaction with total light, photoperiod, and growth regulators	Criley (1975)
Bedding plants	—	10–16	0 to 3° C higher on cloudy days, 5° C on clear days, finish at 7–10° C	Carpenter (1976)
	— 21	—	Germination for Ageratum, Allyssum, Aster, Begonia, Calendula, Celosia, Cosmos, Dahlia, Gaillardia, Impatiens, Lobelia, Marigold, Portulaca, Salvia, Zinnia	Tayama et al. (1975)
	— 18	—	Germination for Centaurea, Coleus, Pansy, Phlox, Verbena	Tayama et al. (1975)
	— 16	—	Germination for Hollyhock	Tayama et al. (1975)
	— 13	—	Germination for Sweet Pea	Tayama et al. (1975)
	3+–5+	4–10	Day temperatures refer to increase above night for cloudy or clear days for Alyssum, Calendula, Larkspur, Lobelia, Snapdragon, Stock, Pansy. Higher night is optimum	Dietz (1976)
	3+–5+	10–16	Same as above for Ageratum, Aster, Begonia, Centaurea, Coleus, Dianthus, Geranium, Nierembergia, Petunia	Dietz (1976)
	3+–5+	13–18	Same as above for Browallia, Celosia, Dahlia, Impatiens, Marigold, Salvia, Verbena, Zinnia	Dietz (1976)
	3+–5+	16–18	Same as above for Kochia	Dietz (1976)
	—	10–13	Optimum night temperature until transplanted for Alyssum, Calendula	Dietz (1976)
	—	16	Optimum night temperature until transplanted for Aster, Balsam, Browallia, Coleus, Dusty Miller, Geranium, Impatiens Marigold, Petunia, Portulaca, Salvia, Verbena, Zinnia	Dietz (1976)

Table 4-1 (continued)

Plant	Temperature		Remarks	Ref.
	Day	Night		
	—	10	Optimum temperature until transplanted for Dianthus, Larkspur, Lobelia, Pansy, Snapdragon, Stocks	Dietz (1976)
	—	13	Optimum temperature until transplanted for Centaurea, Dahlia, Nierembergia, Phlox	Dietz (1976)
Caladium	— 21	—	Storage in dry air prior to planting	Ball (1975)
	— 27–29	—	Soil temperature for forcing. Minimum 18	Ball (1975)
	—	21–24	Growing temperatures, night	Ball (1975)
Calceolaria	— 21	—	Germination	Ball (1975)
	—	9–10	Growing temperature, 3–5° C higher day	Ball (1975)
Carnation	16, 18–19	10–11	For plants in soil, heat to 16, begin cooling at higher day temperature range	Holley (1971)
	17, 19–21	11–12	For plants in soil with CO_2. Heat to 17, begin cooling at higher day temperature range	Holley (1971)
	18, 21–22	12–13	For plants in inert media with CO_2. Heat to 18, begin cooling at higher day temperature range. All preceding values for Colorado, fall, winter, and spring	Holley (1971)
	18, 21	12	Young plants, with the lower day value for winter, Colorado	Seeley (1961)
	14, 18	10–11	Old plants (1 year +), with lower day value for winter, Colorado	Seeley (1961)
Chrysanthemum	18	16	Minimum night and maximum day for year-around	Ball (1975)
	17–18, 18–21	17	Lower day range for cloudy days, higher for clear, for potted Chrysanthemums	Ball (1975)
Daffodils	— 9	—	Storage for precooled bulbs	DeHertogh (1973)
	— 13–16	—	Storage for bulbs not ready to be precooled or are not to be precooled	DeHertogh (1973)
	— 1–10	—	Precooling temperature, starting at 10, ending at 1, depending upon flowering period	DeHertogh (1973)
	—	13	Forcing temperature	DeHertogh (1973)

Easter Lily	—	1–3	—	Precooling for minimum of 5 weeks	Hastings (1975)
	—	2–17	—	Rooting temperatures, starting at 17 and reducing to 2, depending upon variety	Hastings (1975)
	18, 21		16	18 maximum day temperatures until after flower buds formed, force at 21, do not go below 16 night	Hastings (1975)
Foliage plants	—		16, 18, 21	Lowest for holding, 18 for rapid growth up to 21	McColley (1975)
Fuchsia	—	18–24	13	Night 13° C, germination at 18–24	Ball (1975)
Geranium	—	18–24	12–13	Rooting temperatures, higher ranges in root medium	Rathmell (1971)
	16–18		13–16	Minimum night 13, usually 16 to finish, depending upon age	Mastalerz (1961)
	18, 21		16–17	Use 18 for cloudy days, 21 for clear days, without CO_2	Stinson (1971)
	21, 24		18	Use with CO_2, 21 cloudy days, 24 clear days	Stinson (1971)
	—	24	18	Rooting temperature with forcing night temperature	Ball (1975)
Gloxinia	—		18–21		Ball (1975)
Hyacinth	—	9–10	—	Storage temperature for prepared bulbs for early forcing	DeHertogh (1973)
	—	16–18	—	Storage temperature for regular bulbs	DeHertogh (1973)
	—	9	—	Rooting, time depending upon flowering period	DeHertogh (1973)
	—		16–23	Forcing temperature, depending upon flowering period, highest when early	DeHertogh (1973)
Hydrangea	—		13–16, 21	Force at 21, reduce to 13–16 after flowers visible	Shanks (1975)
	—	0–2	—	Cold storage combined with defoliation to overcome dormacy. Long-term storage	Shanks (1975)
	—	4–7	—	Cold storage temperature best for early forcing	Shanks (1975)
	—		18	Less than 18 for flower development	Shanks (1975)
Iris	—	32	—	For 10 days to accelerate flower bud formation	Ball (1975)
	—	9	—	Precooling temperature, length of precooling depending upon flowering period	Ball (1975)
	—		10–14	Forcing temperature, depending upon variety and flowering period	Ball (1975)
Kalanchoe	—	21	16	Germinate at 21, grow at 16 night	Ball (1975)
Poinsettia	—	20, 21	—	Temperature for propagation, lower value for multiflower	Shanks (1975)
	—		16–17	For standards, day temperature up to 8 higher, until mature. Hold at 14	Shanks (1975)
	—		18, 20	Higher value for multiflowering until lighting discontinued. Lower value to finish	Shanks (1975)

Table 4-1 (continued)

Plant	Temperature		Remarks	Ref.
	Day	Night		
Rose	21, 24 27–29	16	Lowest day value for cloudy days, 24 for clear days, highest range for CO_2 injection	Mastalerz (1969)
	22, 27	17	Heat to 22 and begin cooling at 27, Colorado	Goldsberry and Holley (1966)
	22, 24, 30	17	Heat to 22, Set ventilation at 30 between November 1 and May 1, when using CO_2. Use 24 in summer or without CO_2	Hanan (1973)
Snapdragon	— 18–21	—	Germination	Langhans and Maginnes (1962)
	—	16, 10–16	16 during first month, range is to finish crop at lower temperature	Langhans and Maginnes (1962)
	— 21	—	Germination	Ball (1975)
	— 16	—	To transplanting	Ball (1975)
	13, 16–18	10	Use 13 for cloudy days, higher day values for clear days	Ball (1975)
Tulips	— 17	—	Storage temperature for precooled bulbs prior to stage "G" of initiation	DeHertogh (1973)
	— 13–16	—	Storage for non-precooled bulbs prior to planting	DeHertogh (1973)
	— 7, 9	—	Precooling temperatures depending upon flowering period	DeHertogh (1973)
	— 1–20	—	Rooting, prior to forcing, starting at higher value	DeHertogh (1973)
	—	16–18	Forcing temperature, depending upon flowering period, avoid more than 3° C higher day temperature	DeHertogh (1973)
Vegetables:				
Lettuce	16–18–21	2–10–13	For fall heated crop, 21 for bright weather, reducing to 18, then 16. Night temperature 13 to rosette, 2 after rosette, England	Large (1972)
	13, 16	13	Winter crop, during early stages, ventilate at 16, heat to 13, England	Large (1972)
	21, 18, 16, 13	10	Winter crop, ventilate at 21 and heat to 16 with CO_2, heat to 13 and ventilate at 18 without CO_2, England	Large (1972)
	10–16	7	Winter crop, at heart formation, ventilate at 16, heat to 10, England	Large (1972)

	16–21		13	Spring crop, young plants, England	Large (1972)
	10–16		7	Spring crop, drop to night temperature gradually, heat to 10, ventilate at 16, raise by 3 if using CO_2	Large (1972)
Tomato	21–22	24–26	14–16, 16–17	Using higher ranges for bright days, lower ranges for dull days, Ohio	Brooks (1973)
	—	29	—	Germination	Tayama et al. (1975)
	—		16	Temperature until transplanting	Dietz (1976)
	—		13–18	Temperature to produce transplants for garden, 18 optimum	Dietz (1976)
Others	—	35	—	Germination temperature for corn, cucumber, squash, watermelon	Tayama et al. (1975)
	—	32	—	Germination temperature for muskmelon	Tayama et al. (1975)
	—	29	—	Germination temperature for cabbage, eggplant, pepper	Tayama et al. (1975)
	—	27	—	Germination for cauliflower	Tayama et al. (1975)
	—	24	—	Germination for lettuce, onion, parsley	Tayama et al. (1975)
	—	16–21	—	Germination for celery, allow night temperature to fluctuate to 16	Tayama et al. (1975)
	3+–5+		4–10	Production of garden transplants for onion, grow at 7 until transplanting, 3–5° C higher during days	Dietz (1976)
	3+–5+		7–13	Production of transplants for cabbage	Dietz (1976)
	3+–5+		13–18	Production of transplants for eggplant, pepper. Grow at 16 until transplanted	Dietz (1976)

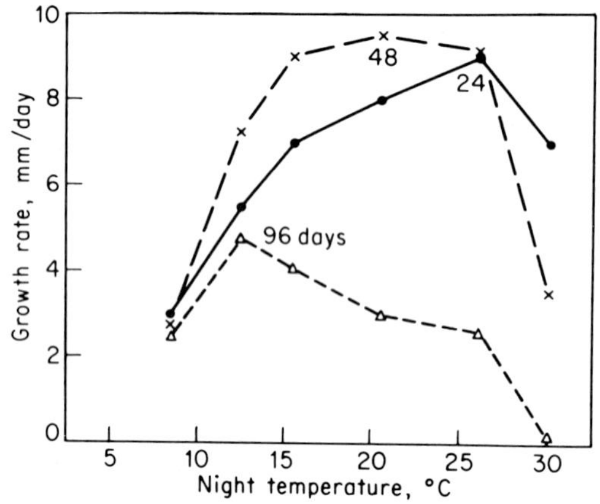

Fig. 4-14. The effect of night temperatures on elongation of pepper plants as a function of age of the plant. Optimum temperatures usually decline as plants age as noted in the figure where maximum elongation occurs at 25° C for plants 24 days old as contrasted to 12° C for those 96 days old (Dorland and Went, 1947)

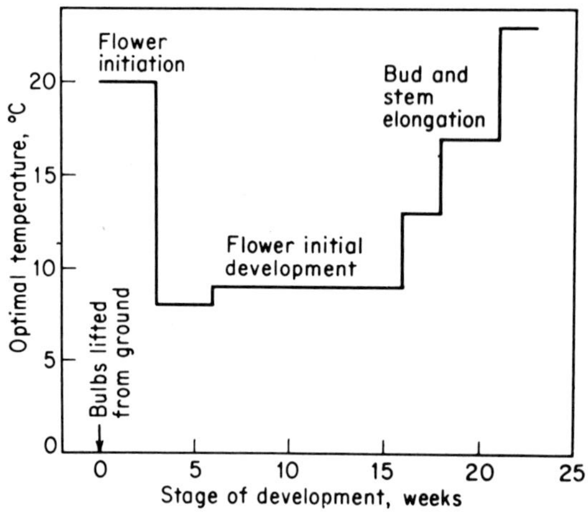

Fig. 4-15. Optimal temperature for continued development and growth of tulips at different developmental stages (Hartsema et al., 1930)

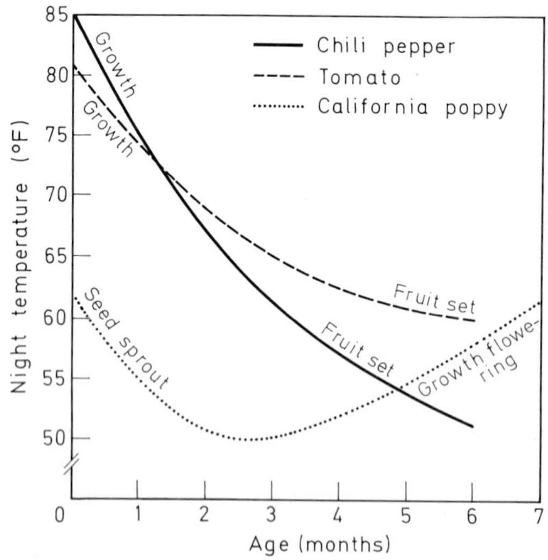

Fig. 4-16. Variation in optimum night temperatures of different species as a function of age and stage of growth (Went, 1957a)

4.3 Controlling Greenhouse Temperatures

The greenhouse operator of the 21st century has become more of a scientist than ever before and thus requires more accurate control of temperatures for optimum production of crops. In developing a desirable year around environment for plant growth, heating and ventilation requirements must be considered. The type of crop, its age or degree of maturity and the desired date of marketing all influence the range and duration of day and night temperature levels used to control the greenhouse environment.

Walker (1965) assumed the environment within a producing greenhouse facility as a "steady-state condition". The heat and moisture balance relationships, under such a condition, are influenced by solar energy, ventilation, plant transpiration, heating methods, and numerous conduction losses (Fig. 4-17). The general principle is expressed as:

$$q_f + q_i + q_a + q_r = q_c + q_t + q_p + q_g + q_e + q_s \, .$$

The heat factors involving plant respiration (q_r), photosynthesis (q_p), and losses to the ground are generally negligible and thus seldom considered. The relationship of ventilation and infiltration (q_e and q_s) and the moisture flow within a facility all influence the energy balance.

The following sections attempt to aquaint the reader with methods, principles, and equipment used to maintain the temperatures within a greenhouse environment.

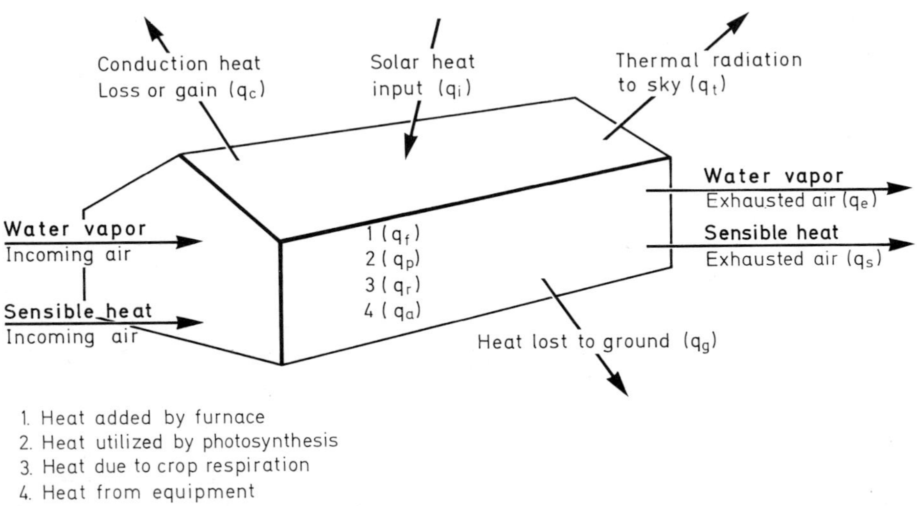

Fig. 4-17. Heat quantities considered in the greenhouse heat budget (Ashrae, 1974)

4.3.1 Heating

The success of a greenhouse manager lies, in part, with his ability to maintain desirable and uniform temperatures during the heating season. A properly designed, installed, and maintained heating system should meet this requirement and be functional for many years.

The following terms are presented so the reader will understand the discussion. Most of them are adapted from Ashrae (1963).

Convection:	The transfer of heat from one place to another by the motion of air or water.
Conduction:	The flow of heat from a point of higher temperature to that of a lower temperature.
Radiation:	The continual emission of energy from the surface of all bodies.
Condensation:	The process of changing a vapor into a liquid when heat is extracted. Example, when the conduction of heat occurs through a greenhouse cover in cold weather condensing water on the inside surface.
C factor:	Thermal conductance, surface to surface; BTU h^{-1} ft^{-2} $°F^{-1}$.
K factor:	Thermal conductivity, BTU h^{-1} ft^{-2} $°F^{-1}$.
U factor:	The overall heat transfer coefficient BTU h^{-1} ft^{-2} $°F^{-1}$.
Infiltration:	Air flowing into a structure through cracks.

4.3.1.1 Heat Loss

The greenhouse, as a structure, contributes little to the ability of a grower to maintain a stable environment. The designs and sizes of structures used in commercial production influence the temperature and air circulation patterns. Carpenter and Bark (1967) demonstrated how high temperature air accumulates in the greenhouse ridge, and heat is lost through the cover by conduction. Roberts (1969) reported that the loss of heat at night in a polyethylene covered house represented almost 39% of the total loss, providing no condensate was present. The presence of condensate on a poly cover has been reported to decrease heat loss by 10–15% (Walker and Walton, 1969). Infiltration of cold air through improperly fitted ventilators, louvers, and cracks in the cover all contribute to increased heat requirements (Morris, 1959; Walker and Duncan, 1973a). Whittle and Lawrence (1961) reported that the overall heat transfer coefficient measured in several glasshouses was approximately 1 BTU h^{-1} ft^{-2} $°F^{-1}$ and a wind velocity of about 15 mph doubled the heat consumption. Tristan (1973) found that a 15 mph wind increased the heat loss in greenhouses with a corrugated FRP roof and glass side walls by 40%. Heat is also lost through improper combustion in heating equipment. Whittle and Lawrence (1961) reported heat losses up to 16% through the chimneys.

The major heat loss is through the greenhouse structure and covers. Depending on temperature, 30–210 and 15–110 BTU h^{-1} ft^{-2} of greenhouse ground area covered can be lost, respectively, from a single and double glazed polyethylene house (Ashrae, 1974).

The ability of greenhouse structure parts and covers to transmit heat can be represented by a heat transmission coefficient, commonly known as the *U* factor. Hart (1950) computed a *U* value for greenhouse glass, exposed to wind of 1.15. Sheldrake and Langhans (1962) calculated the *U* factor for plastic film covered houses. A single layer of 10 mil cellulose acetate with a 2 mil polyethylene liner 2 in from the outside layer, had a value of 0.705.

Liu et al. (1969) measured the heat loss through common greenhouse covers and calculated larger *U* values than those discussed above. The factor for glass was 63% greater and for a single layer of polyethylene, 52% larger than the value obtained by Sheldrake. It is interesting to note that the *U* factor for corrugated FRP panels was 12% greater than for glass.

There appears to be some discrepancy in the *U* values, and it is hoped that recent research can determine which factor is more realistic. It should be noted, however, that the *U* value of 1.13 for glass has been used successfully for more than 25 years.

4.3.1.2 Computing Heating Requirements

The heat requirements for most greenhouse structures are based on a formula developed by Gray (1956) for glass covered house calculations.

$$H = \mathrm{UA}(t_i - t_0),$$

where:

H = heat required per hour in BTUs per h;
U = a constant for glass;
A = the exposed surface area in square feet;
t_i = inside temperature in ° F, and;
t_o = outside temperature in ° F.

1. *National Greenhouse Manufacturing Association Standard.* The above formula has been modified and incorporated into the greenhouse heating standards developed and adopted by the National Greenhouse Manufacturers Association (NGMA). The formula included in the standard is:

$$[A_1 + (A_2 \times R)] \times \Delta T \times G \times W \times C = \text{BTU Heat Loss}$$

A_1 = Square feet of exposed glass area;
A_2 = Square feet of exposed wall area (other than glass);
R = Resistance of curtain wall to transmission of heat (expressed in relation to transmission through glass) (Table 1);
ΔT = Highest temperature to be maintained in greenhouse as compared to outside design temperature (in other words—temperature difference);
G = Coefficient of transmission of glass (Table 2);
W = Wind factor (Table 3);
C = Construction factor (Table 4).

Table 1. "R" factor. (These are factors for "curtain" wall materials normally used on greenhouses)

3/8 in	Corrugated asbestos cement board wall	0.94
4 in	Poured concrete wall	0.76
6 in	Poured concrete wall	0.67
8 in	Poured concrete wall	0.60
4 in	Concrete block wall	0.58
8 in	Concrete block wall	0.46
8 in	Brick wall	0.43

Table 2. Coefficient of transmission of glass. "G" factor

Temperature difference between inside and outside design temperatures (ΔT)	Coefficient of transmission
50°	1.09
55°	1.10
60°	1.11
65°	1.12
70°	1.13
75°	1.14

Table 3. Wind factor—"W". If wind velocity exceeds 15 miles per hour average during the heating season, the heat loss calculation should be increased by 4% for each 5 miles per hour over 15 as below

Wind velocity	"W" factor	Alternate "W" factor
15 mph or less	1.00	1.10
20 mph	1.04	1.14
25 mph	1.08	1.18
30 mph	1.12	1.22
35 mph	1.16	1.26

Table 4. Construction factor—"C". The condition and type of a greenhouse directly affects the heat loss. Consult this table for the factor which best describes the greenhouse on which you are calculating the heat loss

All metal (good tight glass house—20 or 24 in glass spacing)	1.08
Wood and steel (good tight glass house—16 or 20 in glass spacing, metal gutters, vents, headers, etc.)	
Wood houses (glass houses with wood bars, gutters, vents, etc. up to and including 20 in glass spacing)	
Good tight houses	1.00
Fairly tight houses	1.13
Loose houses	1.25
Fiberglass covered wood houses	0.95
Fiberglass covered metal houses	1.00

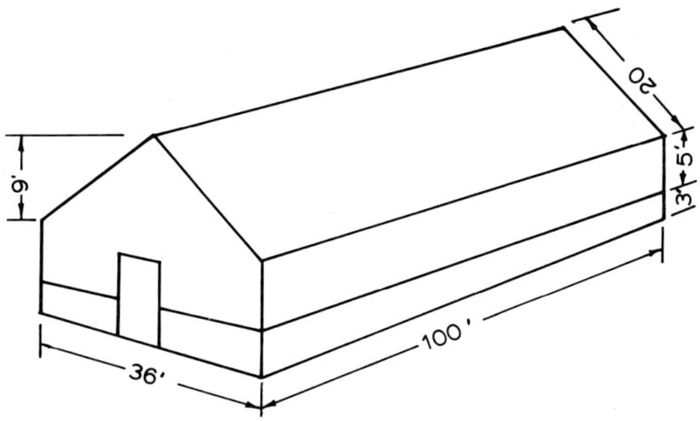

Fig. 4-18. Greenhouse facility with transite foundation walls

The above calculations appear to be complicated and some designers feel there is no need to be so accurate, but they deserve some comments: (1) The "R" factor (Table 1) is seldom needed in FRP covered commercial greenhouses because the plastic panels are generally extended into the ground (Fig. 3-20). (2) The "G" factor (Table 2) is for glass but can be safely used for FRP and single polyethylene covers and 0.70 for double layered covers. Walker and Duncan (1973a) and NGMA recommend that design temperatures be based on a temperature approximately 15° above the lowest figure ever recorded in a geographical location. (3) The average maximum wind for a given area should be used to compute the "W" factor (Table 3). The NGMA standard points out that wind speeds greater than 15 mph can often lower the greenhouse temperature $\frac{1}{2}$° F for every mph over 15. (4) Cracks such as found in lapped glass and fiberglass panels generally freeze shut when outdoor temperatures reach approximately 15° F. The alternate "W" factor is recommended when high design temperatures are used in the greenhouse, or when warm air is blown against the greenhouse walls, keeping the ice melted. (5) The combination of construction factor "C" (Table 4) and factor "G" is the equivalent of the "*U*" factor described at the beginning of the section. (6) Infiltration must also be considered. Gray (1956) discussed the calculations required to determine infiltration which were based on the fact that 0.018 BTU is required to raise the temperature of one cubic foot of air 1° F. The BTUs required to balance the infiltration loss are: 0.018 × structure volume × the design temperature (Table 2) = BTUsh^{-1}. Roberts (1969) simply adds 10% of the basic BTU heat loss figure as the infiltration factor. The greenhouse dimensions used to calculate the sample problem are shown in Figure 4-18.

2. *Other Calculations.* (a) Nomograph. A rapid method of determining heat loss has been presented by Walker and Duncan (1973a). The factors needed for entering the nomograph, Figure 4-19, are frame perimeter, length of greenhouse and temperature difference (design temperature). (b) Ground area relationship. A very rapid estimate of heat loss can be extracted from Figure 4-20. The only information required is the outside temperature and ground area covered by the greenhouse structure.

TABLE I - FRAME PERIMETER (FEET)

WIDTH (W) (FEET)	12 / 6 / 6' — W	W/2 — W	13/21 — W	12 / 6 / 6' — W	12 / 6 / 6' — W
12	26	19	20		
16	30	25	28		
20	34	31	36		
24	39	38	40	40	
28	44	44	48		
32	48	50	56	48	
36	53	56	64		54
40	57	63	72	56	
48				66	66
56				76	
60				80	78
64				84	
72				94	93
80				102	
84					108
96					120
108					135
120					147

TABLE II - EQUIVALENT LENGTH FOR END WALLS (FEET) (ADD TO LENGTH OF HOUSE)

WIDTH (W) (FEET)	12 / 6 / 6' — W	W/2 — W	13/21 — W	12 / 6 / 6' — W	12 / 6 / 6' — W
12	8	6	7		
16	9	8	9		
20	10	10	11		
24	11	12	13	10	
28	12	14	15		
32	14	16	17	11	
36	15	18	20		11
40	16	20	22	12	
48				13	12
56				14	
60				15	13
64				16	
72				17	14
80				18	
84					15
96					16
108					17
120					18

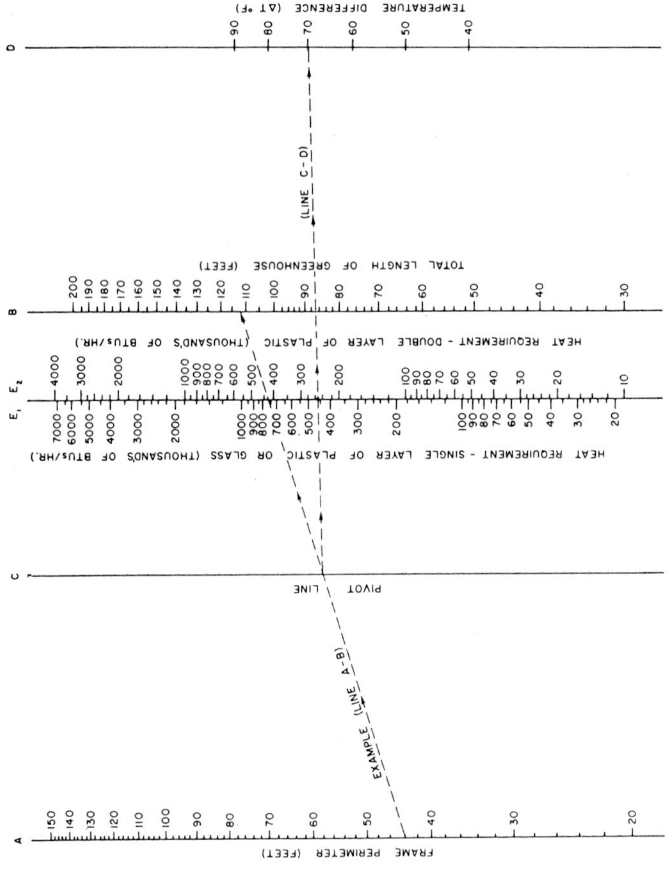

HOW TO USE THE GRAPH AND DETERMINE HEAT REQUIREMENT

1. FROM TABLE I, FIND FRAME PERIMETER OF HOUSE. LOOK UNDER SHAPE OF HOUSE AND WIDTH. MARK FRAME PERIMETER ON LINE A OF THE GRAPH.
2. FROM TABLE II, FIND EQUIVALENT LENGTH OF HOUSE TO BE ADDED FOR END WALLS. ADD THIS TO THE ACTUAL LENGTH OF THE HOUSE. MARK THE TOTAL LENGTH ON LINE B.
3. DRAW A STRAIGHT LINE FROM THE POINT MARKED ON LINE A TO THE POINT MARKED ON LINE B.
4. DETERMINE DIFFERENCE IN OUTSIDE AND INSIDE TEMPERATURE AND MARK ON LINE D. SEE TABLE III FOR NORMAL WINTER DESIGN TEMPERATURE.
5. FROM THE POINT WHERE LINE A-B CROSSES LINE C, DRAW A STRAIGHT LINE TO POINT ON LINE D, LINE C-D.
6. READ HEAT REQUIREMENT ON LINE E OR E_1 WHERE LINE C-D CROSSES.

EXAMPLE:

1. GREENHOUSE (GABLE TYPE) 28 FEET x 100 FEET. OUTSIDE DESIGN TEMPERATURE 0°, INSIDE DESIGN TEMPERATURE 70°
2. FRAME PERIMETER - 44 FEET. (TABLE I)
3. EQUIVALENT LENGTH 12 FEET, TOTAL LENGTH 112 FEET. (TABLE II)
4. TEMPERATURE DIFFERENCE 70°. (70° - 0°)
5. FOR SINGLE LAYER OF PLASTIC OR GLASS, HEAT REQUIREMENT - 460,000 BTU/HR. FOR DOUBLE LAYER OF PLASTIC, HEAT REQUIREMENT - 250,000 BTU/HR.

COOPERATIVE EXTENSION WORK IN AGRICULTURE AND HOME ECONOMICS
DEPARTMENT OF AGRICULTURAL ENGINEERING
COLLEGE OF AGRICULTURE
UNIVERSITY OF KENTUCKY &
U.S. DEPARTMENT OF AGRICULTURE COOPERATING

HEAT REQUIREMENT FOR GREENHOUSES

Designed By G.H.J. & J.N.W.	Sheet 1 of 2
Checked By: H.H. G.A.D.	Plan Number:
Drawn By [illegible]	KY. 11.811-4
Date: JAN 65 REV. 1-71	

Fig. 4-19. Estimating heat requirements for greenhouses (Walker and Duncan, 1973a)

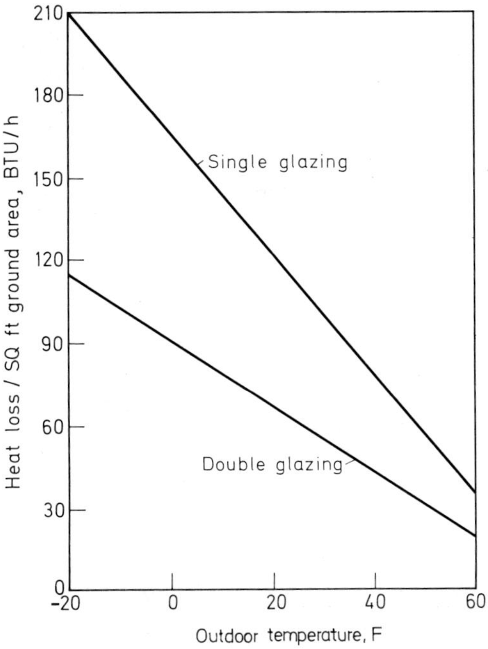

Fig. 4-20. Heat loss from a single and double polyethylene covered greenhouse with a wall-to-floor area ratio of 1.5 (Ashrae, 1974)

4.3.1.3 Sample Heating Problem

$$[A_1+(A_2\times R)]\times \Delta T\times G\times W\times C \qquad = \text{BTU heat loss h}^{-1}$$

A_1 = Exposed greenhouse cover area (FRP)

2 side walls 100′ × 5′ × 2	= 1000
2 end walls 36′ × 5′ × 2	= 360
2 gable ends 36′ × 4.5′ × 2	= 324
Roof 100′ × 20′ × 2	= 4000
Total	5684 ft^2

A_2 = foundation wall area

2 side walls 3′ × 100′ × 2	= 600
2 end walls 3′ × 36′ × 2	= 216
Total	816 ft^2

R = (Corrugated asbestos cement)	= 0.94
ΔT = (Temperature difference)	= 60° F.
G = (Coefficient of transmission)	= 1.11
W = (Wind factor, 20 mph)	= 1.04
C = (Construction factor, metal, and FRP)	= 1.00
[5684.0 + (816 × 0.94)] × 60° F × 1.11 × 1.04 × 1.00	= BTU heat loss h^{-1}
6451 × 60 × 1.11 × 1.04 × 1.00	= 446 822 BTU heat loss h^{-1}
Infiltration 10% additional	= 44 682
Total heat loss	= 491 504 BTU h^{-1}

4.3.2 Heating Equipment

The principles involved in the use of heating water in an open copper kettle with a four inch brass distribution pipe running the length of a Roman greenhouse; or the use of a wood or coal furnace with a brick horizontal chimney constructed through a house for heat transfer purposes (Taft, 1926), are still used in greenhouses today. However, the techniques of distributing heat have been constantly revised through the years.

4.3.2.1 Boiler Types

Most larger commercial greenhouse ranges are heated with some type of boiler system. Boilers may be classified in two general groups (Carrier, 1966).

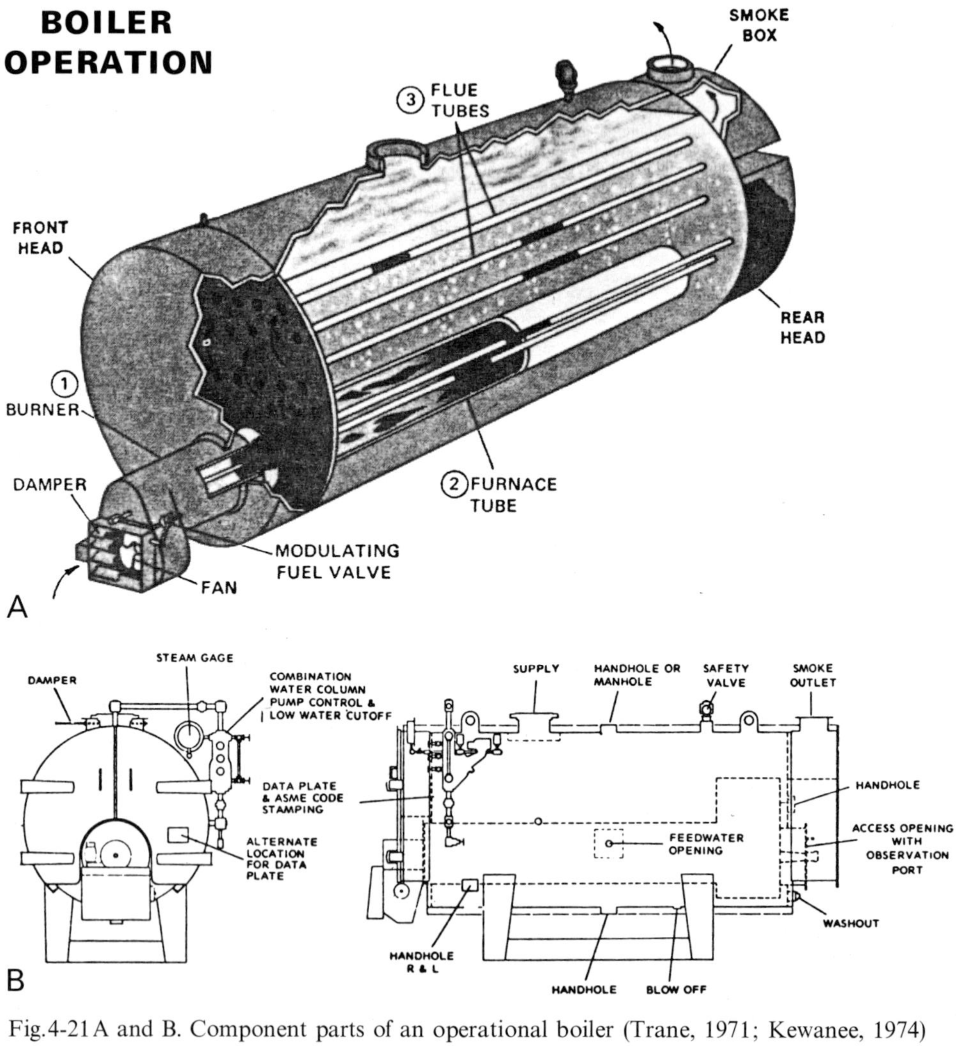

Fig.4-21 A and B. Component parts of an operational boiler (Trane, 1971; Kewanee, 1974)

Section Cast Iron. Suitable for steam generation or hot water heating. A steam fitter is required to assemble the sections, but the overall cost is usually less than steel boilers of the same capacity. If more boiler capacity is required, additional sections may be added. They are normally limited to 15 lbs in $^{-2}$ steam pressure. Any fuel can be used.

Steel Firebox Boilers, Fire-tube or Water-Tube. The fire tube steel boilers have their combustion gases passing through tubes surrounded by circulating water. It is a packaged system having all components including burner, boiler, controls, and auxilliary equipment attached. Fuels can only consist of oil or gas or combinations. The component parts can be seen in Figure 4-21.

The water-tube steel boilers have combustion gases circulating around the tubes and water passing through the tubes. Packaged units are designed principally for oil or gas or combinations, but could be adapted to solid fuel. In the era of limited fuel sources, it would be wise for the greenhouse manager to consider a type of boiler that can burn all fuels, with only slight modifications. Even though a coal stoker does not fit the lifestyle of many greenhouse operators, it would be wise for anyone contemplating expansion to consider a fire box type installation.

Boilers are also classed as natural or forced draft types (Trane, 1971). A natural draft system involves the difference in weight of the column of flue gas inside the chimney and an equal column of air outside. The "lift" created by the hot, expanded flue gases create a negative pressure within the combustion system and induces a draft, bringing in outside air.

The forced draft system has a blower in front of the burner which produces a slight positive pressure in the stack. The air moving through the combustion chamber can be easily controlled in a forced draft system and permits a smaller diameter and shorter stack.

Boilers, especially forced draft types, are fitted with an ignition and combustion flame control system. Fuel injection, purging, pilot lights, ignition, and shut down are all controlled automatically. If any phase malfunctions, alarms sound and the unit shuts down.

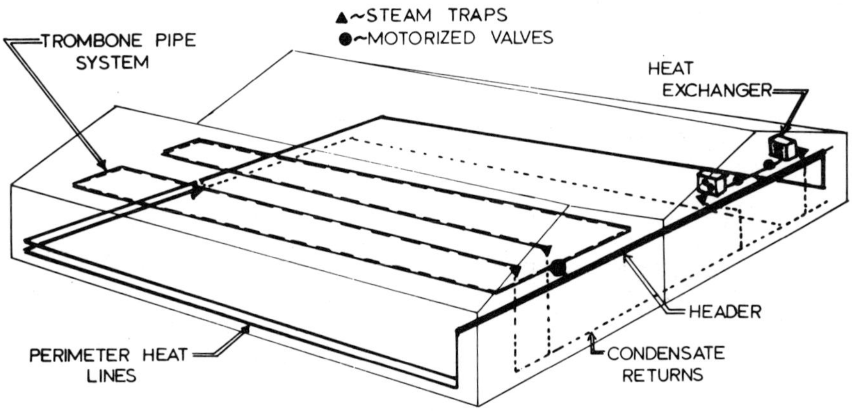

Fig. 4-22. Simple steam distribution system with return lines. *Left house:* pipe design; *right house:* heat exchanger system

4.3.2.2 Boiler Designs

Boiler designs are further catagorized as low or high pressure steam and gravity or accelerated hot water.

Low pressure boiler systems operate at 15 lbs in^{-2} pressure. Steam is carried by "mains" from the boiler to the outlying houses and returned in the form of condensate (Fig. 4-22). Steam pressure in the lateral heat pipes is approximately 5 lbs in^{-2} which provides a pipe temperature approaching 108° C (227° F). Steam traps allow condensate and not free steam to enter the "return" lines, where it usually flows by gravity to a condensate tank and is pumped back into the boiler.

High pressure boilers generate steam from 80 to 100 lbs in^{-2} and the pressure is reduced to the desired levels in adjacent houses. Pipe-steam heating systems provide nonuniform temperatures in greenhouses unless arrangements are properly designed.

Gravity flow hot water systems have been improved by incorporating pumps to aid in water circulation. Many greenhouses constructed in the early 1900 s used a gravity hot water system, but they left much to be desired in the control and distribution of heat. All pipes had to be properly sloped and large amounts of radiation surface were required because of the 180° F water temperature.

High temperature hot water systems operate from temperatures of 180–400° F and pressures of 15–233 lbs in^{-2}. For greenhouse use, water temperatures usually

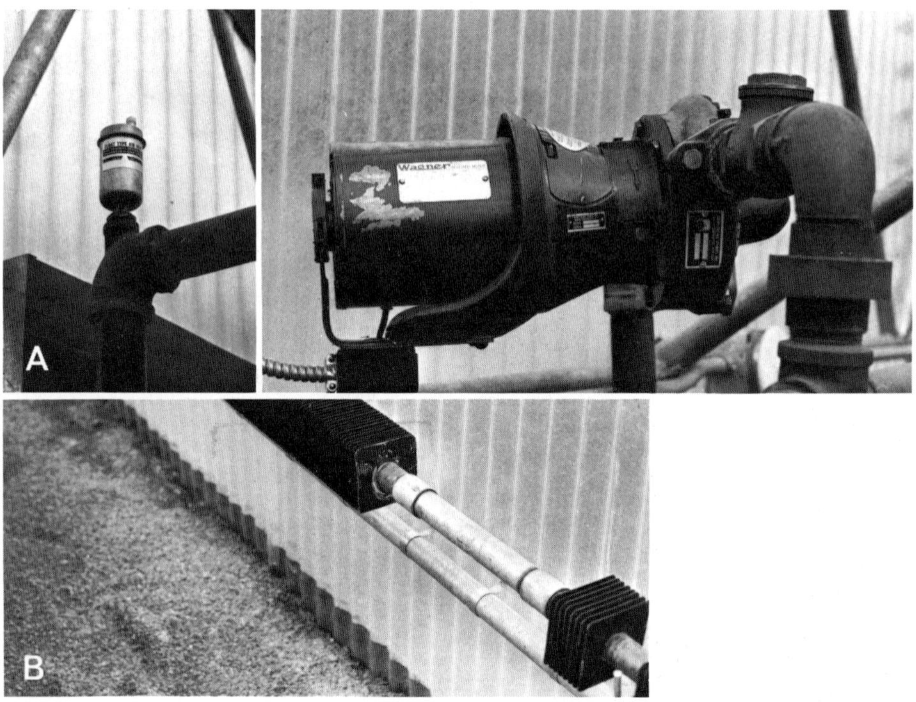

Fig. 4-23. (A) Pump and air vent used in hot water heating systems. Pump is in lieu of automatic valves. (B) Finned steel pipe. (C) Steam traps used at the end of steam lines or downstream from heat exchangers

Fig. 4-23 C

Table 4-2. Conversion factors for boiler ratings (Kewanee, 1972)

1 Boiler Horsepower = 34.5 lbs of steam (from and at 100° C)
1 Boiler Horsepower = 29.6 lbs of steam (10 lb gauge pressure)
1 Boiler Horsepower = 33472 BTU per h
1 Boiler Horsepower = 139 ft^2 of steam radiation
1 Boiler Horsepower = 223 ft^2 of hot water radiation
1 ft^2 equivalent direct radiation (steam) = 240 BTU per h
1 lb of steam (from and at 100° C) = 970.2 BTU
1 lb of steam (10 lb gauge-feed water 15.5° C) = 1131 BTU
1 Therm = 100000 BTU per h

range from 230° to 240° F. Pumps (Fig. 4-23), instead of automatic valves, are installed in key locations and are operated by thermostats. Some hot water boilers can be a source of enough steam for soil pasteurization purposes.

4.3.2.3 *Boiler Ratings*

Until recent years, the size of a boiler has been denoted by boiler horse power (BHP), which was arbitrarily set as equal to the evaporation of 10.0 lbs of water per hour from an initial temperature of 100° F, to steam at 70 lbs in $^{-2}$ pressure. It was later defined as the evaporation of 34.5 lbs of water per hour from a temperature of 212° F into dry steam at the same temperature. Standardized ratings have been established and boiler capacities are now expressed in ft^2 of steam or hot water radiation and in BTU per h (Table 4-2; Steam, 1955; Carrier, 1966; Kewanee, 1972).

4.3.2.4 *Operating the Boiler*

The boiler is probably the most important piece of equipment in the greenhouse and thus requires considerable care. The following areas must be considered: (Kewanee, 1974).

1. Pressure and temperature controls—keep pressures and temperatures only high enough to carry the load. Excessive pressures are wasteful.

2. Flue cleaning—depends on fuel used. If combustion is complete, flues will not require frequent cleaning. If the stack temperature is 100° F higher than it was when the boiler was new and firing at the same rated output, clean it!

3. Combustion air—must be readily available. If boiler room is tight, an opening to the outside should be made.

$$\frac{\text{BHP}}{23} = \text{ft}^2 \text{ of opening required.}$$

4. Water levels—must be observed in the water guage glass as a matter of habit. Low water will prevent ignition. The low water cut off system should be connected to an alarm.

5. Blow off—a means of getting rid of rust colored water and sediment. Do not drain clear water.

6. Water treatment—chemicals added to boiler water to protect against corrosion and pitting. A boiler water of pH 11.0 minimizes corrosion. Sodium sulfite is one of the materials used to combat oxygen. Note that chromates are toxic and should not be used where steam pasteurization is employed. Sodium chromate would give a yellow color to the boiler water which can be observed in the guage glass (Boiler, 1972).

7. Scale—not common in low pressure boilers unless excess makeup water is required. Blow down and yearly "washing out" at inspection time will control most scale problems, especially where hard water is used. Scale 1.5 mm thick on any heating surface will cause a 11 to 13% decrease in BTU output (Kewanee, 1972).

4.3.2.5 Distribution of Heat

The standard method of distributing heat from steam or hot water has been through black steel pipe. Heating installations of today often consist of pipe, finned pipe (Fig. 4-23) and/or forced air heat exchangers (Fig. 4-24). Heat exchangers are rated in BTU h^{-1} and can be easily adapted to steam or hot water

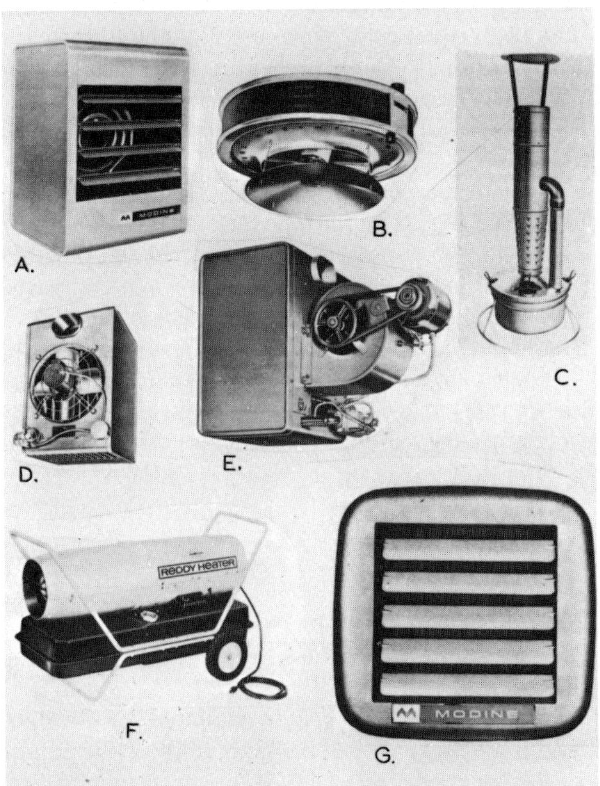

Fig. 4-24A–G. Unit heaters used in the greenhouse. (A) Electric, 17000–85000 BTU h^{-1}, (B) steam or hot water, vertical unit; (C) smokeless—portable radiant oil heater; (D) propeller type gas fired unit; (E) blower type gas fired unit; (F) forced air fuel oil or kerosene portable; (G) steam or hot water, horizontal unit (non oil heaters courtesy Modine Manufacturing Corp.)

Table 4-3. BTU emission rate per lineal foot per hour of pipe in 70° F still air. (Adapted information from IBG, Sun-Fin Company and Sterling Radiation Company)

	Steam	Hot water[a]		
	215°	180°	200°	220°
$1^1/_4''$ aluminum pipe 4″ fins	1440	1138	1264	1390
2.0″ aluminum pipe $4^1/_2''$ fin	1260	995	1106	1216
$1^1/_4''$ galvanized pipe with round galvanized fins	820	570	710	860
2″ galvanized pipe with round galvanized fins	920	630	790	970
$1^1/_4''$ black pipe with $3^1/_4''$ fins	960	660	830	1010
2″ black pipe with $4^1/_4''$ fins	1200	830	1030	1260
$1^1/_4''$ black steel pipe	203	112	158	190
2″ black steel pipe	271	152	190	230

[a] Ratings applicable to water flow rates of three or more ft per s.

systems. When modulating valves are used in conjunction with heat exchanger units, excellent temperature levels are usually maintained. Finned pipe is available with steel or aluminum fins (Table 4-3) and often provides the answer for "hard to heat" areas related to conventional piping. Steel pipe for heating is still popular in many parts of the world. Hoare and Morris (1956a) found that the use of 4 in pipe for hot water heating was not justified and recommended the 1 in and $1\text{-}^1/_2$ in diameter pipes as a replacement. The smaller pipe reduced the thermal inertia which enabled better control of temperatures at all levels of heat input. The available radiation from black pipe is based on the pipe surface area (Table 4-3). Hoare and Morris (1956b) reported that aluminum paint on black pipe reduced the effective emissivity 50% and the total heat output 25% when compared to unpainted pipe.

The proper distribution of steam heat is directly related to the working condition of steam traps (Fig. 4-23). Each heat exchanger or pipe "run" should be "trapped" individually in order to approach constant temperature levels. Even though traps require little maintenance, they should be checked periodically and cleaned.

4.3.2.6 Gas Fired Unit Heaters

The early 1950s brought a revolution to greenhouse heating systems with the introduction of the gas fired unit heater (Fig. 4-24D). Growers found the cost of installation of these units was approximately 50% less when compared to the traditional boiler and pipe system.

Many growers obtain the lowest priced unit heater available and seldom consider the material content of the heat exchanger or the quality of the motor used to drive the fan. Heaters with stainless steel heat exchangers and burners

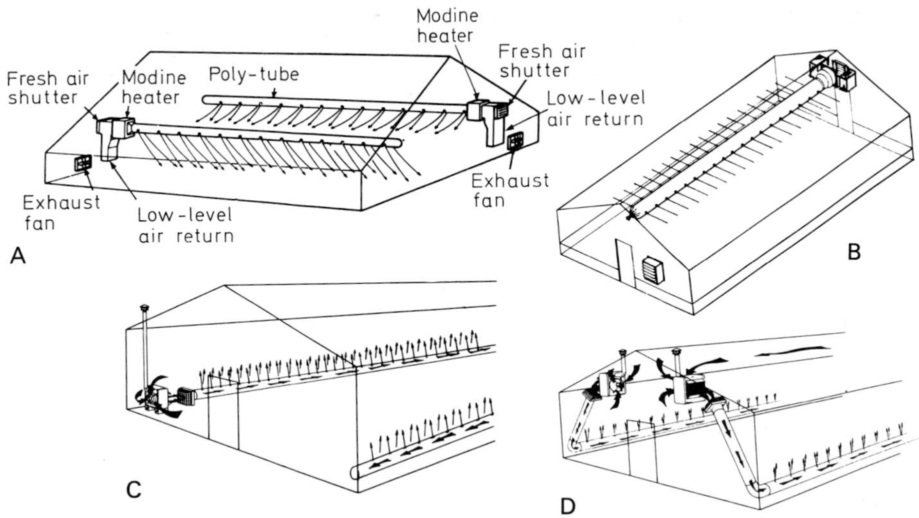

Fig. 4-25 A–D. Forced air distribution systems for heating. (A) Modine Flora-Guard System; (B) Acme Fan-Jet System; (C) Acme ground level system; (D) Acme downflow system. (Courtesy Modine Manufacturing Company and Acme Engineering)

have been introduced (Bishop, 1954), and can be obtained with output ratings ranging from 24000 to 320000 BTU h^{-1}. Totally enclosed, continuous duty, ball bearing motors should be considered for all unit heater installations.

Earlier installations of gas fired unit heaters often created temperature differences of 5–15° F in many greenhouse sections. The advent of the poly-tube, to distribute the warm air, reduced greenhouse temperature differential problems by providing proper circulation patterns.

The Modine Manufacturing Company designed a combination heating and ventilating system, Figure 4-25 A, which is used for distributing both fresh and heated air as well as recirculating purposes. Another proven system, Figure 4-25 B, designed by the Acme Engineering Company (Controlled Environment, 1975), can be adapted to a number of heating and ventilating systems. The latest introduction of heating and circulation equipment by Acme is the perimeter Fan-Jet system (Fig. 4-25 C and D).

a) Combustion. Combustion in gas fired unit heaters is often a problem, especially in very small greenhouses. During winter months when all cracks are frozen shut, there is insufficient air (10 ft^3 h^{-1} of air per ft^3 h^{-1} of natural gas) for complete combustion and plant damage often occurs (see Chap. 8). Hanan (1972) found that the ethylene content of greenhouse atmospheres increased when polyethylene air distribution tubes were not used. The air movement created by the individual unit fans apparently contributed to improper air circulation patterns, leading to incomplete combustion.

b) Gas heater check lists (1970) are an important asset to the greenhouse manager. When a boiler is used for heating, only one burner, stack, and pilot light system is involved. A 100000 ft^2 greenhouse range would have approximately 50 gas fired unit heaters to maintain. Check list considerations are:

1. Clean accumulated dust and dirt around burner section.

2. Adjust flame. It should be blue in color. Any yellow flame indicates an improper air-gas mixture. To prevent incomplete combustion, at least one square inch of free air inlet should be used per 2000 BTU per h input.

3. Clean all orifices. Make sure they are present and the proper type when installing new units.

4. Flues should be properly connected to the draft hoods. Misfits or cracks can supply toxic gases.

5. Check all wiring and pipe connections. Vibrations created by unbalanced fan blades can cause electrical wires to break and pipes to crack, allowing fuel leaks.

4.3.2.7 Infrared Heaters

Some interest has been shown in the use of infra-red heating systems in greenhouses. There are two general types (Ashrae, 1963).

Radiation heaters with surface combustion, have a ported or metallic screen burner face, where the gas-air mixture burns. It heats to incandesence, releasing heat by radiation. Such units have a surface temperature of approximately 1600° F, are usually unvented with rated inputs up to 50000 BTU h^{-1} per unit. These units, to our knowledge, have not been used extensively in plant environments.

Internally-fired heaters consist of heat exchangers with an exposed surface radiating heat at surface temperatures of nearly 800° F. Reflectors are generally installed above the units to direct radiation downward. Units of this type are available with input ratings up to 125000 BTU h^{-1}.

In 1963, we evaluated an infrared vacuum combustion, gas heating system (Fig.4-26). The system consisted of a combustion chamber and approximately 100 ft of 2 in pipe for radiation purposes. A vacuum pump pulled the products of combustion through the pipe and exhausted them outside at the end of the run. Modifications of the system were needed and no definite recommendations were made.

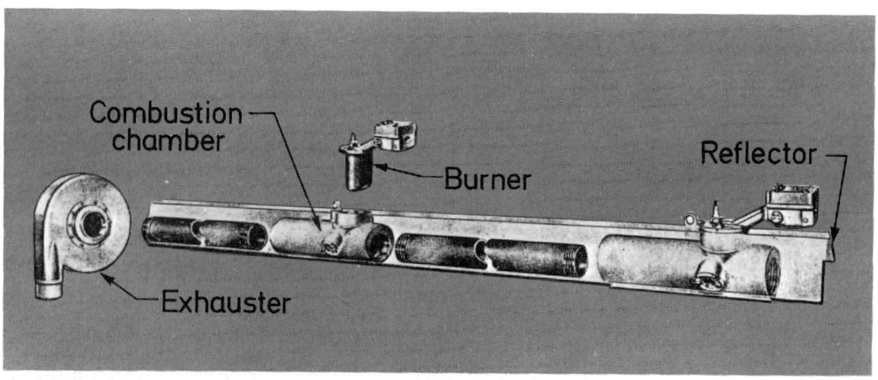

Fig.4-26. Infrared (vacuum combustion) gas heater. Each unit is rated at approximately 40000 BTU h^{-1} (Courtesy Roberts Gordon Corp.)

4.3.2.8 *Characteristics of Fuel*

The greenhouse manager must always be aware of the "heart" of any heating system, the fuel. As the price of fuels continues to climb, and their source becomes questionable, the manager will have to make some hard decisions—namely, raise the selling price of his product to keep pace with inflation; switch to a lower priced fuel; modify the heating system and structure to lower fuel bills or close the doors.

1. Composition. The composition of fossil fuels varies with source, type, and degree of refinement. Natural gas, when mainly composed of methane and ethane, is termed "dry" gas. If other gases such as carbon dioxide, hydrogen sulfide, and nitrogen are present, it is considered a "wet" gas. Further differentiation is made when large quantities of hydrogen sulfide are present; it is then termed a "sour" gas. When only small amounts are present, is it classed as a "sweet" gas (Macrae, 1966; Natural, Gas, 1971).

There have been undocumented reports that possibly sulfur dioxide or hydrogen sulfide damage can occur when sulfur vaporizers, used for mildew control on roses, are operated in houses where gas fired unit heaters are used.

Incomplete combustion and/or improper draft conditions can cause ethylene formation and ultimately plant damage. While many plants are sensitive to ethylene, they are not damaged by prolonged exposures to natural gas (Gustafson, 1944).

Some growers use gas fired unit heaters for CO_2 production during daylight hours. Care should be taken to see that *complete* combustion occurs and no cracks are present in the heat exchanger.

2. Combustion. Combustion of any fuel is directly related to the fuel bill. Therefore, the manager should be acquainted with the general factors involved in combustion, in anticipation of at least a 15% savings in fuel. The following information is partially reproduced from the Kewanee System Data Bulletin (1972).

"Today's practical fuels are composed of carbon, hydrogen, and foreign material. The carbon and hydrogen when chemically combined with the oxygen in the air during the combustion process produce heat and light. The foreign materials, sulfur and noncombustibles as water, nitrogen and ash, serve no useful purpose. The sulfur combines with the oxygen to form sulfur dioxide which has an affinity for water. When sulfur dioxide is dissolved in water, highly corrosive sulfurous acid is formed. It is this sulfurous acid that corrodes smoke pipe, chimneys and furnace, and boiler interiors. Fuel specifications include a clause specifying the maximum sulfur content for this reason. Also, fuel refiners recognize this fact and protect the consumer by keeping the sulfur content at a practical low limit.

"The rate of heat release of a combustion process depends upon the rate at which combustion occurs. This depends upon the rate at which the air is supplied to the reaction. Inasmuch as the various fuels, namely the solid, the liquid and gaseous fuels, contain the desirable combustible elements, carbon and hydrogen, in varying combinations together with the foreign material, the oxygen requirements will vary considerably.

"Complete combustion occurs when all the combustible elements of the fuel combine with oxygen. This is accomplished practically by supplying air in excess of that theoretically required to completely burn the combustibles in the fuel. The more intimate the mixing of the fuel and the air, the less the excess air required to obtain complete combustion. The intimacy of this mixing and thus the completeness of combustion is dependent upon the design of the burner and the combustion chamber.

"When carbon combines with oxygen in the presence of sufficient air, carbon dioxide is formed and heat is liberated. When carbon unites with oxygen in the presence of an insufficient amount of air, carbon monoxide and some carbon dioxide are formed and much less heat is liberated. If the amount of air is too little, some of the carbon will not burn at all, giving a smoky fire. Combustion in this case is said to be incomplete and the difference in the heat released is called the loss due to incomplete combustion.

"The hydrogen in the fuel combines with the oxygen in the air to form water vapor and produce heat.

"Practical combustion requires excess air to assure complete combustion; however, the greater the excess air, the greater the heat carried away by it with the products of combustion. Also, the higher the temperature of these gases leaving the stack, the greater the heat loss.

"From this it can be seen that an analysis of the flue gas and its temperature would be of value to adjust a burner to operate at optimum efficiency. An analysis of the flue gas would establish what percent carbon dioxide is being formed and thus the completeness of combustion. The carbon monoxide percentage will further tell how completely the fuel is being burned. The oxygen percentage will show how much excess air is being delivered to the combustion process. Knowing the amount of excess air being supplied, the loss due to heating this air can be calculated. If an analysis shows the presence of both carbon monoxide and excess oxygen in the flue gas it can be suspected that the fuel and the air are not being mixed intimately enough or not enough time is being allowed to burn the fuel completely.

"The percentages by volume of carbon dioxide, carbon monoxide and oxygen can be determined with an Orsat flue gas analyzer. There are kits available for the determination of carbon dioxide alone. To go with these kits, there are usually graphs or tables to determine the combustion efficiency from the carbon dioxide reading and the flue gas stack temperature. By properly interpreting the reading from either of these instruments, scientifically accurate adjustments can be made on the burners."

3. Equipment Efficiency. All types of heating equipment have established minimums for burning fuel. The American National Standards Institute has developed minimum standards for each piece of gas fired heating equipment. A boiler or gas fired unit heater installed outside a building has a 5% "jacket" loss and cannot be less than 75% efficient. A unit operating inside utilizes the jacket loss and therefore must operate at a minimum efficiency of 80%, which is considered for computations:

Sea level name plate

Rating:	300000	BTU	
	0.80		
	240000	BTU	corrected input rating.

4. *Fuel Heat Value.* The heat value of fuels, Figure 4-28, varies with composition as well as the altitude where they are being used. Using natural gas as an example and disregarding composition, a cubic foot of natural gas at sea level contains approximately 1000 BTUs. At 5000 ft elevation, a cubic foot of this gas would contain 840 BTUs. The difference may be explained by using a one cubic foot elastic container filled with natural gas at sea level (14.70 lbs in $^{-2}$). When the container is moved to an elevation of 5000 ft (12.39 lbs in $^{-2}$), it will expand because of the pressure difference.

$$\text{Thus:} \quad \frac{12.39}{14.70} = 0.84 .$$

When one cubic foot of gas is removed from the expanded container at 5000 ft elevation, it will contain 840 BTUs.

$$\text{Thus:} \quad \begin{array}{ll} 1000 & \text{BTU—sea level} \\ \underline{\quad 0.84} & \\ 840 & \text{BTU—5000 feet elevation} . \end{array}$$

At 5000 ft elevation a gas fired unit heater 300000 BTU h^{-1}, which has a natural draft or gravity vent system, will have a derated input of 240000 BTU h^{-1}. Whereas a forced draft unit with adequately sized blowers may have the capability of the full sea level rating. The relationship of equipment efficiency and heat value can be expressed:

$$\text{Rated input} = \frac{\text{actual output}}{\underset{(80\%)}{\text{efficiency of unit}} \times \underset{(80\%)}{\text{altitude correction}}}$$

$$= \frac{\text{output}}{0.64}$$

$$300000 = \frac{X}{0.64}$$

$$\text{output} = 192000 \;\; \text{BTU h}^{-1} .$$

5. *Fuel Costs.* It is often difficult to change the type of fuel being used. The individual contemplating expansion, especially with boilers, should prepare his heating system to use both solid and liquid fuels. Walker and Duncan (1973 a) prepared graphs showing the annual cost to heat a greenhouse (Fig. 4-27), and also the costs of various fuels per equivalent heat produced (Fig. 4-28). From these graphs, the manager not only can estimate his fuel costs, but follow the effects of price changes on the value of heat received from the various materials.

4.3.3 Greenhouse Temperature Patterns

Greenhouse growers prior to the mid-1900s were well aware of the temperature patterns in their houses. They knew a certain area was colder or warmer than

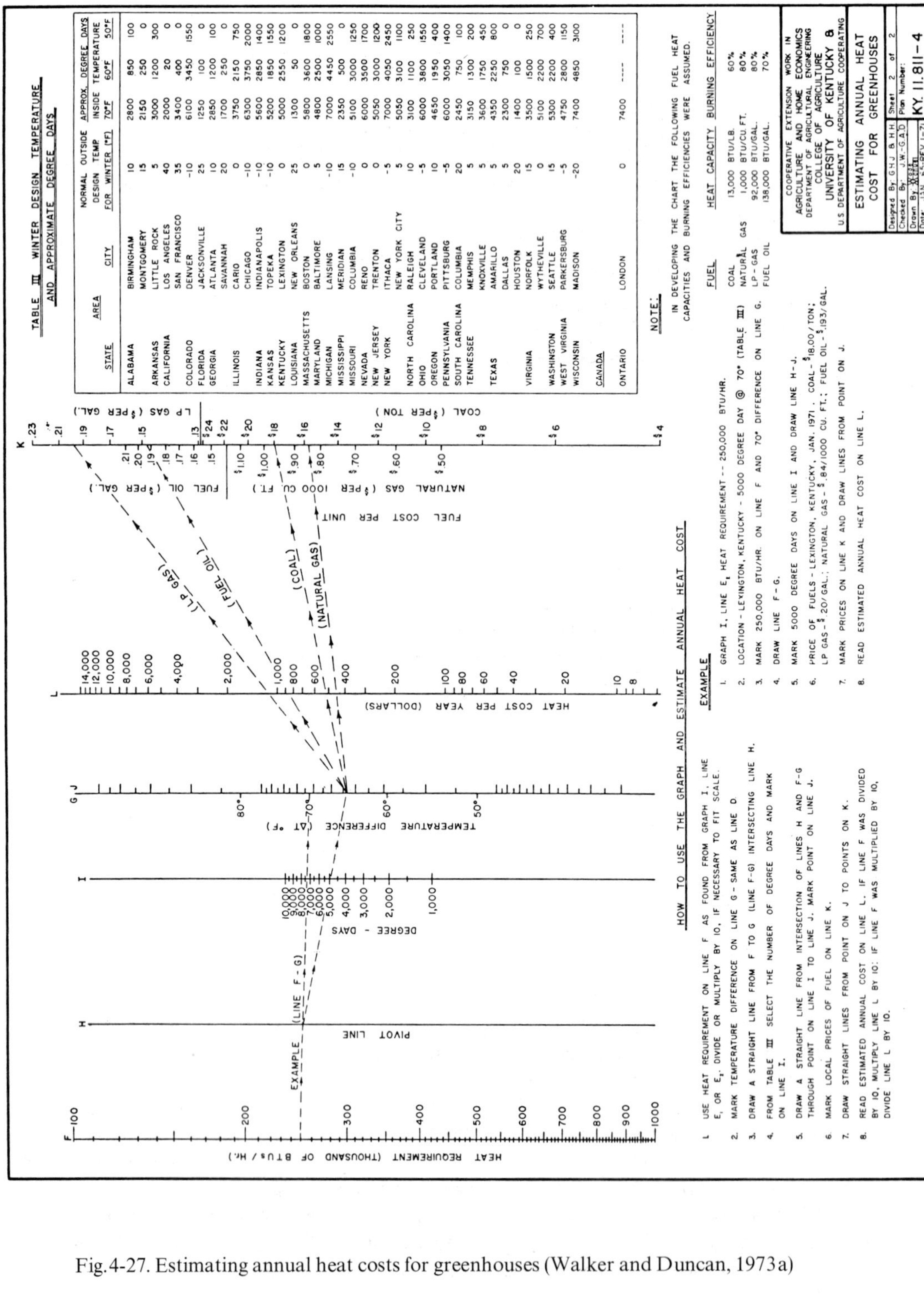

Fig. 4-27. Estimating annual heat costs for greenhouses (Walker and Duncan, 1973a)

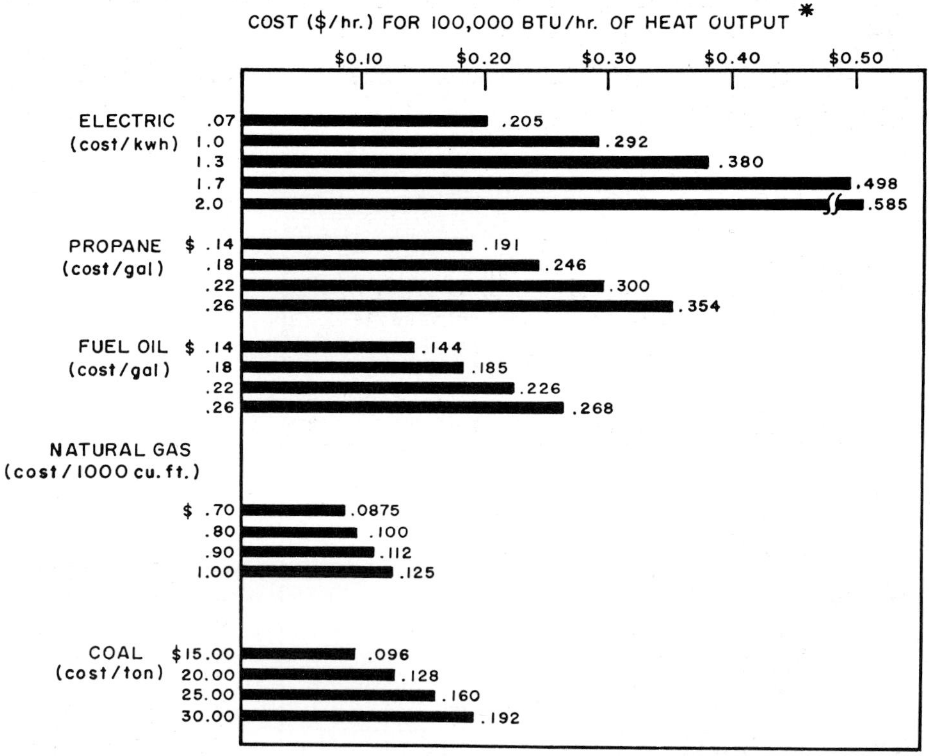

*FUEL HEAT VALUES AND COMBUSTION EFFICIENCIES ARE AS FOLLOWS:

FUEL	ESTIMATED EFFICIENCY PERCENT	HEAT VALUE (BTU PER UNIT)
COAL (BITUMINOUS)	60	13,000 BTU/lb.
NATURAL GAS	80	1,000 BTU/cu. ft.
PROPANE	80	92,000 BTU/gal.
FUEL OIL, NO. 2	70	138,000 BTU/gal.
ELECTRICITY	100	3,413 BTU/kw. hr.

Fig. 4-28. Comparison of fuel costs for equivalent heat production (Walker and Duncan, 1973a)

others and placed their plants accordingly. The so-called modern grower has gone along with the new heating and air circulation systems in anticipation that uniform temperature distribution was being achieved.

Hanan (1958) used circulating fans in small 15′ × 18′ steam heated compartments while conducting temperature research on carnations. Vertically, he found almost a 1° F rise in temperature per foot above the ground. With the circulating fans off, there was a 10° F differential within a six foot vertical distance.

Outstanding contributions on greenhouse temperature patterns came from Carpenter and Bark (1967a). Using thermocouples, they measured vertical and horizontal air temperatures in greenhouses heated with several systems:

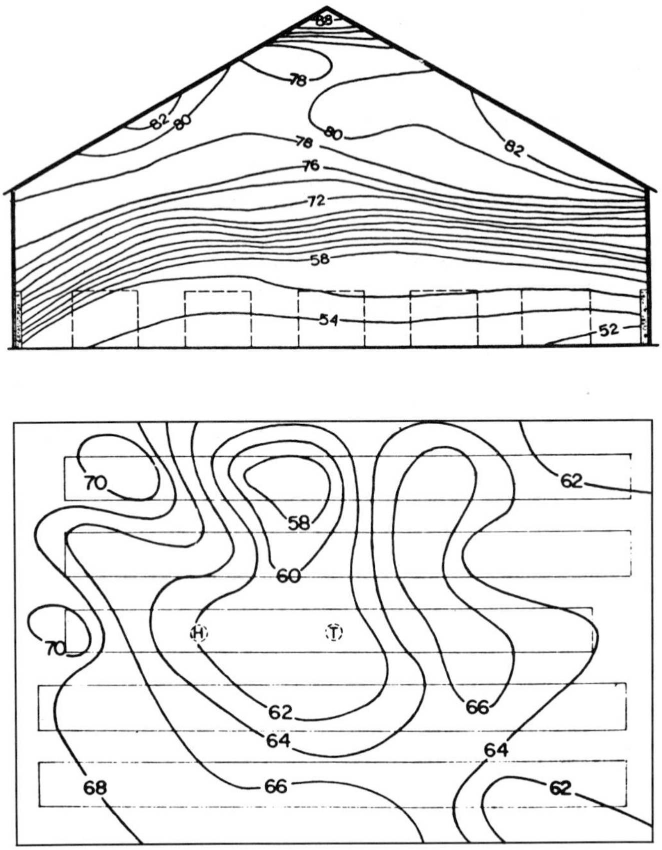

Fig. 4-29. *Top:* Vertical temperature patterns produced by a unit heater. Note isotherms above bench. *Bottom:* Horizontal temperature produced with simultaneous operation of vertical unit heater "H" and air circulating fan "T". Note disorganized warm and cool spots (Carpenter and Bark, 1967 c)

4.3.3.1 Vertical Discharge Units

Figure 4-29 shows temperature gradients of 20° F from soil level to the top of the canopy of 42 in tall plants. The coolest air was found under the operating unit heater and the highest temperatures at plant level were found at the greenhouse perimeter (Carpenter and Bark, 1967 c).

4.3.3.2 Horizontal Unit Heaters

Horizontal unit heaters were evaluated (Carpenter et al., 1970) as to the temperature patterns created by a single and dual system (Fig. 4-30). It was found that unit heater fans should be operating continuously at night when heat is needed, result-

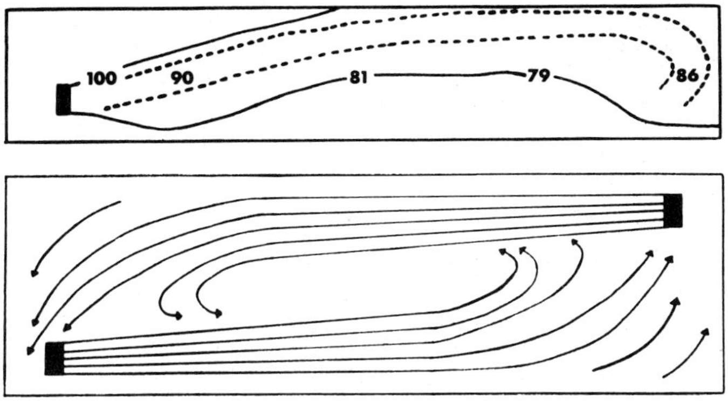

Fig. 4-30. *Top:* Temperatures at a 6-ft height in heating a 75-ft greenhouse with one horizontal unit heater. The warm airstream gradually rises as it moves the greenhouse length, turning downward near the opposite end wall and returning to the heater at plant height. *Bottom:* A circular air pattern was developed in a 30 × 100-ft greenhouse with two horizontal discharge heaters located five feet from the opposite end walls when the forced airstreams were complementary (Carpenter et al., 1970)

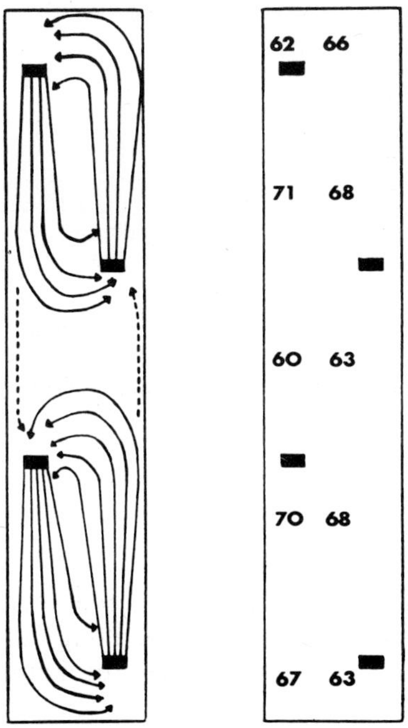

Fig. 4-31. Greenhouse temperatures during heating a 37 × 160-ft greenhouse with four horizontal unit heaters. Heaters were placed 10 ft from each end wall at an 8-ft height, with the center heaters 60 ft from the end walls, also at 8 ft. *Left:* The air circulation pattern; *right:* the temperature at one foot above the floor (Carpenter et al., 1970)

ing in uniform air temperature patterns. Improperly placed units created conflicting air circulation patterns (Fig. 4-31). The authors suggested placing the two center units opposite each other creating one continuous circular pattern. It was concluded that horizontal unit heaters can heat uniformly if a single air circulation pattern is developed for the whole greenhouse at plant height.

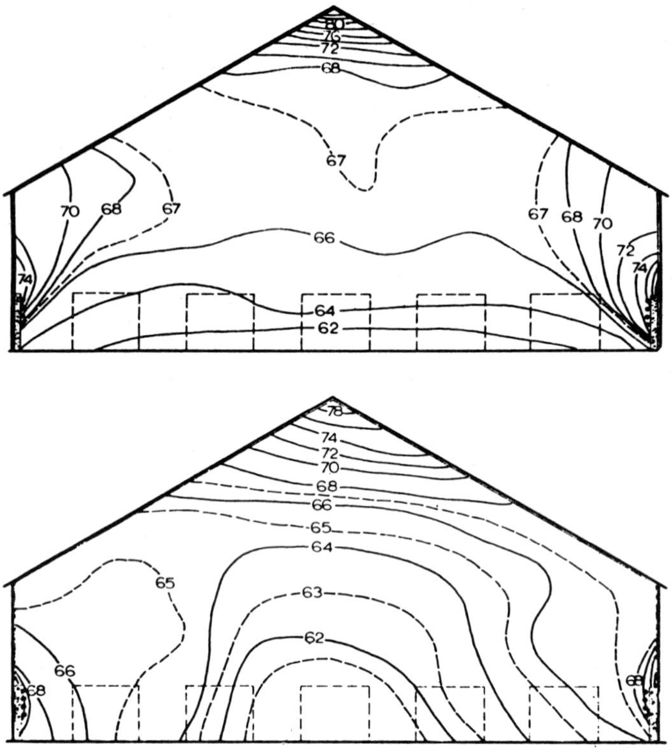

Fig. 4-32. Vertical temperature pattern produced by steam coils along sidewalls. *Top:* Without air circulation fan; *bottom:* with air circulation fan. The combination of turbulators with steam side coils has modified the temperature patterns of each system (Carpenter and Bark, 1967b)

4.3.3.3 Steam Pipe Systems

Steam pipe systems created the most uniform temperature patterns, (Fig. 4-32 and 4-33) in greenhouses. As warm air rises, it creates a canopy over the plants (Gray, 1956; Carpenter and Bark, 1967b). Even though the temperature pattern is relatively uniform, high temperature air accumulates in the gable, creating a condition of high heat loss during cold weather. Many of the older greenhouse heating pipe designs included a trombone or horizontal pipe arrangement over the plants, at eave height, to assure an adequate blanket of heat (Fig. 4-22).

Some pipe heating systems have been designed with pipe on the side walls, over the crop and under or beside benches. Such a system makes more efficient use of heat but still lacks a totally uniform distribution of air temperatures (Carpenter and Bark, 1967e).

4.3.3.4 Air Mixing Fans

Air mixing fans of the vertical movement type, have not effectively mixed the air and created uniform temperatures when the heat is on (Carpenter and Bark,

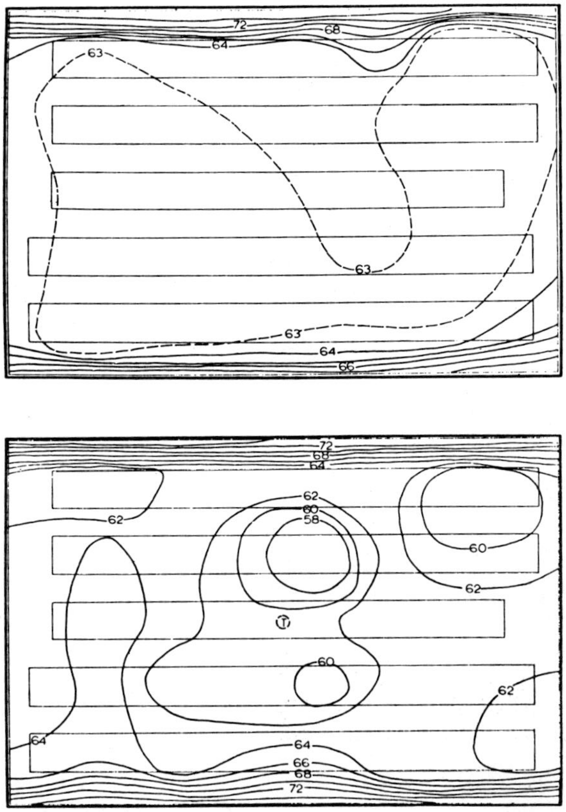

Fig. 4-33. Horizontal temperature pattern produced by steam coils along sidewalls. *Top:* Without air circulation fan; *bottom:* with air circulation fan T. Dominance of the steam sidewall coils along the greenhouse perimeter and turbulator in the other areas is apparent. A greater temperature gradient has resulted from the combination of the two systems (Carpenter and Bark, 1967b)

1967d, e). Unpublished studies showed that centrally located turbulator fans created cold areas in the center of the greenhouse with warmer air at the perimeter. Carpenter and Bark (1967e) (Fig. 4-33) observed similar conditions as well as poor circulation patterns when fans were located beneath bench coils.

4.3.3.5 Poly-Tube Circulation Systems

Poly-tube circulation systems appear to provide the best distribution of warm air in greenhouses. A definite relationship exists between the tube size, number and size of holes, and the velocity and amount of air required to heat or cool a greenhouse (Edlin, 1963). Bark and Carpenter (1969) found relatively uniform temperatures throughout the greenhouse, except immediately under the unit heater (Fig. 4-34).

Duncan and Walker (1973) investigated the use of single and dual poly tubes in greenhouses and found that two or more tubes should be used in houses 30 ft or more wide.

The use of forced air through poly tubes for heating purposes is being developed rapidly (Fig. 4-35). Further considerations on heating and cooling designs will be discussed later in this chapter.

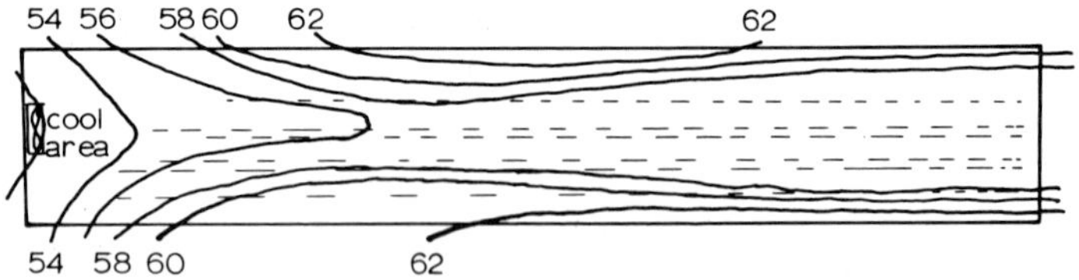

Fig. 4-34. Horizontal temperature pattern produced at bud height (4 ft) by horizontal unit heater with the polyperforated tubing running the length of the greenhouse. Note relatively cool area produced near the heater during the "on" portion of the heating cycle (Bark and Carpenter, 1969)

Fig. 4-35. Poly-tube heating system for tomatoes (Courtesy Ickes Braun Glasshouses)

4.3.4 Heat Conservation

The energy crisis of the mid-1970s was felt around the world by the greenhouse industry as the price of fuels were doubled, tripled, and in some instances quadrupled within months. Grimmer (1975) evaluated the operating costs of a large group of Michigan greenhouse growers and found that 7.95% of all costs of operation was attributed to heating their facilities. A pot plant grower found that the cost of heating his poinsettia crop had increased 40% from 1972 to 1974 (Voigt, 1976). Voigt, however, questions the relationship of fuel costs to the yield and selling price of the products and thus indicates energy problems may be confused with marketing techniques.

No matter how one looks at the fuel situation, it is one of the roles of management to keep all components of overhead in perspective and heat conservation is becoming a major factor.

4.3.4.1 Management Practices

Most greenhouse operators are aware of the need to conserve heat, yet there are several who are not realistic and fail to consider various aspects of heating. Record keeping as discussed in Chapter 11 will quickly bring the relationship of heating costs into perspective with profit for any greenhouse manager.

The type of structure, temporary or permanent, added to a greenhouse range should also be considered. Wright (1921) indicated there was little difference in the surface area of three small greenhouses and a single house providing they cover the same ground area and have an equal roof pitch. He further stated that the environment in the larger single house could be maintained easier. Stickler (1975) calculated the ratio of the above ground surface area and the ground area covered by several greenhouses in West Germany. He reported that individual houses are cheaper to construct, but a four-sectioned ridge and furrow greenhouse requires 23.5% less heat than four single section houses of the same type covering the same ground area.

Faulty steam traps, with an $^1/_8$ in orifice, will use 35 lbs of steam per hour, and on a 40 h per week basis, this represents 7 tons of coal per year (Winspear, 1974). Any steam leak of appreciable size contributes to higher fuel costs.

Unpublished data (Goldsberry, 1973), obtained in a commercial greenhouse range, indicated that 22.5% less energy was received under dirty glass than that transmitted through cleaned glass during a 14 day period in late fall. Dirty glass or weathered FRP covers can cause continued heat requirements on some mornings, and earlier operation during late afternoons due to low solar energy transmission conditions. The greenhouse cover should be thoroughly cleaned when heat is required so the benefits of solar heating can be realized.

Steam used for soil pasteurization is often wasted by poor installation of steam covers or application methods. When steam is applied on top of a growing medium, approximately 8–10 lbs of steam per ft^2 of soil area are needed to raise the temperature adequately at a depth of six to nine inches (Winspear, 1974). Four pounds of steam are effective if buried perforated pipes and a good steam cover are used.

The greenhouse manager can quickly determine other areas of heat or fuel conservation if he will only take time to evaluate his production programs. One method is to lower growing temperatures as described in Growers Talks (Harrison, 1973):

"Average monthly outside temperature, 20° F:
    If you cut 65 to 60°, you save 11%.
    Cut 65 to 55°, you save 22%.
    Cut 65 to 50°, you save 33%.
"Average monthly outside temperature of 32° F:
    If you cut 65 to 60°, you save 15%.
    Cut 65 to 55°, you save 30%.
    Cut 65 to 50°, you save 45%.

"Outside temperature, 40° F:
If you cut 65 to 60°, you save 20%.
If you cut 65 to 55°, you save 40%.
"Outside temperature, 48° F:
If you cut 65 to 60°, you save 29%.
Cut 65 to 55°, you save 58%."

One should be aware that decreased growing temperatures usually delay crop maturity, therefore, the overall saving may be substantially decreased.

4.3.4.2 Heat Blankets

Numerous programs have been undertaken by greenhouse researchers to evaluate methods of heat conservation using supplementary interior or exterior covers. West Germany researchers (Stickler, 1975) have evaluated several greenhouse insulation systems. One system, a solid ceiling, of inflated polyethylene tubes (Fig. 4-36), reduced the night time heat loss by 35%. Researchers at Great Britain's National Institute of Agricultural Engineering evaluated numerous materials for potential "heat blankets". They found that the plastic blackout material, normally used for photoperiod control reduced the heat requirement 35% with an outside wind speed of 2–3 $m\ s^{-1}$ (Reducing Glasshouse, 1975). Other studies show that an aluminum polyester reduced heat loss by 50% when compared to 41% for a clear polyethylene (NBM, 1975).

Tristan (1975) installed an aluminized polyester type fabric inside a 32 × 41 ft greenhouse (Fig. 4-37). The blanket, attached to a Simtrac System (Trade Name), was opened and closed on alternate nights throughout the winter months. The amount of condensate obtained from the heating pipes in the greenhouse was

Fig. 4-36. Inflated "spare-thermal" system of poly tubes which makes a false ceiling in a German greenhouse (Stickler, 1975)

Fig. 4-37. Aluminum lined heat blanket in position for heat conservation studies (Tristan, 1975)

correlated with outside temperatures and wind speed to determine the heat savings potential of the cover system. The condensate units (0.5 lb of water = 1 unit) used by Tristan can be converted to BTU h^{-1} based on the conversion factor of 1 lb water evaporated at 212° F is equal to 970.0 BTU. The temperature of the condensate water at the collecting station in his study was 175° F, thus, each pound of water measured at the meter equals 1007.9 BTU. Tristan found that 76.9% of the heat loss from a greenhouse was during the night, which closely agrees with the data obtained by Barnard (1954) and Christensen (1970). The maximum heat savings created by the blanket application on any one night was 55% with no wind and an average of 23% for the duration of the experiment (86 days). The temperatures during this period ranged from −18° F to +57° F. Tristan found that a blanket was extremely beneficial during periods of cold outside temperatures with moderate or low speed winds (Fig. 4-38). It should be noted, however, that higher velocity winds in mountain regions are usually associated with higher temperatures. Thus, the differential between inside and outside temperatures may be reduced from 70° F to 5° or 10° F in a matter of minutes.

Rebuck et al. (1976) conducted experiments using two different materials including the same one used by Tristan. Their data (Fig. 4-39), taken during short periods of no wind and outside temperatures ranging from 28° F to 50° F, showed a total energy reduction potential of 57%.

Some researchers have advocated the use of an exterior curtain above the greenhouse roof to reduce the heat loss. Whittle and Lawrence (1961) found that such a system was not sufficient to reduce heat consumption unless a still air space was present.

The installation of curtains or "blankets" to reduce the thermal transmittance in greenhouses during cold nights has proven worthwhile in regards to lower fuel

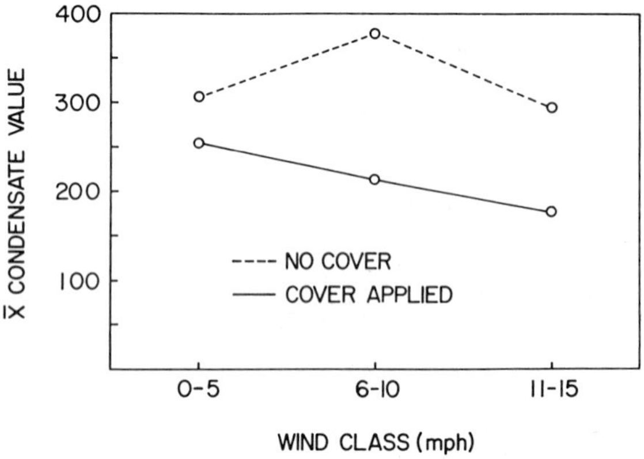

Fig. 4-38. Mean condensate produced, based on grouped wind speeds (Tristan, 1975)

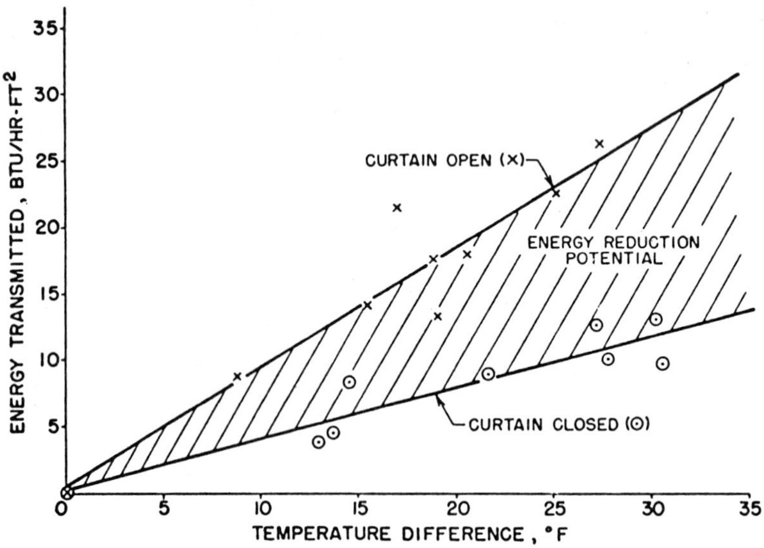

Fig. 4-39. Effects of interior, aluminum-lined curtain on greenhouse heat loss (Rebuck et al., 1976)

bills. New greenhouse facilities should be designed so photoperiod equipment can be used for a dual purpose—black cloth and heat blankets.

The greenhouse manager who does not use photoperiod equipment will need to determine if a heat retention system is feasible. Will a 20% per year savings on fuel warrant the installation? Or can the 20% be obtained in other areas of effective management, at a lower cost (Voigt, 1976)? What if the heat blanket is in use and unexpected heavy snows occur?

4.3.4.3 Double Layer Structures

The air inflated double film structure described in Chapter 3 definitely has a place in greenhouse heat conservation. Sheldrake (1971) observed a 40% savings in fuel

Fig. 4-40. Triple poly vinyl chloride covering over a winter crop in Japan (Nakata, 1972)

with a double layer when compared to a single layer. The British National Institute of Agricultural Engineering (IRG, 1975) has observed a heat savings of 39% with 5.0 m s^{-1} wind speed and 29% at 2.2 m s^{-1}. Axlund et al. (1974) also observed a 40% heat savings during a four month period in San Diego, California. Their work also showed that the distance between the two film layers is not directly related to insulation efficiency. A structure with a two inch spacing between the plastic film layer had a slightly higher heat loss than did the house having the film layers up to eight inches apart.

Other double layer type coverings have been considered for greenhouses. Some have promise (Fig. 4-40). Others are a gamble. Every insulation system must be evaluated for light transmission characteristics. There is a definite area of economics where the sacrifices of light for heat savings can be a demonstration of poor greenhouse management.

4.4 Cooling Greenhouses

Prior to the mid 1950s, little was mentioned in texts regarding principles, methods, and equipment used in ventilating greenhouses. Little did Bailey (1900) realize that some of his 100-word description of ventilation would be the goal of engineers 50 years later. He wrote of ventilation:

"Theoretically, it is employed also for the purpose of introducing chemically fresh air, but with the opening and shutting of doors, and the unavoidable leaks in the house, it is not necessary to give much thought to the introduction of mere fresh air. Ventilating reduces the temperature by letting out warm air and letting

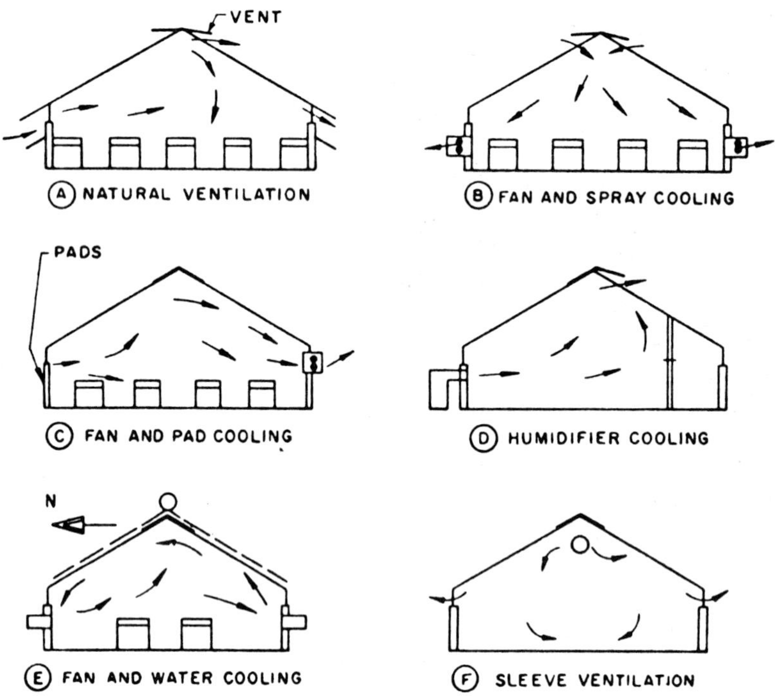

A

Fig. 4-41. (A) Methods of ventilating and cooling greenhouses (Ashrae, 1974). (B) Equipment for operating ventilators; *A* Rack and pinion system; *B* chain-pull vent machine (open worm and gear box); *C* folding or scissor vent arm (courtesy National Greenhouse Co.). (C) Motorized vent unit

in cool air. The air should be admitted in small quantities and at the greatest distance from the plants in order to avoid the ill effects of drafts on the plants. Many small openings are better than a few large ones. Ventilate on a rising temperature."

Until the late 1940s, most greenhouse ventilators, Figure 4-41 A, were operated manually. During periods of cold weather, the greenhouse grower was continuously adjusting top vents to not only maintain the temperature but to try and retain control of the moisture level of the atmosphere. As technology improved and a better controlled environment was desired, the ventilator equipment, Figure 4-41 B, was modified so thermostatically controlled reversible motors could be incorporated to open and close the vents at predetermined temperatures. Similar hand operated equipment is still found in older greenhouses today and some growers have it installed in new facilities for emergency use.

Cooling greenhouses manually with ventilators involves the principle of warm air rising. The air in a greenhouse is generally warmer than that outside and when the top vents are opened, the warm air escapes. If side vents are present, they can be opened and a circulation pattern established. One of the major assets of a good grower, yesterday as well as today, is the ability to maintain an "optimum" growing temperature.

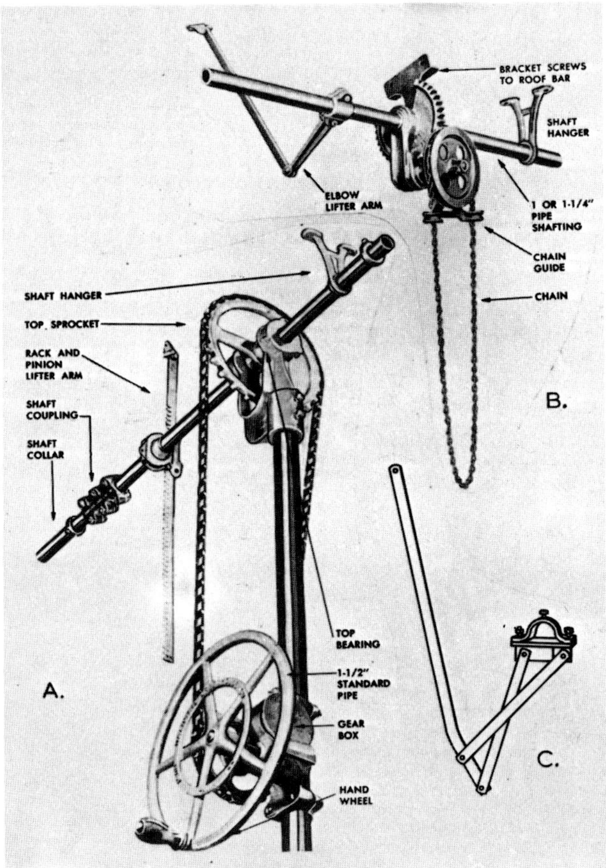

B

Fig. 4-41B and C

C

In 1954, DeWerth and Taska (1954, 1956) pioneered a new method of ventilating greenhouses. It could be classed as forced ventilation but is commonly known as fan and pad cooling.

The basis for cooling greenhouses is centered around the greenhouse energy balance, Figure 4-17. The solar energy input is the largest contributing factor. On a clear summer day approximately 280 BTUs per square foot per hour reaches the earth. In some areas of high water vapor or pollution, as little as 200 BTUs per hour may be received. When insolation reaches a greenhouse, the majority is transmitted inside and is trapped because the cover is opaque to the long wavelengths reradiated from surrounding plants, soil and equipment within the greenhouse, thus the greenhouse effect.

Using Figure 4-20, one can see that more energy is generally received on any clear day if the outside temperature is above 30° F, than is lost through the structure. Therefore, some type of ventilation is required. On a clear 25–30° F day, in Fort Collins, Colorado, latitude 40°30′, elevation of 4780 ft, there is often

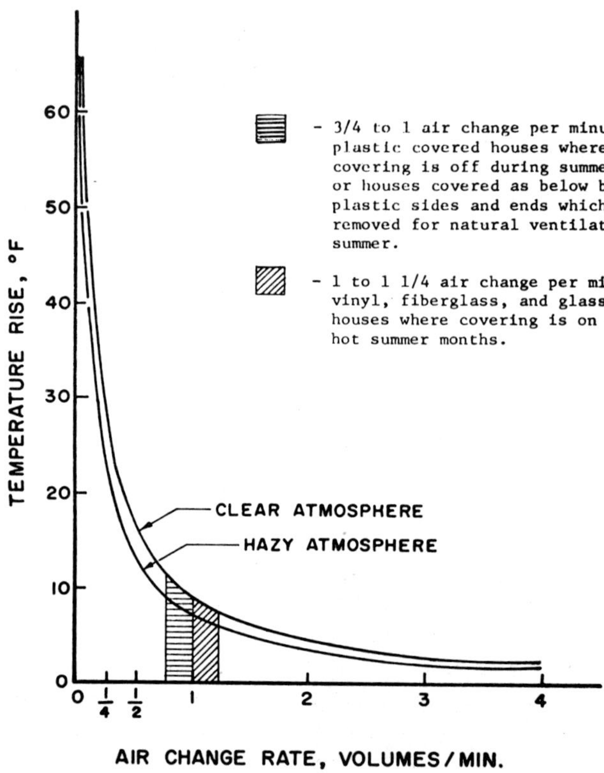

Fig. 4-42. Relationship of air change rate to temperature rise in greenhouses on a clear sunny day. Note: Fans should be sized to provide these air exchange rates at .10 in static pressure (water column) with A.M.C.A. certified ratings to insure adequate fan delivery (Walker and Duncan, 1973b)

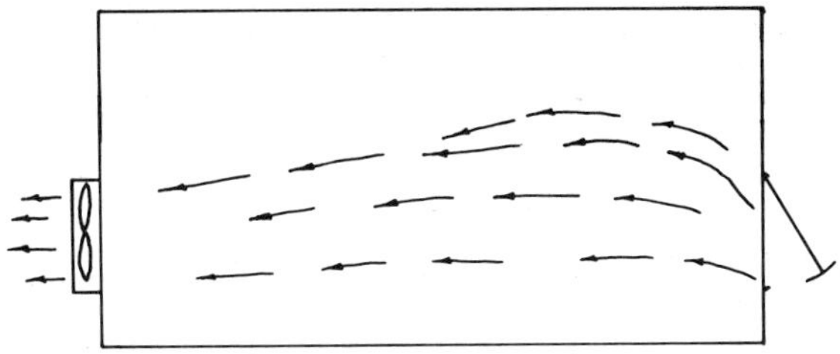

A

Fig. 4-43. (A) Side view of air moving through a greenhouse from fan to pad end. (B) *Left:* Aluminum pivot louver for exhaust fan. *Right:* Counterbalanced louver

enough solar energy reaching the greenhouse to cause the heat gain to equal or exceed the heat loss.

Ventilation must also be considered as one method of controlling greenhouse humidity, which is discussed in Chapter 5.

4.4.1 Ventilation Criteria

4.4.1.1 Air Movement

Walker and Duncan (1973b) determined ventilation requirements by using two factors; one, the volume of the greenhouse and two, the air changes desired. Adequate ventilation can generally be achieved using one air change per minute (Fig. 4-42), providing a negative pressure type (exhaust fan) system is used. The older topside vent system could only be expected to provide one air change per minute under the most unusual circumstances. The system that has proven most effective, Figure 4-43, does not actually change the total volume of air in the house, but moves it in a pattern designed to maintain the most desirable plant environment. The faster air is changed in a greenhouse, Table 4-4, the lower the inside temperature rise.

Table 4-4. Air change rate and inside temperature rise for greenhouse ventilation (Walker and Duncan, 1973b)

Air change rate, volumes/min	Inside temperature above outside
1/2	16–18
3/4	11–13
1	8–10
2	4– 6

Example: When the outside air temperature is 80° and the ventilation capacity provides one air change per min, the inside temperature could be expected to be between 88° and 90° F.

Augsburger et al. (1975) consider additional design factors which correlate the ventilation needs to climatic environments. They have determined that a basic air flow rate 8 CFM per ft^2 is adequate to cool a moderately shaded greenhouse, having a maximum inside light intensity of 5000 ft-c at sea level.

4.4.1.2 Altitude

For elevations above 1000 ft, the basic air flow rate is adjusted for elevation, light intensity, greenhouse temperature differential and distance between the incoming air ventilator and exhaust fan. The following discussion of these factors is taken, in part, from the Greenhouse Climate Control Handbook (Augsburger et al., 1975). Air is less dense at higher altitudes and since the heat removal capacity of air depends on its weight and not its volume, elevation should be considered in design calculations. A greater volume of air will be needed at higher elevations, thus an *elevation factor* (F_{elev}) has been developed based on the local mean barometric pressure (BP).

$$F_{elev} = \frac{29.92}{BP}.$$

4.4.1.3 Solar Input

The type of cover, geographical location of the structure and the degree of roof shade determines the amount of solar heat received in the greenhouse. The light intensity should be measured at high noon in midsummer under a clean cover. The interior light measured in foot candles (ft-c) and the light *intensity factor* (F_{light}) is:

$$F_{light} = \frac{\text{ft-c}}{5000}.$$

4.4.1.4 *Temperature Differential*

Augsburger et al. (1975) have determined that cooled air passing through a greenhouse will pick up heat, which increases the temperature at least 7° F by the time it is exhausted. Based on the fact that the temperature increase of the air from the incoming side of the greenhouse to the exhaust side is inversely proportional to the airflow rate, the *temperature increase* factor (F_{temp}) may be adjusted using the formula:

$$F_{\text{temp}} = \frac{7.0^\circ}{\Delta T^\circ}.$$

Hanan (1970) found that the air moving 41 ft in a glasshouse, on a 60–65° F day in March exceeded an 18° F differential. The differential on a 55° F day in a 105 ft fiberglass covered house was 10.5° F.

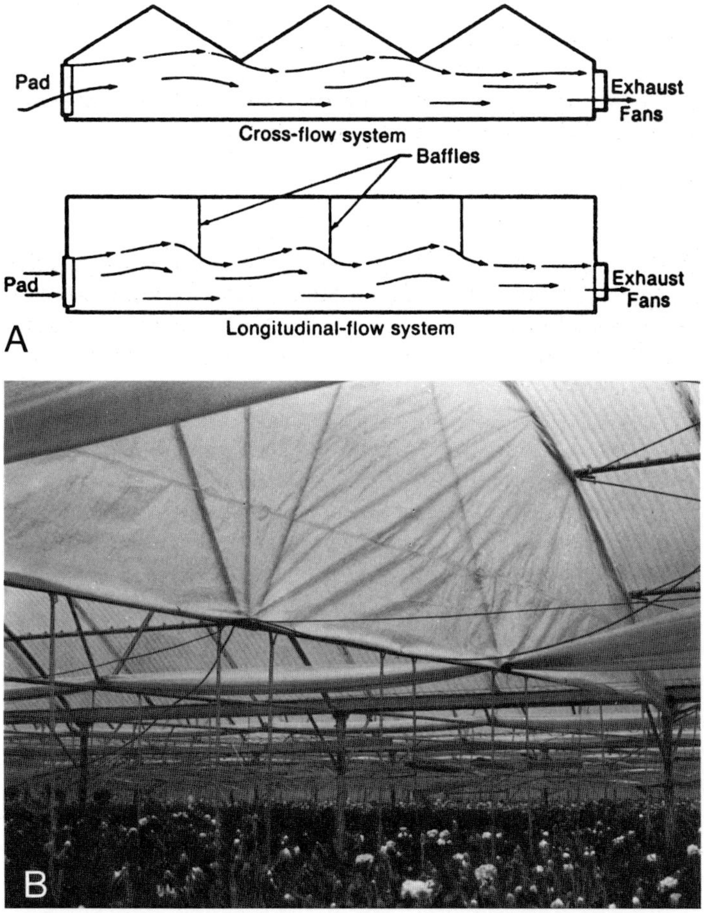

Fig. 4-44. (A) Cross and longitudinal flow designs (Augsburger et al., 1975). (B) Clear plastic baffle attached to a truss

Fig. 4-45. Plastic tube used for winter ventilation—inflated by negative pressure created by exhaust fans

Short air flow distances, less than 100 ft, must be considered somewhat differently than a long pull. Experience has shown that a greater volume of air needs to be moved when air is moved distances less than 100 ft. Thus, the *velocity* adjustment factor (F_{vel}) has been established where the air inlet to fan distance (D) is less than 100 ft (Augsburger et al., 1975).

$$F_{vel} = \frac{10}{\sqrt{D}}.$$

4.4.1.5 Greenhouse Air Leakage

The ability to move air through plant canopies is related directly to the tightness of the greenhouse. If glass is broken in the roof area, excessive opening of panel laps are present or poorly fitted joints occur, air will be pulled into the greenhouse through every crack, as a negative pressure is created by the exhaust fan. The majority of greenhouse ventilation design programs include fans that will move the required volume of air at 0.1 SP (inches of water, static pressure).

4.4.1.6 Airflow Systems

If a greenhouse is extra tight, the airflow from the inlet side of the house to the exhaust fan will be relatively well confined (Fig. 4-44 A). When air is pulled the length of a house, it is wise to consider the use of plastic baffles attached to trusses every 30–40 ft (Fig. 4-44 B). If raised benches are involved, a plastic baffle, under the benches, near the air inlet end will help confine air movement to the plant canopies. The maximum length air can be effectively pulled across a greenhouse is

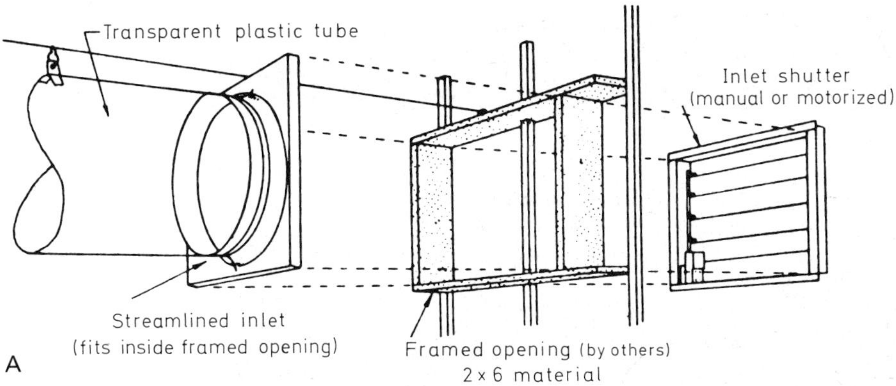

A

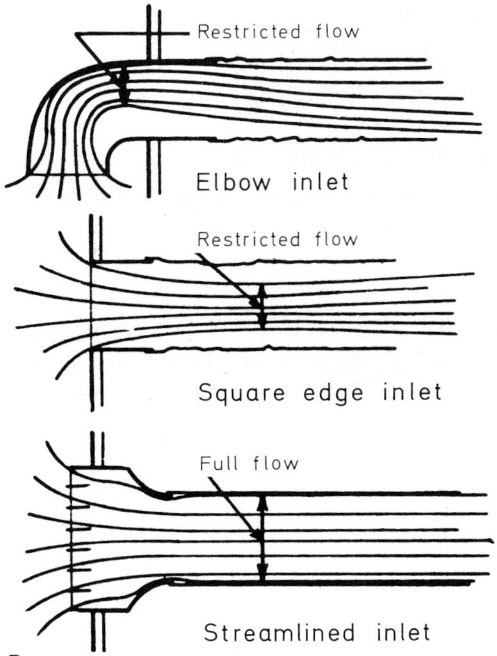

B

Fig. 4-46. (A) Simple poly tube attachment system. (B) Fresh air inlet for plastic tube. A streamlined inlet orifice will increase the capacity of the system by approximately 50% (Acme, 1963)

based on the potential rise in temperature of the air, but is generally limited to 200 ft.

4.4.2 Winter Ventilation

In the early 1960s, an automatic winter ventilation system was introduced in the greenhouse industry. Prior to this time side ventilators were opened and exhaust fans energized during near freezing outside temperatures and the plants nearest the openings were chilled and often frozen. The new winter ventilation system consisted of a perforated plastic tube plenum (Fig. 4-45) that distributed the cold air in the upper greenhouse atmosphere without plant damage.

Fig. 4-47. Motorized louver in series with the first stage cooling fan or energized by a step controller as the first stage of cooling where a poly tube pressurizing fan is used

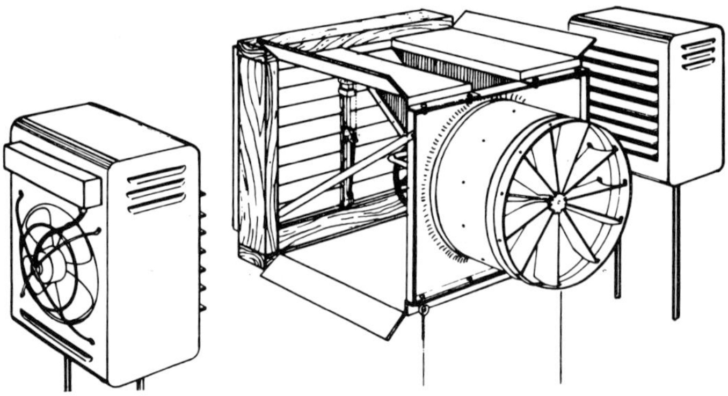

Fig. 4-48. Pressurizing fan system that can introduce fresh air, distribute heat, or recirculate the greenhouse air. (Courtesy Acme Engineering)

4.4.2.1 Plastic Tube Systems

The exact inception of the use of poly (polyethylene plastic) tubes in the United States is unknown, but it was possibly due to, Homer Hill, who had visited Europe in 1961 and noted their potential use in the greenhouse for cooling purposes. The negative pressure created by an exhaust fan in a closed house, during a cold day, could be put to use and might prove to be of value during the winter.

The first poly tubes (often called wind socks or convection tubes) were connected to an adaptor (Fig. 4-46) which allowed outside air to enter a closed greenhouse and inflate the tube due to a negative pressure created when the exhaust fan was energized. The air inlets to the tubes are generally controlled by a motorized louver, which is energized at the same time the first stage exhaust fan is started, Figure 4-47, unless a pressurizing fan and controller are used.

4.4.2.2 *Poly Tube Uses*

In recent years, many refinements have been made to make the tube system more versatile. The Acme Corporation (Augsburger et al., 1975) integrated a fan system which kept the tube inflated. The pressurizing fan recirculates the greenhouse air and/or introduces outside air when the louvers are opened (Fig. 4-48). The latest system adapts the initial cooling, recirculation and heating phases (Fig. 4-25).

The pressurized tube system can also be used to reduce the relative humidity in greenhouses, especially at night. When cooler air is brought into the greenhouse through the poly tube, it has a low water vapor content and when it's temperature is raised, the water vapor holding capacity is increased and thus has drying ability. Greenhouse humidity can be controlled relatively well by introducing fresh air periodically, warming it and exhausting the air with the higher water vapor content, all within minutes.

4.4.3 Computing Ventilation Needs

Based on the various ventilating criteria described previously, the following examples are presented.

4.4.3.1 *Exhaust Fan System*

1. Formula:

Total $ft^3\,min^{-1} = L \times W \times FR_{Air} \times (F_{House}$ or $F_{Vel})$

L = Length of houses

W = Total width

FR_{Air} = Basic flow rate of air (8 $ft^3\,min^{-1}\,ft^{-2}$ of ground covered)

F_{House} = $F_{Elev} \times F_{Light} \times F_{Temp}$

F_{Vel} = Air velocity adjustment factor. To be considered in place of F_{House} if fan to pad distance is less than 100 ft. Use largest factor

2. Data:

L = 150 feet

W = 2×42 ft = 84 ft

FR_{Air} = 8 $ft^3\,min^{-1}\,ft^{-}$

Elevation = 4850 ft

Maximum expected interior light = 8000 ft-c

Maximum temperature variation = 8° F

3. Calculations:

$$\begin{aligned} ft^3\,min^{-1} &= 150 \times 84 \times FR_{Air} \times (F_{House} = F_{Elev} \times F_{Light} \times F_{Temp}) \\ &= 150 \times 84 \times 8 \times (F_{House} = 1.20 \times 1.60 \times 0.88) \\ &= 150 \times 84 \times 8 \times 1.69 \\ &= 170\,352\ ft^3\,min^{-1}. \end{aligned}$$

The total air flow requirement for this 150 × 84 ft ridge and furrow house is 170352 ft^3 min^{-1}. A minimum of two fans, and preferably three, should be used in the end of each house.

4.4.3.2 Nomograph Calculation

The same size structure can be used to calculate the ventilating requirements with a nomograph (Fig. 4-49). The criteria used to enter are minimal and result in a lower airflow requirement.

4.4.4 Air Cooling Systems

4.4.4.1 Evaporative Pads

The pad and fan method of cooling greenhouses revolutionized the industry. Prior to its use, many growers left their greenhouses vacant in the summer because of excessive temperatures. If plants were grown, the greenhouses had to be heavily shaded in order to achieve good quality and production and provide decent working conditions for employees.

Evaporative pad cooling is one of the most inexpensive methods of lowering the temperature inside any building. It does, however, increase the humidity which is often undesirable. It is very impractical to consider cooling greenhouses with refrigeration systems as it will require approximately two tons of refrigeration per 750 ft^3 of greenhouse to maintain desirable growing temperatures. Such a system of cooling would require the total volume of air to be cooled and not just the area immediately surrounding the plants.

The reader should become well acquainted with the section on "wet and dry bulb temperatures" and "controlling water vapor content of air" in Chapter 5. In brief, the wet bulb temperature (Fig. 4-50) indicates the potential inside air temperature that can be obtained by evaporative cooling (Augsburger et al., 1975). If air outside is 35° C (95° F) and a relative humidity of 50% exists, the wet bulb temperature will be 26° C (79° F) (Fig. 5-10). Thus, the air coming into the greenhouse through the pad system will approximate 27° C (80° F) or slightly higher (Fig. 4-50).

a) Aspen excelsior pads (Fig. 4-51) have been the primary material for evaporative cooling systems. The excelsior is made by shredding *Populus tremuloides*, which is made into mats by enclosing it in a cloth netting. The mats are then placed in a framework forming a solid pad area.

The cooling efficiency of aspen pads is dependent on several variables including density, thickness, salt accumulation, dirt and age. Studies have found that pad thickness was important and that a 1½ in thickness was superior to a 2-in heavier material for cooling efficiency. The thinner pad has been unacceptable, however, because of thin spots. In 1960, Holley and Manring observed the cooling efficiency of new and one year old aspen pads. New pads lowered the air temperature 7–8° F.

b) Other cooling pad materials include impregnated paper, "hog hair" and aluminized paper. Where neoprene covered hog's hair has been used, some grow-

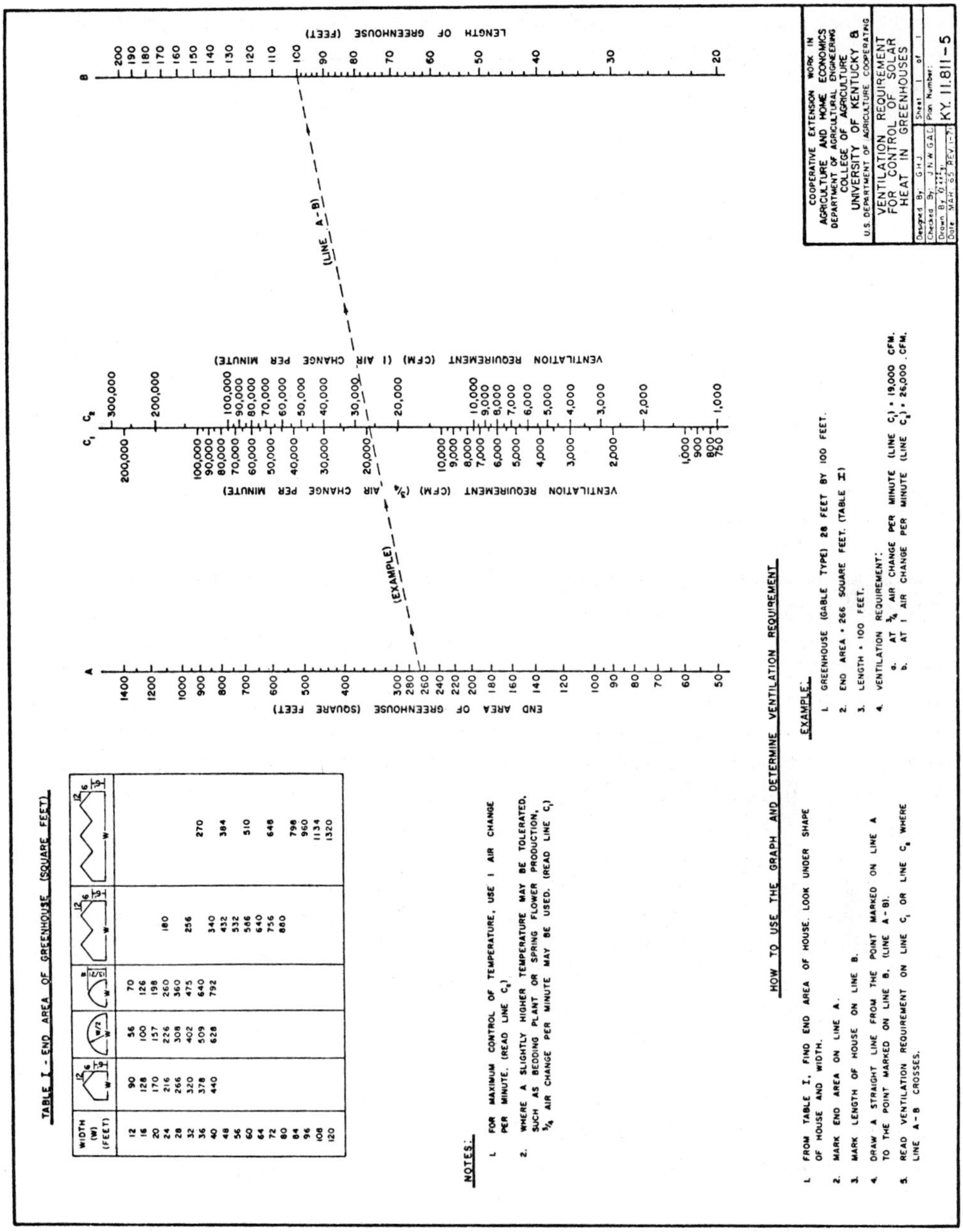

Fig. 4-49. Ventilation requirements for control of solar heat in greenhouses (Walker and Duncan, 1973b)

ers have reported good results with pads up to two years, provided that excessive salts do not accumulate.

Goldsberry (1967) evaluated the effectiveness of two pad types of different ages (Table 4-5). The age of the aspen pads did not appreciably affect the cooling efficiency of airflow providing they remained uniformly thick, did not sag or

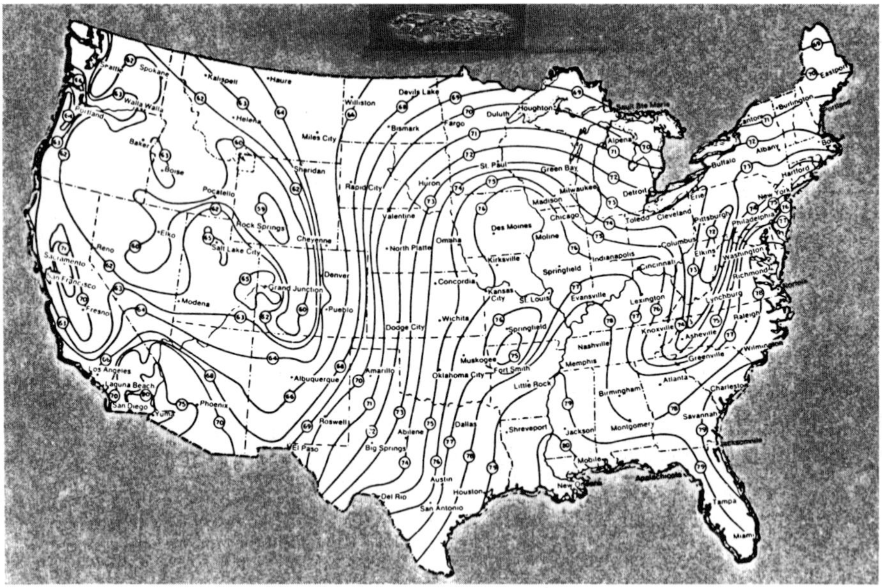

Fig. 4-50. Summer wet-bulb temperature data. The wet-bulb temperatures shown will be exceeded not more than 5% of the total hours, during June to September inclusive, of a normal summer (Augsburger et al., 1975)

Fig. 4-51. Aspen excelsior evaporative cooling pad system

accumulate excessive dirt and salts. The two year old aspen pads did start decomposing. The aluminized paper pad provided relatively good cooling, but the airflow was reduced.

The latest addition to the selection of cooling pads is the impregnated rigid paper material (Fig. 4-52). It is self-supporting and requires only a superstructure with anti-rot salts, wetting agents and fungicides impregnated in a crossfluted paper material. Heins (1974) indicated that the 12-in thick Cel-dek (Trade Name)

Table 4-5. Effects of age on aspen excelsior and aluminized paper evaporative cooling pads (Goldsberry, 1967, unpublished)

	Al-paper		Aspen		Al-paper new	Aspen new
	1 yr	2 yr	1 yr	2 yr		
Mean air flow through wet pad-ft min^{-1}	288	241	364	360	282	350
Temperature reduction through pad °F	7.7	8.9	9.48	8.84	8.2	11.5
Salt concentration Mhos $\times 10^{-5}$ specific conductance	24	13	39	63	3	4

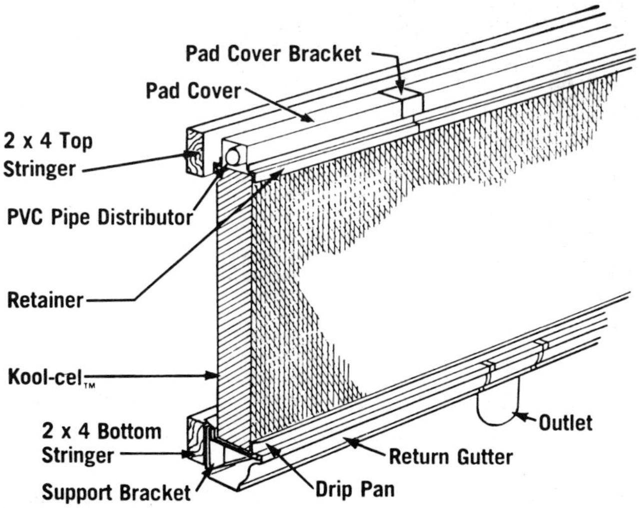

B

Fig. 4-52. (A) Kool-cell installation at Colorado State University. (B) Installation material for Kool-cell. (Courtesy Acme Engineer) paper pads provided 6% more cooling efficiency than the standard aspen pad. The newer Kool-cel (Trade Name) is only four inches thick and can be adapted easily to aspen pad installations. The manufacturers feel this new pad type should last several years, eliminating the labor of changing pad materials each year.

4.4.4.2 Cooling Pad Design

The cooling pad system needs to be properly designed in order to cool a house to within 3–4° of the outside wet bulb temperature. The size of the pad system is determined by the total cubic feet per minute of air flow required for forced ventilation (Table 4-6). The most uniform distribution of air coming through a cooling pad is accomplished with an installation the full width of the greenhouse. Most greenhouses will require about one foot of pad height for every 20 feet of pad to fan distance (Augsburger et al., 1969).

In some instances the pad to fan distance can be so great that the height of the eave may limit the height of the pad system. In such cases, pads have been installed in an accordian manner and hung either vertically or horizontally (Fig. 4-53). New greenhouse facilities should be designed so an effective pad area can be installed with fan to pad distances no greater than 200 ft.

a) Exterior-Interior Pad System. There are aerodynamic and structural advantages and disadvantages for having pad systems installed in or outside the greenhouse. Interior pads with the ventilators on the outside produce less turbulent air flows down stream from the pad. However, lower temperatures near the pad area often affect plant growth in adjacent benches. Inside pads present few problems during winter months and can be rapidly put in use during periods of warm weather. The vent opening only needs to be equal to 1/3 of the area of the pad.

Exterior pad systems, constructed in conjunction with a pad house, present some freezing problems during winter months (Fig. 4-51). However, the pad house does act as a mixing and tempering chamber if utilized effectively. The longitudinal flow system is used and pad cooled air can enter the poly tubes while the vents are closed, providing the second stage of cooling.

The type and design of ventilation used inside the greenhouse are important. When a sash vent, Figure 4-54A, is opened 2–3 in, the incoming air is directed upward. The full effect of the exterior cooling pad is not evident until the ventila-

Table 4-6. Square feet of aspen cooling pad required per CFM of air flow requirement. (Adapted from Augsburger et al., 1975)

$ft^3\ min^{-1}$ required	Square feet aspen pad required	$ft^3\ min^{-1}$ required	Square feet aspen pad required
5000	30	60000	360
7500	45	80000	480
10000	60	120000	720
15000	90	150000	900
20000	120	180000	1080
40000	240	210000	1260

Fig. 4-53 A and B. Increasing effective pad area by (A) horizontal accordian system and (B) vertical accordian system

tors are wide open. The use of opposed blade dampers as the interior ventilator, Figure 4-54 C, provides a uniform distribution of air within the same height as the plant canopy.

b) Cooling Pad Components. Most pad systems should be designed to provide at least $^1/_2$ gal water per min per linear foot of the pad system length. This amount not only provides enough water for wetting the pad, but provides more than that evaporated.

Conventionally, water for pads is applied through a pipe at the top of the pads (Fig. 4-55). Some greenhouse operators have also used low pressure mist nozzles with relatively good results.

Fig. 4-54 A–C

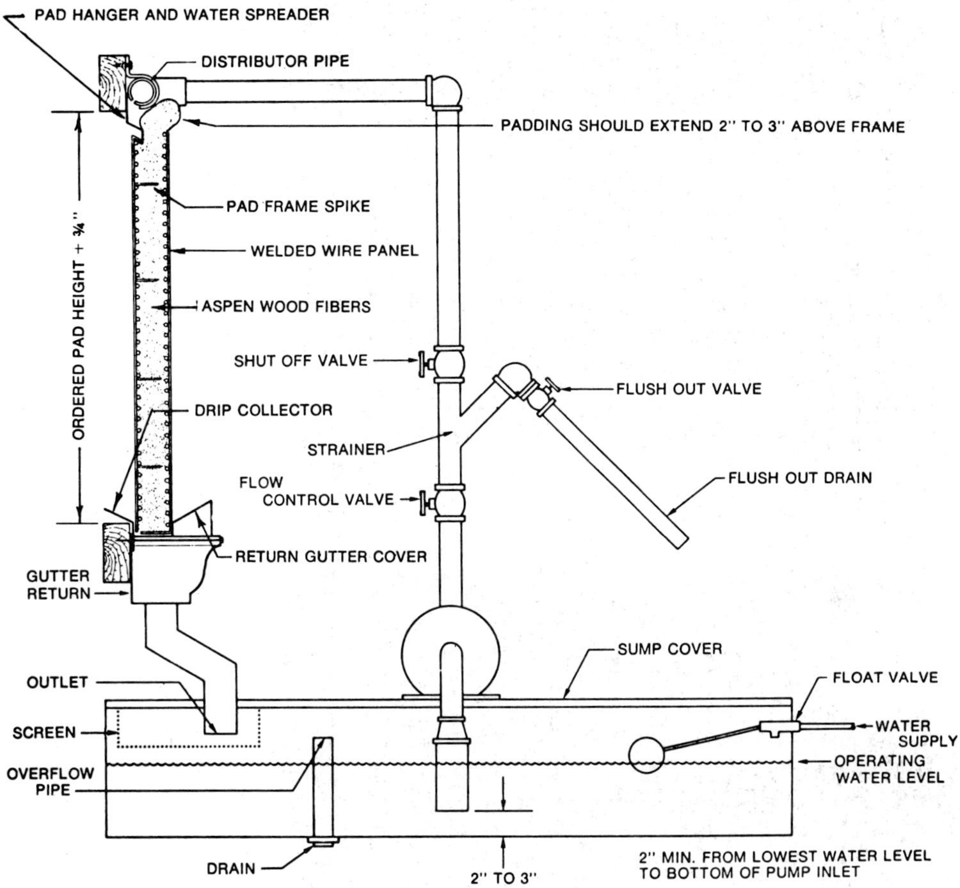

Fig. 4-55. Typical evaporative cooling pad system. Aspen pad system. (Courtesy Acme Engineering)

It is desirable to recirculate water used for cooling, therefore, a sump for water storage is used and water pumped to the cooling pads. Depending on the type of pad, a sump capacity of $^1/_2$ to $^3/_4$ gal of water per square foot of pad area will be required. The water flow to the pad area should be controlled.

The make-up water requirement for the pad system will depend on the water vapor content of the incoming air, and may require up to one gallon per minute per 100 ft^2 of pad.

A clean system can be maintained if care is taken to flush and clean out the system periodically and keep it covered.

◀ Fig. 4-54A–C. Cooling pad ventilators. (A) Top hinged sash. (B) Center pivot sash. (C) Opposed blade dampers

4.4.5 Mist Cooling

Some greenhouse environments are more conducive to mist cooling than others. In 1957, Holley found that some rose varieties performed better under mist than when cooled with a fan and pad system. The rose house with top ventilators fully opened, shaded roof and a high pressure mist system in operation, maintained temperatures 8° F above those achieved in a second fan and pad cooled house. Carpenter and Willis (1957) observed growth responses of chrysanthemums, snapdragons, and carnations when cooled by mist, fog and fan and pad systems. The fan and pad cooled house provided the best environment for the growth of all plants. When mist is applied in a greenhouse, the air circulation must be such that it flows from the mist nozzles to the plant growing area (Walker and Cotter, 1967).

It is possible that growers converting to foliage plant crops should consider mist cooling instead of a fan and pad system. Some foliage plants do not perform well in temperatures normally produced by fan and pad systems. Many species of foliage plants are burned when they are exposed to temperatures above the air stream created by the fans, and a mist cooling should provide a better environment.

4.4.6 Cooling by Shading

The application of a shading compound to the outside of a greenhouse to help control summer temperature and decrease light intensity has been a common practice for years. Gray (1948) reported that a 50% reduction of light, by shading a greenhouse, lowered the average inside temperature at least 6° F. The application of shading compounds to the greenhouse cover can often be more economically harmful to a crop than a grower realizes. The temperature may be lowered during periods of high insolation, but also intensity may be reduced unnecessarily on cloudy days. Fixed types of shade to aid summer cooling or lower light intensities are not desirable in most instances.

Several researchers and greenhouse operators have tried blinds, movable shading cloth, colored water, and many other nonfixed materials. To date, no one has provided a proven system.

The future of greenhouse shading will no doubt hinge on a variable shade system that can control the transmitted light energy. Spencer et al. (1974) have described experiments that can increase the total radiation for a summer season from 65 to 100% over fixed shading methods. They point out that fixed shade is usually applied in proportion to the maximum energy expected during summer.

When fixed shade, such as commercial greenhouse compounds or diluted interior latex paint are applied to a greenhouse, the color and density will determine it's effectiveness of decreasing temperatures or light intensity. Goldsberry and Wolnick (1966) showed that darker colors of shading compounds absorbed infrared and caused heat build up below the glass (Fig. 4-56). It is recommended that only white materials be used. A very fine spray is most desirable and coverage and density can be easily controlled.

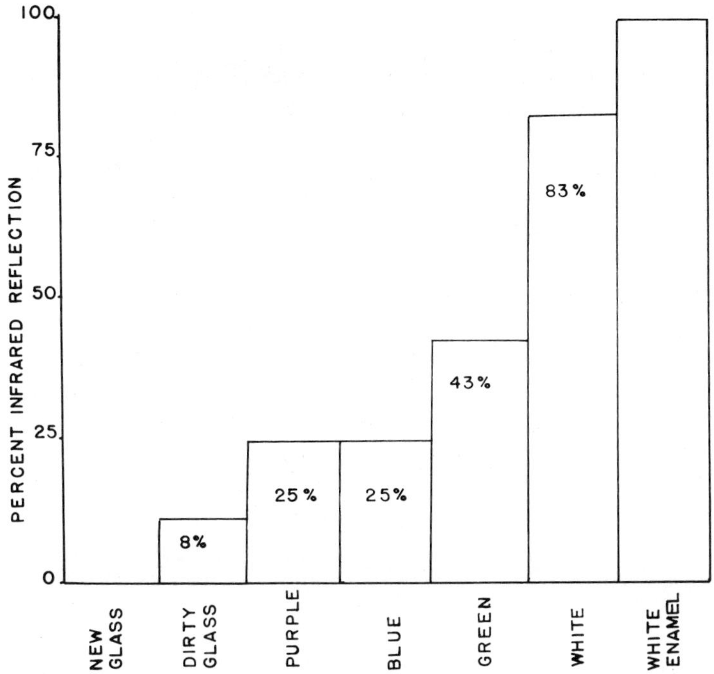

Fig. 4-56. Percentage of infrared reflected by various shading compounds on glass (Goldsberry and Wolnick, 1966)

4.4.7 Maintaining Ventilating and Cooling Equipment

Another one of the merits of a good greenhouse manager is to be aware of mechanical problems such as noticing the sound when a bearing is "going out" on a fan or pump or, seeing a bent ventilator arm and fixing it before a vent is broken.

4.4.7.1 Exhaust Fan

The exhaust fan is a piece of equipment that needs to be checked two or three times a year. The fans designed for greenhouse use are considered as nonloading and at 0 pressure, they have maximum flow like any other type fan, but over their operating range the motor is protected. Greenhouse fan motors should be totally enclosed to eliminate moisture problems. The major requirement for the maintenance man is to see that each fan accomplishes 100% of its intended work load. Each fan is rated for voltage and amperage. It will only be making money if it is operating at the proper amperage. Belts should be kept tight and changed when they start to wear. A temperature rise due to one broken belt could lead to a disaster. A variable pitch sheave (pulley) and clamp-on ammeter (Fig. 4-57 A) are the equipment required to keep an exhaust fan working properly. The variable

A

B

Fig. 4-57. (A) Variable pitch pulley and clamp on ammeter. The ammeter is clamped around a single electrical wire going to the fan. The variable pulley is then changed to provide the proper amperage rating. (B) Sagging aspen pad

pitch sheave permits speed variations of approximately 30% and motor amperage can be set exactly. It is recommended that all exhaust fans be installed with variable pitch pulleys.

4.4.7.2 Pad Systems

The evaporative cooling pad also requires maintenance. The aspen pads tend to sag if not installed properly and if the proper weight material is not used. They can also have thin sections that are relatively inefficient (Fig. 4-57 B).

The accumulation of salts and algae is more of a problem in some areas than others. Algaecides may need to be added periodically where pads are not allowed to dry out before the exhaust fans are de-energized.

The water supply for evaporative cooling pads should be as salt free as possible. If possible, do not use nutrient treated water which has been used for watering plants. Excessive salt build up can be partially removed by periodically "hosing" the pads with a medium pressure nozzle.

4.4.7.3 Other Equipment

Plastic tubes, louvers, and ventilators all require some attention periodically. When plastic tubes are dirty, they obstruct light transmission, and if they are in a cultured plant house, dust could carry diseases. Louvers on exhaust fans tend to get bent and worn. A louver that does not close in the winter could cause some

Table 4-7. Basic temperature regime for cool greenhouse crops

	Day	Night
Heat to:	60° F	54° F
Cool to:	65° F	65° F

frozen plants. The ventilator on the cooling pad system may not close tightly on a cold fall evening and the water supply pipes could freeze. Maintenance is an important concern of a manager.

4.5 Sensing and Controlling Temperature

Every greenhouse has unique environmental conditions which are created by the house design, geographical location, wind patterns or other factors considered in Chapter 3. No one is capable of manually controlling the equipment to overcome some of the heating and cooling requirements in a greenhouse and thus provide a near "optimum" environment for plant growth.

4.5.1 Knowing Your Temperatures

Without a doubt, the professional grower can walk into a greenhouse and "feel" the temperature and know if it is too high or low. Many growers depend on a thermometer that is hanging in the center of the greenhouse (Fig. 4-58 A) and is perhaps viewed three or four times a day and maybe once before bedtime. The column of the common alcohol thermometer will often separate when left in the sun and the temperature reading maybe 5–15° or more incorrect. Hanan (1975) found that some new alcohol thermometers varied 3° F from the true temperature. He recommended that all types of thermometers be calibrated before they are used in the greenhouse, and periodically thereafter.

The best method of determining the true temperatures within a greenhouse is to use a recording thermometer (Fig. 4-58 B). Every grower should take time to make 24 h temperature checks throughout his range. When temperatures are checked, the sensing elements should be at plant height and not two or three feet above or below the crop.

4.5.2 Temperature Sensors

Some greenhouse operators use a single electric thermostat (Fig. 4-58 A) to control 10000 ft^2 of ridge and furrow greenhouses. Experience has shown that most commercially available high voltage bimetal thermostats are unsuitable for green-

Fig. 4-58. (A) Unprotected thermometer and thermostat hanging in the center of a greenhouse. (B) Portable 24-h recording thermograph. Adequate for checking night temperatures, but should be placed in an aspirated box during daylight hours while recording

house use. They usually arc or stick and are not sensitive enough. Some thermostats are too sensitive and easily vibrated, which causes relays to "chatter" and fuses to be "blown" on equipment.

Excellent greenhouse temperature control has been achieved with pneumatic thermostats. Some systems require sophisticated accessories which are often costly but can control all phases of greenhouse environment.

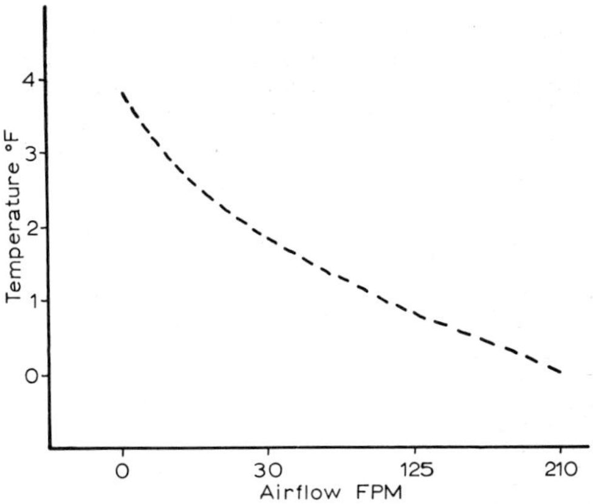

Fig. 4-59. The temperature differences between greenhouse air and the sensing element, due to air velocity through a wooden thermostat holder. Air speed should be at least 125 ft min^{-1} across instruments (Goldsberry, 1963)

Electronics has recently provided sensors that are very useful in the greenhouse. A single thermistor has almost become standard equipment in new greenhouse facilities.

4.5.2.1 Holders

Thermometers, thermostats, and other sensing elements are often placed in undesirable holders. In the less modern greenhouse, they might be found attached to a board or an outside wall, in a box hanging in the center of the house or in an insulated holder. Goldsberry (1963) showed that temperature differences up to 5° F could be expected in an assortment of typical nonaspirated holders used in greenhouses.

a) Aspirated: A second phase of Goldsberry's study determined that relatively true air temperatures could be obtained within enclosed thermostat shelters by moving air through them. An air flow of approximately 125 fpm must pass across the sensing elements before a true air sample is available (Fig. 4-59).

Several aspirated holders are now on the market (Fig. 4-60). Some have been designed so air flows across the motor before it crosses the sensor, thus false readings occur. Others lack proper airflow. Holders should be painted white or have a reflective surface. The sensors should not be in direct contact with the outside portion of the holder due to possible heat conduction through the wall. A wooden aspirated box (Fig. 4-61) was designed for use which has provided reasonable temperature control.

b) Placement: The sensing holders should be placed in a representative area of the greenhouse. If air is being properly circulated or moved through the greenhouse, Goldsberry (1963) found that an aspirated holder could be placed from three to six feet above the ground in a fan and pad cooled carnation house and comparable temperature readings recorded.

The sensing holder should be designed to be moved up and down vertically, placed in an aisle near the crop, and not hang from the superstructure if bimetal or mercury thermostats are used.

Fig. 4-60 A and B. Aspirated thermistor or thermostat holders used with modern control system

4.5.3 Greenhouse Temperatures

Most greenhouse crops are grown in temperatures ranging from 50° F to 70° F. In many instances, the inability to cool a greenhouse allows the temperatures to rise beyond desirable levels. Some greenhouse managers often overwork their cooling and heating equipment, and in many instances have them working against each other. The basic temperature levels should be kept as simple as possible.

A discussion is in order regarding the above basic temperatures. The majority of growers who reset individual thermostats, morning and night, can never be within 25% of the desired temperature. Secondly, most automatic controllers are designed to immediately lower the greenhouse temperature upon switching to nighttime temperatures. It would seem more appropriate to let the night temperature "coast" down to the "to heat" level from 60° or 65° F day settings. The geographical location and time of year will determine the degree of modifications one might consider in temperature regimes.

4.5.4 Automatic Controls

Automatic control systems are an asset to the greenhouse manager. One sensing unit can control all of the environmental equipment, plus provide a convenient

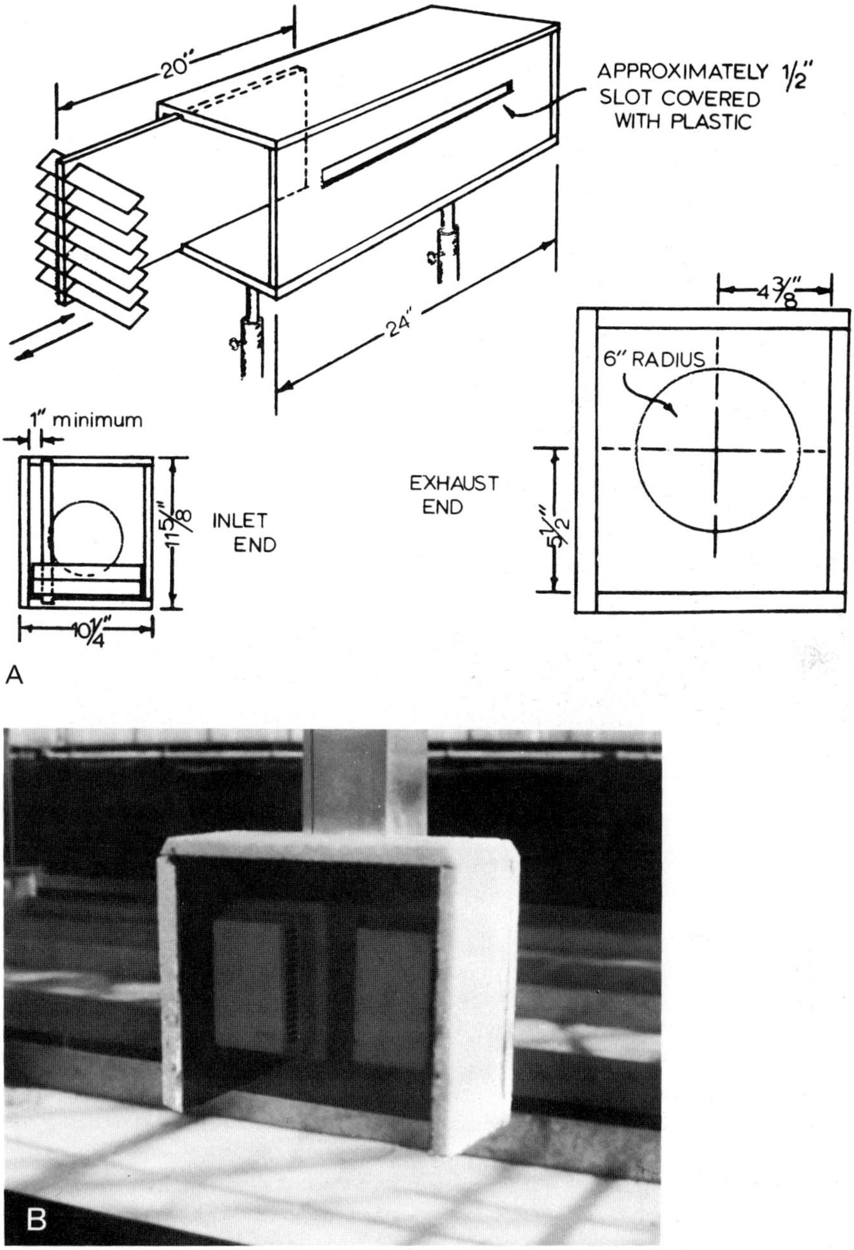

Fig. 4-61. (A) Aspirated control box designed to eliminate heat conduction and reflection (Goldsberry, 1963). (B) Insulated holder, but no air movement

high and low temperature alarm system. The newer environmental control units, Figure 4-62, are of solid state construction and are composed of a sensing thermistor, amplifier, potentiometer, and adjustable "turn on" and "off" points and relays. The equipment is not infallible and should be calibrated periodically and the resulting greenhouse temperatures checked with recorders.

Fig. 4-62 A and B. Environmental control panels commonly used in the greenhouse. (A) The revolutionary electrical unit of the late 1960s. (Courtesy Wadsworth Electric; Denver, Colorado.) (B) The latest in solid state electronics

4.5.4.1 Staging

The value of automatic controllers is found in their ability to provide the desired environmental conditions, allowing only small temperature changes that will not harm a plant's physiological condition or cause a physical distortion. The controller is designed to provide a range of temperatures that can be easily changed, yet is highly versatile. It is impossible to maintain exact greenhouse temperatures

Table 4-8. Recommended staging of environmental control equipment in a greenhouse requiring a single controller

Heating stages ↑	Stage 5—low temperature alarm
	Stage 4—empty or other heaters
	Stage 3—second unit heater per house on
	Stage 2—one unit heater per house on
	Stage 1—minimal heat such as perimeter steam line
0	"neutral," "null," or "set point" (no heat or cooling-only positive air circulation system on)
↓ Cooling stages	Stage 1—louver behind poly tube opens, CO_2 off
	Stage 2—one exhaust fan in each house on
	Stage 3—vent opens slightly—poly tube system off
	Stage 4—second exhaust fan on—vent opening
	Stage 5—all fans on
	Stage 6—empty
	Stage 7—high temperature alarm

in most instances. Automatic controllers are designed to energize cooling and heating equipment in stages, so that minimal temperature changes occur (Table 4-8).

The staging recommendations are somewhat different than those used by the major greenhouse manufacturers. The following points should be considered:

1. CO_2—de-energized when fresh air is first introduced, unless higher CO_2 levels are required (CO_2 off dusk to dawn).

2. Cooling pads—the evaporative cooling pads should be controlled by the outside temperature. In Colorado, pads should be functional when the outside temperature reaches 55° F any time during day light hours. A time clock is used for primary pad control one hour after sunup and one hour before sundown during the summer and approximately 0900 to 1600 through the winter months.

3. Perimeter heat—on many occasions, only a small amount of heat is required. Experience has shown that perimeter steam lines are most desirable as a first stage of heat. Steam or hot water heat exchangers used in conjunction with poly tubes and activated by modulating valves also provide excellent first stage heating systems.

4. Alarm systems—the high and low temperature alarms are indispensable. Every greenhouse should have temperature alarms that are activated by both AC and DC voltages. Alarm systems for the boiler operation, water levels in sump pump pits, walk-in refrigerators, and many other important aspects of greenhouse operation should be installed.

5. Key equipment—the devices important to the staging of all greenhouse equipment are found in Figure 4-63. Every greenhouse manager should become acquainted with common electrical relays, fuses, switches, wiring, and parts used to maintain the greenhouse environment.

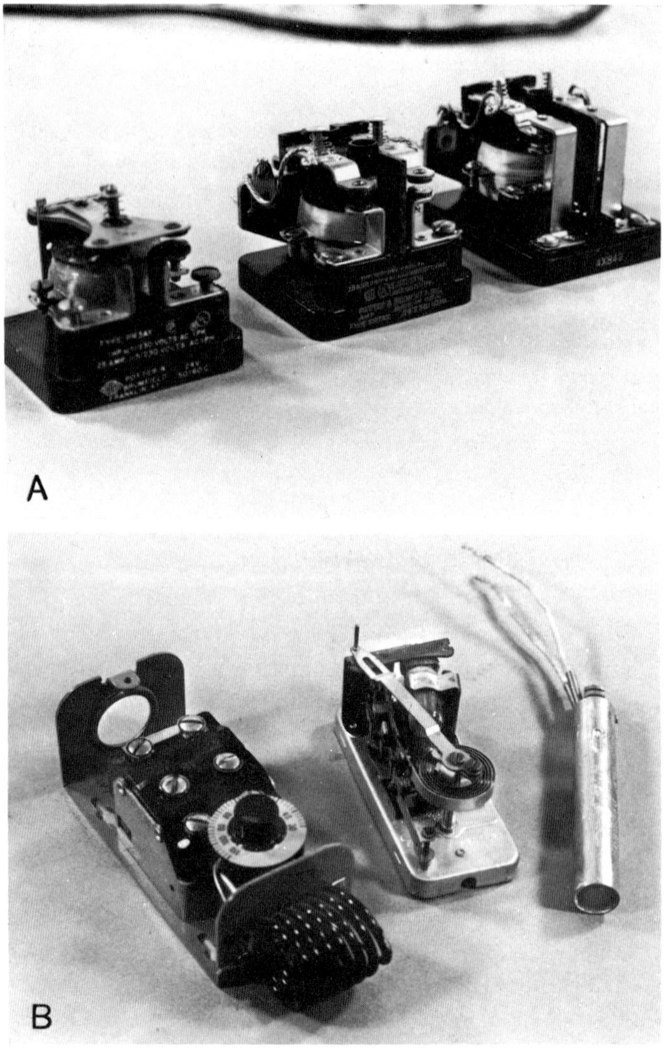

Fig. 4-63 A–D. Key components that are needed for the proper control of greenhouse environment. (A) Relays (*left* to *right*): Single pole single throw (SPST), double pole single throw (DPST), double pole double throw (DPDT). Relays which are generally energized with low voltage and control higher voltage circuits. (B) Thermostats (*left* to *right*): expanding gas type, mercury (24 V only), and bimetal expansion. (C) Circuit breakers: heater sensitive breaker and simple fuse. Can also be used as quick disconnects. (D) Modulating steam valves; *left*: electric; *right*: pneumatic

4.5.5 Making Poly Tubes Work

Temperatures were recorded in a commercial greenhouse where a forced air poly tube circulation system was in use. A thermometer was adjacent to the thermostat in the center of the house and the grower felt uniform temperatures were being

Fig. 4-63 C and D

achieved. The temperature records, Figure 4-64, showed that a very uniform temperature was obtained near the thermostat but those at the ends of the house were highly variable. It is apparent the poly tubes, holes, and heaters were not "matched".

4.5.5.1 *Poly Tube Design*

A poly tube used for cooling may not meet the needs for a heating program, or visa versa, thus one should be aware of certain criteria. Carpenter and Hawkins (1970) developed the following conclusions from their research:

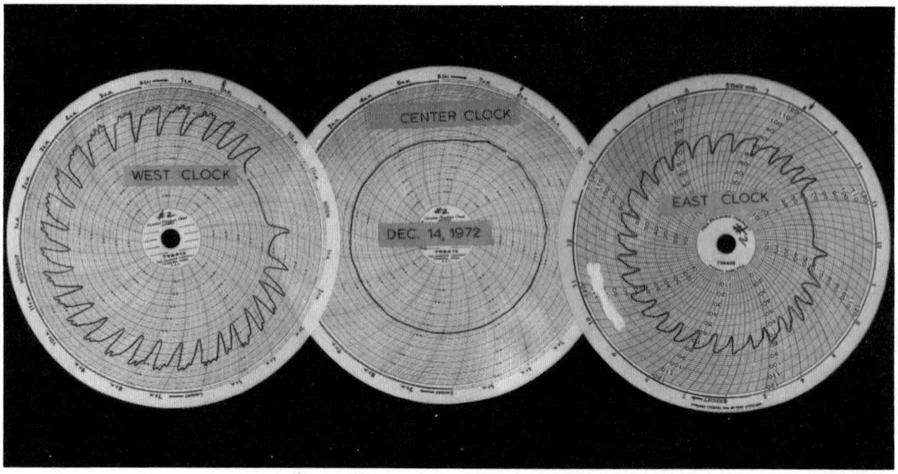

Fig. 4-64. Temperature variations recorded in a commercial greenhouse where double poly tubes were installed on heaters located in opposite ends of the greenhouse. Center recorder was next to thermostat. High temperature delay was observed on a poinsettia crop at each end of the house

1. Holes in tubes up to 2 in diameter have little effect on fan and duct performance.
2. When holes are uniformly spaced, the volume and velocity of air discharged from the holes near the closed end are twice as great as that from holes near the fan.
3. A long duct should have the same number of holes as a short one, for a given air flow, when tubes are less than 100 ft in length. Tubes longer than 100 ft in length should be larger in diameter.
4. An air straightener should be placed down stream from the propeller to counteract circular air motion in the tube.

Duncan and Walker (1973) described additional poly tube requirements. Discharge holes should be positioned at 4 and 8 "o'clock" on the tube for most crops. In some instances, the holes should direct the air horizontally or slightly upward. It is further recommended that tubes be large enough to force air down near the floor and have a continuous air circulation of approximately 30% of the house volume.

4.5.5.2 Sizing Poly Tubes for Heating and Cooling

The following information by Parsons (1974) is provided to acquaint the reader with the principles and requirements for designing poly tubes that will provide more uniform temperatures within the greenhouse environment:

"Most greenhouse operators buy pre-designed, punched tubes; this method is recommended because it is usually more convenient. But some growers who have special applications or interests prefer to design their own ventilation or heating systems and punch the tubing themselves.

To Size Tubing To Distribute Heat From a Space Heater

1. Calculate total heat needed.
   a) Find total surface area (A) of greenhouse walls and roof.
   b) Decide minimum allowable inside temperature (T_1).
   c) Determine minimum outdoor temperature (T_0) in the area.
   d) Calculate total heat loss (Q) in BTU per hour by the following formula:

   $$Q = 1.5\,\mathrm{A}(T_1 - T_0)\,.$$

   When double-layer plastic is used to cover the greenhouse, use the formula $Q = 1.0\,\mathrm{A}(T_1 - T_0)$.
   For glass-covered houses: $Q = 1.08\,\mathrm{A}(T_1 - T_0)$.
   e) Choose number of heaters needed.
2. Choose a fan that delivers 2000 cubic feet per minute per 100000 BTU h^{-1} of heater capacity at a static pressure of 0.15 inch water gauge (w.g.), or:

   $$\mathrm{ft}^3\,\mathrm{min}^{-1} = 0.02\,Q\,,$$

   where $\mathrm{ft}^3\,\mathrm{min}^{-1}$ = fan capacity needed in cubic feet per minute.
3. Select tubing to fit fan diameter chosen. Remember, tubing is flat when purchased and so the purchase or "lay flat" width is:

   $$\text{Lay flat width} = 1.57\,(\text{Tube Diameter})\,.$$

4. Decide number (N) of outlet holes required. The holes are usually in pairs and two feet apart, and so the number of holes equals the tube length.
5. Calculate hole diameter by formula:

   $$\text{Hole diameter} = \sqrt{0.19\,\frac{\mathrm{ft}^3\,\mathrm{min}^{-1}}{N}}\,.$$

6. Punch outlet holes centered one-third of the tube width from the edge of the flattened tube for a discharge angle of 60° from the vertical direction. Holes centered one-fourth of the tube width from the tube edge discharge at an angle of 45° from vertical.

To Size Tubing for Ventilation Only

1. Calculate greenhouse volume and divide by 6 to find the air flow rate desired. This air flow provides 1 air change per house and so temperature inside the greenhouse on a sunny day will be 15° F hotter than the outside temperature.
2. Choose a fan that delivers this air volume at 0.15 inch static pressure.
3. When outside air temperatures exceed 75° F, wall-mounted exhaust fans are used instead of fan tubes for ventilation. Wall fans should change the air in the greenhouse once a minute.
4. Follow steps 3 through 6 in Section "To Size Tubing to Distribute Heat From a Space Heater."

To Size Tubing for Multiple-Duct Systems

1. Decide number of branch tubes to be installed for each heater.
2. Calculate total heat capacity and $\mathrm{ft}^3\,\mathrm{min}^{-1}$ needed for the system.
3. Calculate the air flow in each branch tube by dividing the total air flow from the heater by the number of branch tubes.

   $$(\mathrm{ft}^3\,\mathrm{min}^{-1})_b = \frac{\mathrm{ft}^3\,\mathrm{min}^{-1}}{N_b}\,,$$

where: $ft^3\ min^{-1}{}_b$ = air flow in the branch tube,
$ft^3\ min^{-1}$ = total air flow from heater,
N_b = number of branch tubes.

4. Calculate the branch tube diameter (D_b)

$$D_b = \sqrt{0.15\,(ft^3\ min^{-1})_b}\,.$$

5. Decide number (N) of outlet holes required.
6. Calculate hole diameter in inches by:

$$\text{Hole diameter} = \sqrt{0.19\,\frac{(ft^3\ min^{-1})_b}{N}}\,.$$

7. Calculate main duct diameter (D_m) by:

$$D_m = \sqrt{0.1\,(ft^3\ min^{-1})}.$$

Example of Heating System Design

A 24 × 100 ft greenhouse with 9 ft side wall is to be heated.
Outside minimum temperature: 25° F
Inside minimum temperature: 55° F

Calculations

1. Area:

Side wall area	
2(9 × 100) =	1800 ft²
Roof area	
2(12.65 × 100) =	2600 ft²
End wall area	
2(24 × 9 + 12 × 5) =	552 ft²
Total surface area	4952 ft²

If greenhouse is in the center of a series of ridge and furrow greenhouses with common sidewalls, only the roof and end walls lose heat and need to be included in the area calculations. In that case, the total area is the roof area (2600 square feet) plus end wall area (552 square feet), or 3152 sq. ft.

2. Heat needed: $Q = 1.5(4952)(55\text{-}25) = 223000$ BTU h^{-1}.
For flexibility, choose two heaters with 110000 BTU h^{-1} output feeding into one fan and tube.
3. Fan capacity: (220000) (.02) = 4400 $ft^3\ min^{-1}$ at 0.15 inch static pressure. A 1/3 horsepower, 24-inch-diameter fan can be used.
4. Tubing size: Lay flat width = (1.57) (24) = 37.7 inches, so buy a 38 inch wide plastic tube.
5. Hole spacing: Assume tube is 90 feet long and pairs of holes are spaced 2 feet apart, so 90 holes are punched in the tube.
6. Hole diameter:

$$0.19\,\frac{4400}{90} = 3\ \text{in}\,.$$

7. Discharge angle: 60° from verticle is desired, so holes are centered (1/3) (38) = 125/8 inches from edge of flat tubing."

4.5.6 Temperature Control with Infrared Thermometers

Controlling plant temperature indirectly by controlling air temperature leaves some things to be desired. Ideally, the grower is most interested in the average plant temperature which varies, as pointed out earlier, with radiation, wind velocity, water loss, etc. Attempts have been made in the past to control plant temperature by inserting sensitive elements into leaves, buds and stems, and operating on the signal these devices generate. Such systems suffer from the fact that they measure temperature at one point, and may not be representative of actual canopy temperature.

With the development of remote sensing, infrared thermometers, a method has been developed by which a large area of vegetation can be scanned, and the temperature is integrated to provide an output that is a good average of the total canopy. Such systems have been tested with the output from the infrared thermometer being used to control the heating and cooling system. Theoretically, the system eliminates problems that arise with improper thermostat location, improper aspiration and shielding, and the fact that plant temperature can often differ markedly from the air temperature. There are a number of difficulties: (1) We do not know the proper plant temperature required. All of our recommendations are based on empirical recommendations for air temperature, which vary according to climatic location, greenhouse structures and cultural practices. (2) The apparatus is expensive, and inquiries to manufacturers have not yet resulted in interest to develop the infrared thermometer as a reasonably priced temperature control system. (3) There is insufficient information to say that such control will provide any renumeration to the operator in terms of higher production and quality. Colorado State University's tests indicate, on the average, one to two degree variations between air temperature and average canopy temperature. There does not appear to be any advantage in regard to fuel savings by using the infrared thermometer. The question is largely the requirement for improved technology and cheaper equipment before suitable large-scale tests can be carried forward.

4.6 Interrelations of Temperature and Greenhouse Production

The amount of solar energy that is used to heat a greenhouse soil is directly related to the soil moisture content, duration of radiation contact, and soil area exposed. As plants grow and gradually shade the soil, the soil temperature in raised benches will approach the night time temperature and remain there until a cultural or environmental conditions change it.

The greenhouse grower of late 1800s and early 1900s had hot water pipes under raised benches and next to the sides of ground beds which kept the soil relatively warm. The mid-20th century greenhouse operators have basically eliminated any ground level heating equipment.

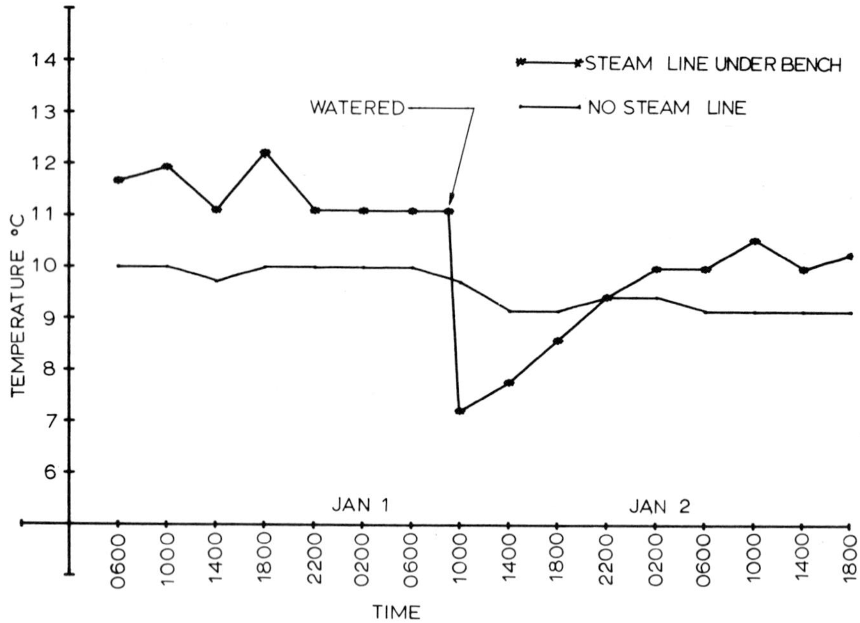

Fig. 4-65. Two-hour mean temperatures of ground bench soils used to grow carnations and watered with 43° F water (Goldsberry, 1962, unpublished)

4.6.1 Soil Heating

Several researchers and growers have been questioning the feasibility of maintaining a certain soil temperature during plant growth. Canham (1970) indicated that soil temperature is limiting for some crops. In 1974, Goldsberry (unpubl.) grew plants, *Lilium longiflorum*, with the pots submerged in sand that was heated to 65° and 70° F. A highly significant linear relationship occurred. Plants in the 70° and 65° F treatment flowered in seven and four days faster, respectively, than those grown at 60° F. One of the agronomic crops that was used to determine the effects of soil heating was grass (Rykbost et al., 1975). By increasing soil temperatures 10° C, the yields of fescue were 19% and sudan grass 50% above those produced in unheated soils. In unpublished results, Goldsberry (1976) found that the fresh weight and rate of growth of several crops, including radishes, lettuce, cauliflower, and Swiss chard were significantly greater in a 60° F ground bed than in an unheated bed.

The benefits for heating soil are not limited to temperature, but include more rapid decomposition of organic matter and ultimately higher CO_2 concentrations (Kruyk, 1976).

There is no doubt that minimal heating of the soil would affect the greenhouse heat budget in a small way. However, where soil temperature is raised, the air temperature of some crops may be reduced and over all fuel savings achieved.

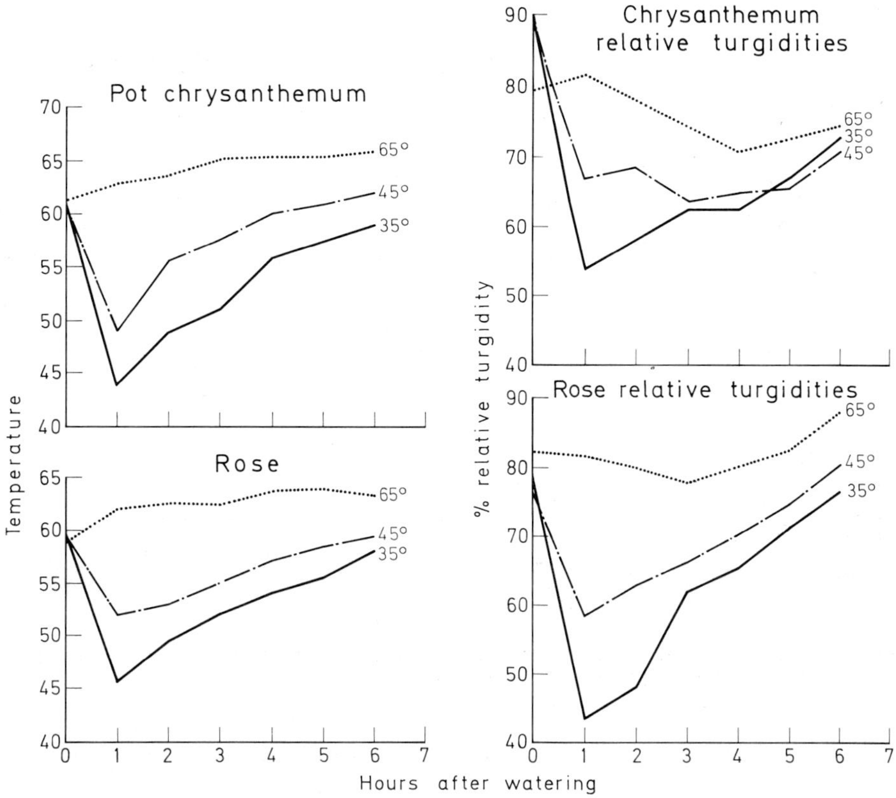

Fig. 4-66. Effects of irrigation water temperature on roses and chrysanthemums (Carpenter and Rasmussen, 1970). *Left:* Changes in soil temperature. *Right:* Changes in relative turgity of plants

4.6.2 Warm Water

The temperature of greenhouse soils can be partially controlled in winter months by using warm water for irrigation purposes. When greenhouse carnation benches are saturated with 43° F irrigation water, the soil temperature drops abruptly (Fig. 4-65).

It is also apparent that water temperature has a rapid effect on the turgidity of plants (Fig. 4-66). Carpenter and Rasmussen (1970) reduced greenhouse soil temperatures with cold irrigation water in midwinter which in turn reduced rose and chrysanthemum turgidities, tended to partially or completely close stomates, and decreased plant heights and fresh weight.

The modern day grower in colder climates should not only consider warming the media in which plants are growing but the use of warm water for irrigation. The pot plant grower may not be able to warrant heating his pots, but he could economically warm the irrigation water to the temperature range of the day time environmental conditions.

References

Acme Engineering and Manufacturing Corporation: A new concept in automatic greenhouse ventilation. Form C637, 1963

Anonymous: New blind material may be money saver. The Grower, November 29, 1975

Ashrae: Guide and Data Book. Fundamentals and Equipment. New York: Am. Soc. Heating, Refrigeration, Air-Conditioning Engineers, Inc., 1963

Ashrae: Handbook and Product Directory, 1974 Applications. New York: Am. Soc. Heating, Refrigeration, Air-Conditioning Engineers, Inc., 1974

Augsburger, N. D., Bohanon, H. R., Calhoun, J. L.: The Greenhouse Climate Control Handbook. Muskogee, Oklahoma: Acme Engineering and Manufacturing Corp., 1969

Augsburger, N. D., Bohanon, H. R., Calhoun, J. L., Hildinger, J. D.: The Greenhouse Climate Control Handbook. Muskogee, Oklahoma: Acme Engineering and Manufacturing Corp., 1975

Axlund, D. S., Besemer, S. T., Brown, A., Jr.: Greenhouse Insulation Experiment. San Diego Gas and Electric Co., Nov. 11, 1974

Bailey, L. H.: Greenhouse. Cyclopedia of American horticulture. New York: MacMillan, 1900

Ball, G. J.: The Ball Red Book. Geo. J. Ball, Inc., 1975, 501 pp

Ball, V.: Roof shading is automated. Grower Talks **38**(2), June (1974). West Chicago, Ill: Geo. J. Ball Inc., 60185

Bark, L. D., Carpenter, W. J.: Understanding the air circulation patterns produced by the horizontal unit heater/convection tube systems. Florist's Review, No. 1493 (3713), pp. 18–19, 60–61, Jan. (1969)

Barnard, G. K.: Panel method of thermal insulation. Comm. Gr., No. 3031, p. 180 (1954)

Bishop, J. R.: Advanced concepts in greenhouse heating. Orchid Digest. **17**(4), July-August (1954)

Boiler Water Compounds: Greenhouse Notes. The Penn. State Univ. Extension Service, August, 1972

Brooks, W. M.: Growing greenhouse tomatoes in Ohio. Columbus, OH: The Ohio State Univ. Bull. SB-19, 1973, 55 pp

Canham, A. E.: How important is soil temperature? The Grower, October 10, 1970

Carpenter, G. A., Hawkins, J. C.: The design of permeable ducts and their application to the ventilation of livestock buildings. In: Poly-tube Heating-Ventilation Systems and Equipment. Duncan, G. A., Walker, J. N. (eds.). Univ. Kentucky, Dept. Agr. Eng. Bull. AEN-7, September, 1973

Carpenter, W. J.: Temperature patterns and greenhouse heating. Penn. Flower Growers, Bull. **221**, 6–8 (1969)

Carpenter, W. J.: Environmental factors: temperature, light. In: Bedding Plants. Mastalerz, J. W. (ed.). Univ. Park, PA: Penn. State Flower Growers, 1976, 515 pp

Carpenter, W. J., Bark, D. L.: Temperature patterns in greenhouse heating. Part 1. Introduction. Florists Rev. Jan. 26 (1967a) pp. 17–19

Carpenter, W. J., Bark, D. L.: Temperature patterns and steam coils. Florists Rev. Feb. 2 (1967b) p. 43–45

Carpenter, W. J., Bark, D. L.: Vertical discharge unit heaters. Florists Rev. Feb. 9 (1967c) pp. 21–22

Carpenter, W. J., Bark, D. L.: Effects of air circulation. Florists Rev. Feb. 16 (1967d) pp. 28, 92–95

Carpenter, W. J., Bark, D. L.: Summary and discussion. Florists Rev. Feb. 23 (1967e) pp. 98–101

Carpenter, W. J., Mecklenberg, R. A., Carlson, W. H.: Greenhouse heating with horizontal discharge unit heaters. Florists Rev. June 11 (1970)

Carpenter, W. J., Rasmussen, H. P.: Temperature effect of greenhouse water on rose and chrysanthemum turgidity, stomate behavior, growth, and flowering. J. Am. Soc. Hort. Sci. **95**(5), 578–582 (1970)

Carpenter, W. J., Willis, W. W.: Comparison of low pressure mist atomized fog and evaporative fan and pad systems for greenhouse cooling and plant response. J. Am. Soc. Hort. Sci. **70**, 490–500 (1957)

Carrier System Design Manual: Part 8. Auxiliary Equipment. Syracuse, N.Y.: Carrier Air Conditioning Co. 1966

Christensen, S. A.: Reducktion of natligt varmetab. Saertryk af Gartnertidende. **86** (48), 821–823 (1970)

Controlled Environment Equipment for Greenhouses: Muskogee, Oklahoma: Acme Engineering and Manufacturing Corp. Form C 29 D 875, 1975

Co Ray Vac.: Form 1158 B. Buffalo, N.Y.: Roberts Gordon Appliance Corp., 1963

Criley, R. L.: Effects of light and temperature on flower initiation and development. In: Growing Azaleas Commercially. Richmond, Calif.: Univ. Ca. Publ. No. 4058, 1975, 108 pp

DeHertogh, A.: Holland Bulb Forcer's Guide. Mich. Agr. Exp. Sta. and Netherlands, Flower-Bulb Inst. Netherlands 1973

DeWerth, A. F.: A practical method of cooling greenhouses. Texas Agric. Expt. Sta. Prog. Rpt. 1729 (1954)

DeWerth, A. F., Jaska, R. C.: Greenhouse cooling. Texas Agric. Expt. Sta. MP-163 (1956)

Dietz, C. F.: Sowing schedules. In: Bedding Plants. Mastalerz, J. W. (ed.). Univ. Park, PA.: Penn. State Flower Growers, 1976, 515 pp

Dorland, R. E., Went, F. W.: Plant growth under controlled conditions. Am. J. Botany **34**, 393–401 (1947)

Duncan, G. A., Walker, J. N.: Poly tube heating-ventilation systems and equipment. Univ. Kentucky. Dept. Agr. Eng. Bul. AEN-7, September (1973)

Edlin, R. E. Jr.: Greenhouse Temperature Control with Forced Air Circulation. M.S. Thesis, Colorado State University, May, 1963, p. 54

Gas Heater Check List: Colo. Flower Growers' Assoc. Bull. 247, December, 1970

Geiger, R.: The Climate Near the Ground. Cambridge: Harvard Univ. Press, Translated from Das Klima der bodennahen Luftschicht. Brunswick, Germany: Vieweg, 1965, 611 pp

Goldsberry, K. L.: Controlling greenhouse temperatures. Colo. Flower Growers' Assoc. Bull. **165**, December, (1963)

Goldsberry, K. L., Holley, W. D.: Six year evaluation of environment on yield and quality of greenhouse roses. Colo. Flower Growers' Assoc. Bull. **191**, 1–3 (1966)

Goldsberry, K. L., Wolnick, D.: Evaluation of greenhouse shading compounds. Colo. Flower Growers' Assoc. Bull. **190**, 1–3 (1966)

Gray, H. E.: Summer cooling of greenhouses. N.Y. State: Flower Growers Bull. **33**, 8 (1948)

Gray, H. E.: Glasshouse heating. Booklet Cornell Univ. Agr. Exptl. Sta. Bull. **906**, 54 (1953)

Gray, H. E.: Greenhouse Heating and Construction. Chicago, Ill: Florists' Publishing Co. 1956, p. 78

Grimmer, W. W.: Greenhouse Cost Accounting. Kenosha, Wisconsin: Gateway Technical Inst., 1975

Gustafson, F. G.: Is natural gas injurious to flowering plants? Plant Physiol. **19**, 551–558 (1944)

Hall, A., Hanan, J. J.: Measurement of total light energy in a carnation bench. Colo. Flower Growers' Assoc. Bull. **308**, 2 (1976)

Hanan, J. J.: Air movement and temperature control. II. Colo. Flower Growers' Assoc. Bull. **103**, 1–4 (1958)

Hanan, J. J.: Influence of day temperatures on growth and flowering of carnations. Proc. Am. Soc. Hort. Sci. **74**, 692–703 (1959)

Hanan, J. J.: Temperatures in carnation flowers. Colo. Flower Growers' Assoc. Bull. **226**, 1–3 (1969)

Hanan, J. J.: Temperature and humidity observations in CSU greenhouses. Colo. Flower Growers' Assoc. Bull. **238**, 1–4 (1970)

Hanan, J. J.: Use of natural gas for CO_2 production in greenhouses. Colo. Flower Growers' Assoc. Bull. **262**, 1–5 (1972)

Hanan, J. J.: Ventilation temperatures, CO_2 levels, and rose production. Colo. Flower Growers' Assoc. Bull. **279**, 1–5 (1973)

Hanan, J. J.: Accuracy of common thermometers. Colo. Flower Growers' Assoc. Bull. **298**, 1–3 (1975)

Hanan, J. J., Heins, R.: Effect of plant density on two years of carnation production. Colo. Flower Growers' Assoc. Bull. **302**, 1–3 (1975)

Hanan,J.J., Moon,D., Holley,W.D.: Summer quality in carnation benches. Colo. Flower Growers' Assoc. Bull. **276**, 1–2 (1973)
Harrison,M.: How much does temperature dropping save? Grower Talks Supplement. West Chicago, Il.: Geo. J. Ball, December, 1973
Hart,S.A.: Studies of heating, cooling and humidity control within greenhouses. PhD dissertation, Cornell University, N.Y., 1950
Hartsema,A.M., Luyten,I., Blaauw,A.H.: The optimal temperatures from flower formation to flowering of Darwin tulips. II. Proc. Koninkl. Ned. Akad. Wetenschap. **27**, 1–46 (1930)
Hastings,R.: Lilies for Easter pots. In: The Ball Red Book. West Chicago, Il.: Geo. J. Ball, 1975, pp. 375–385
Heins,R.: Evaporative pad evaluations. Colo. Flower Growers' Assoc. Bull. **294** (1974)
Hoare,E.R., Morris,L.B.: The heating and ventilation of glasshouses. Farm Mechanization, **8**, 7 (1956a)
Hoare,E.R., Morris,L.B.: The heating and ventilation of glasshouses. J. Inst. Heating and Ventilation Engineers Britain, April, 1956b, pp. 1–34
Holley,W.D.: Rose cooling can be overdone. Colo. Flower Growers' Assoc. Bull. **87**, 1–2 (1957)
Holley,W.D.: Some responses of carnations to temperature. Colo. Flower Growers' Assoc. Bull. **254**, 1–5 (1971)
Holley,W.D., Manring,J.: Greenhouse Cooling Efficiency. I. Density of evaporative pad. Colo. Flower Growers' Assoc. Bull. **126**, 1–2 (1960)
Kewanee Boiler Care and Operation: Form KBM-507. Kewanee, Il.: Kewanee Boiler Corp., June, 1974
Kewanee System Data: Bulletin SD-002. Kewanee, Il.: Kewanee Industries, Inc., December 1972
Kruyk,P.A.: Soil heating makes a come-back. The Grower, January 24, 1976 (Suppl.)
Langhans,R.W., Maginnes,E.: Temperature and light. In: Snapdragons. Langhans,R.W. (ed.). Ithaca, N.Y.: N.Y. State Flower Growers Assoc., 1962, 93 pp
Large,J.A.: Glasshouse lettuce. London: Grower Books, 1972, 181 pp
Larson,R.A.: Propagation. In: Growing Azaleas Commercially. Richmond, Calf.: Univ. CA. Publ. No. 4058, 1975, 108 pp.
Laurie,A., Kiplinger,D.C.: Commercial Flower Forcing. Philadelphia: The Blakiston Co., 1948, 550 pp
Liu,R.C., Bailey,W.A., Klueter,H.H., Krinzek,D.T.: Problems in measuring heat losses through greenhouse covering materials. In: Proc. 9th Natl. Agr. Plastics Conf., October, 1969, pp. 198–212
Love,J.W.: Vegetative growth. In: Growing Azaleas Commercially. Richmond, Calf.: Univ. CA. Publ. No. 4058, 1975, 108 pp
Macrae,J.C.: An introduction to the study of fuel. New York: Elsevier, 1966
Mastalerz,J.W.: Stock plant culture. In: Geraniums. Mastalerz,J.W. (ed.). Univ. Park, PA: Penn. State Flower Growers, 1961, 100 pp
Mastalerz,J.W.: Environmental factors—light, temperature, carbon dioxide. In: Roses. Mastalerz,J.W., Langhans,R.W. (eds.). Univ. Park, PA: Penn State Flower Growers, 1969, 331 pp
McColley,R.: Tropical foliage plants. In: The Ball Red Book, pp. 319–329. West Chicago: Geo. J. Ball, Inc., 1975
Morris,L.G.: Principles of glasshouse heating. Agriculture **66**, 403–407 (1959)
Nakata,M.: Development of special PVC films for agricultural use in Japan. Plasticulture Bull. **15**, 19–27 (1972)
Natural gas. Encyclopedia of Science and Technology. McGraw Hill **9**, 6–8 (1971)
Parsons,R.A.: Sizing perforated polyethylene tubes for greenhouse ventilation and heating. Univ. of Calif. Flower and Nursery Report, November, 1974
Post,K.: Florist Crop Production and Marketing. New York: Orange Judd, 1950, 891 pp
Rathmell,J.K. Jr.: Growing and finishing flowering plants. In: Geraniums. Mastalerz,J.W. (ed.). Univ. Park, PA: Penn. State Flower Growers, 1971, 350 pp
Rebuck,S.M., White,J.W., Aldrich,R.A.: Internal curtains reduce energy requirements in greenhouses. Penn. Flower Growers Bull. 289, 1, 6–8 (1976)

Reducing Glasshouse Heat Losses. G.B.: Natl. Inst. Agr. Eng. Glasshouse Dep. Rep. May, 1975

Roberts, W.J.: Heating and Ventilating Greenhouses. Booklet of Coop. Ext. Serv. Rutgers Univ., August, 1969, p. 20

Rykbost, K.A., Boersman, L., Mack, H.J., Schmisseur, W.E.: Yield response to soil warming agronomic crops. Agron. J. **67** (6), 733–737 (1975)

Seeley, J.: Temperature and splitting. In: Carnations. Langhans, R.W. (ed.). Ithaca, N.Y.: N.Y. State Flower Growers Assoc., 1961, 107 pp

Shanks, J.B.: Hydrangeas. In: The Ball Red Book. West Chicago, Il.: Geo. J. Ball, Inc., 1975

Sheldrake, R.: Air makes the difference. Am. Veg. Grower. **19 (1)**, 28–34 (1971)

Sheldrake, R., Langhans, R.W.: Heating requirement of plastic greenhouses. Proc. Am. Soc. Hort. Sci. **80**, 666–669 (1962)

Spencer, D.L., Smith, T.F., Daunicht, H.J.: Variable shading for greenhouses. Speech presented at Am. Soc. of Mechanical Engineers Winter Meeting, November 17–22, 1974

Stålfelt, M.G.: Der Gasaustausch der Moose. Planta **27**, 30–60 (1937)

Steam, its generation and use. New York: The Babcock and Wilcox Co., 1955

Stickler, M.P.: The use of plastics for heat insulation in greenhouses. Plasticulture **25**, 41–53 (1975)

Stinson, R.F.: Environmental factors—temperature, light, carbon dioxide. In: Geraniums. Mastalerz, J.W. (ed.). Univ. Park, PA: Penn State Flower Growers, 1971, 350 pp

Taft, L.R.: Greenhouse Construction. New York: Orange Judd, 1926

Tayama, H.K., Kiplinger, D.C., McDowell, T.C., Staby, G.L., Powell, C.C., Lindquist, R.K.: Tips on growing ornamental and vegetable bedding plants. The Ohio State Univ. MM-265 Ext. Bull. 1975, 26 pp

The Trane Company: Boilers for steam and hot water. LaCrosse, Wisconsin 56401, 1971

Tristan, J.: Greenhouse heat conservation. M.S. thesis. Colo. State Univ., 1975

Voigt, A.O.: Are growers confusing energy problems with problems caused by poor marketing? Florists Rev. **159** (4094), pp. 23, 78–80 (1976)

Walker, J.N.: Predicting temperatures in ventilated greenhouses. Transactions Am. Soc. Agr. Engineers **8** (3), 445 (1965)

Walker, J.N., Cotter, D.J.: Cooling of greenhouses with various water evaporation systems. Paper presented at summer meeting of Am. Soc. Agr. Engineers, 1967

Walker, J.N., Duncan, G.A.: Estimating greenhouse heating requirements and fuel costs. In a report of the Coop. Ext. Service, Univ. Kentucky, Bull. AEN-8, 1973 a, p. 5

Walker, J.N., Duncan, G.A.: Estimating greenhouse ventilation requirements. Univ. Kentucky Dept. Agr. Eng. Bull. AEN-9. March 1973 b

Walker, J.N., Duncan, G.A.: Air circulation in greenhouses. Univ. Kentucky Dept. of Agric. Eng. Bull. AEN-18. July 1973 c

Walker, J.N., Walton, L.: Effects of condensation on greenhouse heat requirement. In ASAE Paper, Univ. Kentucky 69–416 (1969)

Went, F.W.: The role of environment in plant growth. Am. Sci. **44**, 378–398 (1956)

Went, F.W.: Climate and agriculture. Sci. Am. **852**, 3–11 (1957 a)

Went, F.W.: The Experimental Control of Plant Growth. Waltham, Mass.: Chronica Botanica, 1957 b, 343 pp

Whittle, R.M., Lawrence, W.J.C.: Part V—The heat consumption of glasshouses. J. Agr. Eng. Res. **5** (4), 399–405 (1960)

Wilsie, C.P.: Crop Adaptation and Distribution. San Francisco: W. H. Freeman, 1962, 184 pp

Winspear, K.W.: Glasshouse engineering research the energy crisis. Speech given at the National Glasshouse Energy Conference. Littlehampton, Eng. October 15 (1974)

Wright, W.J.: Greenhouses. New York: Orange Judd, 1921

Young, F.D.: Nocturnal temperature inversion in Oregon and California. Monthly Weather Rev. **49**, 138–148 (1921)

Chapter 5 Water

5.1 Introduction

Water use by greenhouse plants is directly related to available sunlight. However, less than two percent of the water that enters the roots remains in the plant. Most water passes through the water conducting tissue and evaporates into the air. For example, one kg of dry alfalfa represents 800 l of water (Maxwell, 1965). One carnation flower can use one gallon of water by the time it is cut (Hanan and Jasper, 1967). Rutland (1972) found that single stem, flowering snapdragons, in 6-inch pots, could transpire nearly 25 g of water per hour. With greenhouse crops, increasing yield requires more water (Hanan, 1972). In spite of high water requirements (in excess of 6 acre-ft per year for carnations), the economic return of greenhouse crops usually reduces cost for water to an insignificant factor in greenhouse culture (Hanan, 1968) (Fig. 5-2). Greenhouse culture is one agricultural system that can compete directly with industrial and domestic water users for available water supplies.

To move this water, a difference in pressure must exist between the supply (soil) and the "sink" (atmosphere around the leaves). It is this pressure differential

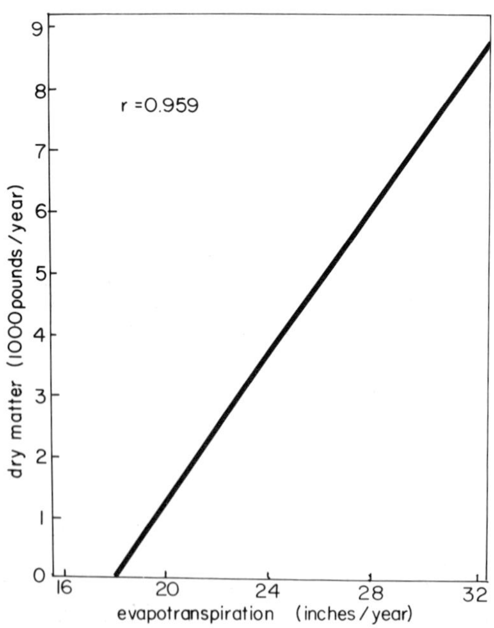

Fig. 5-1. Relationship between crop yield and water use. (After Allison et al., 1958)

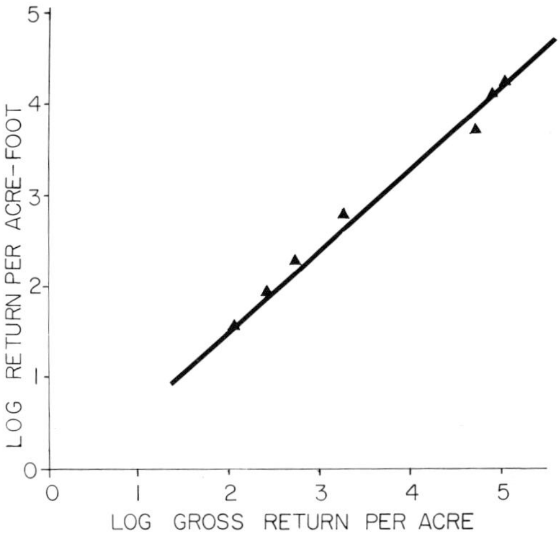

Fig. 5-2. Relationship between gross return per acre for various irrigated crops and return per acre-ft of water applied. Highest returns per acre-ft of water are for carnations, snapdragons and greenhouse tomatoes. Lowest returns are for alfalfa and sugar beets, followed by potatoes and field tomatoes. Values in axis of graph represent number of zeros after one. e.g. 1 = $ 10, 2 = $ 100, 3 = $1000, etc. (Hanan, 1968)

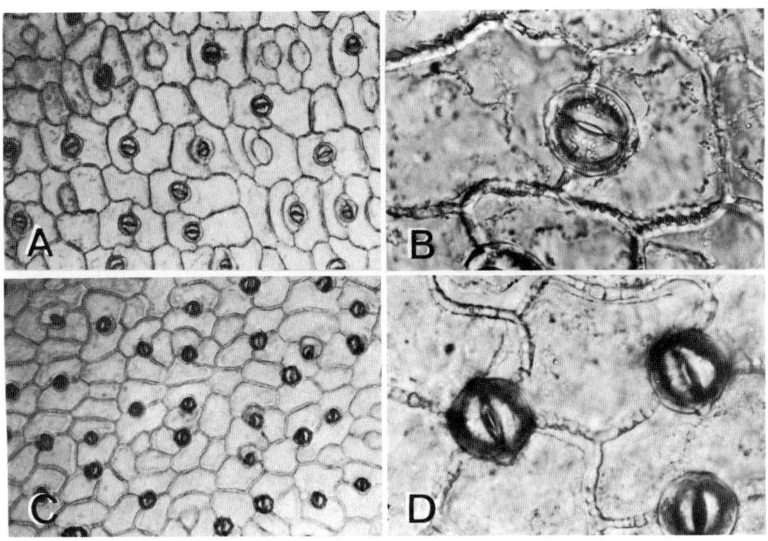

Fig. 5-3 A–D. Surface photograph of carnation stomata grown under two different irrigation regimes. (C) Note smaller stomata and closer arrangement of pores. This and heavy ring around the stomata in (C) and (D) indicate changes due to higher water stress induced by less frequent irrigation as compared with (A) and (B) taken from carnations grown in an inert medium. (From Hanan and Kowalczyk, 1968)

that intervenes in plant growth. The actual suction, or potential (Ψ_T), is regulated by: (1) species; (2) previous history; (3) flow rate; (4) supply rate; and (5) demand rate. Any restriction of water supply will increase the potential (more negative); an increase in demand (evaporation-transpiration) will do the same. Finally, an increase in flow rate through the plant will increase the gradient. Factors (3), (4), and (5) can be controlled in greenhouses by manipulating soil and irrigation practices, and controlling plant temperature and humidity. The effect of previous treatment is illustrated by Fig. 5-3, which gives an example of the adjustment a carnation undergoes when exposed to different water stresses. Higher stress usually increases cell wall thickness, reduces cell size, and can seriously affect basic physiological processes of the plant. Horticultural terminology refers to this as "hardening". A more recent term important to foliage plants is "acclimatization". Or, the modification of the plant to withstand conditions in the customer's home.

5.2 Terminology

Water stress, or internal plant water potential, is designated by the Greek letter Ψ_T, and can be described by a simple equation:

$$\pm \Psi_T = -\Psi_m - \Psi_o + \Psi_P$$

where Ψ_T = total plant water potential (stress)
Ψ_m = potential or stress due to matric forces
Ψ_o = potential resulting from osmotic forces
Ψ_p = the pressure potential resulting from turgor pressure within the plant.

Definitions of these terms are provided in the appendix. Essentially, matric forces (Ψ_m) are those resulting from the attraction of internal plant surfaces, colloids, etc. for water. In well-watered, herbaceous plants, it is often ignored. The osmotic potential (Ψ_o) arises from the fact that the internal cell sap contains materials in solution. The cell walls are more permeable to water molecules than to the materials (solutes) inside the cell. Water will move from a location where it is more dilute into the cell. This ability can be expressed in terms of potential or suction a cell can exert to imbibe water. The pressure potential (Ψ_p) is often referred to as "turgor" pressure, and is usually positive, tending to prevent water from moving into the cell.

The two extreme cases are: (1) when the Ψ_p is zero, and $\Psi_T = \Psi_o$. The plant is wilted, and (2), $\Psi_p = \Psi_o$, $\Psi_T = 0$, and the plant is fully turgid. The value of Ψ_T represents the internal water status, or the stress to which a plant is being subjected. A grower can modify Ψ_T through his cultural practices to influence water supply and demand. Plant water stress varies with soil moisture content, the physical properties of the soil, humidity, sunlight, temperature, wind velocity, soil volume, soil temperature, CO_2 concentration and other factors. As water movement through plants requires a pressure difference, the internal water stress will not be the same everywhere in a plant, but varies continuously with changes in the factors mentioned above.

Stress can be stated in energy terms (Joules, ergs). Usually employed are pressure terms such as "bars", "atmospheres", "pounds per square inch" or "kg per cm^2". "Bars" are most common and one bar is approximately equivalent to 14.7 lb in^{-2}. A value of -10 bars can be thought of as the ability of a cell, or

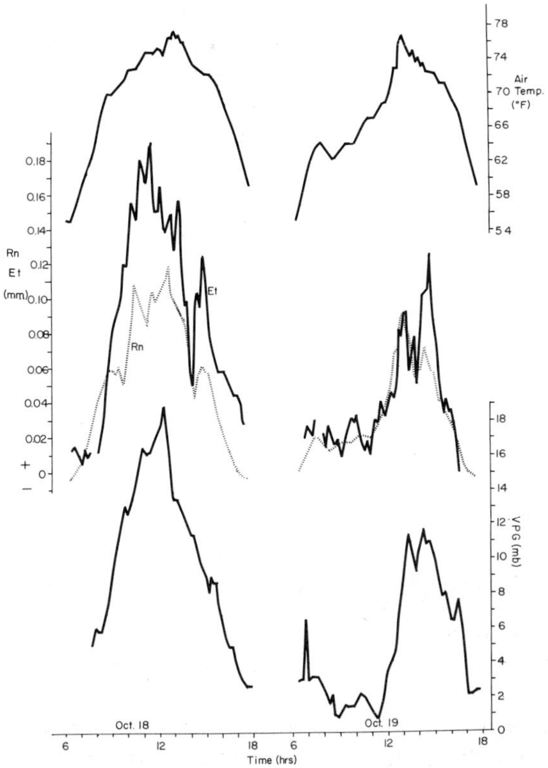

Fig. 5-4. Time course of water loss (evapotranspiration) from carnations in soil, compared with temperature and vapor pressure gradient between leaf and air in glass-covered greenhouse. *Rn* net radiation; *Et* evapotranspiration; *VPG* vapor pressure gradient, leaf to air (Hanan, 1969, unpublished)

plant, to extract water from the soil. To prevent water from moving, the potential of the soil water must be equal to or more negative than -10 bars.

During the day, water stress will be at some pressure less than that when the plant is fully turgid, as water is moving through it. The grower is dealing with a dynamic, constantly changing system, indicated to some extent by Figure 5-4. The final effect will depend upon how well the grower has managed his environment to provide optimum conditions for the longest period.

5.3 Effect of Water Stress

In general, reducing water supply by: (1) reducing irrigation frequency, or (2) increasing environmental demand, will reduce growth. An example is provided by Hanan and Jasper's work on carnations (1967). Figure 5-5 shows that different irrigation regimes will influence branching, stem length, quality and yield. Water stress may have far-reaching effects (Fig. 5-6), influencing keeping life and resistance to air pollution.

However, if a crop is planted single stem (chrysanthemums or snapdragons), yield is fixed. It can be desirable to reduce water application or decrease humidity in order to obtain a more compact flower head, or to reduce brittleness. The same

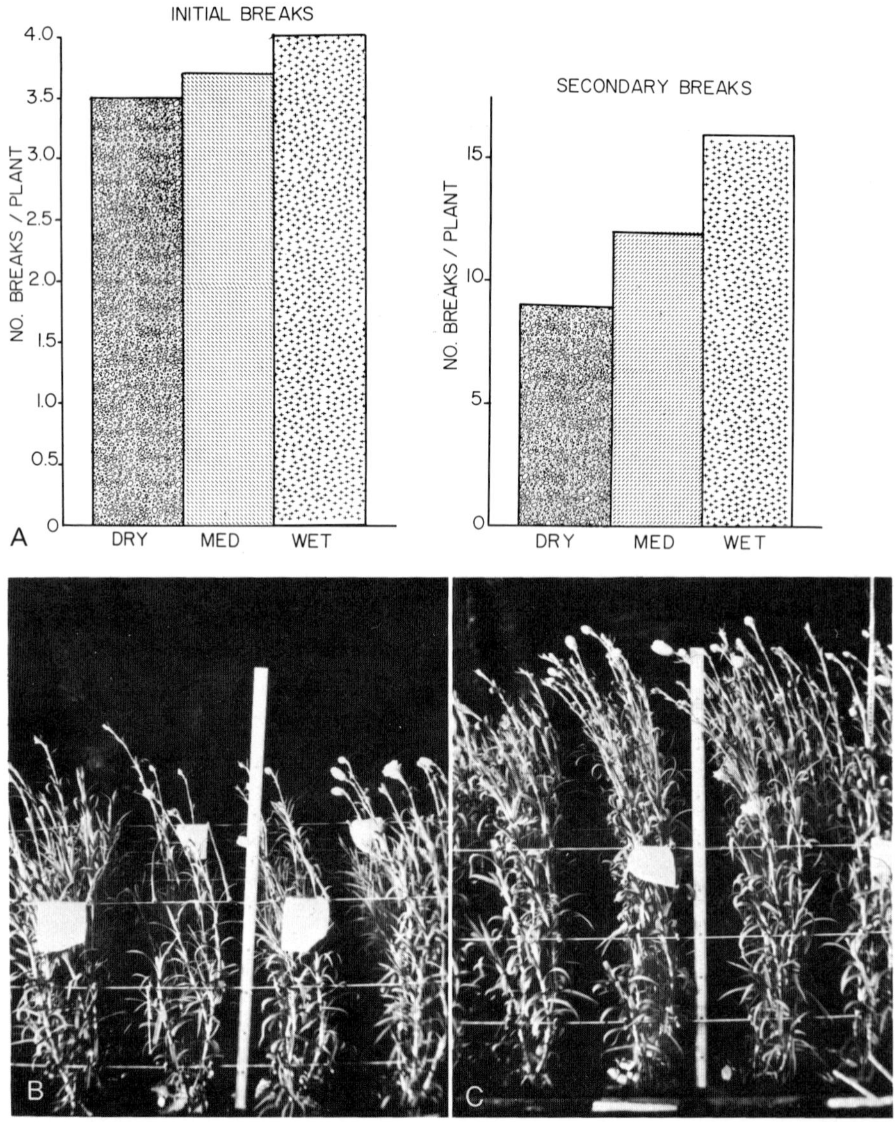

Fig. 5-5 (A) Effect of irrigation regime on number of breaks produced by carnations grown in soil. Plants single-pinched, and resulting initial breaks. Secondary breaks are those resulting after cutting the first flower (Hanan and Jasper, 1967). (B) and (C) Effect of irrigation regime on carnations grown in gravel. Plants shown were grown in $^1/_2$ to $^3/_4$-in aggregate: (B) irrigated once daily in the winter season, (C) 4–6 times daily. The effect of irrigation regime is more pronounced in an inert root medium (Hanan, 1969)

may be necessary with foliage, bedding, or potted plants where some judicious stress will improve ability to withstand transplanting shock, or the new environment provided by the customer. The exact effect of stress will depend upon how large the stress is, how long stress is imposed, how the plant was treated previously (past history), species, and stage of growth. With continuously flowering

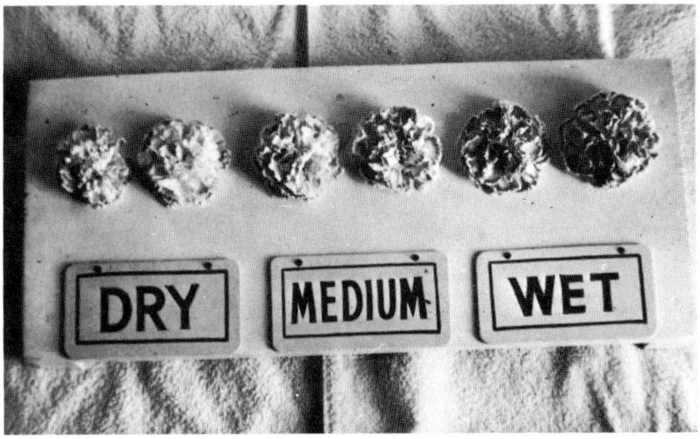

Fig. 5-6. Effect of irrigation regime on dye uptake in carnations (Hanan and Jasper, 1967)

plants, undue stress at any stage may severely restrict stem length and branching. For single cropping, high stress at flower initiation may reduce flower number and size.

The general effect of stress on some physiological processes in herbaceous plants was provided by Boyer (1970a, b) (Fig. 5-7). Stresses of less than −4 bars were sufficient to reduce elongation, and beyond −8 bars, photosynthesis was decreased. These are not extreme when, as a general rule, a soil suction of −15 bars is often accepted as the value where most field plants will permanently wilt. Internal water potentials approaching −15 bars have been measured by Hanan (1969) in carnations, and by Aikin (1974) in roses. Increasing stress will often reduce photosynthesis to such an extent that solar energy cannot be efficiently utilized (Boyer, 1970b).

If low stress is necessary for maximum growth, it does not follow that zero stress is best. Heydecker et al. (1970) showed that abnormal growth could result where there was no pressure gradient from the leaf to the air. Another problem at low stress conditions is increased damage from foliar disease such as *Botrytis* or mildew (Winspear et al., 1970). If there is no stress, it is highly unlikely that a plant is actively producing food. If there is sufficient energy (sunlight) to manufacture sugars, there will be sufficient energy to evaporate water and induce stress.

5.4 Control of Water Demand

There are several ways to control water demand, or reduce the transpiration rate in greenhouse plants:

I. Reduce vapor pressure gradient between leaf and air.
   1. Increase relative humidity.
      a) Add water to the air.
      b) Reduce air temperature.
   2. Reduce leaf temperature.
      a) Syringe or mist with water.
      b) Reduce light intensity by shading.

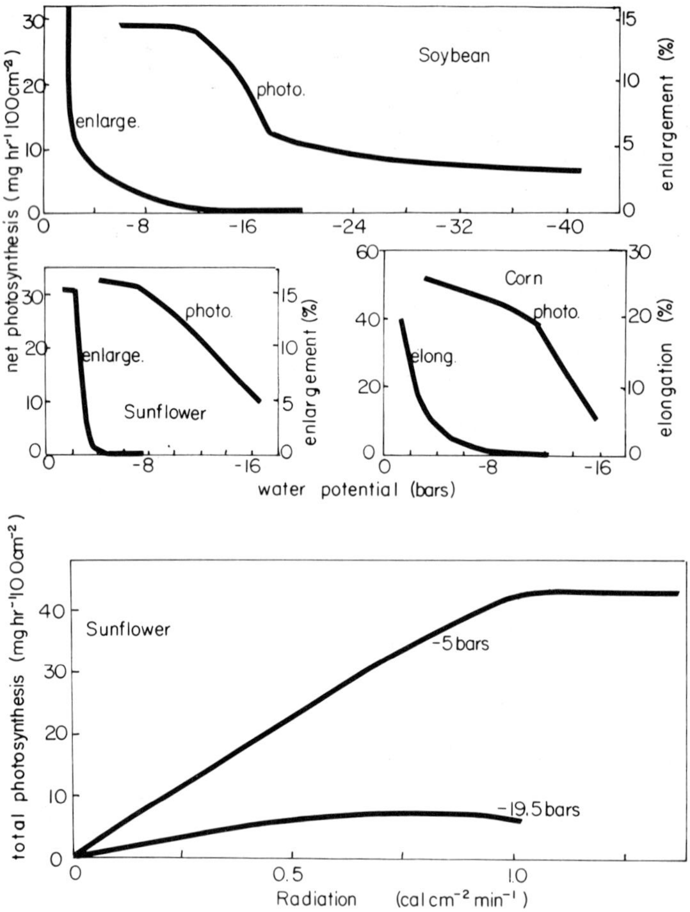

Fig. 5-7. Effect of internal plant water stress (potential) on photosynthesis and size increase in soybean, corn and sunflower. Bottom graph shows the relationship between photosynthetic rate and radiation intensity for two different plant water potentials. (Redrawn from Boyer, 1970a, b)

II. Increase resistance to water vapor flow from leaf to air.
   1. Reduce stomatal aperture by.
      a) Application of a chemical to close stomates.
      b) Reduce stomate opening by increasing CO_2 concentration.
   2. Increase length of diffusion path-length from leaf to air by reducing air velocity.

This outline shows that water loss is determined by: (1) the difference in water vapor concentration between inside the leaf and outside; and (2), by the resistance to movement of water molecules from inside the leaf to outside. The resistance varies according to the length of the path which water molecules must traverse, and the size of the stomate opening.

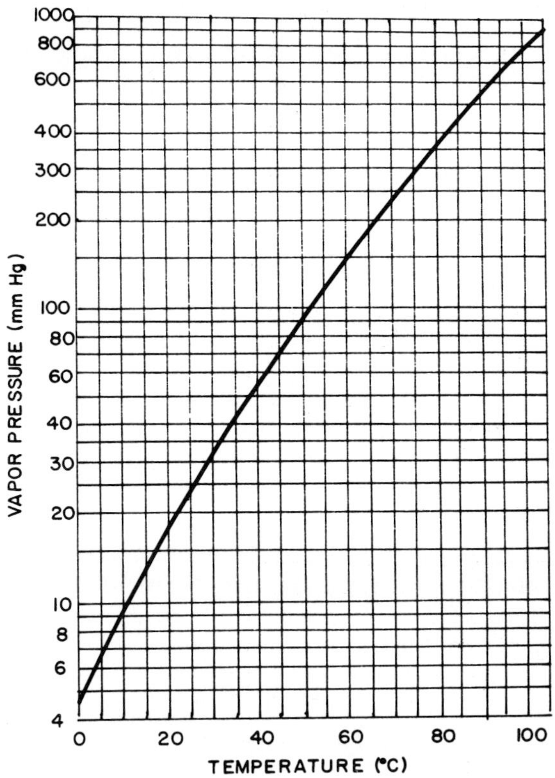

Fig. 5-8. Relationship between air temperature and saturation vapor pressure (List, 1966)

5.4.1 Behavior of Water in Air

Water molecules will escape from a water surface and return to it. There is a constant exchange between liquid water and the air above a water surface. If a pan of water is enclosed in an air-tight box, at a constant temperature, water molecules will escape from the liquid until they are so concentrated that the rate of return equals the rate of escape. The air is saturated, and the number of molecules is a direct function of the temperature. This concentration can be expressed as vapor pressure, dew point, wet bulb temperature, absolute humidity or specific humidity. For all practical purposes, the behavior of water molecules in other gases, will be independent of any other molecules, oxygen or nitrogen, that may be present. The total of all the concentrations of gases present (vapor pressure) will equal the total atmospheric pressure—which at sea level is 760 mm Hg or 14.7 lb in^{-2} (see appendix for units and conversions). The relationship between temperature and vapor pressure (e), when the air above the water is saturated (e_s), is illustrated in Figure 5-8. For example, at 10° C, e_s is about 10 mm Hg (13.3 mb or 0.19 lb in^{-2}); at 80° C, however, e_s will be about 350 mm Hg.

With freely watered, nonwilted plants encountered in the greenhouse, the internal spaces of leaves will be saturated (100% relative humidity, RH). The internal leaf, vapor pressure can be determined if leaf temperature is known (Fig. 5-8). If the vapor pressure outside the leaf is constant, then lowering the leaf temperature will lower the internal vapor pressure, and water loss will be less. This basic principle explains why a leaf can loose water to air that is at 100% RH. If leaf temperature is higher than air temperature, both places at 100% RH, then vapor pressure will be higher inside the leaf and water molecules will diffuse outward. It is on this principle that mist propagation is so successful. Leaf temperature is reduced and the outside air is maintained close to 100% RH.

Relative humidity as a means to indicate drying power of the air is misleading. Relative humidity is the ratio between actual vapor pressure and the vapor pressure of water in air if the air is saturated at the same temperature ($e_a/e_s \times 100 = \mathrm{RH}$). If temperature changes, the saturation pressure will change (e_s), and RH will change. The actual amount of water in air (e_a) does not have to change.

5.4.2 Measuring Water Vapor in Air

There are many ways to express water content of air, and there are many devices to measure water content of air. By far the most common are "psychrometric" devices that measure the temperature difference between a wet bulb thermometer, the bulb enclosed in a wetted cover, and a similar dry bulb thermometer. By using tables, the thermometer readings can be translated directly to RH, actual water

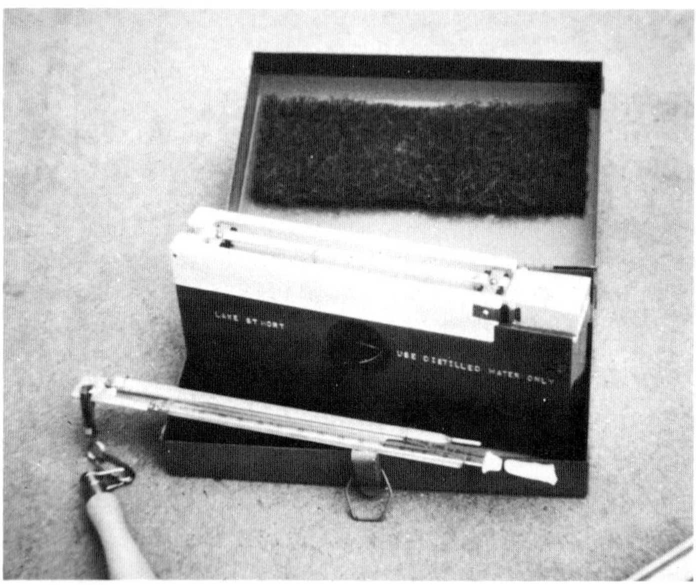

Fig. 5-9. Two types of psychrometers for measuring relative humidity. To achieve adequate ventilation, lower psychrometer is whirled. Cased psychrometer contains battery-powered fan to achieve ventilation, and is less susceptible to breakage

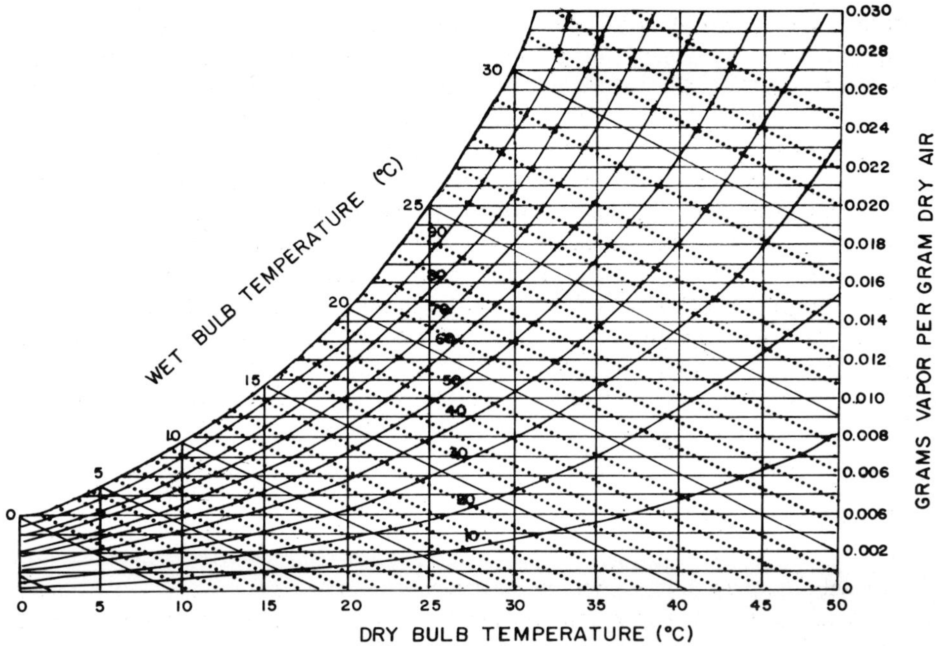

Fig. 5-10. Psychrometric diagram showing relationship between wet bulb and dry bulb temperatures, relative humidity and quantity of water in dry air. Wet bulb temperatures run diagonally toward lower right-hand corner of diagram. Intersection of vertical dry-bulb and diagonal wet-bulb temperature lines gives relative humidity (List, 1966)

content, or dew point. Other types that measure changes in length (hair, paper) can be purchased to directly read RH. But, these are seldom accurate to within ±20%. Psychrometers, such as the two shown in Figure 5-9, can be accurate to within ±5%. Simple sling type psychrometers can be purchased for as little as $ 10.00.

According to Zwart (1966), these precautions should be used when determining water content with psychrometric thermometers: (1) the thermometers should be matched; (2) the air speed over the thermometer bulbs should be above 3 m s^{-1}; (3) the wick on the wet bulb should be clean and wetted with distilled water; and (4) the thermometer bulbs should be shielded from radiation such as that from the sun. Readings require at least one minute, or a series of readings. Figure 5-11 shows two control devices (humidistats), using hair or paper to sense a humidity change. Both are subject to errors from dirt, or in the case of hair elements, exposures to relative humidities below 15% or above 85% can cause permanent shifts in calibration. Readings from wet and dry bulb thermometers can be converted to percentage relative humidity or absolute humidity through the use of charts such as Figure 5-10, or by means of tables. If the air is cooled below the wet bulb temperature, it will reach a temperature at which moisture will condense from the air. This is the "dew point" temperature, which determines, for example, whether condensation will occur on surfaces such as leaves.

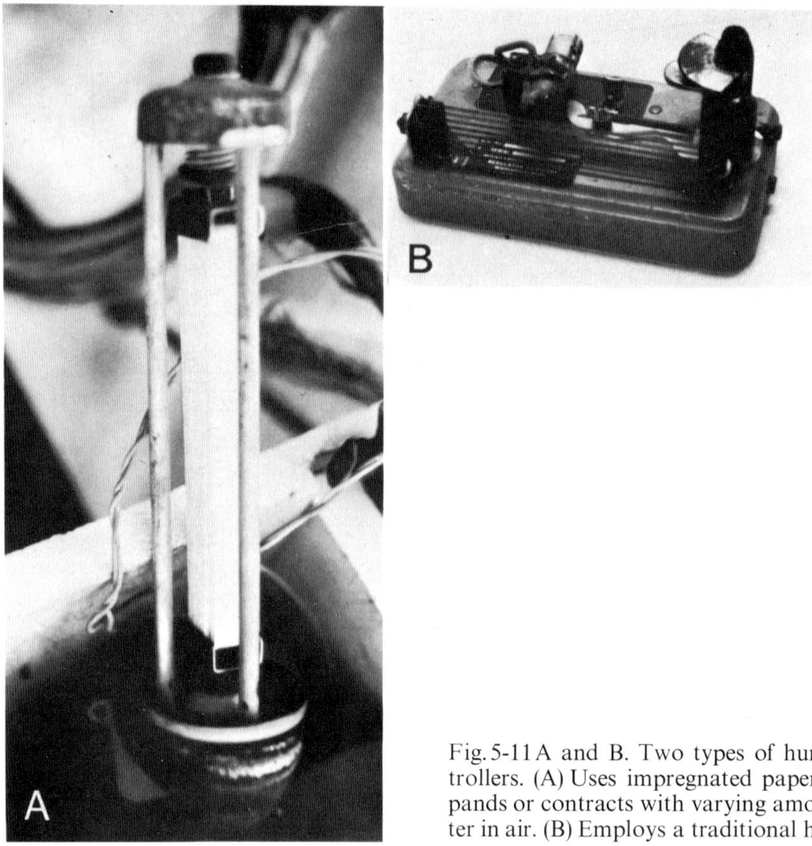

Fig. 5-11 A and B. Two types of humidity controllers. (A) Uses impregnated paper which expands or contracts with varying amounts of water in air. (B) Employs a traditional hair element

5.4.3 Importance of Wet and Dry Bulb Temperatures

In addition to indicating the drying power of air, wet bulb temperatures will show the theoretical maximum a greenhouse can be cooled using evaporative pads. If the wet bulb is 10° C below the dry bulb temperature, then the air can be cooled a maximum of 10° C. A wet and dry bulb measurement inside and outside the greenhouse will indicate pad efficiency. If the dew point is known, the grower knows how much the vegetation temperature can be reduced before free water will be present. Many foliage diseases are influenced by humidity. Some species will not tolerate free water, others are relatively tolerant. From the standpoint of cooling and disease control, knowledge of humidity and its control can save money.

5.4.4 Controlling Water Vapor Content of Air

Constant temperatures in greenhouses are seldom achieved. Furthermore, water is always being added to the air from transpiring plants. Any temperature de-

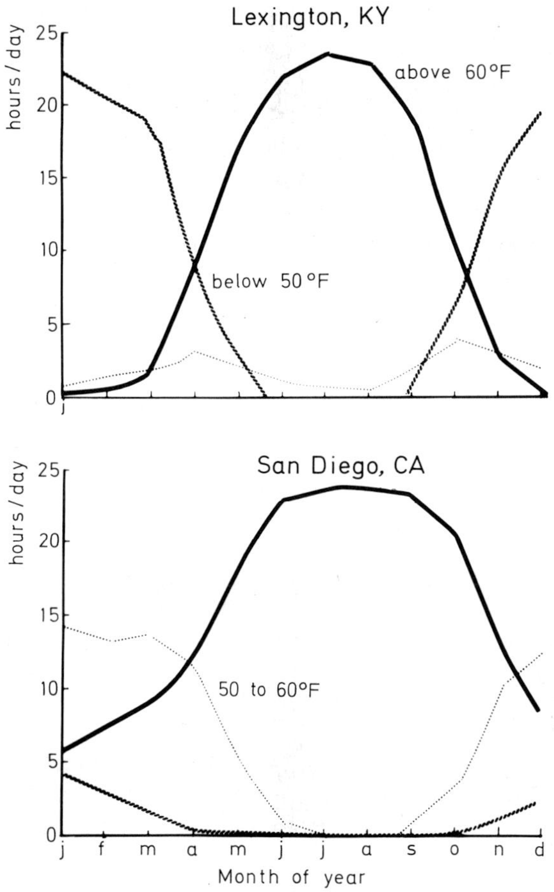

Fig. 5-12. The mean hours per day when temperatures are 60° F and above, below 50° F, and between 50 and 59° F for a ten-year period in San Diego, California, and a five year period in Lexington, Kentucky. (After Cotter and Walker, 1968)

crease, as often occurs in the evening, will tend to increase humidity. Cotter and Walker (1968) studied various climatic conditions which are illustrated in Figure 5-12. They assumed that heating is required when outside temperatures are below 50° F, and cooling is required above 60° F. The graph provides the average hours per day for each of the three conditions (below 50, above 60, or between 50 and 59° F) for Lexington, Kentucky, and San Diego, California. If heating is required, the exterior greenhouse surfaces will be cool, and condensation will occur on those surfaces. This causes dehumidification, and will be increased with rapid air movement. If ridge ventilators are open, interior heating will force some of the moisture laden air from the greenhouse. This, combined with a rise in temperature, is commonly used in the late afternoons and early evening hours to prevent 100% RH, or condensation on the foliage. It is a proven method for disease control.

At outside temperatures above 60° F, ventilation will usually occur, and inside air can be exchanged for dryer outside air. If RH is high (Lexington), large air exchanges can be permitted. At San Diego, or similar dry regions, high rates of air exchange may severely stress crops. During the transition period (50–59° F), positive humidity control is minimal, and is the danger period. Night firing during warm periods has been practiced for years as a means to reduce humidity during the critical transition period.

The degree of humidity control varies with greenhouse type and heating system. If natural gas is burned to supply CO_2, large quantities of water will be added to the greenhouse atmosphere. The tighter the greenhouse, the higher the humidity. Condensation on interior surfaces may cause a severe drip problem (Goldsberry, 1970) which can increase spread of disease.

5.4.4.1 *Increasing Humidity*

In arid climatic regions, or with some species, it is often desirable to increase humidity. With evaporative-pad cooling systems, lowering of the dry bulb temperature will generally raise RH above 70–80% (Hanan, 1970). This is usually sufficient for crops such as carnations or chrysanthemums. The flow of cool, humid air over leaf surfaces will tend to reduce the vapor pressure gradient further. Depending upon local conditions, the water flow to pad systems may be shut off shortly before exhaust fans cease to operate for the evening. Otherwise, RH may increase to 100%.

Water can be added to air by a number of mechanical means. Davidson (1953) originally developed the high pressure (150–500 lb in^{-2}) mist system for cooling. This system is still preferred where evaporative-pad cooling might be excessive. High pressure mist is largely employed to control humidity where a minimum of free water is desired on the foliage. Cheaper, low pressure (30–70 lb in^{-2}) systems are more common. Figure 5-13 shows a variety of nozzles for high and low pressure systems. In any case, high quality water is required to avoid salt deposition on leaves with consequent damage and unsightly salt formation. With fine mists, filters as in Figure 5-14 are required. Reciprocating pumps for high pressure mist are not desirable as excessive vibration can dislodge dirt in the lines and clog nozzles. Rotary, high pressure pumps (Fig. 5-15) are available, which can be connected in series with a time clock and humidistats to control misting.

Mist propagation systems, as devised by Langhans (1954, 1955), are usually low pressure. As free water is maintained on the cuttings during the day, humidistats may not function properly. Devices that sense the amount of water by change in weight of a simulated leaf, or conductivity of a water film are available. The timing cycle can be set by cyclic timers as shown in Figure 5-16.

High pressure mist systems are expensive. Oil burner nozzles as in Figure 5-13 will cost more than $ 1.50 each. Rigid copper lines with sweat fittings are preferred. Depending upon pressure, such nozzles will disperse 0.3–0.6 gal h^{-1}. Low pressure systems will often utilize galvanized steel or plastic supply lines, with nozzles ranging in price from 50 cents to more than $ 1.00. Nozzle arrangement will depend upon objective, local conditions, and arrangement of benches and greenhouse structure. Above propagation benches, nozzles spaced every three to four feet will usually be satisfactory. For misting roses, two lines in a 30 ft-wide house with nozzles spaced 15–20 ft can be adequate if the nozzle has a sufficiently wide pattern.

Fig. 5-13A–D. Selection of low and high pressure mist nozzles. (A) Fourth from left in the picture is a high pressure, oil-burner nozzle, operating best at pressures in excess of 150 lb in^{-2}. The others are operated at pressures less than 100 lb in^{-2}, and are commonly employed for mist propagation where the excess water deposited on foliage is usually desired. (B) Low pressure, rotating sprinkler nozzle. (C) and (D) Two types of low pressure, circular spray nozzles

Fig. 5-14. Filtering system for high pressure mist system. Filter capable of removing particles down to 150 μ diameter. All mist systems should be protected, and require water with few salts

Fig. 5-15. Example of a high pressure, rotary mist pump. Suction line enters top of pump from right. Water is taken from a sump, maintained at constant water level by float valve. Outlet on left includes a pressure gauge, pressure-regulating valve and filter. Note that a flexible hose connects pump to piping system. This reduces vibration of main lines which tends to knock loose debris, clogging mist nozzles

5.4.4.2 Other Methods to Reduce Water Demand

The common method to reduce transpiration is to increase humidity as described. By syringing, or misting, the solar energy will be utilized to evaporate free water from the leaf, rather than raising leaf temperature and increasing water loss. Temporary, low pressure mist systems to prevent wilting until new transplants are established are a common practice.

A second method is to reduce sunlight deliberately by shading plants or applying shade to the greenhouse roof. Prior to mist propagation, cuttings were

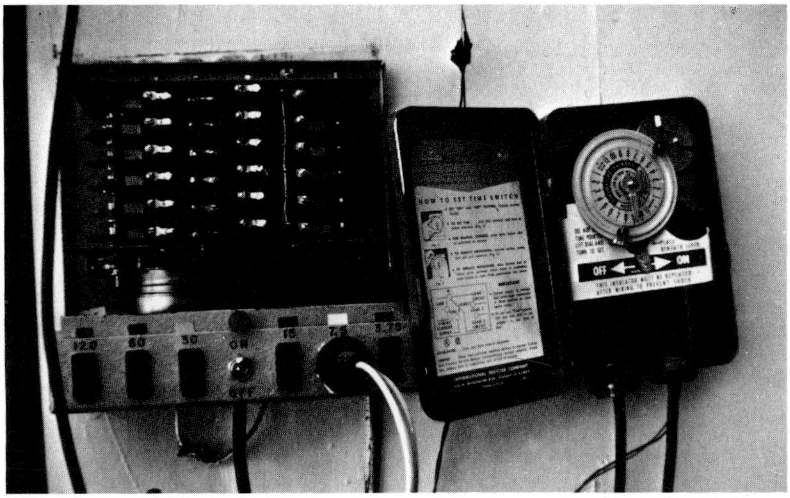

Fig. 5-16. Example of automatic mist control system for mist propagation. *Right:* 24-h clock turns system on in morning and off in evening. *Left:* clock controls time mist is on, and interval between cycles

often covered with paper. Obviously, this reduced the rate of root production, and increased disease under the cover. Still another method to reduce water loss is to decrease air velocity. This will have the effect of increasing the distance water molecules must diffuse from the leaf into the air. If stomata are closed (high diffusion resistance), water loss will be lower. Waggoner and Zelitch (1965) and Gale and Hagan (1966) discussed the use of antitranspirants which restrict water loss by closing the stomates, or by a membrane impermeable to water vapor on the leaf. Effectiveness of antitranspirants will depend upon conditions that tend to increase water loss. For practical purposes, chemical regulation is not often practiced as the effects are not sufficiently consistent, and there can be phytotoxic effects.

Increasing CO_2 concentration will also close stomates. As the resistance to CO_2 uptake in the leaf is higher than resistance to vapor movement, a slight closure is likely to affect water loss more than photosynthesis (Gale and Hagan, 1966). According to de Wit (1958), CO_2 diffusion is only controlled by the stomates when they are nearly closed.

5.5 Control of Water Supply

Supplying water to plants probably accounts for most problems in greenhouse culture. There are factors of water quality, soil structure, soil placement, fertilizer application, application method, and amount of water required which determine actual supply to root systems.

5.5.1 Behavior of Water in Greenhouse Soils

5.5.1.1 Principles

The ability of a soil to retain water against gravity, or the negative suction by plants, is a function of particle size and arrangement. Under all conditions, the root medium must supply water with minimum restriction, and without substantially interfering with oxygen requirements of the root system. As a soil dries, the suction required to remove water from the soil pores, and from the surfaces of individual particles, increases. The relationship between force to remove water and soil moisture content is sometimes referred to as a "soil-water characteristic curve" as illustrated for several soil types in Figure 5-17. A sandy soil (Curves *A* and *B*), loses most of its water at a relatively low suction, and once this occurs, very slight changes in moisture content result in radical changes of suction. On the opposite extreme is a clay soil (Curve *F*) where moisture content can be high even at high suctions. Clay soils can be utilized in greenhouses, but must be modified substantially in order to avoid deficient aeration, and allow rapid water penetration. Characteristics of plants grown in a clay soil versus a sandy soil will be different. The higher suctions required to extract water will usually mean a slower growth rate and harder plants caused by the greater water stress in the plant. Sands or gravels may have freely available water at low suctions with maximum aeration, but reserves are low, and extreme deficits can be reached in very short intervals. The grower's objective is often to provide a compromise between extremes as a safety factor.

For two typical greenhouse soils, the relationship between solids, water and air is shown in Figure 5-18. The peat-perlite mixture will hold less water at high suctions (more negative) than the 1-1-1 soil mixture when expressed as a percent of total soil volume. Actual total water is higher for the soil. The bottom graph in Figure 5-18 is particularly important for it shows the rate of water movement in

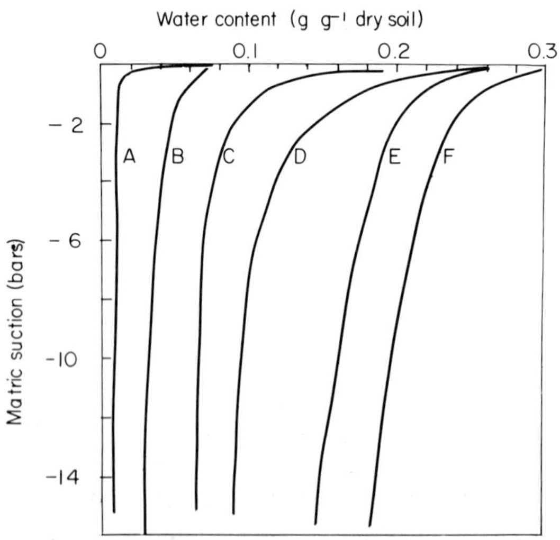

Fig. 5-17. Soil water characteristic curves for various soils, relating moisture content with matric suction. *A* Coarse sand; *B* coarse sandy loam; *C* very fine sandy loam; *D* clay loam; *E* silty clay loam; *F* clay. (From Taylor and Ashcroft, 1972)

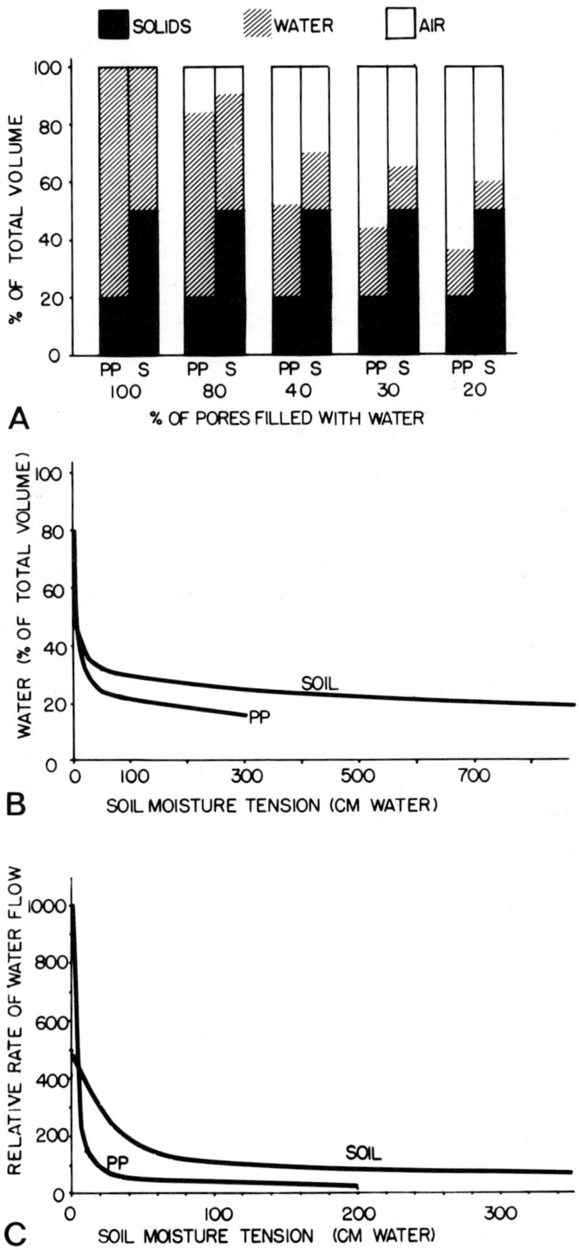

Fig. 5-18 A–C. Diagrammatic representation of relationship between solids, water and air in two typical greenhouse substrates; (A) 1-1 peat-perlite mixture and a 1-1-1 mixture of soil, sand and peat moss. (B) Soil water characteristic curves; (C) rate of water flow (capillary conductivity) of each at different suctions and moisture contents (Hanan, 1964, unpublished)

response to a suction exerted by the plant. This is sometimes referred to as "capillary conductivity", and can directly influence growth. For example, in the peat-perlite mixture, newly transplanted snapdragon seedlings can be severely stressed even though water can be squeezed from the mixture by hand. At fairly low suction, conductivity is nearly zero as contrasted to a finer textured 1-1-1 mix.

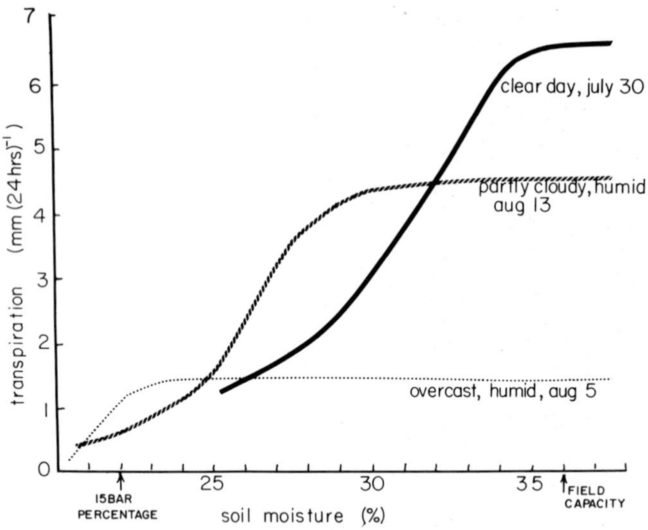

Fig. 5-19. Actual transpiration rate of corn crop as function of soil moisture content. (Redrawn from Denmead and Shaw, 1962)

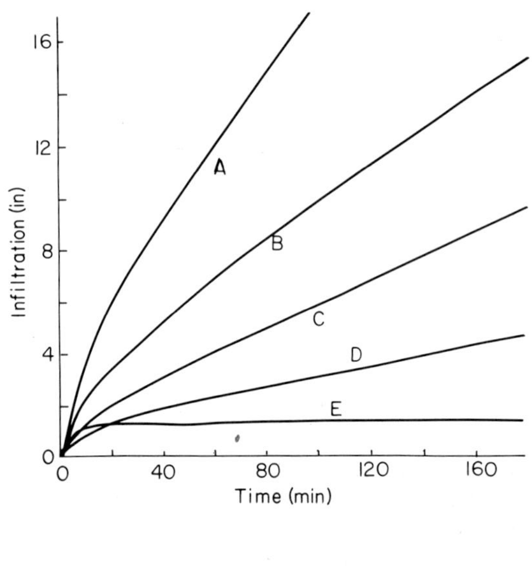

Fig. 5-20. Infiltration rates for various soil types. *A* Honeoye gravely silt loam; *B* Aiken clay loam; *C* Crown sandy loam; *D* Palouse silt loam; *E* Cecil clay loam. (Redrawn from Taylor and Ashcroft, 1972)

If the plant is large, with a rapidly growing root system, this effect of water movement is not often important. The interrelationship between suction, moisture content and conductivity is indicated for a real situation in Figure 5-19. Under low water demand (Aug 5), the soil can become fairly dry before any effect on transpiration is observed. An adequate water supply is available without very high stress. As transpiration increases (Aug 13 and Jul 30), drying so reduces water conductivity that demand can no longer be met, and transpiration decreases with possible wilting. The factor of conductivity explains why a soil can be

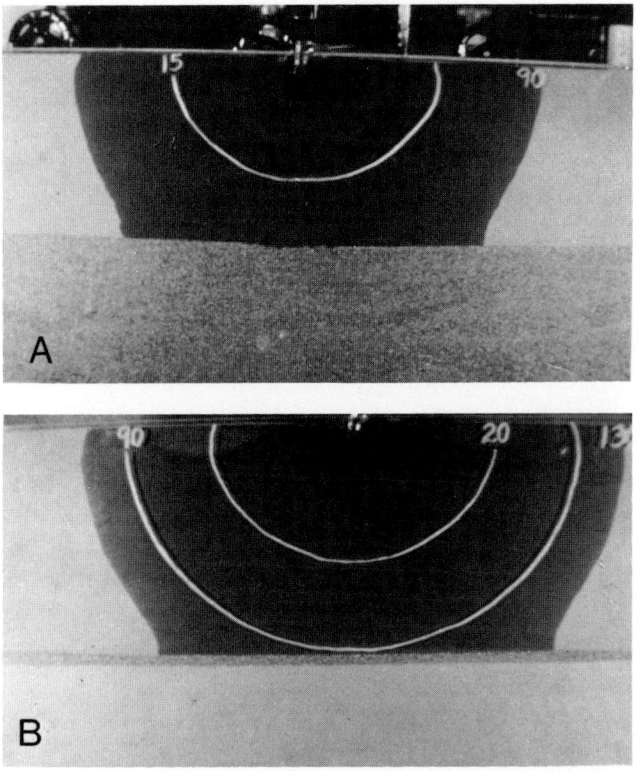

Fig. 5-21 A and B. Penetration of water into soil from point source at surface. (A) Soil overlies more permeable sand layer; (B) sand lens is placed between two layers of soil. In neither case will water penetrate into coarser mixture until pressure in upper layer at interface exceeds zero. *Numerals:* time required for wetting front to move from application point. (Courtesy of W. H. Gardner, Washington State Univ.)

"wet" but plants wilt on a sunny day; or, the soil is "dry" and plants remain turgid on a cloudy day.

Conductivity should not be confused with water infiltration. A particular soil (peat-perlite) may have very low conductivity but very high infiltration when watered. It is common to measure infiltration rates in greenhouse soils exceeding 60 in (150 cm) per h. Compare this value with those provided for some common field soils in Figure 5-20. With the high fertilization rates encountered in greenhouse soils, it is vitally important that infiltration be as high as possible so soluble salts can be controlled when necessary, and drainage is rapid. Unfortunately, modifying soils for maximum infiltration reduces conductivity. For most field soils, according to Gardner and Ehlig (1962) resistance to water movement does not become important until soil suctions are less than -0.6 bars. As will be noted, most greenhouse soils are irrigated before -0.3 bars is reached.

No root medium, in which there is a growing plant, can be maintained at a constant moisture level less than the maximum water holding capacity of the soil. If water is applied to a soil surface, it will move in a uniform wetting front through

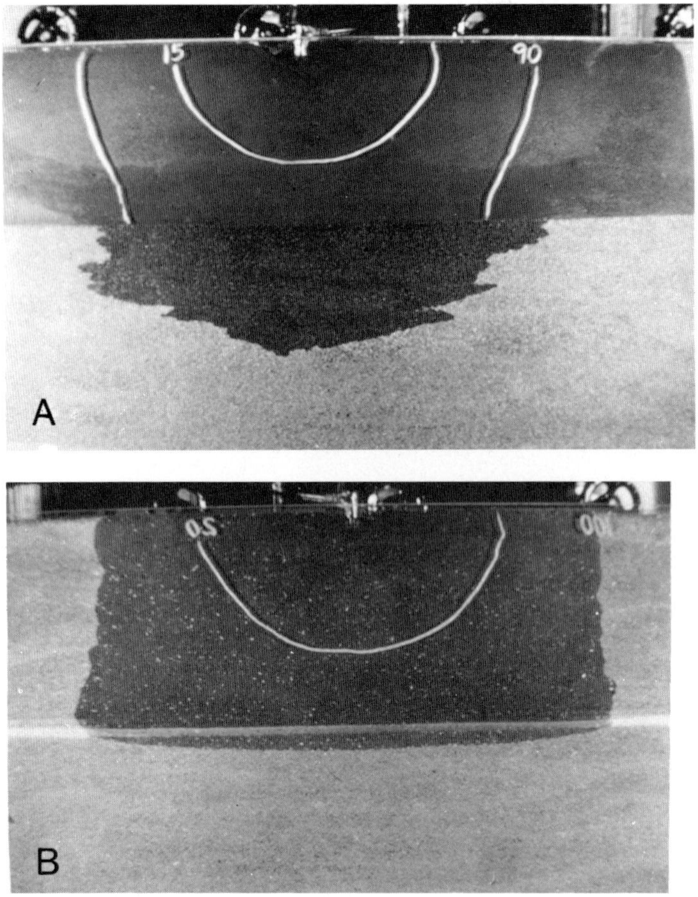

Fig. 5-22 A and B. Penetration of water into soil from point source, overlying a clay. (A) Water spreads through sandy loam rapidly, and penetrates more slowly into clay. (B) Water movement is very rapid through sand, but penetrates clay lens, which acts as hard pan, very slowly. (Courtesy of W. H. Gardner, Washington State Univ.)

the soil mass as shown in Figures 5-21 and 5-22. After the soil has been brought to its maximum capacity, evaporation from the surface and extraction by plants, will cause a gradual suction increase. Therefore, plants will be subjected to a constantly varying moisture stress. There are only two points growers can control: (1) the minimum suction represented by the maximum moisture capacity of the substrate after drainage; and (2) the maximum suction at which the soil is watered. Russell (1959) and Hagan and Vaadia (1961) pointed out that this cyclic process accounts for the many contradictory effects observed in practice and in research. Plants do not respond equally to all values of soil moisture content and soil moisture suction. Each day, as the sun rises, transpiration causes a suction gradient to be established in the plant (water stress), which causes water to move from the soil into the root system. At night, with closed stomates, plant stress

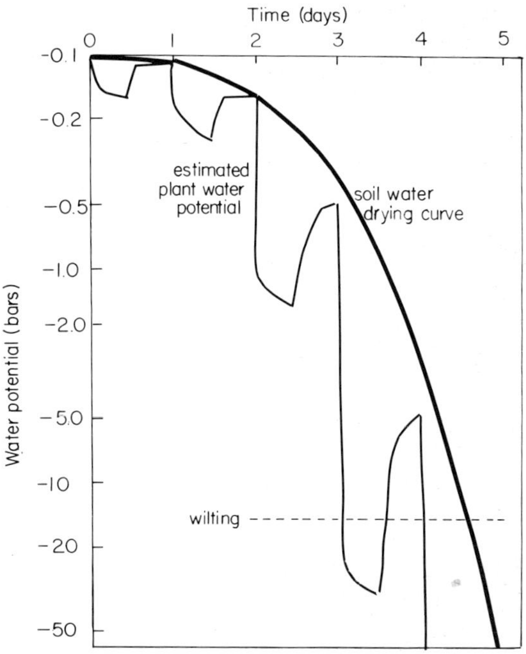

Fig. 5-23. Estimated internal plant water stress over several days as the soil dries out. (From Taylor and Ashcroft, 1972)

decreases. As the soil dries, increasing suctions are caused in the plant, eventually reaching a point where the remaining soil water is held with such tenacity, that the supply is insufficient and wilting occurs (Fig. 5-23). It is the grower's objective to maintain the range of suction as low as possible, commensurate with adequate aeration and plant quality.

A typical situation of recyling moisture suctions for a greenhouse crop of snapdragons is shown in Figure 5-24. Even if a crop is watered at some designated soil suction, the actual conditions to which the plants are subjected can vary drastically. When a grower states that he is running "wet", he means one of two possibilities: first, the drying rate is so slow (Crop 1 in the winter, Fig. 5-24) that the moisture content remains high for long intervals; or, secondly, he irrigates very frequently (both Crops 1 and 2), and he prevents suction from reaching extremes. Even with frequent irrigation (Crop 2, summer), high stress can be caused, and the relative situations of "wet" and "dry" must be related to the season and size of the plants.

5.5.1.2 Factors of Placement

Soil handling in greenhouses is complicated by location: (1) the operator might grow directly in the ground, using the soil native to the location; (2) he may produce in benches; or (3) he can produce his crop in small containers. The last two can be described as freely draining with small volume and shallow. Soil behavior will be different in each case, and cultural practices must be modified accordingly.

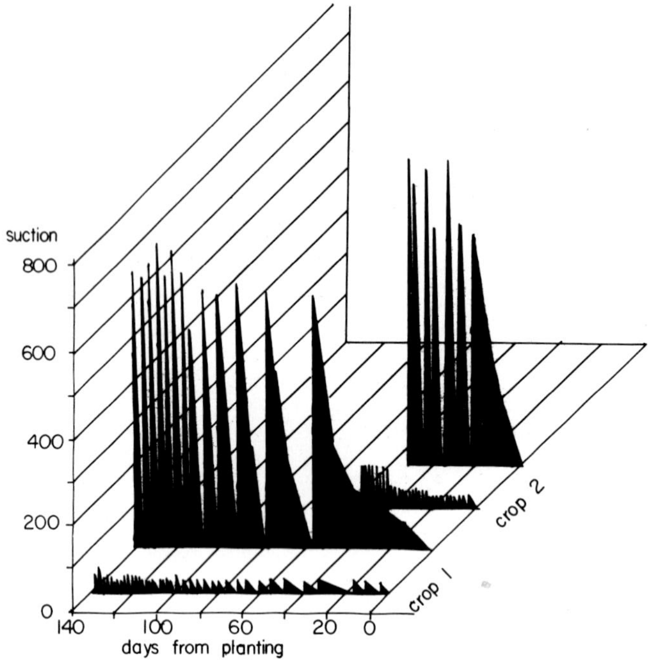

Fig. 5-24. Course of matric suction in greenhouse soil planted to single-stem snapdragons and grown at two different times of year and watered at two different suctions. *Crop 1* was benched in late fall, *Crop 2* in early summer. In each crop one treatment (high suction) was watered on the day when suction reached 600 cm water, the other on the day when suction reached 30 cm water (Hanan, 1963, unpublished)

In the case of ground locations, movement of water upon irrigation can proceed as indicated in Figures 5-21 and 5-22. The volume for root growth will be unrestricted—unless the density of the soil is too high (1.8 g cc^{-1} or higher). Unless the upper layers are modified, infiltration rates will be slower (Fig. 5-20), and time between waterings will be longer. In deep soils, the wetting front continues to move even after water to the soil surface is shut off. When, for all practical purposes, this wetting front has ceased to move, a field soil is said to have reached its "field capacity", or its maximum moisture holding capacity. For many soils, soil suctions at this point range between 200–300 cm water. Because field soils are heavier (higher bulk density), with smaller pore size, they must be run "dryer" in order to avoid deficient aeration.

When placed in a shallow, freely draining, restricted location, the maximum moisture holding capacity of a soil changes dramatically. When watered, it is as if the downward water movement encountered a restriction as shown in Figure 5-22. For water to move from the soil into free air (drainage), the water suction in the bottom of the soil layer must be greater than zero. When drainage is complete, suction in the bottom of the layer will be zero, and the suction at the upper surface will be equal to the depth of the soil if suction is expressed in terms of a water column. An example of actual behavior is provided in Figure 5-25. If a soil is 5 in

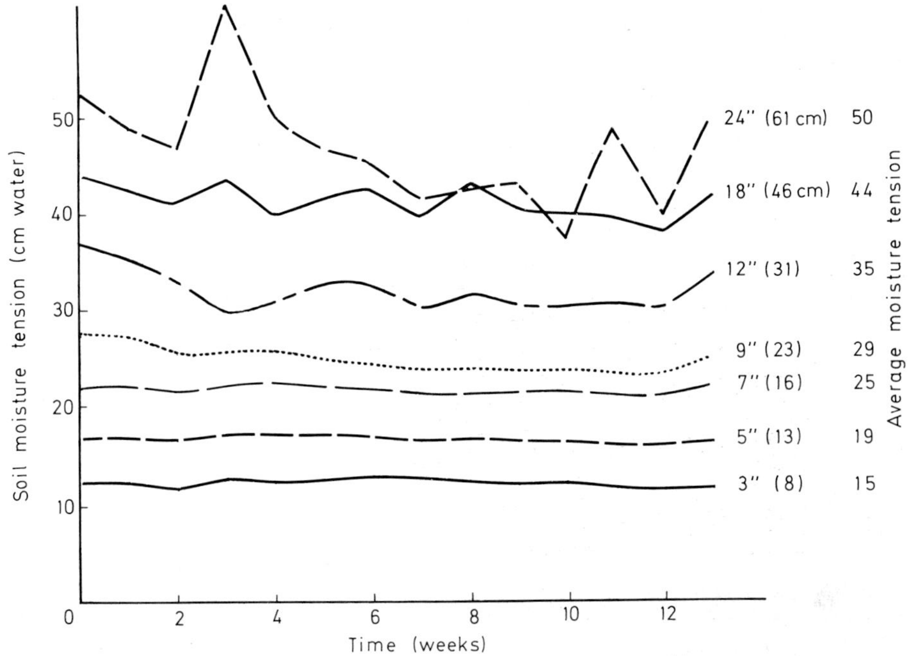

Fig. 5-25. Effect of soil depth on suction in upper layers. *Curves:* soil suction with time during growth of single-stem snapdragons in soils ranging from 3 in (8 cm) to 24 in (61 cm) in depth. In 24-in depth, soil layer begins to approach behavior of ordinary field soil (Hanan and Langhans, 1964)

deep (13 cm), the suction in the upper surface will be 13 cm water when drainage stops. This suction will represent maximum moisture holding capacity, and the soil layer will become wetter toward the bottom. If most field soils are brought directly into the greenhouse, without modification, serious aeration problems usually develop. This principle is known as Richard's "Outflow Law" (Richards, 1950), and the maximum water content is sometimes referred to as "bench capacity" (Hanan and Langhans, 1964) or "pot capacity" (White, 1965). As long as water flow from the undersurface is not restricted, the same principle applies to a bench, or pot, or market-pac.

Some growers, attempting to dry soils out faster, reduce soil depth. In fact, reducing soil depth merely leaves the wettest, bottom layer in place, and can compound the original problem. Reducing depth also reduces an already limited soil volume for water storage and root growth.

5.5.2 Measuring Moisture Content of Soils

Most growers still depend upon visual observation and touch for determining time to water. This procedure requires experience. The environment is not always uniform in a greenhouse. One end, or side, of a bench may dry out earlier than

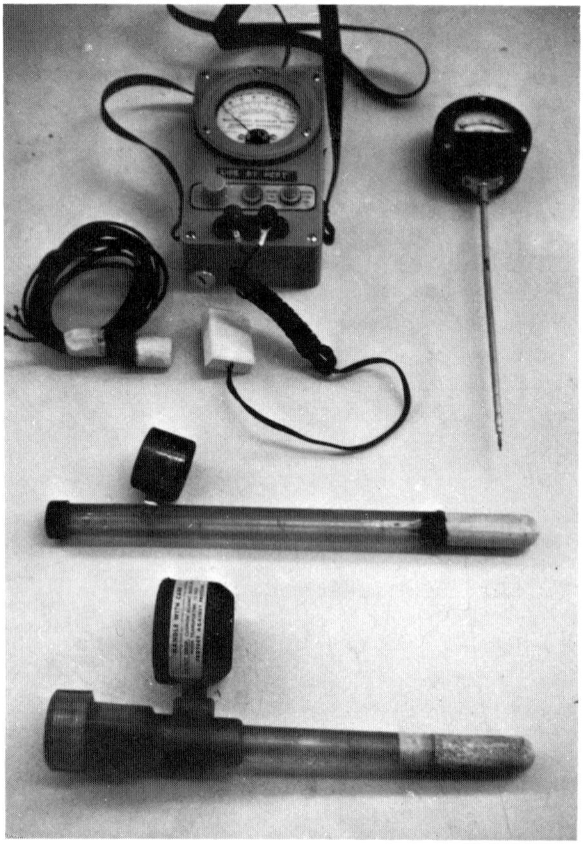

Fig. 5-26. Common devices for estimating soil moisture content. Two devices in lower part of photograph are porous, clay cup tensiometers, adequate to suctions of about 800 cm water. Electrical device in upper part, with two styles of plaster blocks is a Bouyoucos bridge, suitable for suctions in excess of 1 bar. Narrow probe on right is another electrical device seldom seen in practice, due to the fact that soluble salts greatly influence its reading

other parts of the bench. Potted plants close to exhaust fans can dry faster than those close to evaporative pads. There can be differences in the potting mixture that cause some pots, or locations, to dry out before others. There is not, at present, a good, inexpensive method that will tell a grower the moisture content of his root medium for every location. Any device must be supplemented with observation and experience.

A few instruments that can aid the grower are shown in Figure 5-26. Porous cup tensiometers are common, and can be used to start irrigations automatically (Post and Seeley, 1947; Wells and Soffe, 1961). They sense soil moisture in the immediate vicinity of the cup. Tensiometers can be made in sizes small enough to fit potted plants, or made longer to measure moisture in deeper layers of field soils. Tensiometers are water-filled, and as they depend upon the continuity of the water films in the cup and in the soil, they must be checked regularly for air and refilled. Suction devices are inadequate in gravels or very porous soils. They are not accurate above 0.8 bar (800 cm water) as air enters the cup pores. In Colorado, the usual recommendations are to water between 300–500 cm water suction (Holley, 1966).

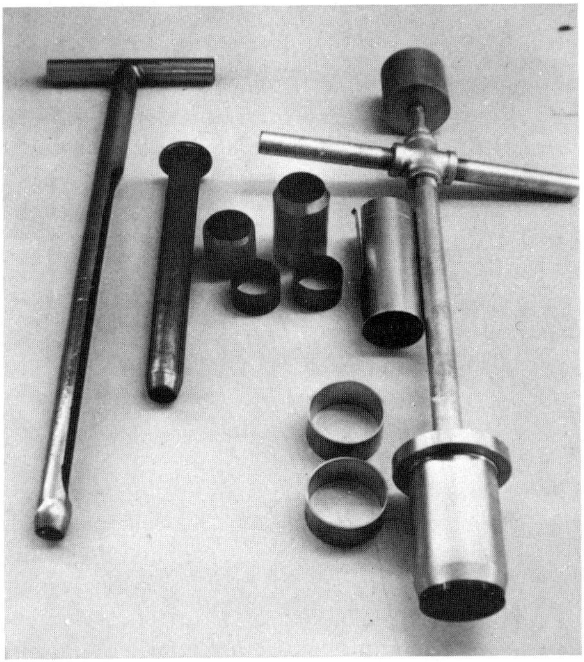

Fig. 5-27. Various types of tools for taking soil samples for analysis. Two on the far left are primarily for taking profile cores for nutrient analysis where disruption of structure is of no consequence. Large core sample on right is used for field soils where a core is to be examined for moisture content and pore volume. Smaller tubes are useful for greenhouse soils where it is desired to remove an undisturbed sample

Plaster of paris blocks (Fig. 5-26), as devised by Bouyoucos and Mick (1940), are seldom seen in greenhouse culture. They work best at suctions in excess of 1000 cm water (1 bar). Long before this suction is reached, growth will be seriously reduced in common greenhouse root media. Electrical conductivity systems are influenced by variations in salt content of the soil and water. A wet reading can be obtained in a soil with high total salts when actually the soil is fairly dry. Plaster of paris, or similar systems, as used by Bouyoucos, avoid this problem by insuring that the electrodes are surrounded by a saturated salt solution.

Actual water content of a soil can be determined by weighing a sample and drying it to determine weight loss. As most field soils have a particle density of about 2.65, and a bulk density 1.0 or more expressing weight loss as a percentage of the dry weight approximates actual water content. The method is unreliable with most light greenhouse soils with high organic matter. Moisture contents of 600% are not unusual when expressed as a percentage of dry weight. An undisturbed core sample of known volume must be taken and the difference in weight upon drying expressed in g or ml per volume of sample. Examples of different soil samplers are provided in Figure 5-27.

The problem of measuring soil moisture can be evaded by ensuring a medium so porous that excessive irrigation will not damage the root system. An inert medium such as gravel permits timed waterings. The reduced volume of small containers also allows timed schedules; in both cases, the frequencies changing from season to season. There are systems that will accumulate the amount of solar energy and automatically start watering. But, the soil must be such that occasional overwatering will not injure the plants.

5.5.3 Water Quality

Distilled water provides maximum flexibility in fertilization practices, fewest problems with excess salts, or irrigation system clogging, and maximum growth. But, distilled water is expensive, and the water available can contain various salts in solution which force growers to modify practices. Figure 5-28 shows typical carnation growth resulting from irrigation with water having a total salt concentration in excess of 3000 μmhos/cm^{-1}. Production in this example was economical due to factors of production costs and return from sales. Such growth would be unacceptable in other areas of the world.

Water quality is particularly important in arid regions where good water can be difficult to obtain, expensive, or in short supply. Under Colorado conditions, the cost per carnation flower was about 0.02 cents in 1967 (Hanan and Jasper, 1967). If systems must be installed to improve quality, the cost can increase markedly.

According to authorities (Thorne and Thorne, 1951; Richards, 1954), water with a total electrical conductivity (EC) below 750 μmhos/cm^{-1} should not cause problems. Irrigation waters that have been successfully used have had ECs below 2250 μmhos/cm^{-1}. The upper limit, with automatic fertilization injection, is considered to be 3000 μmhos/cm^{-1} (Schekel et al., 1968). At these upper levels, surplus water must usually be applied as noted in Table 5-1 to prevent additional salt build-up in soils. With fertilizer injection, the initial salt concentration should not exceed 1000 μmhos/cm^{-1} (Schekel and Hanan, 1971).

The effect of salt level is largely osmotic. The higher the concentration, the greater the tendency for water to move out of the plant. The osmotic suction

Fig. 5-28. General appearance of a carnation crop grown with high salt water in soil. Crop planted in June, flowering for Christmas. Location: the Mediterranean. Grower stated that return was economic

Table 5-1. Leaching requirement of soils as related to salt content of irrigation water. Leaching requirement is expressed as percentage of applied water that must be leached through rootzone to maintain maximum values of electrical conductivity given below (Richards, 1954)

EC of irrigation water (mmhos cm^{-1})[a]	Leaching requirement for individual maximum values of EC in the drainage water (mmhos cm^{-1})			
	4	8	12	16
0.1	2.5%	1.2%	0.8%	0.6%
0.25	6.2	3.1	2.1	1.6
0.75	18.8	9.4	6.2	4.7
2.25	56.2	28.1	18.8	14.1
5.0	—	62.5	41.7	31.2

[a] 1 mmho = 1000 μmhos.

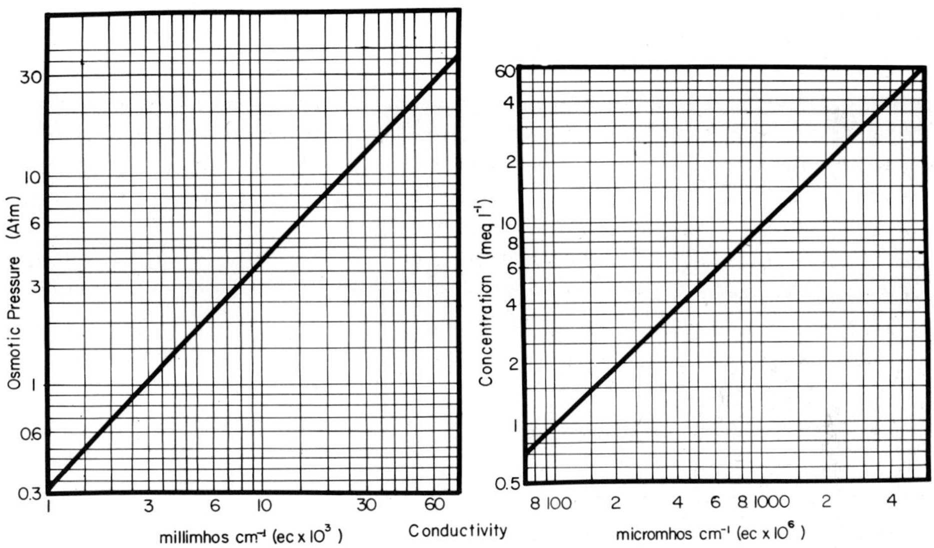

Fig. 5-29. Relationship between electrical conductivity of a solution (EC) and osmotic pressure (potential) *(left)* and concentration *(right)*. (From Richards, 1954)

(potential) is added to the soil matric suction to obtain the total soil water potential or suction; this value must be less negative than the plant water stress if water is to move into the plant. Some plants will survive and reproduce at high salt levels. This would mean, however, that carnations are salt tolerant (Schekel, 1970). Actually, there are no greenhouse crops that can be thought of as salt tolerant in view of the economic return required. Figure 5-29 shows the general relationship between EC, osmotic pressure and salt concentration. A value of 3000 μmhos/cm^{-1} represents an osmotic force of one bar, and a total ion concentration of 25 meq l^{-1}. While this is not particularly high, one should remember that as a soil dries, following irrigation, the salt concentration will increase.

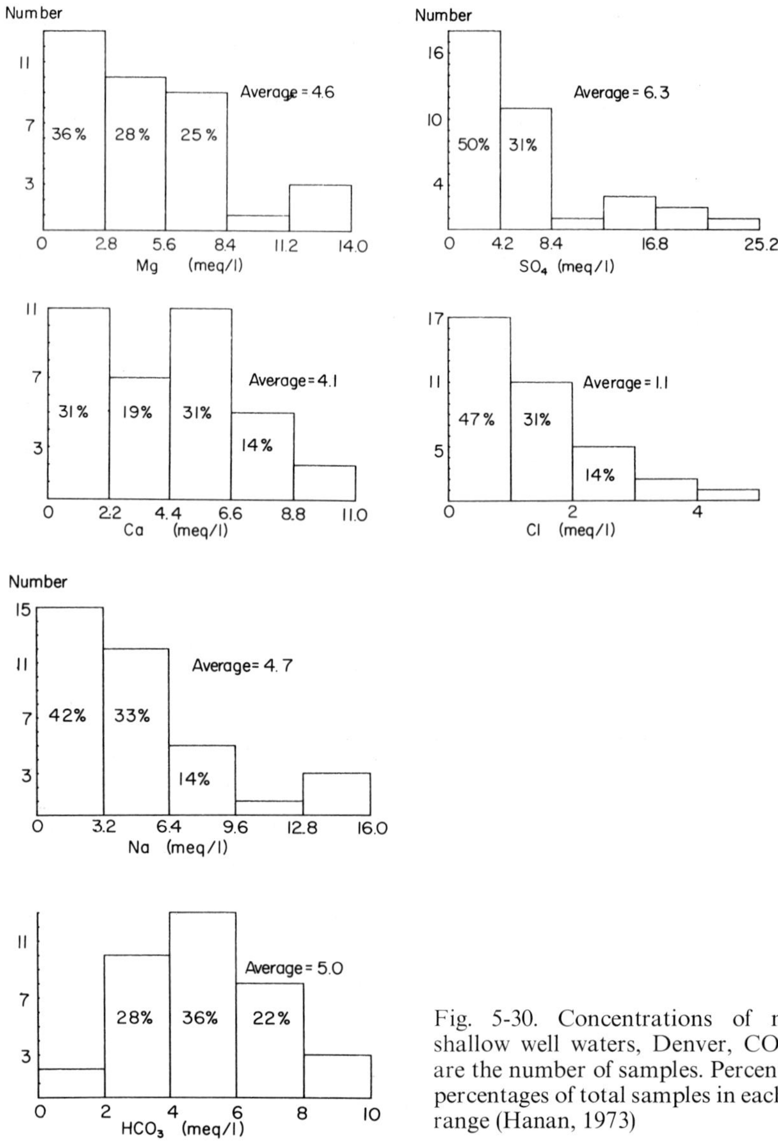

Fig. 5-30. Concentrations of major ions in shallow well waters, Denver, CO. Vertical axes are the number of samples. Percentage figures are percentages of total samples in each concentration range (Hanan, 1973)

In addition to total salt concentration in water, the grower needs to know concentrations of specific salts that make up the total. Sodium, sulfate, magnesium, calcium, carbonate, bicarbonate, chloride, and boron should be determined in a water analysis. Traces of potassium and nitrates suggest shallow wells, contaminated by leaching from field soils. Hanan (1973) made an analysis of several shallow wells in Colorado, and Figure 5-30 shows the concentrations of important ions. If the makeup of the irrigation water is known, the grower will modify his cultural program to avoid duplication, reduce salt level, and eliminate possible soil structural problems.

Table 5-2. General rules-of-thumb for assessing suitability of irrigation water on the basis of individual ions, and possible modifications (Schekel and Hanan, 1971)

1. Calcium	Do not use irrigation water with calcium in excess of 10 meq l^{-1} (200 ppm). Concentrations in excess of 6 meq l^{-1} will prevent phosphorous injection with irrigation, and increase precipitation in irrigation lines
2. Chloride	Should not exceed 4 meq l^{-1} (140 ppm). Requires the use of ammonium in fertilizer solutions for carnations and roses
3. Bicarbonate	Usually not a serious problem below 6 meq l^{-1}, depending upon pH. Will cause precipitation in irrigation system if not corrected by acidification. Not above 3 meq l^{-1} for roses
4. Magnesium	Should not exceed 3 meq l^{-1}. Adds to the precipitation problem
5. Sodium	May go as high as 8 meq l^{-1} (184 ppm) as long as potassium is maintaned equal in concentration. Causes serious problems with soil structure, and requires increased calcium application to soils. The latter is no problem with inert media
6. Sulfate	May be as high as 10 meq l^{-1} (480 ppm) although 1 meq l^{-1} is adequate for carnations

Schekel and Hanan (1971) presented some general rules-of-thumb that may be used to assess water suitability (Table 5-2). For example, calcium and magnesium (hard water), in combination with bicarbonate, will generally precipitate. Emitters on trickle systems can be blocked by precipitate, and it may not be possible to add phosphorous in the injection system. Sodium will require increased calcium fertilization to prevent deflocculation of soil structure. If Ca and Mg are already present, they need not be added by injection, but ammonium fertilization may have to be increased.

Trickle, or drip, systems may also be blocked by waters containing iron or sulfur (Ford and Tucker, 1975). In the latter, traces of oxygen in the system will cause sulfur slimes to grow in the emitters. If bicarbonate cannot be eliminated by acidification, or sulfur and iron plugging reduced, then an irrigation system is selected that is less prone to plugging. Bodies of surface water such as lagoons, lakes or drainage canals are likely to offer serious problems from plugging by molds or algae.

5.5.4 Irrigation Systems

The number and types of irrigation systems for greenhouse culture are nearly infinite. The grower has to select from a number of conflicting claims the system which will perform best for his particular conditions. The selection will depend upon the type of soil, species, cultural method, water quality, and relative costs. Figure 5-31 provides a few examples of available systems.

5.5.4.1 Hand Watering

Hand watering is still employed where the number of units (area to be covered) is small, the labor cheap, or if there is a large diversity in unit size and species and there is a rapid turnover in the production space. The initial investment is least of any system. There are a number of disadvantages: (1) it is labor-intensive and not economic in developed countries where labor is expensive; (2) good watering requires considerable training and reliable labor; (3) overhead flooding, and force of moving water, breaks down soil structure, leading to crust formation, and reduced infiltration and oxygen supply; (4) the end of a hose is one of the best ways to spread disease. Invariably, hose ends will be contaminated unless employees are specifically trained in sanitation practices.

5.5.4.2 Subirrigation

Wetting the soil from below has been used on a small scale for several years. The effect of a constant water table is illustrated in Figure 5-32. For ordinary soils, suction will be maintained near bench capacity when 7 in or less in depth. If much over 7 in, capillarity is usually insufficient to prevent severe drying of the upper soil surface. The desirability of constant water table varies considerably with soil type (Cook et al., 1953) as would be expected. Caparas (1955) found that constant water table resulted in flowers with the lowest keeping life. The distance of the water table below the soil surface must be adjusted for the particular soil type.

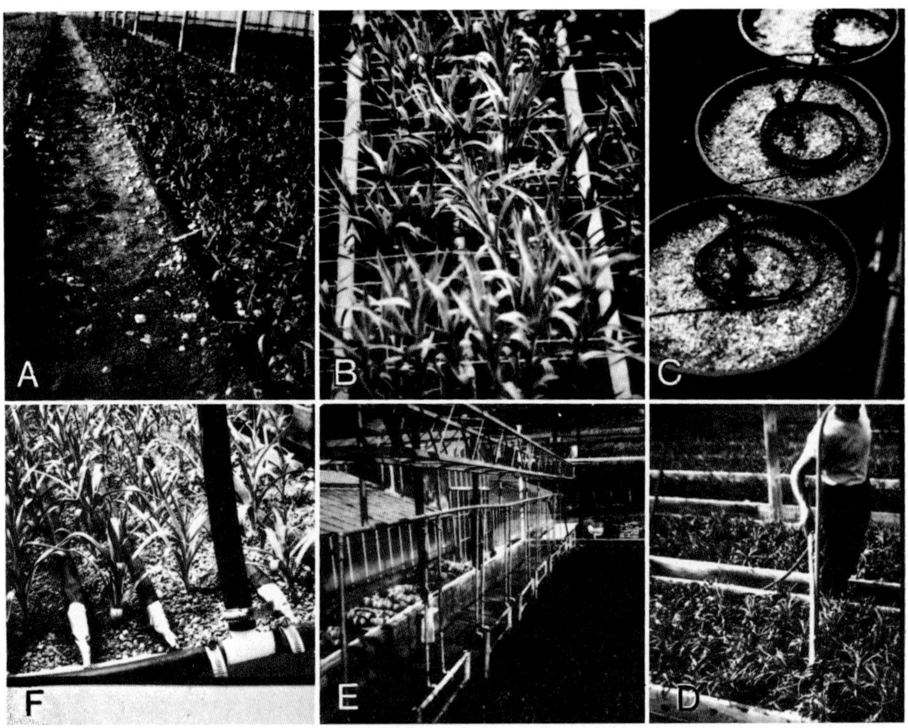

Fig. 5-31 A–F. Various types of watering systems: (A) 180° peripheral gates irrigation system on gravel; (B) porous "Viaflo" tubing; (C) ooze tubes; (D) hand-watering with a water breaker; (E) overhead trolley system carrying spray nozzles at bench level; (F) double wall "Chapin" drip tubes

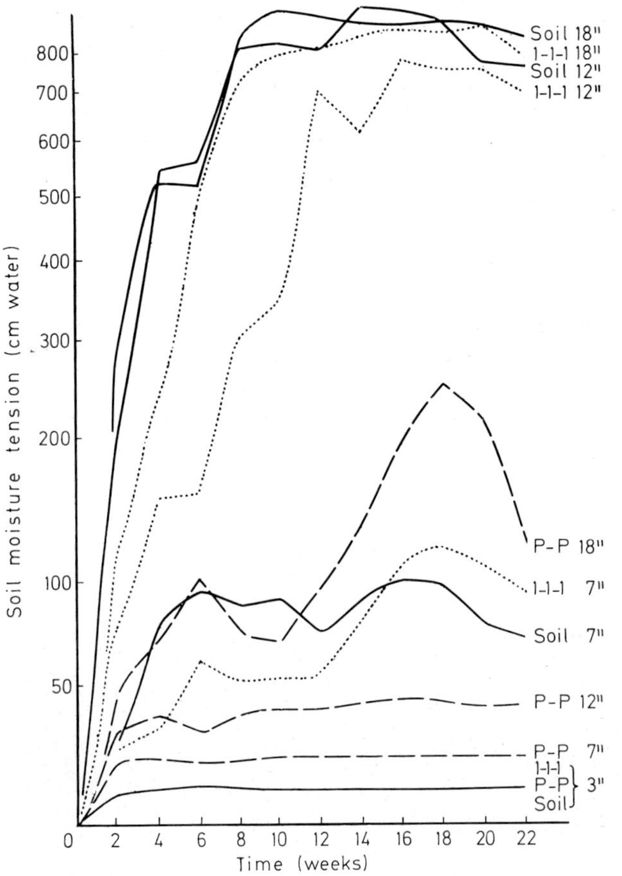

Fig. 5-32. Effect of constant water tables on soil suction in the upper layers of various soils and soil depths. Figures in inches refer to the soil depth above the water table. *P-P:* 1-1 mixture of peat and perlite. *1-1-1:* A 1-1-1 mixture of sand, peat moss, and soil. *Soil*: field soil without modification (Hanan and Langhans, 1964)

Constant water tables, and similar methods of subsurface application, would not be suitable for water supplies containing salts. Surface evaporation can cause salt accumulation to toxic levels. There is, however, very low labor input, water is applied uniformly; but, initial capital investment can be high. Post (1950) described various systems in detail.

A variation of subirrigation has received increasing attention. This is referred to as "capillary irrigation", and was studied by Seeley as early as 1948. It has been the subject of several papers by British workers (Wells and Soffe, 1962; Ministry of Agriculture, 1962). Europeans introduced the "capillary mat" in the 1960s. Water is applied to the mat, and water is drawn into pots by capillary action. There are many variations, ranging from mats constructed of rags to very fine sand. A fairly level bench is required, but it does not have to be watertight. Water quality should be high, with a uniform potting soil, and with fertilization practices modified to incorporate slow-release nutrients, or periodic, over-head feeding.

Hydroponic systems, employing a coarse gravel, are often injected from below and flooded (Hanan and Holley, 1974). As with constant water level methods, a watertight bench is usually required. A modification by Holley avoids this requirement when using inert media.

5.5.4.3 Surface Application Systems—High Volume

High volume, surface systems are often employed for bench crops, crops grown in the ground, and occasionally for potted crops. An overhead system may physically damage the crop, as well as spreading disease organisms. Some species will tolerate overhead watering better than others. In general, however, the less water placed directly on foliage, the fewer the problems that will result. Good overhead systems will apply water directly above the soil surface in a spray to avoid unnecessary foliage wetting. A common system is the "gates", 180°, peripheral system. Initial investment is intermediate to constant water tables and some of the newer trickle systems. Figure 5-33 shows some of the nozzle types, and Figure 5-34 is a nomograph for calculating flow rates and pipe size for different nozzle spacing and bench lengths.

Gates, or similar, systems have been found most useful for woody crops grown in benches, or in the ground, where the water force is unlikely to cause physical damage. With succulent, herbaceous, species, water pressure at the nozzle is critical, and should not exceed 2.5–4 lb in^{-2}. If used for a hydroponics system

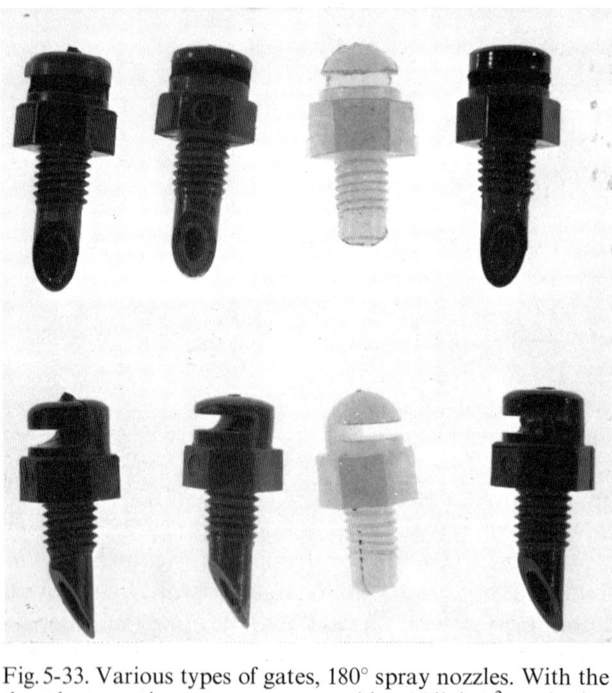

Fig. 5-33. Various types of gates, 180° spray nozzles. With the exception of the white nozzles, the others require pressures approaching 15 lb in^{-2} to obtain an adequate 180° spray pattern. Narrow orifices at the outlet may cause serious plugging if the water supply contains debris. (After Hanan and Duke, 1970)

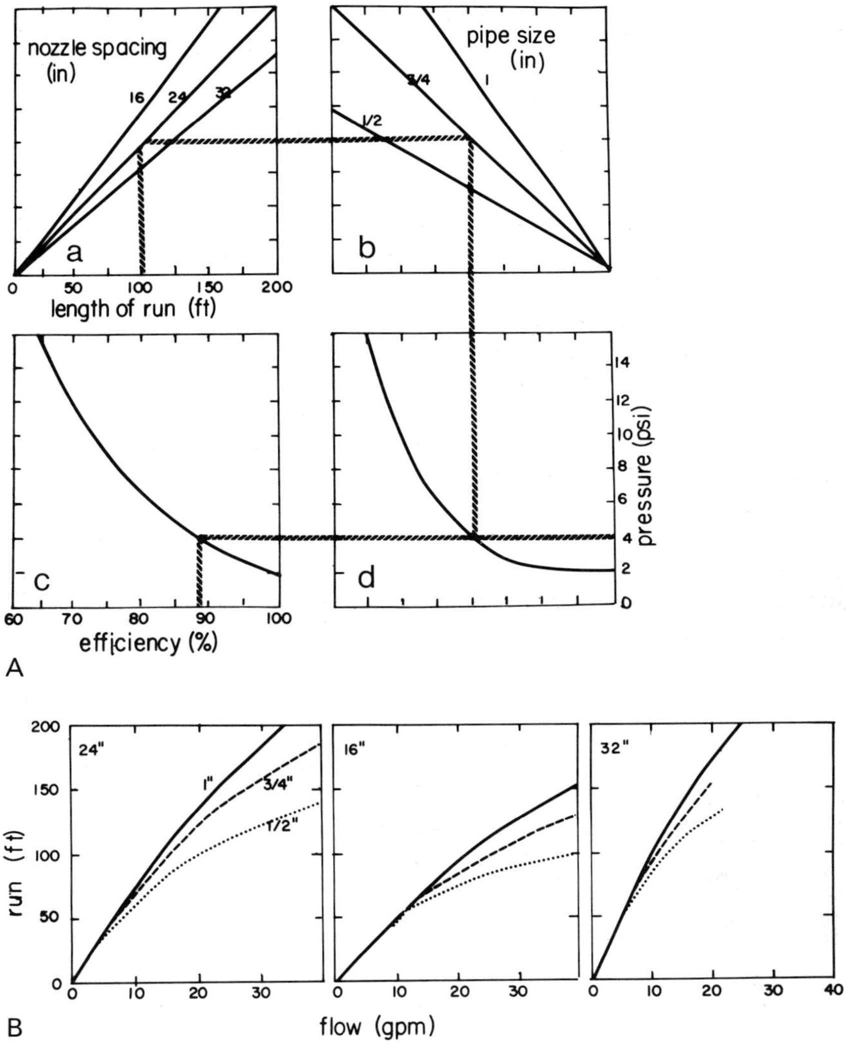

Fig. 5-34 (A) *a–d* Nomograph for determining pipe size and nozzle spacing for various lengths for the gates irrigation system. Data are for the white nozzle shown in Figure 5-33. The maximum pressure at a gates nozzle should never exceed 4 lb in^{-2}. The hatched line shows the pressure, nozzle spacing, length of run and efficiency of the system when using a pressure of 4 lb in^{-2} at the last nozzle from the supply inlet. By trying various combinations, the best pipe size and nozzle spacing can be obtained for a particular length. (B) *From left to right:* the total water flow to a bed for pipe diameters of 1, $^3/_4$, and $^1/_2$ in at nozzle spacings of 24, 16, and 32 in as a function of the length of the run. (From Hanan and Duke, 1970)

(Hanan and Holley, 1974), it is often better to shift to a trickle system. Where hard water is the only available supply, there will be less problem with precipitation and blockage with the gates system. For common soil depths (6–7 in), sufficient water can be applied in 5 min, about 2–4 quarts per ft^2. As salt concentration increases, this amount may have to be doubled (Table 5-1). For inert media, the cycle can be reduced to less than one min.

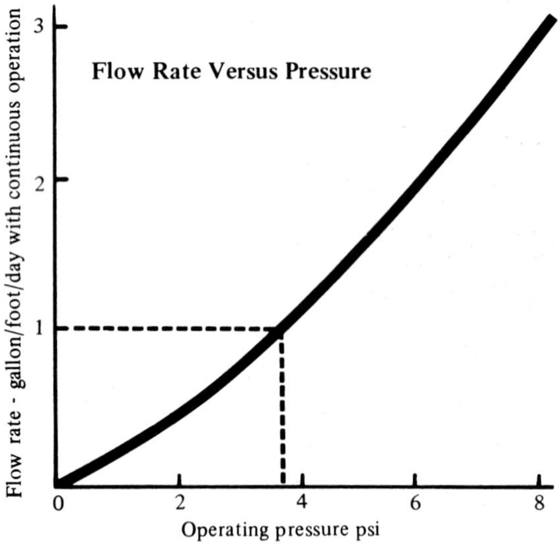

Fig. 5-35. Flow rate versus pressure for "Viaflo" tubing. (From Mitchell et al., 1974)

5.5.4.4 Low Pressure, Low Volume Systems

The number of trickle irrigation systems has exploded in recent years. Water is applied at low rates over long periods. There is no free water standing on the surface. Outstanding advantages are: (1) no water on the foliage; (2) less water loss from evaporation; and (3) reduced salinity problems with high salt water. These systems may not operate adequately if the major salts include bicarbonates, or there is sulfur (H_2S) or iron (Fe^{+2}). They can be cleaned with hydrochloric acid, but this is expensive and dangerous. Adequate filtration is required. Typical flow rates range from one to 3 gallons per h at pressures less than 1–4 lb in^{-2}. In some instances, the systems have watered for 24 h periods. Some systems distribute water along their entire length, or "emitters" are incorporated at regular intervals in the water line. Due to the low volume, much larger areas can be irrigated simultaneously. Care must be exercised to ensure sufficient volume to operate fertilizer proportioners.

The ability to apply water more or less continuously tends to move salts away from the soil surface, with continual dilution. Figure 5-35 correlates water delivery with pressure for one of the many low volume types available.

Drip or ooze systems are common for pot watering. A small tube or emitter is placed in each pot, and the system operated from a central valve. As with capillary mats, uniform potting and soil is desired so that all pots can be wetted to the same moisture content. These systems can be used in combination with almost any type of fertilizer program.

5.5.5 Automation

The distribution system used to apply water uniformly and adequately to the soil is most critical. Success is most dependent on reliable distribution. When install-

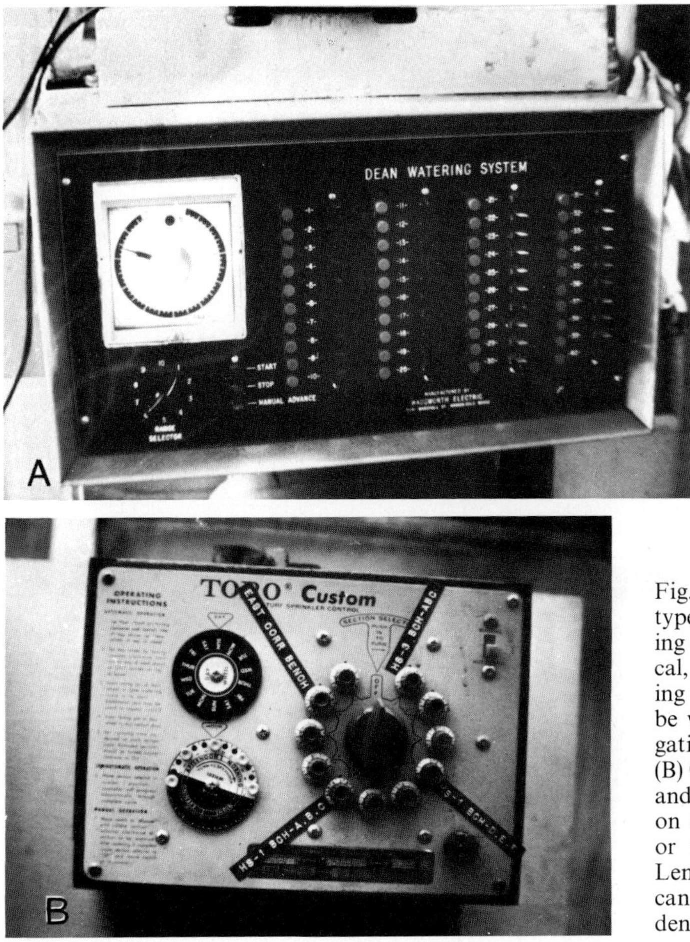

Fig. 5-36 A and B. Two types of automatic watering controllers. (A) Electrical, each button representing one or more benches to be watered. Length of irrigation is set by the left dial. (B) Operates hydraulically, and can be preset to water on specified days and one or more times each day. Length of irrigation period can be regulated independently for each unit

ing new systems, growers should balance costs against objectives, and test the proposed system on a limited basis until they are familiar with it under their conditions. But, the decision must still be made as to "when" and for "how long". In most ranges, the grower, on the basis of observation and representative tests, often sets automatic systems as shown in Figure 5-36, that permit valves for designated areas to be turned on for intervals determined by the operator. The use of materials to indicate stomate size as a means to schedule irrigations has been investigated by Ofir et al. (1968) and Shillo and Halevy (1964). Unfortunately, by the time a significant change may be determined, high stress probably is already affecting growth (Hanan, 1972).

For pot plants, weighing devices have been devised that automatically initiate and terminate a watering cycle on the basis of weight change of a representative plant. Or, a device can be used that will automatically terminate the watering cycle (Fig. 5-37). Problems of overwatering can be bypassed to some extent by choosing a medium that cannot be overwatered—such as gravel, perlite, volcanic

Fig. 5-37. One type of irrigation timer for pot watering. Watering cycle must be initiated by operator, but shuts off automatically when weight of water in cup exceeds a preset limit. (Courtesy of J. W. White, Penn. State Univ.).

ash, etc. These can be programmed by a simple timing device to irrigate a section once to several times daily.

Despite these considerable advances, soils usually require visual inspections to be made at regular intervals. Greenhouse environments are seldom uniform enough to permit a large section to be watered on the basis of a single spot inspection. Tensiometers can be used as a tool to aid in decision making, or, there are several solar radiation integrators on the market that will measure available light and indicate time to water. These are not effective unless the root medium is such as to allow occasional overwatering. There is no system to replace completely value judgments of an experienced grower. Hydroponic systems offer problems peculiar to themselves that can outweigh problems with conventional irrigation.

References

Aikin, W. J.: Photosynthesis in roses. III. The nature of rose water stress and its effect on CO_2 uptake. Colo. Flower Growers' Assoc. Bull. **291**, 1–3 (1974)

Allison, F. E., Roller, E. M., Raney, W. A.: Relationship between evapotranspiration and yields of crops grown in lysimeters receiving natural rainfall. Agron. J. **50**, 506–511 (1958)

Bouyoucos, G. J., Mick, A. H.: An electrical resistance method for the continuous measurement of soil moisture under field conditions. Mich. State College, Agr. Expt. Sta. Tech. Bull. **172**, 38 pp. (1940)

Bower, C. A., Wilcox, L. V., Akin, G. W., Keyes, M. G.: An index to the tendency of $CaCO_3$ to precipitate from irrigation waters. Proc. Soil Sci. Soc. Am. **29**, 91–92 (1965)

Boyer, J.S.: Differing sensitivity of photosynthesis to low leaf water potentials in corn and soybean. Plant Physiol. **46**, 236–239 (1970a)

Boyer, J.S.: Leaf enlargment and metabolic rates in corn, soybean, and sunflower at various leaf water potentials. Plant Physiol. **46**, 233–235 (1970b)

Caparas, J.T.: Basic methods of irrigating greenhouse carnations. M.Sc. Thesis, Colo. State Univ., Ft. Collins, CO. 1955, 97 pp.

Cook, R.L., Erickson, A.E., Krone, P.R.: Soil factors affecting constant water level sub-irrigation. Proc. Am. Soc. Hort. Sci. **62**, 491–496 (1953)

Cotter, D.J., Walker, J.M.: Strategy of relative humidity and temperature control in greenhouses. In: Proc. 8th Natl. Agr. Plastics Conf. San Diego: Univ. of Calif. 1968, pp. 45–50

Davidson, O.W.: High pressure sprays for greenhouse air-conditioning. N.J. Plant Flower Growers' Assoc. Bull. **3(4)** (1953)

Denmead, O.T., Shaw, R.H.: Availability of soil water to plants as affected by soil moisture content and meterological conditions. Agron. J. **54**, 385–390 (1962)

Ford, H.W., Tucker, D.P.H.: Blockage of drip irrigation filters and emitters by iron-sulfur-bacterial products. HortScience **10**, 62–64 (1975)

Gale, J., Hagan, R.M.: Plant antitranspirants. Ann. Rev. Plant Physiol. **17**, 269–282 (1966)

Gardner, W.R., Ehlig, C.F.: Impedance to water movement in soil and plant. Science **138**, 522–523 (1962)

Gates, D.M.: Transpiration and leaf temperature. Ann. Rev. Plant Physiol. **19**, 211–238 (1968)

Goldsberry, K.L.: Condensate in plastic greenhouses. Colo. Flower Growers' Assoc. Bull. **244**, 1–2 (1970)

Hagan, R.M., Vaadia, Y.: Selecting measurements and data in water-soil-plant relations experiments. UNESCO Arid Zone Res. **16**, 331–336 (1961)

Hagan, R.M., Vaadia, Y., Russell, M.B.: Interpretation of plant responses to soil moisture regimes. In: Water and Its Relation to Soils and Crops. Advan. Agron. **11**, 77–98 (1959)

Hanan, J.J.: Water utilization in greenhouses: Alternatives for agriculture in arid regions. In: Proc. 3rd Ann. Water Res. Conf. San Francisco. Urbana, Ill: Am. Water Res. Assoc. 1968, pp. 160–169

Hanan, J.J.: Inert media and irrigation frequency. Colo. Flower Growers' Assoc. Bull. **227**, 1–2 (1969)

Hanan, J.J.: Temperature and humidity observations in CSU greenhouses. Colo. Flower Growers' Assoc. Bull. **238**, 1–4 (1970)

Hanan, J.J.: Repercussions from water stress. HortScience **7**, 108–112 (1972)

Hanan, J.J.: Denver water quality. Colo. Flower Growers' Assoc. Bull. **280**, 1–3 (1973)

Hanan, J.J., Duke, H.R.: The Gates-type greenhouse irrigation system: Design and problems. Colo. Flower Growers' Assoc. Bull. **245**, 1–4 (1970)

Hanan, J.J., Holley, W.D.: Hydroponics. Colo. State Univ. Agr. Exptl. Sta. General Series Bull. **940**, 21 pp. (1974)

Hanan, J.J., Jasper, F.D.: Water utilization by carnations. Effect of three irrigation regimes on growth and flowering. Colo. Flower Growers' Assoc. Bull. **204**, 1–4 (1967)

Hanan, J.J., Kowalczyk, R.A.: Observations on carnation stomata. Colo. Flower Growers' Assoc. Bull. **214**, 1–4 (1968)

Hanan, J.J., Langhans, R.W.: Control of moisture content in greenhouse soils. Agron. J. **56**, 191–194 (1964)

Heydecker, W., Pareek, O.P., Sivanavatam, T.: Major effects of relative humidity on plant performance. In: Proc. 18th Intern. Hort. Cong., Tel-Aviv, Israel Min. Agr. 1970, Vol. I, pp. 96–97

Holley, W.D.: Watering carnations. Colo. Flower Growers' Assoc. Bull. **130**, 1–4 (1966)

Hsiao, T.C.: Plant responses to water stress. Ann. Rev. Plant Physiol. **24**, 519–570 (1973)

Langhans, R.W.: Mist propagation and growing. N.Y. State Flower Growers' Bull. **103**, 1–3 (1954)

Langhans, R.W.: Mist equipment for starting roses. N.Y. Flower Growers' Bull. **116**, 3–4 (1955)

List, R.J.: Smithsonian Meterological Tables, 6th ed. Smith. Misc. Coll. 144, 1966, 527 pp.

Maxwell, J.C.: Will there be enough water? Am. Sci. **53**, 97–102 (1965)

Ministry of Agriculture: Automatic watering of pot plants with the N.I.A.E. capillary bench. Minis. Agr., Fisheries and Food., Great Britain, 1962, pp. 2–6
Mitchell, W. H., Stevens, R. F., Williams, T. H.: Farm and garden crop production with trickle irrigation. Coop. Ext. Serv. Bull., Univ. Del. Apr. 1974, 4 pp.
Ofir, M., Shmueli, E., Moreshet, S.: Stomatal infiltration measurements as an indicator of the water requirement and timing of irrigation of cotton. Exptl. Agr. **4**, 325–333 (1968)
Post, K.: Florist Crop Production and Marketing. New York: Orange-Judd 1950, 891 pp.
Post, K., Seeley, J. G.: Automatic watering of roses. 1943–1946. Proc. Am. Soc. Hort. Sci. **49**, 433–436 (1947)
Richards, L. A.: Laws of soil moisture. Am. Geophys. Union **31**, 750–756 (1950)
Richards, L. A.: Diagnosis and improvement of saline and alkali soils. U.S. Salinity Laboratory, USDA Agr. Handbook 60, 1954, 160 pp.
Russell, M. B.: Plant responses to differences in soil moisture. Soil. Sci. **88**, 179–183 (1959)
Rutland, R. B.: Transpiration of Antirrhirum majus L. in relation to radiant energy in the greenhouse. HortScience **7**, 39–40 (1972)
Schekel, K. A.: Influence of salinity on nutrient uptake, growth and water relations of carnation. Ph.D. Dissertation, Colo. State Univ., Ft. Collins, 1970, 85 pp.
Schekel, K. A., Hanan, J. J.: Nitrogen sources for carnations and general limits on saline waters. Colo. Flower Growers' Assoc. Bull. **253**, 1–4 (1971)
Schekel, K. A., Hanan, J. J., Holley, W. D.: Colorado water analysis. Colo. Flower Growers' Assoc. Bull. **222**, 2–4 (1968)
Seeley, J. G.: Automatic watering of potted plants. Proc. Am. Soc. Hort. Sci. **51**, 596–604 (1948)
Shillo, R., Halevy, A. H.: Experiments in the irrigation of gladioli according to absorption of viscous fluid through stomata. Israel J. Agr. Res. **14**, 89–95 (1964)
Slatyer, R. O.: The influence of progressive increases in total soil moisture stress on transpiration, growth and internal water relationships of plants. Australian J. Biol. Sci. **10**, 320–336 (1957)
Taylor, S. A., Ashcroft, G. L.: Physical Edaphology. San Francisco: W. H. Freeman 1972, 533 pp.
Thorne, J. P., Thorne, D. W.: Irrigation waters of Utah. Utah State Univ. Agr. Exptl. Sta. Bull. **346**, 1951, 64 pp.
Waggoner, P. E., Zelitch, I.: Transpiration and the stomata of leaves. Science **150**, 1413–1420 (1965)
Wells, D. A., Soffe, R.: A tensiometer for the control of glasshouse irrigation. II. Use of controlling tensiometer. J. Agr. Eng. Res. **6**, 16–26 (1961)
Wells, D. A., Soffe, R.: A bench method for the automatic watering by capillarity of plants grown in pots. J. Agr. Eng. Res. **7**, 42–46 (1962)
White, J. W.: The concept of container capacity. Annual Meeting. Am. Soc. Hort. Sci., Boulder, Colo. August, 1965
Winspear, K. W., Postlethwaite, J. D., Cotton, R. F.: The effectiveness of automatic humidity control in reducing fungus diseases of glasshouse tomatoes. Ann. Appl. Biol. **65**, 1–9 (1970)
deWit, I. C. T.: Transpiration and crop yields. Versl. landbouwk. Onderz, Wageningen **64**, 6 (1958)
Zwart, H. C.: Errors in humidity measurements. Instr. Control. Systems **39**, 95–96 (1966)

Chapter 6 Soils and Soil Mixtures

6.1 Introduction

Plants can be supported and grown in many types of materials. In fact, observation will show that if a root can penetrate the mixture, then most plants can survive. The crux comes with the question of how well the plant grows in the mixture. Once beyond the question of mere survival, the problem of soils and root substrates becomes complex. Soils, as usually encountered, are complex physically and chemically, and can be thought of as a living organism.

Root media provide a number of functions in addition to support for the above ground parts—functions that often appear mutually exclusive. For example, roots respire, requiring oxygen and a means of eliminating CO_2. On the other hand, plants also require water, and the soil must provide water in abundance without adversely restricting aeration. The substrate must also supply adequate nutrition, and the medium must be free of disease organisms.

Complexity is further increased by the numerous methods, application and objectives of the grower. Depending upon the species, and how well the location has been chosen, production may be directly in the ground, as is often the case with roses and carnations or tomatoes. The soil volume is essentially unrestricted. Drainage, leaching, nutrition, modification, and disease control will be different from the more usual case where production occurs in restricted, shallow, soil volumes. Examples of the large variety of situations a grower can encounter are shown in Figures 6-1 and 6-2.

Growers have a number of options. They can select an essentially inert medium with the potential of maximum production—provided they have the competence to handle the requirements of such a medium; or, they can select a soil mixture which provides acceptable growth, but allows a degree of safety against mistakes. If election is made to grow in the ground from the standpoint that initial investment will be less, then site selection becomes extremely important for the grower will be forced to accept whatever is available. If pot or foliage plants are grown, then consumer requirements must be met, and a medium is selected that will provide the consumer with maximum satisfaction.

6.2 Principles

Soil properties are derived from the size of the individual particles composing the soil, the mineral composition of those particles, their structural arrangement, location, and how the soil is handled. If a cube, having sides one centimeter long,

Fig. 6-1A–H. Examples of soil placement conditions a greenhouse grower may encounter. (A) Concrete, V-bottom, ground benches with tile drainage. (B) Raised, concrete, flat-bottom benches. (C) Ground beds with no restriction to root growth, wooden side boards to contain the mulch. (D) Specialized "bow" bench for asexual propagation. (E) Movable, shallow trays for bedding carnations. (F) Six inches of gravel placed directly on the ground with no restrictions to root growth. (G) Roses directly in the ground, no restriction to root growth. (H) Potted chrysanthemums in 6-in pots, placed on moveable benches

Fig. 6-1 (continued)

or a total area of 6 cm^2 is subdivided to obtain the usual particle size (less than 0.002 mm), total surface area will exceed 2.4 ha. This large surface area is the principal difference between gravel substrates, as used in inert or hydroponic systems, and common soils. It accounts in large measure for the fact that clay soils will usually retain more water at higher soil moisture suctions than sandy soils; and that clay or highly organic soils will be more resistant to radical chemical changes. Size alone, however, has more of an effect on water retention properties than on chemical properties. Quartz ground to the dimensions given above has no particular chemical properties.

6.2.1 Chemical

The small clay and humus particles—referred to as "micelles"—have a negative electrical charge. As a result, they will attract positively charged cations such as potassium, sodium, calcium, magnesium and hydrogen as indicated in Figure 6-3.

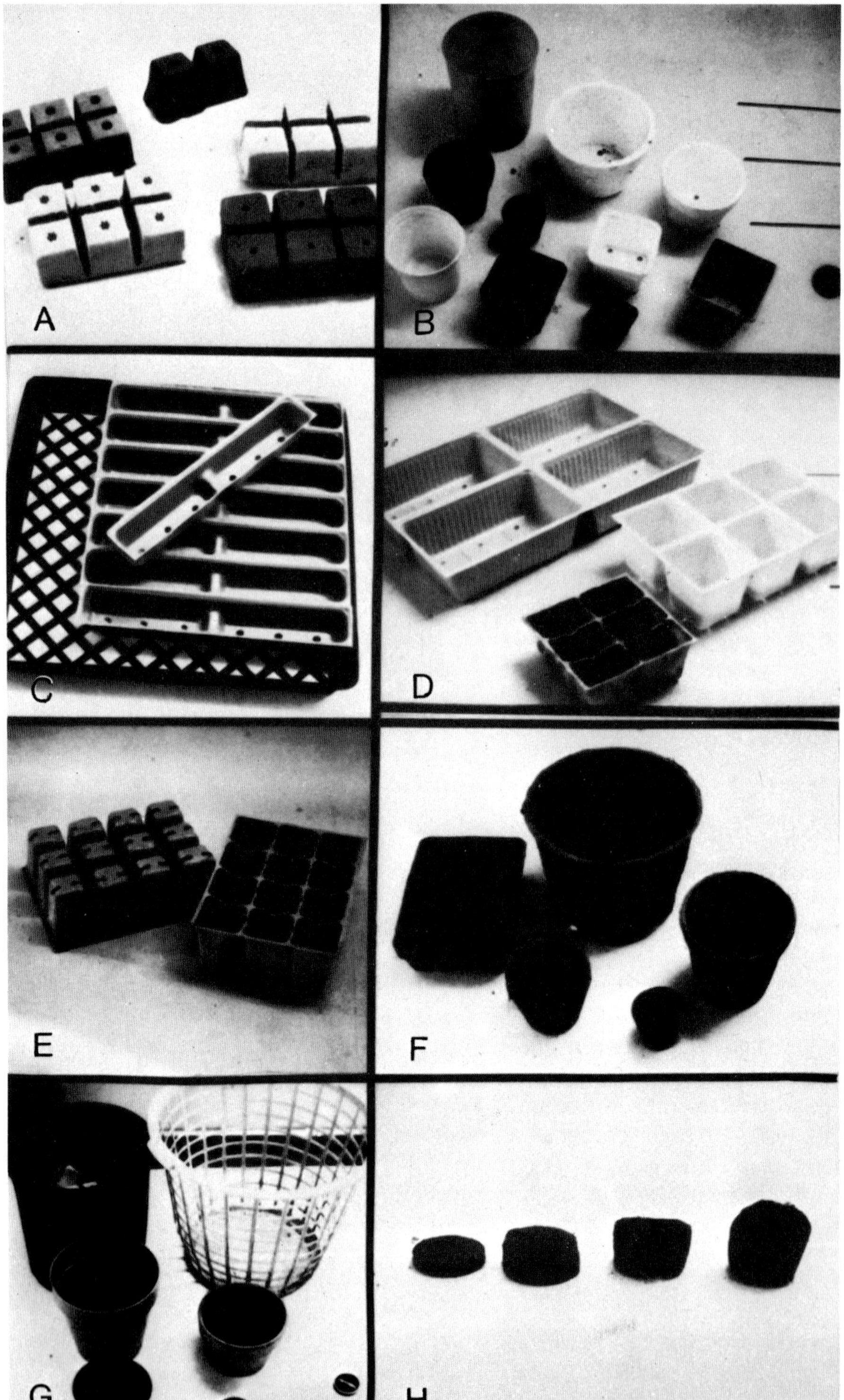

Fig. 6-2 A–H

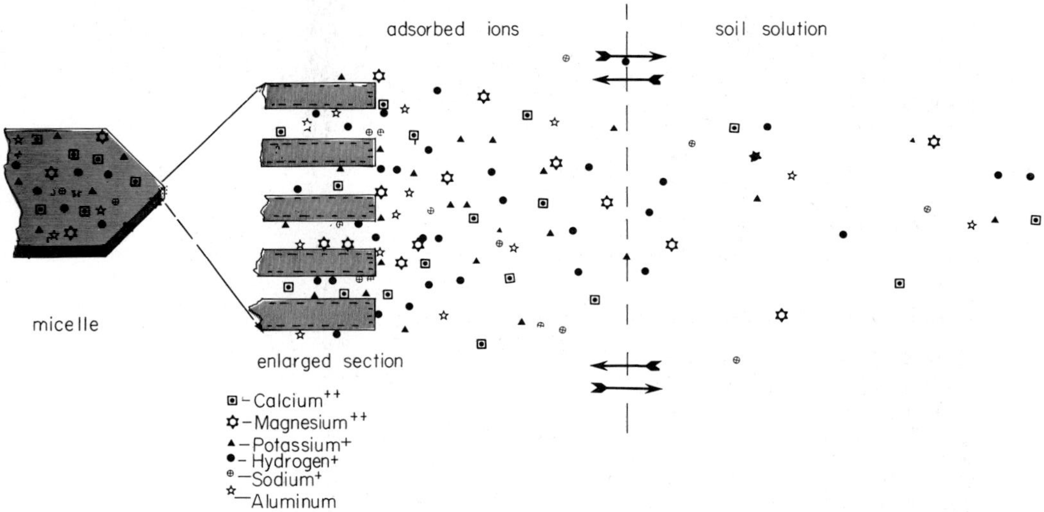

Fig. 6-3. Diagrammatic representation of a clay micelle, greatly enlarged. Cations of calcium, magnesium, sodium, potassium, etc. are adsorbed onto the negatively charged surface of the soil particle. Concentration of ions on the micelle is in equilibrium with concentration of ions in soil solution. Changes in concentration of soil solution will cause release, exchange, or adsorption of ions on the particle surface. Ability of a soil to adsorb ions is expressed as "exchange capacity", and determines in large part the soil's buffering ability

If fertilizer is added to the soil, the concentration of the fertilizer salts increases in the soil solution, and some of the cations are adsorbed on the clay or organic matter. Others are returned to the solution. If concentration of one type of cation is reduced in the soil solution, ions are released from the exchanger, tending to maintain a relatively constant concentration of that cation in solution. Cations are not equally attracted to the clay and organic particles (exchanger), and are usually replaced in the order Na > K > Mg = Ca. We say the soil is "buffered", or is resistant to changes in total salt concentration. This ability to adsorb, or store, ions is measured by the soil's cation exchange capacity (CEC), expressed as milli-equivalents per 100 g of soil. The relative ability of different soils and amendments to store ions is given in Table 6-1. The table shows that gravel or sand media require precision in handling nutrition as there is no buffering action, and small chemical additions will have immediate and marked effect.

6.2.2 Structure

Physical characteristics of greenhouse soils are probably more important than chemical properties. As noted in Chapter 7, greenhouse operators have considerable opportunity for precise manipulation of nutrients. On the other hand, once

---

◂ Fig. 6-2A–H. Various examples of plant containers, plastic, fiber, etc. (A) BR-8 and Oasis blocks for rooting cuttings. (H) Various stages of expansion upon wetting of a Jiffy-7, peat pellet

Table 6-1. Relative cation-exchange capacity of common soil materials. (From Furuta, 1974)

Material		Relative cation-exchange capacity
Clay	Montmorillonite	10
	Illite	3
	Kaolinite	1
Silt		½ or less
Sand		0
Vermiculite		15
Humus		20
Peat moss		14
Redwood sawdust		3

the root medium has been selected, and the plant is established, little or nothing can be done to alter water and aeration relationships. The grower is committed.

The primary soil practicles—clay, silt, humus, sand—determine the soil's texture and basic characteristics of water retention, aeration and chemistry. The root system is in intimate contact with these particles as depicted in Figure 6-4. The fact that some people have estimated root surface area for a single plant in two cubic feet of soil to exceed 6800 ft^2 (Dittmer, 1937), emphasizes the intimacy suggested by Figure 6-4. During the day, when plants are losing water, the liquid must be supplied to the roots at rates sufficient to meet needs, and a large portion of the required nutrients are carried to and into the roots with the water flow.

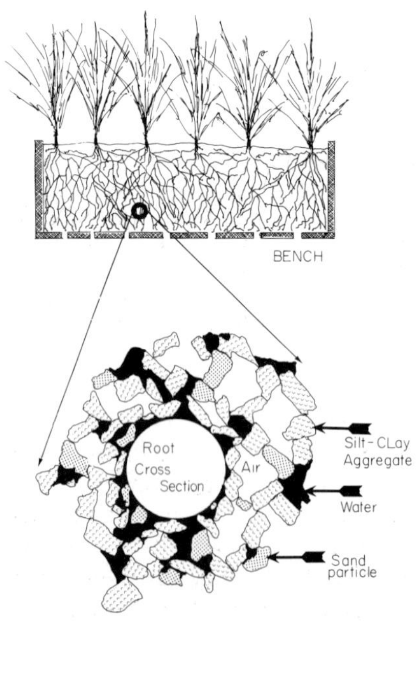

Fig. 6-4. Diagram of soil particle arrangement about a single root. For air to be taken in by the root, molecules must diffuse across the water film which is 10000 times slower than oxygen diffusion in air. The air-filled pores are referred to as the "macropores", "free air pores", or "porosity". Their volume at maximum water content should exceed 20% of the total volume under most conditions. In coarsely aggregated media, the water-filled pores may be discontinuous so that water movement to the root can only occur in the thin films around each particle

Fig. 6-5A–D. Examples of soil structure: (A) Crumb structure of heavy clay; (B) granular structure; (C) subpolyhedric; (D) polyhedric. Other types may be massive, prismatic, columnar, or complex. (From Taylor and Ashcroft, 1972)

Oxygen must be supplied to the roots through the water films although the films may not be continuous. As oxygen diffusion through water is 10000 times slower than through air, the thickness of the water film around the root becomes critical. Fortunately, the soil can be modified to provide a "structure" that permits adequate water and air supply. Water films around the roots can also be changed through irrigation practice, or, in shallow soil layers, by changing the depth. It is the grower's objective to provide a suitable "macrostructure" such that water and

Table 6-2. Effect of adding peat moss to a silt loam on pore space and infiltration. (From White, 1964)

Silt loam-peat moss proportions	Total pore space (%)	% moisture for a volume 17.2 cm deep	% air	Infiltration rate (in per h)
10–0	57.0	43.9	13.1	1.6
9–1	60.7	43.7	17.0	1.8
7–3	64.9	41.0	23.9	15.4
5–5	73.4	47.6	25.8	39.2
3–7	81.1	57.3	23.8	58.4
1–9	91.1	68.6	22.5	60+
0–10	94.4	63.8	30.6	60+

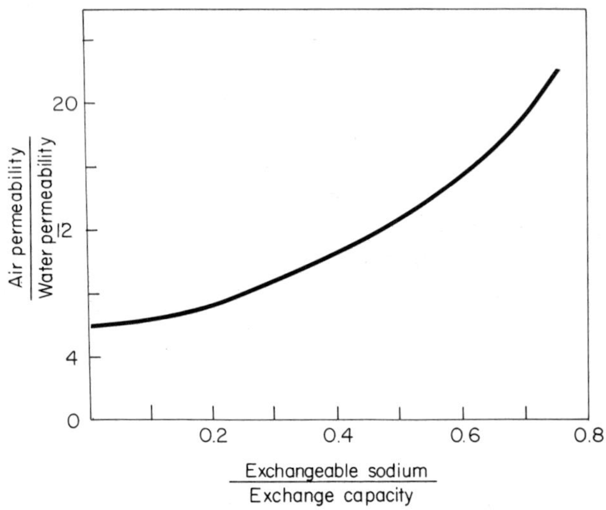

Fig. 6-6. Influence of sodium on air and water permeability ratio in sandy loam soil. (After Reeve, 1953.) Reducing amount of sodium (higher ratio of exchangeable sodium to exchange capacity) increases stability of soil aggregates, and increases soil permeability as indicated by higher ratio of air to water permeability

air move through the soil rapidly, but the aggregates, composed of the primary particles, retain sufficient water and nutrients to meet requirements between irrigations and fertilizations.

Soil aggregation can be changed by adding organic matter, as noted in Table 6-2, or by changing the kinds and amounts of ions adsorbed on the soil micelle (Fig. 6-6). Table 6-2 shows that as the proportion of soil is reduced, total pore space, water retention, air-filled pore space and infiltration rate are increased. Or, if for some reason, a major portion of the exchange capacity consists of sodium, the soil will tend to loose its aggregation, or deflocculate, and the water and air permeability will be reduced. The soil is "puddled". This is a good reason for measuring the sodium hazard of irrigation waters. The fertilization program must take into account the need to keep sodium in the exchange complex from becoming dominant.

6.2.3 Water Relationships

The effect of water on growth, and relationships in the soil were covered in Chapter 5. Briefly, the amount of water retained by a soil will depend upon particle size, structure and depth of the soil if in a shallow layer. In deep field soils, maximum moisture holding capacity (field capacity) is often found at soil moisture suctions of 200–300 cm water. Infiltration is usually slower as indicated in Figure 5-20, and water content will decrease as suction increases in the manner depicted in Figure 5-17. For most field soils under common field conditions, plants will permanently wilt when moisture content has decreased to a point where the pressure required to extract water is less than -15 bars. Actually, plants can extract water from much dryer soils (Slatyer, 1957), but the delivery rate is too slow. With highly modified greenhouse soils, wilting often occurs at much lower suctions, depending upon the evaporative demand surrounding the leaves. This is because the lateral flow from source to "sink" (root surface), or capillary conductivity, in greenhouse soils may be nearly zero at very low suctions—10 cm water suction in peat soils according to Richards and Wilson (1936).

The large soil volume in the ground—on the order of two million pounds for an acre six inches deep—means that very little can be done in a short time to significantly alter physical characteristics. The success of drain lines will depend upon soil depth and texture. Very seldom do growers installing drain lines in deep, clay soils see water running out of them. For water to move from the soil into the drain line requires that the soil water suction be positive at the surface between drain and soil—as described in Chapter 5 for a shallow, freely draining layer. If the soil overlies an impermeable layer, or the water table is high, then drain lines can be useful. A thorough physical and chemical analysis should be made before locating greenhouses if the crop is to be grown directly in the ground. Core samples to determine the soil profile are necessary.

It is seldom that a field soil can be directly placed in a shallow pot or bench and used successfully. Field soils often have bulk densities ranging from 1.0 to 1.4 g cc^{-1}, with commensurate porosities and infiltration characteristics. Under normal handling in the greenhouse, field soils will retain excess water with reduced aeration. The maximum moisture capacity is a direct function of depth. For this reason, considerable effort and expense are gone to to modify the soil for bench or pot culture. Greenhouse soils will have bulk densities ranging from 0.1 to 0.8 g cc^{-1} with total porosities often exceeding 90% and infiltration rates in excess of 60 in per h. Soils with densities in excess of 1.8 g cc^{-1} will virtually exclude root growth (Zimmerman and Kardos, 1961).

Obviously, there should be minimum restriction to the water flow away from the soil undersurface, unless a capillary mat or constant water table is employed. Often, individuals will place gravel, or some other drainage material, in the bottom of pots. This reduces soil depth, and increases water retention. If a proper substrate has been provided, it is much cheaper to forget about drainage material. Greenhouse benches must have sufficient cracks between boards, or holes, for minimum restriction. Lumber will swell, and boards are commonly spaced to leave a 3/8th to ½-in crack after wetting. A good rule of thumb is that drain locations should not be further apart than 6–8 in. Wide bottom boards will restrict drainage as the water must move laterally a significant distance. In ground beds with "V" bottoms, a layer of coarse gravel is often placed around the drain tile.

6.2.4 Aeration

Unless a surface is unwettable, a water film can be found on it. In soils with large surface areas, and where the relative humidity is nearly always 100%, the film is thick, accounting for the fact that clay soils hold considerable amounts of water. Plants grown in heavy soils often have characteristics approaching those subjected to drought. The relationship between thickness of water film and pressure required to remove that film is shown for a glass and mica surface in Figure 6-7. A similar situation exists in soils, and so also will root surfaces often be covered by a layer of water (Fig. 6-4). Oxygen must diffuse across the layer to the root. As soil water content increases, the water layer will tend to become thicker, and oxygen concentration at the root will decrease (Fig. 6-8). If texture and structure of the soil are such that most of the pores remain filled with water after irrigation, oxygen supply will be severely restricted and growth will be reduced. Soil modification provides pores large enough to drain so that air can readily diffuse through the soil, and the water film is thin enough to allow rapid oxygen supply.

As oxygen supply decreases, ability of a root to grow and penetrate the soil decreases (Fig. 6-9). Work by several individuals (e.g. Stolzy et al., 1961; Hanan,

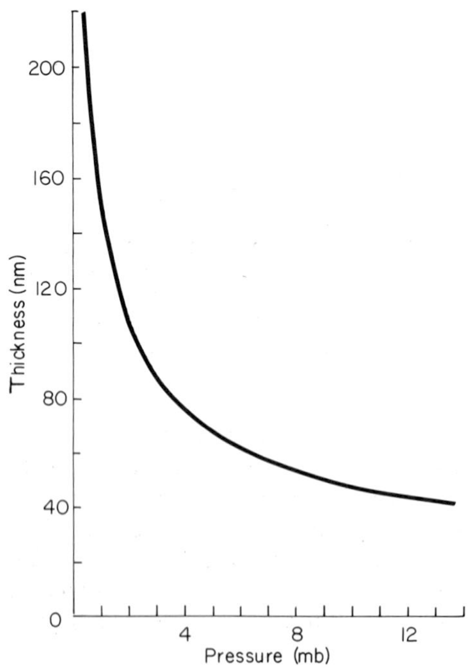

Fig. 6-7. Relationship between thickness of a water film on glass and mica as influenced by pressure. (Adapted from Schofield, 1946.) Similar situation exists in soils. Clay soils with large surface areas will retain considerable water against high suctions even though water films are quite thin

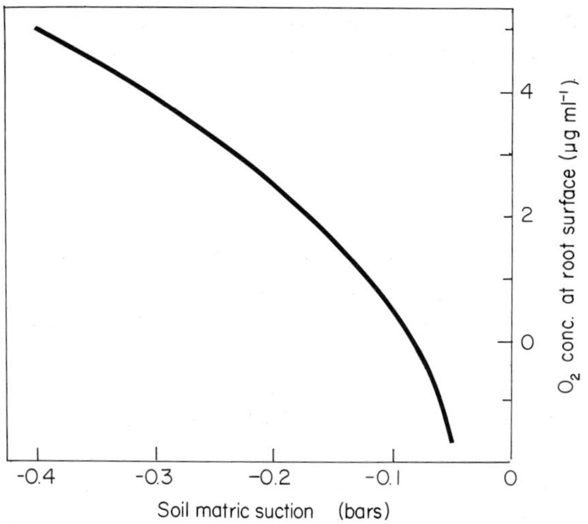

Fig. 6-8. Calculated concentration of oxygen at root surface as determined by soil suction. Values below zero result from assumption that roots consume oxygen at a constant rate regardless of aeration which is obviously erroneous below zero. For practical purposes, critical oxygen concentration is often said to be reached when oxygen is zero at center of root. (From Weigand and Lemon, 1958, as replotted by Taylor and Ashcroft, 1972)

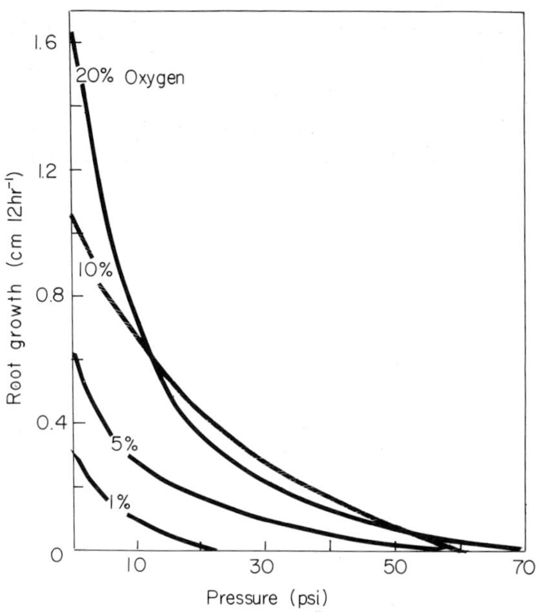

Fig. 6-9. Rate of root growth as influenced by oxygen concentration and applied pressure. Applied pressure has essentially same effect as increased bulk density or impedence to root growth. (After Gill and Miller, 1956)

1964; Hanan and Langhans, 1964) has shown that even with 20% oxygen concentration in the soil air, roots can suffer if the water film is too thick. In properly modified soils, oxygen is seldom deficient in the air-filled pores (Hanan, 1964). In summary, modifying the soil accomplishes three things: (1) it provides sufficient air-filled pores so that air rapidly diffuses through the mixture; (2) it reduces the thickness of the water film around the root so that diffusion of oxygen and CO_2 is not a limiting factor; and (3) it reduces the resistance to root growth as indicated by bulk density.

Once the soil has been modified, every effort should be made to ensure that it retains its structure by eliminating excessive compaction. The irrigation system should not disturb the soil surface. If excessive washing occurs, fine particles will settle out in the upper layers with consequent crusting which impedes water and air infiltration. Growers will sometimes cultivate the surface layers to break up this crust. However, this is wasteful and often hastens reformation of the crust. If the soil is properly constituted, and the irrigation system well chosen and properly handled, there should be no need to cultivate. In instances where there may be a problem, a one-inch mulch of leaf mold, straw, wood chips, corn cobs, etc., on the surface will usually prevent crusting, as well as adding organic matter.

Attempts are sometimes made to "aerate" the soil by forcing air through it, or installing devices to pump air into the mixture at several locations. The value is questionable as oxygen sufficiency is most often dependent upon the water film about the individual root and not the status of oxygen in open pores. If the soil is properly handled, special devices are unnecessary.

There are some disadvantages to modifying soils. Modification reduces capillary conductivity. Small seedlings, or new cuttings, may suffer drought before they can establish themselves with a rapidly expanding root system. Water cannot move easily to a point where it is taken into the plant. Irrigation frequency may have to be increased for a period until the root system is established.

6.3 Soil Modification

No soil remains the same from year to year. Not even inert media will remain the same. Mixtures undergo constant chemical and physical changes. For a pot plant grower, the problem is simplified as long as his potting soil remains uniform from one crop to the next. The greatest danger is that sudden changes resulting from different amendments and basic soils will require different cultural practices. Until the grower becomes familiar with the new root medium, he may have difficulty.

6.3.1 Effect of Soil Amendments

The choice of amendments to modify soils is wide. Examples of various types, and suitable proprietary mixtures, have been examined by White (1974) (Table 6-3) and Johnson (1968) (Table 6-4). An amendment should decrease bulk density, increase total porosity, and increase water and air content. Infiltration rates will rise rapidly, and capillary conductivity decrease (Table 6-2). Amendments can be combined as shown by White (1964) (Table 6-5). As a general rule, the tables show that whatever is added to the basic soil should equal at least one-third to one-half of the final volume if significant changes are to be made. Generally, any mixtures

Table 6-3. Comparative costs and weights of media useful for container crop production—costs from 1973. (Adapted from White, 1974)

Medium	Approx. cost m^3 (\$)	Dry bulk density[a] (g cm^{-3})
Peanut or rice hull	0.82	0.11
Sawdust, bark	1.02	0.21
Topsoil	1.43	0.97
Sand-gravel	2.04	1.62
Waylite	2.45	0.81
Lelite or Haydite	2.86	0.89
Freelite	2.86	0.89
Anthracite refuse	2.86	0.65
Idealite	2.86	0.81
Pumice	2.86	0.49
Scoria	2.86	0.89
1-1-1, soil-peat-perlite	5.72	0.55
U.C. Mix, 50–50	6.74	0.58
Steer manure, dried	7.14	0.13
Organic humus	7.14	0.11
Sphagnum peat moss	7.76	0.10
Perlite	7.76	0.10
Vermiculite (Terralite)	7.76	0.11
Sorbolite	12.25	0.49
Peat Grow-Mix	15.31	0.10
Terragreen-Lusoil-Turface	22.45	0.57
Silvabark	25.52	0.19
Jiffy-Mix	28.58	0.10

[a] See Appendix A for conversions.

which have infiltration rates less than 20–30 in per h, air-filled porosities less than 10% at maximum moisture capacity, or bulk densities exceeding 1.0 g cc^{-1}, are likely to be difficult when in shallow, restricted, freely draining layers such as benches or pots. The same soil may be satisfactory in the ground. The effect of amendments as compared to a straight loam soil at different suctions is shown in Figure 6-10. Unmodified soils usually fail to drain at suctions encountered at bench or pot capacity—around 15–20 cm water suction. All three mixtures in Figure 6-10 had air-filled pore space exceeding 20% at 10 cm water suction.

Sand or gravel is a common additive to field soils or other mixtures. Spomer (1975), however, has shown that small volumes of sand will actually decrease porosity. The final volume will be less than the total volume of the components separately as the sand particles will fill the pore spaces of the soil or peat moss. Spomer has defined the amounts of such materials that will result in the miminum porosity as the "threshhold proportion". If the amount of sand in the final volume continues to increase, porosity will begin to increase (Fig. 6-11). Therefore, heavy materials such as fine sand or gravel should be used with caution, since they do nothing to lighten the mixture (Table 6-6). Sand or gravel should never be added to heavy clay soils in the field.

Table 6-4. Physical properties of amendments, soils, and mixtures. (Condensed from Johnson, 1968)

Material	Bulk density		Moisture capacity[a] (% vol)	Total porosity (%)	Free-air porosity (%)
	Dry ($g\ cm^{-3}$)	Wet ($g\ cm^{-3}$)			
Bark, fir (0–1/8″)	0.23	0.62	38	69.5	31.5
Bark, fir (1/8–5/8″)	0.19	0.34	15	69.7	54.7
Bark, redwood (3/8″)	0.13	0.44	30.8	80.3	49.5
Loam, clay	0.95	1.51	54.9	59.6	4.7
Loam, sandy	1.58	1.95	35.7	37.5	1.8
Peat, sedge, AP4	0.21	0.74	52.3	69.3	17.0
Peat, sedge, BD	0.26	0.95	68.6	77.0	8.4
Peat moss, hypnum	0.19	0.79	59.3	71.7	12.4
Peat moss, sphagnum	0.11	0.70	58.8	84.2	25.4
Perlite (1/50–1/16″)	0.09	0.52	42.6	75.8	33.2
Perlite (1/4–5/16″)	0.10	0.29	19.5	73.6	53.9
Pumice (1/5–1/16″)	0.46	0.87	40.5	62.2	21.7
Pumice (5/16–5/8″)	0.48	0.74	25.5	70.5	45.0
Rice hulls	0.10	0.23	12.3	81.0	68.7
Sand, builders	1.68	1.95	26.6	36.0	9.4
Sand, fine B	1.44	1.83	38.7	44.6	5.9
Sawdust, cedar	0.21	0.60	38.2	80.8	42.6
Sawdust, redwood	0.18	0.68	49.3	77.2	27.9
Vermiculite (0–3/16″)	0.11	0.65	53.0	80.5	27.5
50-50 mixture (clay loam with)					
Peat, sedge	0.66	1.23	56.5	61.8	5.3
Peat moss, hypnum	0.63	1.23	59.9	66.2	6.3
Peat moss, sphagnum	0.55	1.18	61.0	71.0	10.0
Sand, builders	1.28	1.69	40.8	47.0	6.2
Sand, fine	1.32	1.74	41.5	47.4	6.9
50-50 mixture (sandy loam with)					
Peat sedge	0.90	1.41	50.3	53.9	3.7
Peat moss, hypnum	0.94	1.45	49.8	54.2	4.4
Peat moss, sphagnum	0.87	1.41	52.8	59.1	6.3
Sawdust, redwood	0.80	1.33	52.7	62.8	10.1
50-50 mixture (fine sand with)					
Bark, fir (0–1/8″)	0.86	1.23	37.4	54.6	15.2
Bark, redwood	0.94	1.38	43.5	56.8	13.3
Peat, sedge	0.93	1.43	49.0	53.5	4.5
Peat moss, sphagnum	0.75	1.23	47.3	56.7	9.4
Perlite (1/16–3/16″)	0.86	1.29	42.6	52.0	7.6
Pumice (1/16–1/8″)	1.05	1.43	37.5	42.3	4.8
Sawdust, redwood	0.93	1.31	40.5	52.6	12.1
50-50 mixture (peat moss with)					
Perlite (3/16–1/4″)	0.11	0.63	51.3	74.9	23.6

[a] Moisture capacity determined for a column 17.8 cm deep.

Table 6-5. Physical characteristics of some soil mixtures. (Adapted from White, 1974)

Mixture (soil-perlite-peat)	Bulk density (g cm^{-3})	Total porosity (%)	Max. water capacity[a] (%)	Air porosity (vol.-%)	Percolation rate (cm h^{-1})
10-0-0	1.15	57.0	43.9	13.1	4.1
9-1-0	1.15	56.9	42.0	14.9	5.3
9-0-1	1.05	60.7	43.7	17.0	4.6
8-1-1	1.03	61.3	46.0	15.3	6.6
7-3-0	1.03	61.5	41.8	19.7	50.8
7-0-3	0.93	64.9	41.0	23.9	39.1
7-1-2	0.85	67.9	45.6	22.3	35.8
7-2-1	0.90	66.4	44.9	21.5	49.0
6-1-3	0.72	72.5	44.2	28.3	30.0
6-2-2	0.82	69.2	41.2	28.0	31.2
6-3-1	0.86	67.5	43.8	23.7	34.8
5-5-0	0.82	69.3	42.4	26.9	20.3
5-0-5	0.69	73.4	47.6	25.8	99.6
3-7-0	0.68	73.6	39.6	34.0	132.6
3-0-7	0.48	81.1	57.3	23.8	148.3
3-6-1	0.54	78.7	39.5	39.2	108.0
3-1-6	0.45	82.5	53.3	27.2	123.2
2-7-1	0.46	82.1	38.8	43.3	152+
2-1-7	0.38	84.7	63.9	20.8	152+
2-6-2	0.40	84.3	42.0	42.3	152+
2-2-6	0.36	85.8	53.8	32.0	152+
1-9-0	0.40	84.2	40.3	43.9	152+
1-8-1	0.31	87.6	38.1	49.5	152+
1-7-2	0.30	87.9	45.9	42.0	152+
1-6-3	0.29	88.3	43.2	45.1	152+
1-3-6	0.26	89.3	55.9	33.4	152+
1-2-7	0.27	88.6	64.0	24.6	152+
1-1-8	0.27	88.7	64.8	23.9	152+
1-0-9	0.22	91.1	68.6	22.5	152+
0-10-0	0.18	92.4	36.8	55.6	152+
0-9-1	0.17	92.7	38.7	54.0	152+
0-7-3	0.14	93.8	43.5	50.3	152+
0-5-5	0.14	93.4	51.5	41.9	152+
0-3-7	0.12	93.8	52.6	41.2	152+
0-1-9	0.18	89.8	64.6	25.2	152+
0-0-10	0.10	94.4	63.8	30.6	152+

[a] Moisture capacity determined for a column 17.2 cm deep.

6.3.2 Soil Mixtures

Uniformity is a major problem in greenhouse soils. Good field soils are increasingly difficult to find. As a result, there has been a shift to artificial substrates that can be duplicated from year to year. Beginning with the John Innes composts in the 1950s (Table 6-7), there have been an increasing number of mixtures that can be purchased by the grower, or mixed on the premises to his particular satisfaction. The John Innes composts require a good, organic field soil, with consider-

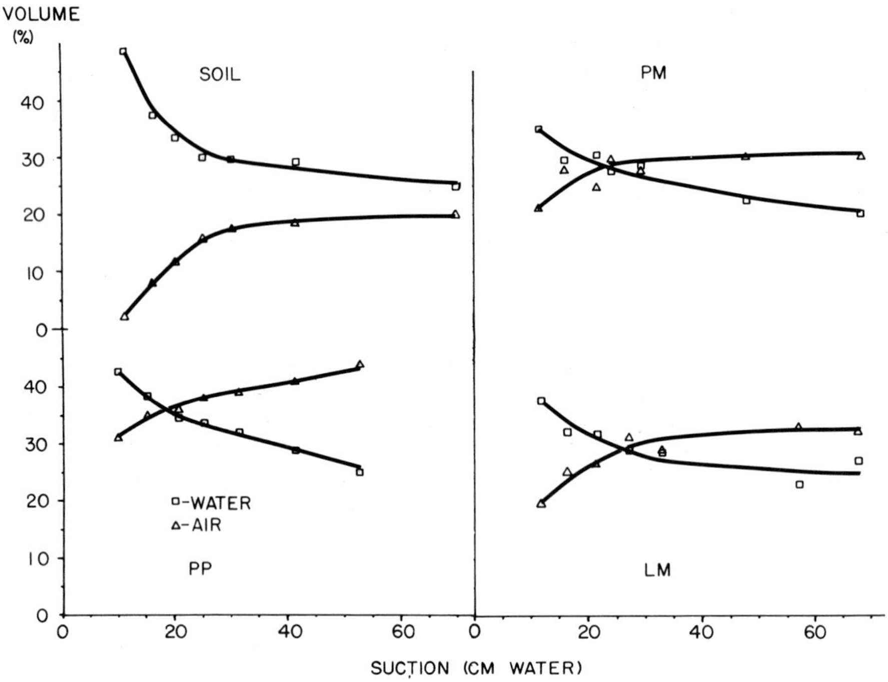

Fig. 6-10. Air and water content of various root media at different soil suctions. *SOIL* silt loam field soil; *PP* 1-1 mixture of peat moss and perlite; *PM* 1-1-1 mixture of soil, sand, and peat moss; *LM* 1-1-1 mixture of soil, sand, and leaf mold (Hanan, 1963, unpublished)

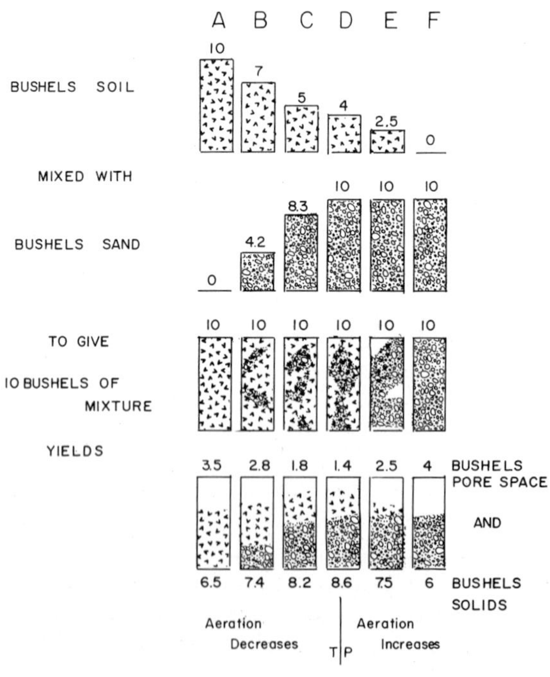

Fig. 6-11. Schematic representation of Spomer's "threshold proportion" (*TP*). Initially adding sand to a soil decreases pore space. *A* through *F* to a point where sand is in larger volume than soil. Thus, in *D*, for example, four bushels of soil mixed with 10 bushels of sand still result in 10 bushels final volume. This illustrates a common danger of adding sand alone, in insufficient amounts, to another medium such as soil or peat moss. As sand fills the soil's pores, sum of sand and soil is less than volumes of sand and soil separately. (From Spomer, 1975)

Table 6-6. Influence of particle size on bulk density in soil mixtures and when sand added. (From Richards et al., 1964)

Soil composition			Bulk density (g cc^{-1})	
Sand (%)	Silt (%)	Clay (%)	Soil alone	Mixed with 60% sand by volume
74	18	8	1.53	1.56
58	25	17	1.45	1.55
43	34	23	1.36	1.52

Table 6-7. John Innes compost. Loam selected should have enough clay to be slightly greasy when smeared without stickiness, from a good pasture, 2–7% humus. This is to be composted in alternating layers of 4–1/2 in loam and 2 in strawy manure, allowing 6 month's time before using. (Adapted from Lawrence and Newell, 1952)

	Ingredients, parts by volume	Fertilizer ingredients, per m^3
Seed compost	2 loam 1 peat moss[a] 1 sand[b]	1.2 kg 20% superphosphate 0.6 kg calcium carbonate
Potting compost	7 loam 3 peat moss 2 sand	1.2 kg 20% superphosphate 0.6 kg potassium sulfate 0.6 kg calcium carbonate

[a] Canadian or German sphagnum or equivalent.
[b] Coarse sand.

able labor in composting. Any time soil is used as part of the mixture, it should have high productivity when in the field. In the late 1950s, the California workers (Baker, 1957) put together the U.C. mixes, using fine sand and peat moss (Table 6-8). Mixtures such as these have no significant nutrition, and can be fertilized with high precision. Any time a soil is the base material, one should test it for nutrients, and experiment on a small scale in order to tailor the fertilization program. The U.C. mixes were shortly followed by the Peat-Lite, Cornell mixes (Table 6-9). Growers may make their own mixes from local materials, taking into account costs and availability. Aside from the characteristics already mentioned, the mixtures should be capable of duplication and free from weeds and disease organisms. Particular care should be given to the possibility of herbicides. There have been instances of damage from amendments placed on ground treated with persistent herbicides, taking soil from roadside ditches that have been sprayed for weed control, or from farmers' fields that have been treated. Chemicals should never be stored in the same area as the materials used for root media.

Peat moss is a major constituent of most soil mixes, but quality varies widely. German, Canadian, and Finnish sphagnum peats have been major sources in recent years. In mountainous regions, there may be small areas of peat. These

Table 6-8. Basic U.C. mixes, using fine sand and peat moss. Sand should consist of particles between 0.5–0.05 mm diameter, with silt plus clay not exceeding 15%, and coarse sand not exceeding 12–15%. Peat moss should be finely ground and pathogen-free, Canadian, German sphagnum or equivalent. (Adapted from Baker, 1957)

Mix	Ingredients, % by volume		Weight ($g\ cc^{-1}$)		Maximum water content, % by volume	Fertilization for indefinite storage amounts per m^3	Approx. cost per m^3, 1957
	Sand	Peat moss	At max[a]	Oven dry			
A	100	0	1.87	1.42	43	227 g potassium nitrate 113 g potassium sulfate 1.1 kg 20% superphosphate 0.7 kg dolomite lime 1.1 kg gypsum	$1.64
B	75	25	1.68	1.22	46	170 g potassium nitrate 113 g potassium sulphate 1.1 kg 20% superphosphate 2.0 kg dolomite lime 0.6 kg calcium carbonate 0.6 kg gypsum	$2.56
C	50	50	1.50	1.01	48	113 g potassium nitrate 113 g potassium sulfate 1.1 kg 20% superphosphate 3.4 kg dolomite lime 1.1 kg calcium carbonate	$3.47
D	25	75	1.06	0.54	51	113 g potassium nitrate 113 g potassium sulfate 0.9 kg 20% superphosphate 2.3 kg dolomite lime 1.8 kg calcium carbonate	$4.32
E	0	100	0.69	0.11	59	170 g potassium nitrate 0.5 kg 20% superphosphate 1.1 kg dolomite lime 2.3 kg calcium carbonate	$5.22

[a] Weight at maximum water content for a 6-inch column of a fine sand of the Oakley series and Canadian peat moss.

should be used with care as the peat may be in an advanced decomposition stage. Some evidence of the original vegetation that formed the peat should be visible in unground peat moss.

Other materials such as wood chips, straw, corn cobs, bark, etc. have been widely used. With precaution, these materials are perfectly suitable with the possible exception of hardwood bark. In using carbonaceous amendments, attention should be given to particle size and the carbon/nitrogen ratio. Decomposition rate increases the finer the particle, and the large microbial population will tie up nitrogen. When decomposition is completed, death of the microbes will release large amounts of nitrogen so the grower, at different periods, may first experience a nitrogen deficiency and later salt damage. General practice is to apply addi-

Table 6-9. Peat-lite, Cornell mixes. (Adapted from Cornell Recommendations-Boodley, Sheldrake and Mott, 1974)

Mixture	Ingredients	Fertilization, per m^3						Wetting agent[b]
		Limestone	20% superphosphate	Calcium-nitrate	Trace[a]	Iron-sulfate	10-10-10	
A	50% sphagnum peat[c] 50% hort. vermiculite	3 kg	1.2 kg	0.6 kg	43 g			64 g
B	50% sphagnum peat 50% hort. perlite	3 kg					4.8 kg	64 g
Foliage plants	50% sphagnum peat 25% hort. vermiculite 25% hort. perlite	4.8 kg	1.2 kg	0.6 kg KNO_3	43 g	16 g	1.5 kg	64 g
Epiphytic mix	33% sphagnum peat 33% hort. perlite 33% douglas fir bark, fine ground, 1/8–1/2″	4.2 kg	2.4 kg	0.6 kg KNO_3	43 g	11 g	1.5 kg	64 g

[a] Fritted trace elements.
[b] Wetting agents such as Aqua-Gro, Ethomid or Triton B-1956, etc.
[c] Canadian or German or equivalent.
Note: Allow for 20% shrinkage.

Table 6-10. Nitrogen requirements and decomposition rates of selected organic amendment sources. (From Johnson, 1968)

Material	Nitrogen requirement (kg m^{-3})			% decomposed in 60 days
	As N	As ammonium nitrate	As blood meal	
Calif. incense cedar	0.5	1.5	3.6	4.9
Cypress sawdust	0.7	2.1	5.4	4.9
Redwood sawdust	0.5	1.6	4.2	5.3
Douglas fir sawdust	0.5	1.5	3.3	11.2
Ponderosa pine sawdust	1.2	3.6	9.0	13.7
White fir bark	1.9	5.7	14.4	7.8
Redwood bark	0.2	0.6	1.8	2.2
Pine bark	1.5	4.5	10.2	—
Sphagnum peat moss	0.04	0.1	0.4	—

tional nitrogen to compensate for the microbial buildup. Johnson (1968) has published nitrogen requirements and decomposition rates of some high carbon, low nitrogen amendments (Table 6-10).

Vermiculite, perlite, manufactured materials such as Corfuna, Floramull, and Styromull are available. The first two are very common, and often employed as a rooting medium in propagation. Vermiculite, however, tends to collapse in a short time and may contain potassium and magnesium. The type used for insulation should not be bought as it is usually coated with an oil. Horticultural perlite is a good, light amendment, but if mixed with peat moss, about one pound of calcium carbonate per square yard should be added as low pH may result in aluminum release with consequent aluminum toxicity to cuttings. A common difficulty with many of these materials is "wetting". Once dry, they are difficult to rewet, and cuttings should never be planted until the mixture has been brought to the proper moisture level. Table 6-9 provides examples of some suitable wetting agents that can be added to the basic mixture to improve wettability. These chemicals have also been employed with success on old bench soils that are highly organic.

Manure, at one time, was a very common amendment. Difficulties arose from variability in chemical and physical characteristics, and the fact that it may contain high numbers of weed seeds that will sprout rapidly after having passed through the animal's digestive system. Fresh manure, aside from aesthetic problems, may release sufficient ammonia to cause phytotoxicity. This is particularly true if the mixture is steamed and then allowed to stand before planting.

6.4 Inert Media

An "inert medium" is a term coined by Holley (1967), referring to the use of material having essentially no buffering capacity, which does not break down, and

Table 6-11. Water holding capacity of some common inert media and soil mixtures per square foot of medium 7–8 in deep (17.8–20.3 cm)

Material	Quarts per ft²	l per m²
Soil-perlite-peat moss		
10-0-0	7.7	78.8
7-3-0	7.3	74.7
5-5-0	7.4	75.7
7-0-3	7.2	73.6
5-0-5	8.3	84.9
3-0-7	10.0	102.3
1-9-0	7.0	71.6
1-0-9	12.0	122.7
0-0-10 (peat moss)	11.2	114.5
0-10-0 (perlite)	6.4	65.5
Rice hulls	2.1	21.5
Sawdust	9.2	94.1
Firbark	2.6	26.6
Clay loam soil	9.6	98.2
Sandy loam soil	6.2	63.4
Inert media[a]		
Granitic sand and gravel	3.3	33.8
Large idealite	1.5	15.4
Regular idealite	1.8	18.4
Fine idealite	4.5	46.0
River sand and gravel	2.0	20.5
Volcanic ash (Scoria)	5.8	59.3

[a] See Table 6-12 for description.

which has no organic matter for sustaining microbial action. If the amount of water per square foot of bench surface, of an inert medium 7 in (18 cm) deep is compared with soil mixes (Table 6-11) one notes that maximum moisture capacity for inert media is about 26% of common soil mixtures and amendments. In other words, the ability to supply water is severely restricted. However, pores in inert media are large and, therefore, there is no restriction to their drainage following irrigation. Thus, watering frequency must be often enough to compensate for lack of reserve.

The use of inert media as discussed by Holley (1967) and Hanan and Holley (1975) differs from conventional hydroponics which can make use of the same media (Table 6-12). As used by Gerike (1940), Steiner (1968) and Sholto-Douglas (1972), classical hydroponics requires a watertight bench, storage tanks for the nutrient solution, and suitable pumps and piping to flood the inert medium in the bench. Capital investment can equal investment for the greenhouse structure. Thus, the reason that hydroponics is not employed on a large scale to any extent in the United States is commonly a matter of installation expense. There are also the problems arising from the fact that there is no buffering capacity and available water storage is low. There is less room for error, and any error is likely to be marked and rapidly visible. The system developed in Colorado makes use of

Table 6-12. Particle size distribution of various inert media

Material	Percent of particles with diameters larger than:				Percent of particles with diameters smaller than 0.5 mm
	5.1 mm	3.1 mm	1.0 mm	0.5 mm	
Granitic sand and gravel	7.1	25.7	75.6	98.2	1.8
Large idealite[a]	100.0	—	—	—	—
Regular idealite[a]	42.3	76.7	98.9	99.4	0.4
Fine idealite[a]	—	—	53.9	94.9	5.0
River sand and gravel[b]	23.7	83.0	95.5	98.8	1.4
Volcanic ash[c]	46.7	62.0	79.1	92.0	7.1

[a] Artificial, light-weight concrete aggregate, made from fired illite shale.
[b] Commonly called "sqee-gee", major portion with particles about 1/4″ diameter.
[c] Commonly called "scoria", prone to crumbling with build-up of fine particles. Water holding capacity exceeds a good greenhouse soil.

Fig. 6-12. (A) Inert medium planted to carnations. Wood benches with a gates peripheral irrigation system. (B) Hydroponics system with chrysanthemums planted in vermiculite and periodically flooded

existing benches, but requires an automatic fertilizer proportioner and a good irrigation system. The water and nutrients are not recirculated, but wasted. Once the nutrition has been established, there is no requirement, however, for periodic analysis to adjust the solution as in the case of regular hydroponics. There is less likelihood that any disease will be spread.

To be suitable as an inert medium, particles smaller than 0.5 mm diameter should not exceed 5% of the total volume (Table 6-12). Otherwise, water holding capacity increases rapidly, and the advantages are reduced. An automatic timing system for irrigation is required, as, in the summer, watering may be required as often as six times each day. This can be decreased in the winter months to once daily and skipped completely on cloudy days. If high volume irrigation systems are utilized, such as the gates, peripheral system, pressure at the nozzles should not exceed 4 lb in^{-2} (see Chap. 5). Benches much wider than 36 in require an additional line down the center. A trickle irrigation line between each row is satisfactory, and does not wet the foliage. Benches on hydroponic systems are usually flooded at similar frequencies.

During the first year of production in inert media, irrigation, and fertilization are particularly critical. As crops are replanted, roots will be left behind gradually increasing organic matter content until the medium is a gravely soil. Watering frequencies will gradually decrease with the increase in water holding capacity. Nutritional balance becomes less critical with the higher exchange capacity. As pointed out in Chapter 5, saline water supplies may require dry fertilization with calcium and phosphorous (see Chap. 7).

There are innumerable variations on conventional hydroponics and inert media. Cooper (1973) has recently described a continuous recirculation system for tomato production without any root substrate whatsoever. Plastic troughs are formed on a slant incline into which the plants are set. The nutrient solution is allowed to flow through the troughs to a holding tank, and then pumped back to the beginning. Tomatoes are sometimes planted on straw bales, or in bolsters of plastic filled with straw or manufactured compost. These systems require different handling techniques, and become a matter of decision for the individual grower. Very often, it is a matter of which system the grower finds most comfortable. However, as yields continue to increase, the root medium becomes an important limiting factor. Inert media, with suitable handling, will provide an additional yield and quality improvement for most bench crops.

6.5 Soil Handling

Soil handling in greenhouses is a major expense. With bench crops, it must be dug up, tilled, fertilized, fumigated or pasteurized, leveled, and planted. Pot plant growers must mix their materials, haul from one location to another, fumigate or pasteurize it, fill the container, plant, and perhaps move the container to its bench location. Various stages of handling are susceptible to mechanization, as for example in Figure 6-13, the grower operator has utilized two surplus concrete mixers for preparing the potting substrate. The mixture is then placed in carts for steaming, and can be moved to the location where potting is done. Europeans have manufactured various types of machines (Fig. 6-14) for filling the container and sometimes making the hole for the new plant.

Devine (1975) has given six principles that should be followed in determining choice of a soil or crop handling system: (1) Will the system eliminate all handling

Fig. 6-13. Surplus concrete mixers installed to mix several cubic yards of potting substrate at a time. Front-end loader charges mixers from center, and unloading occurs onto conveyer belt between the two and under loading hopper

Fig. 6-14. Potting machine. Soil is loaded on left. Workers set pots on conveyor which passes them under loader at right. Such machines will fill more than a thousand pots per hour. A large crew will be required if machine is to be maximally utilized

except essential moves? (2) Will the system reduce loading and unloading to a minimum? (3) Can containers be standardized, with the numbers of standards kept to a minimum? (4) Can the handling system be integrated with the production? (5) Can the equipment and layout be standardized for flexibility in the cropping sense? And (6) is it economically justifiable to mechanize? Machinery of any kind is a substantial investment, so that all factors of labor, replacement, depreciation, utilization, maintenance, etc. must be evaluated before the final decision.

In a study on labor usage for processing one gallon containers for nursery stock, Warneke (1975) found the number of cans per worker per hour to vary from 130 to 180, but the total labor force varied from 4 to 25. When there was no bending over, reaching or shifting about, planting rates were higher. In the processing and planting for 5-gal containers, the containers per manhour varied from 27.65 to 57.14, and maximum productivity was not found in the nursery with the greatest mechanization. Wage incentive was found to provide the highest productivity. The most mechanized nursery had 38% more production per worker than the least mechanized.

Attention to detail should not be forgotten when working with root media. Some mixtures, particularly those incorporating peat moss, can shrink nearly 30%, and allowance must be made for the volume loss. Some fertilizer amendments cannot be stored for long periods as they may break down. Moisture content in handling, steaming and planting is important. If too dry, thorough pasteurization is difficult; if too wet, steaming may cause structural breakdown and increased difficulty in handling. Obviously, slow release fertilizers must be added after steaming. Small details such as tilling benches to their full depth to break up any hardpan at the bottom, good leveling of the surface, and mixing of additives may make the difference between a uniform crop, providing high quality flowers at the time required; or a bench containing spots of delayed and stunted flowers. Uniform filling, packing and planting of containers can determine whether the irrigation system performs adequately and the crop is ready for sale during the period projected in the timing plan. Good soil-handling practices ensure that investment in light, temperature, and CO_2 control will not be wasted.

References

Baker, K.F.: The U.C. System for producing healthy container-grown plants. Univ. Calif., Calif. Agr. Exptl. Sta. Man. 23, 1957, 332 pp

Cooper, A.J.: Rapid crop turn-around is possible with experimental nutrient film technique. The Grower **79**(18), 1048–1052 (1973)

Cornell Recommendations: N.Y. State College Agr. and Life Sciences: Cornell Recommendations for commercial floriculture crops. Part I. Ithaca, N.Y., 1974, 60 pp

Devine, E.S.: Six principles dictate choice of a crop handling system. The Grower **84**, 493–495 (1975)

Dittmer, H.J.: A quantitative study of the roots and root hairs of a winter rye plant (*Secale cereale*). Am. J. Botany **24**, 417–420 (1937)

Furuta, T.: Environmental Plant Production and Marketing. Arcadia, CA.: Cox 1974, 232 pp

Gericke, W.F.: The Complete Guide to Soilless Gardening. New York: Prentice-Hall 1940

Gill, W.R., Miller, R.D.: A method for study of the influence of mechanical impedence and aeration of the growth of seedling roots. Soil Sci. Soc. Am. Proc. **20**, 154–157 (1956)

Hanan, J.J.: Oxygen and carbon dioxide concentrations in greenhouse soil-air. Proc. Am. Soc. Hort. Sci. **84**, 648–652 (1964)

Hanan, J.J., Holley, W.D.: Hydroponics. Colo. State Univ. Exptl. Sta., Gen. Series 941, 1975, 21 pp

Hanan, J.J., Langhans, R.W.: Soil water content and the growth and flowering of snapdragons. Proc. Am. Soc. Hort. Sci. **84**, 613–623 (1964)

Holley, W.D.: Inert media to replace soil? Colo. Flower Growers' Assoc. Bull. **205**, 1–2 (1967)

Johnson, P.: Horticultural and agricultural uses of sawdust and soil amendments. Tech. Bull. Publ. by Paul Johnson, 1904 Cleveland Ave., National City, Calif., 1968, 46 pp

Lawrence, W.J.C., Newell, J.: Seed and Potting Composts. London: Allen and Unwin 1952, 176 pp

Reeve, R.C.: A method for determining the stability of soil structure based upon air and water permeability measurements. Soil Sci. Soc. Am. Proc. **17**, 324–329 (1953)

Richards, L.A., Wilson, B.D.: Capillary conductivity measurements in peat soils. J. Am. Soc. Agron. **28**, 427–431 (1936)

Richards, S. J., Warneke, J. E., Marsh, A. W., Aljibury, F. K.: Physical properties of soil mixes. Soil Sci. **98**, 129–132 (1964)
Schofield, R. K.: Ionic forces in thick films of liquid between charged surfaces. Trans. Faraday Soc. **42 B**, 219–225 (1946)
Sholto-Douglas, J.: Beginner's Guide to Hydroponics. New York: Drake 1972
Slatyer, R. O.: The influence of progressive increases in total soil moisture stress on transpiration, growth, and internal water relationships of plants. Australian J. Biol. Sci. **10**, 320–336 (1957)
Spomer, L. A.: Soil amendment (or why does 4 bushels plus 10 bushels equal only 10 bushels?). Illinois State Flor. Bull. **355**, 7–8 (1975)
Steiner, A. A.: Soilless Culture. In: Proc. 6th Colloq. Intern. Potash Inst., Florence. Berne, Swiss: Intern. Potash Inst. 1968, pp. 324–341
Stolzy, L. H., Letey, J., Szuszkiewicz, T. E., Lunt, O. R.: Root growth and diffusion rates as functions of oxygen concentration. Soil Sci. Soc. Am. Proc. **25**, 463–467 (1961)
Taylor, S. A., Ashcroft, G. L.: Physical Edaphology. San Francisco: W. H. Freeman 1972, 533 pp
Warneke, J. E.: Nursery labor usage for containerized ornamental planting. Univ. Calif. Flower and Nursery Report, Sept.-Oct. 1975, pp. 1–5
Weigand, C. L., Lemon, E. R.: A field study of some plant-soil relations in aeration. Soil Sci. Soc. Am. Proc. **22**, 216–221 (1958)
White, J. W.: The concept of "container capacity" and its application to soil-moisture-fertility regimes in the production of container-grown crops. Ph.D. Dissertation. Pennsylvania State Univ. 1964, 173 pp
White, J. W.: Criteria for selection of growing media for greenhouse crops. Florists Rev. **155 (4009)**, 28–30, 73–75 (1974)
Zimmerman, R. P., Kardos, L. T.: Effect of bulk density on root growth. Soil Sci. **91**, 280–288 (1961)

Chapter 7 Nutrition

7.1 Introduction

7.1.1 Essential Elements

Most plants grown in greenhouses require these elements if they are to survive: carbon (C), hydrogen (H), oxygen (O), nitrogen (N), phosphorous (P), potassium (K), sulfur (S), calcium (Ca), magnesium (Mg), iron (Fe), manganese (Mn), boron (B), copper (Cu), zinc (Zn), chlorine (Cl), and molybdenum (Mo) (Table 7-1). Carbon, hydrogen, and oxygen are supplied with methods not considered here under the heading of nutrition. The remaining elements are usually taken up in the plant through the root system, although certain of them, under special conditions, can be applied through the leaves. The next six elements listed are generally referred to as "macronutrients" as their concentrations in plant tissue are relatively high (Table 7-19), and failure to supply them usually results in readily visible growth reduction. Iron, manganese, boron, copper, zinc, chlorine, and molybdenum are called "micronutrients" as their concentrations in tissue are usually low, and demonstration of their essentiality for growth requires special procedures. For example, excess chlorine is more often a problem than a deficiency. Zinc has become increasingly important as growers shift from galvanized piping systems to plastic irrigation systems and supply piping. Most of the micronutrients can be supplied as contaminants in other fertilizers and chemicals used in greenhouse culture.

Other elements such as sodium (Na), cobalt (Co), silicon (Si), and vanadium (V) have not been shown to be essential for higher plants, although there have been indications of growth enhancement from sodium, and, under certain conditions, sodium, rubidium (Rb), and strontium (Sr) can partially replace the requirement for some nutrients such as potassium. Fluorine (F), iodine (I), and selenium (Se) have been classed as functional nutrients for some organisms, but are seldom dealt with in greenhouse nutrition. The fact that a particular element may be found in a tissue analysis does not mean that it is essential. It is common to find elements in plants because they are present in the soil and are taken up with other elements.

However, we are not concerned with "survival". The macro- and micronutrients must be supplied in adequate amounts, at the proper time, and in the proper fashion if the grower's objectives are to be fulfilled. This is most often maximum yield and quality, although indoor decorative plants are not handled with this objective uppermost. Problems of adequate fertilization are extremely complex. The "nutrient requirement" of plant tissue, as defined by Gerloff (1963),

Table 7-1. Elements important in nutrition, atomic weights, valence, and equivalent weights

Ion	Atomic weight	Charge	Valence	Equivalent weight
Boron (B)	10.82	+	3	3.61
Calcium (Ca)[a]	40.07	+	2	20.0
Chlorine (Cl)[a]	35.46	−	1, 3, 5, 7	35.46 (Usual)
Copper (Cu)	63.57	+	1, 2	31.8
Hydrogen (H)	1.0	+	1	1.0
Iron (Fe)	55.85	+	2, 3	18.6
Magnesium (Mg)[a]	24.32	+	2	12.16
Manganese (Mn)	54.94	+	2, 3, 4, 6, 7	27.47
Molybdenum (Mo)	95.95	+	3, 4, 6	32.0
Nitrogen (N)	14.0	+	3, 5	4.7
Oxygen (O)	16.0	+	2	8.0
Phosphorous (P)	31.0	+	3, 5	10.3
Potassium (K)[a]	39.1	+	1	39.1
Sodium (Na)[a]	23.0	+	1	23.0
Sulfur (S)	32.1	+	2, 4, 6	8.0
Zinc (Zn)	65.4	+	2	32.7
Nitrate (NO_3)[a]	62.0	−	1	62.0
Ammonia (NH_4)[a]	18.0	+	1	18.0
Bicarbonate (HCO_3)[a]	61.0	−	1	61.0
Sulfate (SO_4)[a]	96.0	−	2	48.0
Phosphate (H_2PO_4)[a]	97.0	−	1	97.0

[a] Macronutrients most often existing in the states shown as ions in solution, and most frequently dealt with as such.

Fig. 7-1. Effect of nitrogen supply on lettuce growth. Plant on left had full nitrogen in nutrient solution, gradually decreasing to zero on right

is not the same as that found at maximum yield. For example, Figure 7-1 shows the effect of decreasing nitrogen supply on lettuce growth. Similarly, an obvious deficiency symptom as illustrated in Figure 7-2 means that nutrition has been inadequate for some time. Loneragan and Snowball (1969) found that luxury and deficient concentrations of calcium could exist simultaneously in different parts of

Fig. 7-2. Potassium deficiency in poinsettia

the same plant. Critical and functional nutrient concentrations are not the same. These two investigators suggested that "functional nutrient requirement" be used to indicate the minimal concentration of nutrient within the plant which does not limit growth. While a start has been made in recent years to define minimum critical concentrations (Table 7-19), we shall see in this chapter that greenhouse operators have a long way to go.

7.1.2 Role of Mineral Elements

The action of mineral elements that have been proved essential in plant growth has not been fully defined. However, Evans and Sorger, in 1966, briefly outlined the probable functions:

Nitrogen:	Constituent of amino acids, proteins, nucleic acids, coenzymes, etc.
Phosphorous:	Component of sugar phosphates, nucleic acids, coenzymes, plays key role in energy transfer in plants.
Potassium:	Acts as an activator for a wide variety of enzymes.
Sulfur:	Component of several amino acids and proteins.
Calcium:	Constituent of cell walls, important in some enzyme systems, with deficiency thought to increase chromosome fragmentation.
Magnesium:	Required by a number of enzyme systems, a component of chlorophyll.
Iron:	A component of cytochromes and iron proteins involved in photosynthesis, nitrogen fixation and respiratory processes.

Manganese:	Required for photosynthetic evolution of oxygen, activates numerous enzyme systems.
Boron:	Evidence for involvement in carbohydrate transport.
Copper:	Essential component of several enzyme.
Zinc:	Essential component of several enzymes.
Molybdenum:	Essential for nitrogen fixation, component for enzymes involved in nitrogen transformations.
Chlorine:	Required for photosynthetic reactions involved in oxygen evolution.

Generally, lack of any of these elements will reduce growth, and ultimately cause death if not supplied. Several of the micronutrients can be highly toxic if concentrations are too high. As we will note later, sufficient, and efficient supply of these minerals can be a function of several factors not always under control of the grower.

7.1.3 Expression and Units

There are several ways of expressing nutrient concentrations in fertilizers, soils, solutions and plant tissues. Despite the fact that fertilization is a significant production cost, many of the systems most commonly used are illogical. For example, nitrogen in fertilizer is most often expressed in terms of actual "N", phosphorous as P_2O_5 and potassium as K_2O. Unfortunately, plants absorb N, P, and K from the solution as nitrate (NO_3^-), ammonium (NH_4^+), phosphate ($H_2PO_4^-$ or HPO_4^{2-}), and potassium (K^+). The same situation holds for other essential elements. Plants are capable of absorbing fairly large and complex molecules such as chelates, amino acids, hemoglobin, and other organic compounds. Root systems themselves possess an ion exchange capacity, similar to soil particles. However, uptake of large molecules has not been shown sufficiently important to consider in greenhouse nutrition—perhaps an indication of ignorance. It is safe to assume that the source of an essential mineral (from organic or inorganic materials) is not of direct importance. Elements are absorbed from the soil as ions in solution as indicated above.

One can easily see that a 50 ppm concentration of N or P_2O_5 in solution is not the same as a 50 ppm concentration of NO_3^- or $H_2PO_4^-$. While there are factors that can be used to convert from one to the other (Table 7-6), it can not be done accurately, and the relative proportions of N in NO_3^- versus NH_4^+ can have high impact on growth.

Parts per million (ppm) is most common in the greenhouse industry. But, as a start toward efficient nutrient control, concentrations should be expressed in terms of the ions actually absorbed by roots. The literature is further confused by the fact that agronomists use milliequivalents per liter (meq/l), and most physiologists use molar values [millimoles per liter (mmoles/l)]. Milliequivalents were introduced for practical greenhouse nutrition control by Green (1967a). While there has been resistance to its use, meq/l has a number of advantages: (1) the unit is metric; (2) it takes into account the types of ions actually present in root media and absorbed by plant systems; (3) it allows a uniform system for determining concentrations regardless of the particular situation; (4) as chemical reactions and physical systems operate according to well-known physical principles, the unit encourages more efficient research that can be translated into practical operations. Milliequivalents per liter state in unequivocable terms: the number of bricks of what color one has in his house; rather than the total weight of all bricks of all colors.

7.2 Factors Affecting Nutrition

7.2.1 Ion Uptake

Minerals are generally absorbed by root systems as cations (NH_4^+, K^+, Ca^{2+}, Na^+, Mg^{2+}) or anions (NO_3^-, $H_2PO_4^-$, HPO_4^{2-}, SO_4^{2-}, Cl^-). However, they are not absorbed in equal amounts. Univalent cations are often taken up more rapidly than divalent cations such as Ca^{2+} or Mg^{2+}, there may be competition between ions that inhibits uptake. Depending upon the plant species, K^+ uptake can be lower from solutions containing SO_4^{2-} at high concentrations, but not at low concentrations. Again, Na^+ uptake may be depressed in the presence of SO_4^{2-} regardless of concentration. According to Leggett (1968) the interactions of ions in uptake can be inhibitory, stimulatory or neutral, depending upon concentrations and times.

One detailed study on nutrition of a greenhouse crop has been the series of experiments reported by Green (1967a–c), Hartman (1971a–c), Hartman and Holley (1968), Schekel (1969, 1971), and Hanan (1975). In a new approach to the problem they attempted to learn the interaction of the various ions in an inert substrate where complicating soil factors could be eliminated. For example, Na^+ was generally taken up in accordance with concentration of the ion in solution, whereas SO_4^{2-} was limited regardless of its level outside the roots. Hartman showed that the sum of Ca^{2+} and Mg^{2+} in the plant was maintained more or less constant even though the proportions of Ca^{2+} and Mg^{2+} in solution varied. The ammonium ion (NH_4^+), according to Green, tended to depress K^+ uptake; but, Hartman showed that Cl^- uptake increased with NH_4^+ level in the nutrient solution.

The effects of varying ion ratios can be quite subtle, depending upon environmental conditions of light, water, CO_2 concentration and temperature. Thus, Green found that NH_4^+ was not necessary for carnation growth, but later work showed that requirement for NH_4^+ varied with the season. Schekel's work pointed up the importance of ratios by showing that the effect of increasing salinity could be reduced by making sure the ion ratios were held in their proper proportions. What these ratios may be has not been worked out for most greenhouse crops. Another example is shown in Figure 7-3, where a slight adjustment upward in the potassium level tended to overcome the effect of high sodium in the water supply. The results emphasize that as the genetic potential of a plant is approached, as a

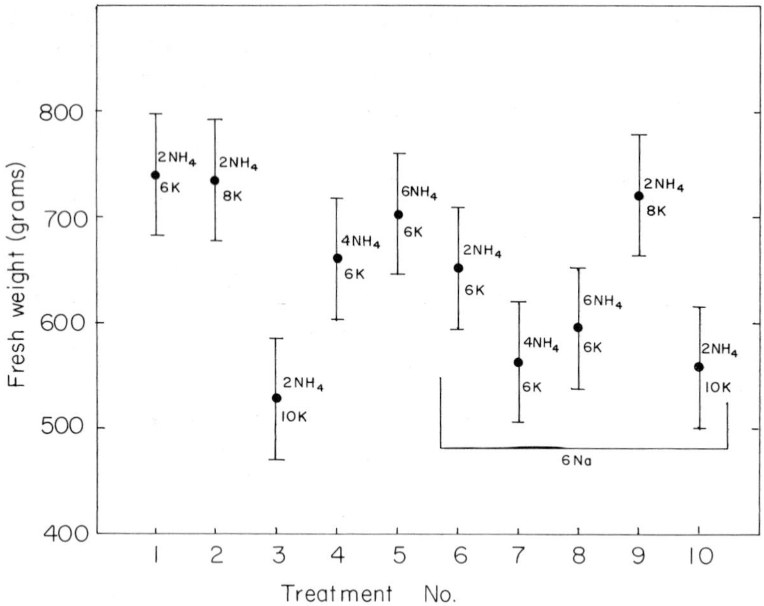

Fig. 7-3. Effect of nutrient solutions on growth of carnations. *Numbers in graph:* concentration in meq/l of NH_4, K, and Na. All other nutrients are held constant (from Hanan, 1975). *Vertical bars:* variability of the average. Those which do not overlap are significantly different from each other

shift is made to inert media, relative proportions of individual ions in solution become more important.

In addition to the interactions of ions, there is also the fact that plants can actively accumulate salts against a concentration gradient. The relative importance of active uptake (the energy expenditure by a plant to absorb an ion) versus passive uptake (absorption with the water flow into the root) has not been fully determined. It appears, that if plants are conditioned to low salt levels, the majority of uptake will occur by active processes. If, however, plants are conditioned to high salts, most of the uptake will be passive—that is, with the transpirational water stream through the root. Whether ion uptake is active or passive, uptake rate is dependent on the fact that the ion must be in a position where it can be absorbed. This supply is dependent upon water flow to the root, and the mobility of the ion in the root medium.

Ion absorption is often restricted to certain portions of the root system, often in the root-hair region or behind the root growing point, and back to the region of the root where surfaces become suberized or internal barriers within the root are formed. One of the conditions for a healthy plant is an actively growing root system. It is illogical to speak of "maintaining" a foliage plant, for example. A growing root system means that not only are new surfaces being formed for ion uptake, but that distance between absorbing surface and water and ion supply is being reduced concomitantly. This factor is seldom considered in greenhouse production. The restricted soil volume in most greenhouse production accounts, in large measure, for higher fertilization rates. As phosphorous is highly immobile in soils, for example, extension of root absorbing surfaces to new supplies will enhance phosphorous availability. It was the conclusion of Broyer and Stout (1959) that the volume of soil explored by roots was one of the important factors in phosphorous availability—as well as renewal from the solid phase. Furthermore, as root density increases individual roots will compete with each other for available supplies. Stevenson (1967) found that top growth of several species increased as the container volume increased. Similarly, Baker and Woodruff (1963) found that 0.02 ppm of phosphorous in the expressed soil solution from pots was required to adequately supply corn. But, under field conditions only 0.001 ppm was required. Even with high rates of fertilization, one or two gallon containers were found to be too small for many experiments. It is not unusual to find large plants growing in very small containers in greenhouse culture, and it is understandable that fertilization and watering rates must be increased proportionally. However, there appear to have been no studies to relate container size to practical application in greenhouses. What research is available is largely empirical, and related only to the particular situation of the experiment.

7.2.2 Ion Supply

It was shown in Chapter 6, that soils possess the ability to store and release several of the minerals important in plant growth. We refer to this ability as exchange capacity expressed in milliequivalents per weight of soil. However, not all ions are equally adsorbed or released from soil particles, and the amounts released or adsorbed will depend upon the chemical composition of individual soil particles, and physical properties such as size and structure. As roots absorb ions from the soil solution, new ions are released from the clay or organic micelles in exchange for others such as hydrogen or perhaps sodium. The minerals dealt with as exchangeable are calcium, magnesium, potassium, sodium, aluminum, and hydrogen. Such ions as nitrates are highly mobile, remain largely in the soil solution, and are, therefore, readily leached from the root medium. Phosphorus, on the other hand, is highly immobile, with a large proportion in insoluble forms. Supply of ions to the soil solution will also depend upon breakdown of the soil minerals such as apatites (phosphorus), feldspars and micas (potassium), and limestones (calcium and magnesium).

It is only rarely that sufficient nutrients are supplied from materials already present in the soil. Some will be supplied from the exchange sites of the soil—if it is continually renewed. Calcium is peculiar in that a large percentage of the exchange complex must be occupied by calcium if plants are to receive adequate calcium. On the average, about 30% of the soil's cation exchange capacity must be occupied by calcium. But, this depends upon the basic soil particle. If the dominant clay mineral is montmorillonite, and the complementary ion is sodium, according to Broyer and Stout (1959), then at least 50% of the exchange capacity must be occupied by calcium. This drops to 20% if the clay mineral is kaolinitic with weakly exchangeable hydrogen as the complementary ion. The lower limit of adsorbed magnesium is about 5% of the total exchangeable capacity, and potassium is about 2%. These soil properties must often be taken into account, for it is possible to cause nutrient deficiencies by introducing potassium to the exclusion of magnesium, or failing to ensure that the exchange complex is adequately supplied with calcium. We emphasize, that for an exchange to occur, a release of a Ca^{2+} ion for example, there must be another ion to exchange for it. This must be supplied either from the soil solution or from the plant root normally as an hydrogen (H^+) ion.

The dominant ions adsorbed on the soil micelles will vary with climate and pH. In humid regions, the cations in descending order will be Al^{3+}, Ca^{2+}, Mg^{2+}, K^+, Na^+, and H^+. For arid regions, the order is usually Ca^{2+}, Mg^{2+}, Na^+, K^+, and H^+. At very low pHs, only hydrogen and aluminum may be present on the exchange sites, and held tightly to resist replacement. At high pH the dominant ion may be calcium or sodium.

Assuming that an ion has been exchanged, or supplied by fertilization, and is available in the soil solution; it must be moved through the water film around the soil particles to the absorbing surface of the root. As ions are absorbed by roots, there will be a concentration gradient established so that ions will tend to diffuse toward the root. Diffusion is generally insufficient to meet plant needs, and most ion transfer takes place with the mass movement of water from the soil to the plant as transpiration proceeds. Thus, the more rapidly water is taken up, the greater the resupply of nutrients. Withholding water by reducing irrigation frequency not only increases the water stress in the plant, but also can restrict nutrient supply. Another factor seldom considered, but obviously important, is "tortuosity" imposed by the textural and structural characteristics of the root medium. If the soil is clay, then diffusion path length will be increased as compared to a sandy soil where the path of ion movement is more direct. Fertilization and irrigation rates must be adjusted accordingly.

It is the objective of the operator to provide sufficient mineral elements to the root medium to meet plant needs from one application to another. He must do this without oversupply or undersupply. The more frequent the application, the lower can be the concentration at any one application. Theoretically, if supply could be continuous, and the proper concentration maintained at the absorbing root surfaces, the actual concentration would be minimum. The system recently devised by Cooper (1974) would appear to be a step in the direction of constant supply at the root surfaces. However, the concentration of nutrients in his recirculating system (Table 7-11) exceeds Hoagland and Arnon's recommendations. Torsten Ingestad, in a discussion several years ago (Ingestad, 1969), suggested that if ions could be supplied to the absorbing surfaces at rates equal to the actual uptake rate; solution concentrations would be extremely low. Practically, the fertilizer program has to be adjusted empirically due to the diversity of conditions under which plants are grown. An indication of this is illustrated in Figure 7-4 where Hartman and Holley (1968) compared carnation growth in soil with a relatively high exchange capacity, and perlite with little or no exchange capacity, and, of course, different moisture holding capacities. An example of the relation between solution concentration and watering frequency is provided in Figure 7-5. With lower than normal fertilizer concentration, Hartman (1971c) could partially offset nutritional deficiency by increasing watering frequency. The effect of low

Fig. 7-4. Effect of root substrate on carnation growth. Ratio of NH_4 to NO_3 was 1, each applied in the nutrient solution at a concentration of 6 meq/l (Hartman and Holley, 1968)

concentrations could not be wholly compensated—probably due to the limitations of the root medium and particular experimental conditions.

The optimal nutrient concentrations for any particular situation will also vary with light and CO_2, with uptake increasing during the summer. With higher light and CO_2 concentrations, there is an increased nutrient requirement. For a given root medium, the optimum nutrient applications will increase as growth increases and as the environment is manipulated to provide more growth (Figure 7-6).

7.2.3 Effect of pH

The pH of a soil solution is a means of stating the hydrogen ion activity, or the acidity or alkalinity of the soil, with a value of 7.0 indicating neutrality. The number value is a negative exponent such as 10^{-1}, 10^{-2}, 10^{-3}, etc. If pH is to be changed, the amounts required will differ depending upon where on the pH scale one is starting. If enough lime is added to neutralize the hydrogen ions in solution, more H^+ ions will be released from the soil colloids, and pH change will be negligible. Not until sufficient liming material has been added to neutralize the soil's reserve acidity, represented by the amount of hydrogen ions adsorbed on the clay or organic micelles, will pH actually rise. We say the soil is "buffered". The lower the exchange capacity of the "soil" (e.g., an inert medium), the quicker any pH change will occur with a given amount of liming or acidifying material.

The major effect of pH is on aluminum and trace element availability. When pH is low in mineral soils, appreciable quantities of aluminum and manganese may be released into the soil solution so that it becomes toxic. As pH increases, precipitation occurs, until at neutrality (7.0), some plant species may suffer deficiency of manganese and iron. Copper and zinc are affected similarly. Boron is more complicated as lime and a rising pH together can cause rapid fixation in

Fig. 7-5. Effect of fertilizer concentration in the irrigation water on growth of carnations. (From Hartman, 1971c)

insoluble forms, but either one alone does not appreciably precipitate the element. Molybdenum availability is significantly dependent upon pH, and it is unavailable in strongly acid soils.

The kind of phosphate ion varies with pH. At low pHs, soluble phosphates may be fixed in unavailable forms with iron and aluminum; and above 7.0, calcium phosphates are formed which are also insoluble. Furthermore, excess calcium can interfere with phosphorous uptake. The activity and types of soil microorganisms will vary with pH, with fungi predominating at values below 5.5. Nitrification and nitrogen fixation take place rapidly only at pHs above 5.5, and bacteria and actinomycetes function better at levels slightly below neutrality. Most elements are most available between pH 6.0 and 7.0, and these are the

Fig. 7-6. Interaction of CO_2 concentration and fertilizer concentration in nutrient solution on carnation growth. (From Green, 1967c)

general recommendations for most greenhouse crops. There are exceptions where much higher acidity levels are desired, as with azaleas and when blueing is required for hydrangeas.

7.2.4 Effect of Salinity

Salinity refers to the total concentration of salts present in the soil solution. This may be expressed in terms of electrical conductivity (mhos), parts per million,

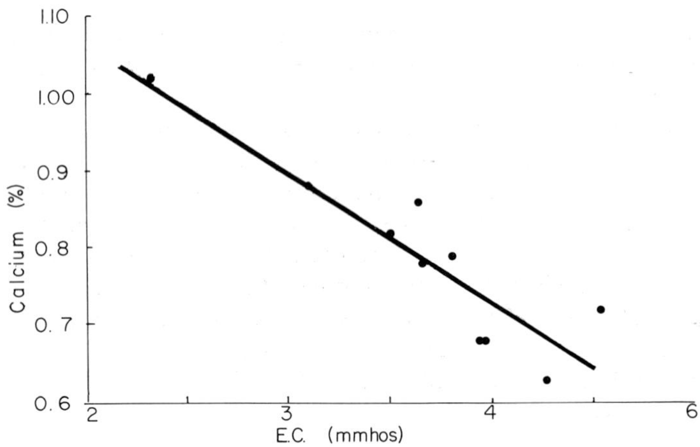

Fig. 7-7. Effect of total salts on calcium tissue levels in carnation. (From Hanan, 1975)

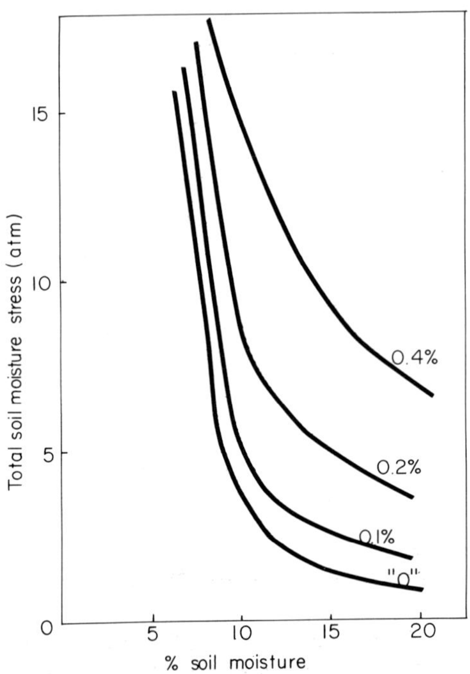

Fig. 7-8. Effect of salinity and soil moisture content on total soil moisture suction (stress). (From Wadleigh and Ayers, 1945)

milliequivalents per liter, moles per l, or grains per gallon. The most general expression is millimhos per centimeter or micromhos per cm.

As mentioned earlier, excess concentrations of individual ions in the soil solution (as may occur when using water containing sodium, bicarbonate, magnesium, calcium or sulfate) may upset desirable ion proportions, and competitively inhibit ion uptake. Although plants will adapt to increasing salinity, if the change is not too rapid, growth will always be less at the higher salinity level, and more so if the ion ratios are not balanced. Higher salt concentrations alone may inhibit ion uptake as illustrated in Figure 7-7.

Probably the greatest effect of salinity on growth is on water availability. Increasing salt concentration increases the osmotic pressure of the soil solution, so that greater negative suction must obtain within the plant if water uptake is to occur. Also, as the soil dries between irrigations, the effect is to further increase salinity as noted in Figure 7-8, where total stress required to extract soil moisture as a function of soil water content is plotted for a number of salt concentrations. It can be seen then that soil structure becomes doubly important under conditions of salinity. Irrigation must be more frequent in order to avoid severe stunting, and this places a greater demand on oxygen availability and on the ability of the soil to drain rapidly.

7.2.5 Summary

The factors influencing nutrition of greenhouse plants can be listed:

1. Plant species
2. Environmental conditions
   a) Total available solar radiation
   b) CO_2 concentration
   c) Temperature
   d) Humidity
   e) Water availability
3. Root medium
   a) Chemical composition
   b) Textural characteristics
   c) Structural arrangement
   d) Soil volume
   e) Exchange capacity
   f) Moisture content
4. pH of soil solution
5. Total concentration of salts in soil solution
6. Relative proportions and absolute concentrations of individual ions
7. Previous history

The uptake of an ion is the sum of several processes: (1) the placement of the ion in an available form in the soil solution; (2) its movement to an absorbing root surface, and (3) its absorption into the plant system. Our knowledge is insufficient to predict with precision and accuracy a particular plant response to fertilization when one or more of the factors is changed. As a result, nearly all practical fertilization studies are empirical; suitable for the particular climatic region in which they are conducted, with certain soil mixtures and types, and under conditions closely approximating the cultural conditions to which the results are to be applied. The greenhouse grower must "fine tune" the fertilization program to his particular conditions.

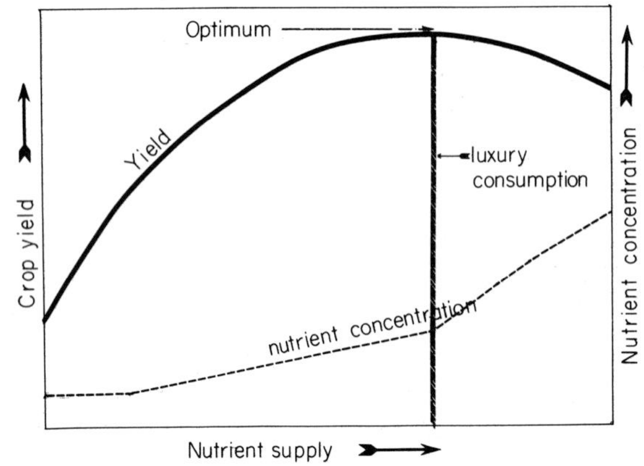

Fig. 7-9. Schematic graph of the manner in which nutrient concentration and crop yield varies with the supply of nutrients. (After Munson and Nelson, 1973)

7.3 Nutrition Control

Despite the intricacies of nutrition we have discussed, numerous fertilization studies result in the curve depicted in Figure 7-9. As fertilization application increases, yield increases to a maximum, beyond which "luxury" consumption begins, and growth may decrease. Overfertilization is common. Not only are production costs increased, but water consumption can increase with needless pollution. In the United States, the greenhouse industry is not sufficiently large to warrant concern, at the present, from governmental agencies. That situation may not remain, and as with other industries, greenhouse production will eventually be subjected to study—if not outright regulation. It is to the manager's best interests to be fully knowledgeable.

7.3.1 Macronutrients

The materials supplying macronutrients are listed in Tables 7-2 and 7-3. Most of these compounds are considered high analysis, mineral fertilizers, as contrasted to organic fertilizers with often low and variable nutrient analyses (Table 7-4). Materials in bulk quantities may differ in analysis, depending upon the manufacturing process. Any fertilizer sold in the United States must have its analysis clearly stated, particularly as to percentages of N, P_2O_5, and K_2O, in that order. Not all possibilities are listed here.

Compounds with solubilities exceeding 10 g 100 ml^{-1} can be supplied in the irrigation system. Their analysis is usually high enough to require caution when applied dry. When applied, many compounds will have an acid or basic reaction, tending to raise or lower the soil pH. This is expressed in terms of equivalent pounds of $CaCO_3$ per 100 pounds of the material. Thus, constant use of $Ca(NO_3)_2 \cdot 4\,H_2O$ will tend to raise pH, and the grower may have to adjust his fertilization program accordingly. Fertilizers are also rated according to their salt index, or the ability to contribute to the osmotic potential of the soil solution.

Table 7-2. Common fertilizers supplying macronutrients required by plants

Compound	Formula[a]	Analysis[b]	Molecular weight	Equivalent weight	Effect on acidity	Solubility[c]		Other materials
						g 100 ml^{-1}	lbs 100 gal^{-1}	
Ammonium chloride	NH_4Cl	25-0-0	53.5	53.5	Acid	39.7	248	—
Ammonium nitrate	NH_4NO_3	33.5-0-0	80.05	80.05	Acid	118.3	991	—
Monoammonium phosphate (Ammophos A)	$NH_4H_2PO_4$	11-48-0	115.04	115.04	Acid	22.7	190	1.4% Ca 2.6% S
Diammonium phosphate	$(NH_4)_2HPO_4$	21-53-0	132.07	66.0	Acid	42.9	359	—
Ammonium sulfate	$(NH_4)_2SO_4$	20-0-0	132.15	66.1	Very acid	70.6	591	24% S
Aluminum sulfate	$Al_2(SO_4)_3 \cdot 18H_2O$	0-0-0	666.45	222.15	Very acid	Soluble		14% S
Calcium sulfate (gypsum)	$CaSO_4 \cdot 2H_2O$	0-0-0	172.18	86.09	Neutral	0.2	1.7	23% Ca 19% S
Calcium nitrate	$Ca(NO_3)_2 \cdot 4H_2O$	15-0-0	236.16	118.1	Basic	102.0	854	17% Ca
Sodium nitrate	$NaNO_3$	16-0-0	85.01	85.01	Basic	73.0	611	27% Na
Urea	$CO(NH_2)_2$	45-0-0	60.06	30.03	Acid	78.0	653	—
Superphosphate	$CaH_4(PO_4)_2$	0-20-0	—	—	Neutral	1.8	15	18% Ca 12% S
Treble superphosphate	$CaH_4(PO_4)_2$	0-42-0	—	—	Neutral	1.8	15	12% Ca
Phosphoric acid	H_3PO_4	0-52-0	98.0	98.0	Very acid	548	4.6×10^3	—
Potassium chloride (muriate of potash)	KCl	0-0-62	74.55	74.55	Neutral	34.7	291	—
Potassium nitrate (saltpeter)	KNO_3	13-0-44	101.1	101.1	Basic	13.3	111	—
Potassium sulfate	K_2SO_4	0-0-53	174.26	87.13	Neutral	6.9	58	18% S
Magnesium sulfate (epsom salts)	$MgSO_4 \cdot 7H_2O$	0-0-0	246.5	123.25	Neutral	71	595	10% Mg 13% S
Magnesium nitrate	$Mg(NO_3)_2 \cdot 6H_2O$	11-0-0	256.4	128.2	Neutral	42.3	354	10% Mg
Calcium carbonate (Limestone)	$CaCO_3$	0-0-0	100.1	50.1	Basic	0.002	0.01	40% Ca
Calcium hydroxide (hydrated lime)	$Ca(OH)_2$	0-0-0	74.1	37.1	Basic	0.19	1.6	60–80% Ca
Potassium monophosphate	KH_2PO_4	8-53-34	120.1	120.1	Basic	33	276	—
Potassium diphosphate	K_2HPO_4	0-41-54	174.2	87.1	Basic	167	1.4×10^3	—
Calcium cyanamide	$CaCN_2$	20-0-0	80.11	—	Basic	Decomposes		—
Sulfur	S	0-0-0	32.1	—	Acid	Insoluble		—

Table 7-2 (continued)

Compound	Formula[a]	Analysis[b]	Molecular weight	Equivalent weight	Effect on acidity	Solubility[c]		Other materials
						$g\ 100\ ml^{-1}$	$lbs\ 100\ gal^{-1}$	
Dolomite (Dolomitic limestone)	$MgCO_3 \cdot CaCO_3$	0-0-0	—	—	Basic	—	—	22% Ca 13% Mg
Basic slag	$5CaO \cdot P_2O_5 \cdot SiO_2$	0-17-0	—	—	Basic	—	—	33% Ca 10% Fe

[a] Formula may differ due to addition of water molecules, or source of material as the case of limestone, superphosphate, etc.
[b] Percentage N, P_2O_5, and K_2O will differ between sources, depending upon manufacturing process. Molecular and equivalent weights are for the formulas as given.
[c] Solubility of materials in water, usually 0° C.

Table 7-3. Examples of complete and special fertilizers supplying macronutrients

Materials	Analysis[a]	Reaction speed	Solubility	Effect on acidity
Complete (examples)	12-12-12	Rapid	Very soluble	Varies
	20-20-20	Rapid	Very soluble	Varies
	25-10-0	Rapid	Very soluble	Varies
	10-10-10	Varies	Slight	Varies
	5-10-5	Varies	Moderate	Varies
MagAmp	7-40-6 (12% Mg)	Slow	Slight	—
Osmocote	14-14-14	Slow	Slight	Varies
	18-9-9	Slow	Slight	Varies
Urea Formaldehyde	38-0-0	Slow	Slight	Acidic

[a] Percentage of elemental N, P_2O_5, and K_2O.

Usually, this ability is compared to an equal weight of $NaNO_3$ as a reference. For example, with $NaNO_3$ equal to 100, KCl has a salt index of 114.3, whereas treble superphosphate has an index of 10.1. Actual values of salt index and equivalent acidity or basicity can be found in reference pamphlets such as the Farm Chemicals Dictionary of Plant Foods (1972).

Fertilizers should be stored in dry and protected locations. Unless coated, pellatized or contained in watertight bags, many will absorb water from the atmosphere. This can cause undesirable caking, and they can actually liquify. Acids, of course, should be stored separately. Ammonium nitrate is used in explosive manufacture, and should be stored protected from possible flame or high temperatures. Some fertilizers should not be mixed, such as calcium cyanamide with manure, superphosphate or ammonium sulfate, ammonium sulfate with calcium nitrate, basic slag or lime, and lime with superphosphate or bone meal. With the exception of those fertilizers listed in Tables 6-7–6-9, most materials should not be mixed with root media unless used immediately.

Highly insoluble fertilizers such as superphosphates, limestone, etc., can be added and mixed in the soil before steam pasteurization. However, slow release compounds such as Osmocote cannot be heated without danger from excessive release and consequent plant damage. This precaution is not applicable to organic fertilizers, with the exception of fresh manure. Aside from odor problems with fresh manure, steaming can cause rapid ammonia release with probable plant injury.

Organic fertilizers are not often employed in greenhouses due to their bulk and variable nutrient content. Availability varies from one place to another. Any organic fertilizer is good if utilized properly with regard to effects on soil physical characteristics as well as nutrient supply. Often, observed increased growth with organic additions results from improved aeration and drainage of the root medium rather than increased nutrient supply.

Urea or urea-formaldehyde and calcium cyanamide are not often employed. During winter conditions, the soil temperature may be too low to allow suitable release from urea. Care must be exercised to see that no biuret is present as this

Table 7-4. Common organic fertilizers

Material	Acidity	Analysis[a]	Remarks
Activated sludge	Acid	*4–6, 2–4, 0*	4 lbs/100 ft^2, 6-in depth, suggested application. (Names: Milorganite, Chicagrow, etc)
Beet sugar residue	Basic	*3–4, 0, 10*	Mostly calcium, traces of micronutrients
Castor pomace	Acid	*5–6, 0, 0*	5 lbs/100 ft^2, 6-in depth, poisonous to animals
Cocoa shell meal	Basic	5% total	Used as a conditioner in some fertilizers
Cocoa tankage	Basic	*4, 1.5, 2*	May contain 20% lime
Dried blood	Acid	*9–14, 0, 0*	2 lbs/100 ft^2, 6-in depth, rapidly available
Steamed bone meal	Basic	*2, 25–30, 0*	4 lbs/100 ft^2, 6-in depth, 20–30% Ca
Fish scrap	Acid	*9, 7, trace*	2–4 lbs/100 ft^2, 6-in depth
Garbage tankage	Basic	variable	
Guano	Acid	*12, 11, 2*	Average also contains 8% Ca. If leached by rain, may have 27% P_2O_5
Hoof and horn meal	—	*13, 0, 0*	4 lbs/100 ft^2, 6-in depth
Kelp (seaweed)	—	*2, 1, 4–13*	2–3 lbs/100 ft^2, 6-in depth
Cottonseed meal	Acid	*7, 2–3, 2*	5 lbs/100 ft^2, 6-in depth
Oyster shells	Basic	*0, 0, 0*	31–36% Ca, liming material where available
Process tankage	Acid	*4–12, 0, 0*	Sometimes used as conditioner in fertilizer
Rapeseed meal	—	*6, 2, 2*	
Rice hulls, ground	—	*0.5, 0.2, 0.5*	Very light and bulky, used as conditioner
Shrimp bran	Basic	*7, 4, 0*	57% Ca
Soybean meal	—	*6, 1, 2*	Residue of soybeans from which oil is removed. Mostly used for feed
Tobacco stems	Basic	*2, 0, 6*	3% Ca + 1% Cl
Animal tankage	Basic	*7, 10, 0*	Principal use in animal feeds
Vermiculite	Basic	*0–1.9, Trace, 1.9–3.1.*	Mostly used as soil conditioner
Wood Ashes	Basic	*0, 0, 5*	23% Ca, 5 lbs/100 ft^2, 6-in depth

[a] Analysis in percentage N, P_2O_5, and K_2O.

Table 7-5. Composition of fresh manure from various sources (Brady, 1974)

Source	Feces/urine ratio	Water (%)	N (%)	P_2O_5 (%)	K_2O (%)
Dairy cattle	80/20	85	10.0	2.7	7.5
Feeder cattle	80/20	85	11.9	4.7	7.1
Poultry	100/0	62	29.9	14.3	7.0
Swine	60/40	85	12.9	7.1	10.9
Sheep	67/33	66	23.0	7.0	21.7
Horse	80/20	66	14.9	4.5	13.2

Table 7-6. Factors for converting weight or percentage of one material to its equivalent when expressed in a different form

Multiply this material	By this number	To obtain this equivalent
N (nitrogen)	4.425	NO_3 (nitrate)
N (nitrogen)	1.286	NH_4 (ammonia)
N (nitrogen)	6.067	$NaNO_3$ (sodium nitrate)
N (nitrogen)	7.218	KNO_3 (potassium nitrate)
N (nitrogen)	4.717	$(NH_4)_2SO_4$ (ammonium sulfate)
NO_3 (nitrate)	0.226	N (nitrogen)
NH_4 (ammonia)	0.778	N (nitrogen)
P (phosphorous)	2.288	P_2O_5 (phosphorous pentoxide)
P_2O_5	0.437	P (phosphorous)
K (potassium)	1.204	K_2O (potash)
K_2O (potash)	0.830	K (potassium)
KCl (muriate of potash)	0.632	K_2O (potash)
K_2O (potash)	1.583	KCl (muriate of potash)
K_2SO_4	0.541	K_2O (potash)
K_2O (potash)	1.849	K_2SO_4 (potassium sulfate)
$NaNO_3$	0.165	N (nitrogen)
$(NH_4)_2SO_4$	0.212	N (nitrogen)
Ca (calcium)	1.399	CaO (calcium oxide)
Ca (calcium)	2.496	$CaCO_3$ (calcium carbonate)
CaO (calcium oxide)	0.715	Ca (calcium)
$CaCO_3$	0.401	Ca (calcium)
Mg (magnesium)	1.658	MgO (magnesium oxide)
MgO (magnesium oxide)	0.603	Mg (magnesium)
S (sulfur)	2.994	SO_4 (sulfate)
SO_4 (sulfate)	0.334	S (sulfur)
CaO (calcium oxide)	1.784	$CaCO_3$ (calcium carbonate)
MgO (magnesium oxide)	2.092	$MgCO_3$ (magnesium carbonate)
$MgSO_4$	0.335	MgO (magnesium oxide)
MgO (magnesium oxide)	2.981	$MgSO_4$ (magnesium sulfate)

material, formed by thermal decomposition, is highly phytotoxic. Calcium cyanamide is poisonous and irritating to the skin. It is now largely employed as an herbicide.

Often there is a need to convert from P_2O_5 to actual P, from Ca to $CaCO_3$, or some other material. A list of conversion factors is provided in Table 7-6.

7.3.2 Micronutrients

There are over two dozen companies in the United States, supplying one or more trace elements in single or combination formulas. Table 7-7 lists a few of the more common single sources that can be obtained. Care is always necessary in application as oversupply can easily lead to plant damage. Quite frequently, most of the micronutrients will be found as contaminants in other fertilizers, in the water supply, or in sufficient quantity in the soil. All of them can be obtained in slow-release frits, or as chelated materials. The latter are organic chemicals in which the essential element is held between two or more atoms of the chelating agent. In the absence of chelates in the soil, iron, copper, manganese, and zinc can all be

Table 7-7. Materials supplying micronutrients

Compound	Analysis	Formula	Comments
Boron (B)			
Borax	11% B	$Na_2B_4O_7 \cdot 10H_2O$	Dissolve in boiling water
Boric acid	17% B	H_3BO_3	Dissolve in boiling water
Boron frits	2–6% B	—	Such as FTE 501, FTE 502, etc. Slow release where boron availability may be a problem
Copper (Cu)			
Copper sulphate	35% Cu	$CuSO_4 \cdot H_2O$	Most common in greenhouses
Basic copper sulfates	13–53% Cu	$CuSO_4 \cdot 3Cu(OH)_2$	
Copper chelates	13% Cu	$Na_2CuEDTA$	Slow release where copper
	9% Cu	NaCuHEDTA	availability may be a problem
Iron (Fe)			
Ferrous sulfate	19% Fe	$FeSO_4 \cdot 7H_2O$	
Ferric sulfate	23% Fe	$Fe_2(SO_4)_3 \cdot 4H_2O$	
Iron frits	Varies	Varies	
Iron chelates	5–14% Fe	NaFeEDTA	May be unavailable at pH's above 7.0. Common as foliar application
	6% Fe	NaFeHEDTA	
	10% Fe	NaFeEDDHA	Better above pH 7.0 then EDTA
Manganese (Mn)			
Manganese sulfate	26–28% Mn	$MnSO_4 \cdot 3H_2O$	
Manganese chelate	12% Mn	MnEDTA	
Manganese chloride	17% Mn	$MnCl_2$	
Manganese frits	10–25% Mn	—	Such as FN 502 and FN 239 B
Molybdenum (Mo)			
Sodium molybdate	39% Mo	$Na_2MoO_4 \cdot 2H_2O$	
Ammonium molybdate	54% Mo	$(NH_4)_6Mo_7O_2 \cdot 4H_2O$	
Molybdic acid	59% Mo	H_2MoO_4	
Molybdenum frits	2–3% Mo	—	
Zinc (Zn)			
Zinc sulfate	35% Zn	$ZnSO_4 \cdot H_2O$	
Zinc frits	Varies	—	
Zinc chelates	14% Fe	$Na_2ZnEDTA$	
	9% Fe	NaZnHEDTA	

converted to insoluble hydroxides or basic salts. Chelates have the ability to keep these elements in an available form. Iron chelates are generally the most important.

7.3.3 Application

We do not attempt to provide specific recommendations for individual crops. As noted in Tables 7-8 through 7-12, the rates are general suggestions. Precise and

Table 7-8. Suggested rates of application of fertilizers as preplant, dry feed or in solution[a]. These rates are general guides only, or as a starting point in figuring fertilizer needs. Refer to recommendations for specific crops

Material	Preplant		Dry application		Liquid application[b]		
	lbs per 100 ft²	lbs per yd³	lbs per 100 ft²	kg per 10 m²	oz per 100 gal	lbs per 1000 gal	g per l
$(NH_4)_2SO_4$	—	—	0.5–1.0	0.2–0.4	50[a]	3.0 (20)[a]	0.4
NH_4NO_3	0.7	0.25	0.5	0.2	20[a]	1.0–4.0 (10)[a]	0.1
$NaNO_3$	—	—	1.0	0.4	50[a]	3.0 (20)[a]	0.4
$Ca(NO_3)_2$	1.0	1.5	1.0	0.4	50[a]	3–6.0 (20)[a]	0.4
KNO_3	0.7	1.0	0.5	0.2	20[a]	3.0–5.0 (10)[a]	0.3
Superphosphate	1.0–5.0	1.0–2.0	1.0–5.0	2.3	—	—	—
Treble superphosphate	—	—	1.3	0.6	—	—	—
KCl	—	0.5	0.5	0.2	20[a]	2.0–6.0 (10)[a]	0.1
K_2SO_4	—	0.5	0.5	0.2	20[a]	1.0–3.0 (10)[a]	0.2
5-10-10	—	—	1.0–3.0	0.9	—	—	—
10-10-10	—	0.5	1.25	0.6	—	—	—
20-20-20	—	—	0.5–0.75	0.3	25[a]	(20–30)[a]	0.2
H_3PO_4	—	—	—	—	50[a]	0.5–1.0	0.3
MagAmp (7-40-6)	10–16	5.0–7.5	10.0–15.0	5.7	—	—	—
Osmocote (14-14-14)	4–10	4.0–10.0	10.0	4.5	—	—	—
$MgSO_4$[c]	—	0.25	0.5–3.0	0.9	20[a]	0.5–2.0 (10–60)[a]	0.5
Dolomitic limestone[a]	2–10	2–10	2–10	3.0	—	—	—
$CaSO_4$[c]	—	1.0–2.0	2–10	3.0	—	—	—
$Ca(OH)_2$[c]	—	—	1.0–2.0	0.9	—	—	—
$CaCO_3$[c]	—	1.0–5.0	5–10	3.4	—	—	—
Sulfur[c]	—	—	2–5	1.4	—	—	—
Boric acid	—	—	0.6 oz	0.017	0.02–0.06	0.2–0.6 oz	0.002
Copper sulfate	—	0.15	0.25–0.5 oz	0.009	0.01–0.05	0.1–0.5 (6.5)[a] oz	0.002
Iron chelate	—	0.1	1.75 oz	0.049	3.6 oz	(2.0)[a] oz	0.270
Manganese sulfate	—	0.05	0.25–0.5 oz	0.008	0.01 oz	0.1 (1.0)[a] oz	0.007
Ammonium molybdate	—	—	3.0–6.0 oz[a]	d	d	d	d
Zinc sulfate	—	0.1	0.25–0.5 oz	0.008	0.01–0.03 oz	0.1–0.3 oz	0.006
Fritted trace	—	2–5 oz	—	—	—	(1.0)[a] oz	—

[a] Values have been compiled from a number of sources: Laurie and Kiplinger (1948); Baker (1957); Cornell Recommandations (1974); Furuta (1974).
[b] These amounts are for single feedings or corrections to soil levels. The lower values are for constant injection.
[c] Amounts of these materials will depend upon pH, see Table 7-14.
[d] Usually requires dilution from a stock solution such as 1 lb in 5 gal water, with 3.75 fl oz of the stock in 50 gal for a 1:200 proportioner, or 1.14 fl oz in 5 gal water to give 1 ppm Mo.

Table 7-9. Parts per million and milliequivalents per liter supplied when one pound or 1 kg of a material is dissolved in 1000 gal or 10 m^3 of water

Compound	Analysis [a]	ppm			Milliequivalents per l [b]		
		lb/1000 gal	kg/10 m^3		lb/1000 gal	kg/10 m^3	
NH_4NO_3	33.5-0-0	40	N	34	1.5	NH_4	1.3
					1.5	NO_3	1.3
NH_4Cl	25-0-0	30	N	25	2.2	NH_4	1.9
		79	Cl	66	2.2	Cl	1.9
$(NH_4)_2SO_4$	20-0-0	24	N	20	1.8	NH_4	1.5
		29	S	24	1.8	SO_4	1.5
$Ca(NO_3)_2 \cdot 4H_2O$	15-0-0	18	N	15	1.0	NO_3	0.8
		44	Ca	37	1.0	Ca	0.8
$NaNO_3$	16-0-0	19	N	16	1.4	NO_3	1.2
		32	Na	27	1.4	Na	1.2
H_3PO_4	0-80-0 [c]	80	P	64	1.2	H_2PO_4	1.0
KNO_3	13-0-44	16	N	13	1.2	K	1.0
		53	K_2O	53	1.2	NO_3	1.0
KCl	0-0-62	74	K_2O	62	1.6	K	1.3
		57	Cl	48	1.6	Cl	1.3
$(NH_4)_2HPO_4$	21-53-0	25	N	21	1.8	NH_4	1.5
		63	P_2O_5	53	1.8	HPO_4	1.5
K_2SO_4	0-0-53	63	K_2O	53	1.4	K	1.1
		21	S	18	1.4	SO_4	1.1
$NH_4H_2PO_4$	11-48-0	13	N	11	1.0	NH_4	0.9
		57	P_2O_5	48	1.0	H_2PO_4	0.9
$MgSO_4$	—	24	Mg	20	2.0	Mg	1.7
		32	S	27	2.0	SO_4	1.7
$Mg(NO_3)_2 \cdot 6H_2O$	11-0-0	13	N	11	0.9	Mg	0.8
		11	Mg	10	0.9	NO_3	0.8
HNO_3 (pure)	18-0-0	21	N	18	1.9	NO_3	1.6
K_2HPO_4	0-41-54	49	P_2O_5	41	1.4	K	1.1
		64	K_2O	54	1.4	HPO_4	1.1
KH_2PO_4	0-53-34	61	P_2O_5	53	1.0	K	0.8
		41	K_2O	34	1.0	H_2PO_4	0.8

[a] Analyses in percentage N, P_2O_5, and K_2O.
[b] Milliequivalents are calculated on the basis of equivalent weight.
[c] Percentage H_2PO_4 in liquid.

accurate fertilization depends greatly upon local climatic conditions, local soil, and local handling procedures. Recommendations can be found in the literature, but invariably these are applicable only to those conditions under which the recommendations were derived. The grower must, through trial and error, find those combinations that best fill his need.

Application concentrations will vary with frequency of application. The closer one moves to an inert medium—such as Peat-Lite mixes—the more precise one can be in supplying only that which is required. Some preplant and mixing suggestions are given in Tables 6-7–6-9. Fertilizers can be applied to the soil

Table 7-10. Example, showing how a constant feed or nutrient solution can be made through the use of meq

Compound	Milliliters of stock solution added to 1 l[a]	Pounds of material in 1000 gal[b]	Milliequivalents added per l of final solution								
			K^+	Ca^{2+}	Mg^{2+}	NH_4^+	Na^+	NO_3^-	SO_4^{2-}	$H_2PO_4^-$	Cl^-
KNO_3	6	5	6					6			
NH_4NO_3	2	1.4				2		2			
$Ca(NO_3)_2$	6	6		6				6			
$MgSO_4$	1	1			1				1		
H_3PO_4	1	0.7								1	
		Total	6	6	1	2	0	14	1	1	0

[a] Stock solutions are made by dissolving the equivalent weight (see Table 7-2) of the compound in 1 l of water. This gives a one normal solution. See Table 7-9, for conversion of one pound of material per 1000 gal into meq/l.
[b] See Table 7-9.
Note that H_3PO_4 contributes one meq/l hydrogen, which is not listed in the table as a separate ion. The total of cations must equal anions (positive-negative) for electrical neutrality.

surface or injected into the irrigation water. They can also be applied foliarly, especially in the case of micronutrients where a rapid change is desired. Iron chelate is commonly sprayed on the foliage. With broad leaved, slow rooting plants, a very dilute nutrient solution is sometimes supplied with the mist system in a propagation bench.

Generally, a regular fertilization program is set up. Over a period, the grower makes applications to the soil before planting, and then carries on a procedure for fertilization throughout the growing cycle. If automatic fertilizer injection is utilized at each irrigation, then no amendments need be made unless some of the various, periodic measurements show a trend beyond desirable limits. For bedding and potted plants, a grower may incorporate fertilizer in the soil mixture before planting (Tables 6-7–6-9), and supplement with injection during growth. If the growing cycle is short, slow release fertilizers may be sufficient. If observations of growth and soil and tissue analyses show undesired trends, then changes are made for the next crop. A good fertilization program will take into account the varying plant requirements as a function of plant size, time of year and cultural procedures. Small seedlings will be started at lower rates and then the rates increased. Application rates will increase in the summer under high radiation, and decrease in the winter.

Table 7-10 shows how the milliequivalent system can be employed to make up a solution for injection through the irrigation system with some device as shown in Figure 7-10. Examples of hydroponic solutions are provided in Table 7-11. Automatic proportioning permits a low level of constant fertilization, and with proper attention to soil and tissue analyses, can be used with considerable precision. However, such devices require periodic checking. Some fertilizers manufactured especially for injection, carry a dye to color the water and to tell the grower

Fig. 7-10 A–G. Several types of fertilizer proportioners for injecting dissolved fertilizers into the irrigation system. (A) Water powered, 1:200 ratio, single-head injector. (B) Electrically operated, 1:200 ratio, single head injector. (C) Modified coal stokers used to proportion fertilizers into a storage tank. (D) Electrically operated, double-head, variable proportioner. (E) Two electrically operated, single-head, variable proportioning injectors. (F) Small, water-powered 1:40 injector. (G) Water storage tank, with fertilizers first mixed and dissolved in the small tank on the right, and then mixed into the large tank

Table 7-11. Nutrient solutions supplying macronutrients for hydroponic systems and automatic injection[a]

Material	Milliequivalents per l				
	Carnations (1)	Roses (2)	Tomatoes or general culture (3)		Tomatoes recirculating solution (4)
Potassium (K^+)	6	4	7	6	6.6
Calcium (Ca^{2+})	6	5	10	8	8.4
Magnesium (Mg^{2+})	1	1	4	4	4.1
Ammonium (NH_4^+)	2	1	—	1	0
Nitrate (NO_3^-)	14	9	15	14	15.0
Sulfate (SO_4^{2-})	1	1	4	4	4.1
Phosphorous ($H_2PO_4^-$)	1	1	2	1	2.4

Sources: (1) Hanan and Holley (1975); (2) Sadasiviah and Holley (1973); (3) Hoagland and Arnon (1950); (4) Cooper (1974).

[a] The solutions given for carnations and roses are acceptable for constant feeding in most soils and inert media. The carnation solution will grow acceptable plants of other species, e.g., snapdragons, chrysanthemums, etc. Hoagland and Arnon provide two solutions that may be used. Hoagland and Arnon's have not been extensively tested for constant feeding in soils. As yields increase, ratios and proportions of ions in solution become increasingly critical. See Table 7-10 for an example of making these solutions.

Table 7-12. Trace element stock solution made by dissolving these materials in the given amount of water[a] (Hoagland and Arnon, 1950)

Material	Amount in	
	1 l	50 gallons (1:200 proportioner)
Boric acid	2.86 g	3.8 oz
Manganese chloride	1.81 g	2.4 oz
Zinc sulfate	0.22 g	0.3 oz (31 grains)
Copper sulfate	0.08 g	0.1 oz (44 grains)
Molybdic acid	0.02 g	12.0 grains
FeEDTA (FE 330 sequestrene)	5.0 g	6.7 oz

[a] 1 ml from the stock solution is put into 1 l to make the final dilution. If the amount under the column heading 50 gal is injected through a 1:200 proportioner, the same dilution will be provided.

that the system is operating. Others install a conductivity bridge to determine salinity. If the conductivity drops below a minimum value, an automatic alarm will sound. Most growers depend upon observation of the water level in the concentrate barrels. This often leads to mistakes.

When using automatic injection, materials such as calcium nitrate should never be placed in the same concentrate barrel with phosphorus or magnesium sulfate. These will precipitate. Trace elements such as boric acid or borax must be dissolved in boiling water before adding to the concentrate. Growers using automatic injection should have a water analysis made before using injection. The ratios provided in Table 7-11 are for "good" waters only. If the water supply already contains calcium and magnesium, then it may not be necessary to add these materials. High bicarbonate waters cause precipitation in the concentrate and in the irrigation system. This can be neutralized with acid, and some growers, to get their chemicals into solution in the concentrate, will add 0.5–1 pound HNO_3 per 50 gal. Others may heat with steam. Solution of many chemicals will remove heat from the water, and in the winter, it is possible to have ice formation in the concentrate when mixing fertilizers. If the water supply already has a high salt concentration, the grower may have to exercise considerable care not to drastically increase salinity by excessive injection.

Not all nutrients move with equal facility through the root medium. Calcium, and phosphorus in particular, are considered immobile. Thus, these materials are often mixed into the root medium prior to planting. Dry applications to the surface are made when necessary. Dry fertilization during crop growth is expensive, unless a mechanical dispenser can be used, as is often the case with ground beds with no sidewalls. Unless properly supervised, workers may improperly distribute the fertilizer, causing plant damage. Slow release fertilizers offer advantages where dry application or injection is inappropriate. These types are very common with bedding and pot plants, and generally serve to bring the crop to completion before nutrition drops too low. There will be differences in availability, depending upon whether the slow-release material is mixed into the soil, or applied to the soil surface shortly after potting. Watering frequency and temperature will also influence availability.

The choice of fertilizer will depend upon cost and plant species. Some foliage plants, for example, are highly sensitive to fluoride. As much of the phosphate rock that goes into production of superphosphate comes from deposits containing fluorine, foliar damage can result if phosphorus is applied. Sufficient boron for a carnation may result in boron toxicity with a chrysanthemum. Again, chrysanthemums are considered heavy feeders, and the fertility level may be much higher than need be for a snapdragon crop. Thus, if a crop requiring low nutrient levels is to follow a crop where soil fertility levels have been built up, the grower may find it necessary to withhold fertilization for a period prior to pulling out the crop. If the decision is made to leach the bench prior to planting a new crop, this should be done before harvest begins on the old crop. Without the evaporating surface area of the old crop to extract water from the soil, it may not dry out rapidly enough to permit the cultural operations necessary before replanting.

7.3.4 Fertilizer Calculations

Calculation of amounts of fertilizers to apply, mixing of fertilizers, and cost comparisons are common. Most of the difficulty comes from the varied units and methods of expression. We have provided a series of common problems here for the student to use as a guide.

1. The grower is using a 1:200 injector and wishes to apply a 200 ppm solution of nitrogen and potassium:

a) Several fertilizers can be employed (see Tables 7-2 and 7-9). For our example, use NH_4NO_3 and KCl.

b) From the above tables, NH_4NO_3 contains 33.5% N and KCl contains 62% K_2O.

c) From Table 7-9, one pound of NH_4NO_3 in 1000 gal will supply 40 ppm. So 200/40 = 5 lbs of NH_4NO_3 in 1000 gal will provide 200 ppm N; and, 200/74 = 2.7 lbs of KCl. Assume that the concentrate tank holds 50 gal. Then 50 × 200 = 10000 gal will be applied when 50 gal of concentrate are injected through a 1:200 injector. So, 10 × 5 = 50 lbs of NH_4NO_3 must be dissolved in 50 gal of water; and 10 × 2.7 = 27 lbs of KCl must be dissolved to provide 200 ppm N and K_2O.

2. Suppose one does not know the amount of K_2O in KCl:

From Table 7-2 the formula weight of KCl = 74.55. From Table 7-1, the atomic weight of *K* = 39.1. So 39.1/74.55 = 0.52 or 52% K. Table 7-6 provides the conversion factor to K_2O, or, 52 × 1.204 = 62% K_2O.

3. You want to supply a 300 ppm solution of N in 10 gal:

We will use calcium nitrate ($Ca(NO_3)_2 \cdot 4\,H_2O$) which contains 15% N. A 300 ppm solution is the same thing as 300 μl in 1 l, 300 mg in 1 kg, or 300 mg in 1 l of water if we assume that 1 l of water weighs 1 kg. A simple proportion can be set up: 1/0.15 = X/300, as one milligram of $Ca(NO_3)_2 \cdot 4\,H_2O$ will contain 0.15 mg N. Or, 300/0.15 = 2000 mg = 2 g/l required. As one gallon = 3.8 l, then 3.8 × 10 = 38 l, and 2 × 38 = 76 g of fertilizer to be dissolved in 10 gal. From the conversion tables in Appendix A; 0.035 × 76 = 2.7 ounces.

4. How many ounces of boric acid should be dissolved in 100 gal to provide a 30 ppm boron solution?

From Table 7-7 boric acid (H_3BO_3) contains 17% boron. Or, 10.82/61.82 = about 0.17. The same procedure for problem 3 can be followed. As a shortcut, we can multiply the percent of any element in a fertilizer by 75 and this will give the ppm of one ounce of fertilizer dissolved in 100 gal water; e.g. 0.17 × 75 = 12.75 ppm. As 30 ppm is desired, then 30/12.75 = 2.4 ounces of boric acid must be dissolved in 100 gal water.

5. Suppose you wish to apply 6 meq/l NO_3 through a 1:100 injector:

a) Potassium nitrate, in solution, will separate into the ions K^+ and NO_3^-. NO_3 can be supplied with this material. One pound of KNO_3 per 1000 gal will supply 1.2 meq/l NO_3 (Table 7-9). Therefore, 6/1.2 = 5 lbs of KNO_3 are required per 1000 gal. If a 20 gal concentrate tank is employed, then 20 × 100 = 2000 gal will be dispensed, or 2 × 5 = 10 lbs of KNO_3 should be dissolved in the concentrate.

b) Another approach for either large amounts or small quantities is to remember that the equivalent weight of KNO_3 = 101.1 (Table 7-2). If 101.1 g are dissolved in 1 l, this provides a one normal solution. That one liter will contain one equivalent or 1000 meq. As there are 1000 ml in 1 l, each ml will contain one meq. If 1 l of a nutrient solution containing 6 meq/l NO_3 is desired, then 6 ml from the one normal stock solution in 1 l of water will provide 6 meq/l.

6. Occasionally the fertilizer is in a liquid form, and it is more convenient to measure volume than weight. For example, the use of phosphoric acid: From a handbook of physics and chemistry, phosphoric acid weighs 1.834 g/ml (density).

Suppose the grower wishes to apply 1 lb of H_3PO_4 in 1000 gal. One pound is equivalent to 453.6 g. So 453.6/1.834 = 247 ml required. From the conversion in Appendix A, 247 ml = 8.4 fluid ounces or slightly more than one cup.

7. A grower wishes to provide 2 meq/l H_2PO_4 in his hydroponic system that uses less than 100 gal of water. This means he may wish to make up a stock solution. However, H_3PO_4 concentrate is 85% pure. How many milliters of concentrate H_3PO_4 should he add to one liter of water to make up a 4 normal strength stock solution? As the equivalent weight of H_3PO_4 = 98, a 4 normal solution will require $4 \times 98 = 392$ g of pure H_3PO_4. However, $100 - 85 = 15\%$ more that must be added to account for the fact that the concentrate is actually 85%, not 100%. So, $392 \times 0.15 = 58.8$, which is added to 392, or 451 g actual weight. As the density is 1.834, then 451/1.834 = 246 ml of the concentrate should be added and brought up to 1 l. Then each milliliter of the stock solution will contain 4 meq of $H_2PO_4^-$.

8. Occasionally a grower may wish to mix his own complete fertilizer. Suppose he wishes a 5-5-5 mixture, using ammonium sulfate (20% N), superphosphate (20% P_2O_5) and muriate of potash (62% K_2O). Therefore, 1000 lbs of the mixture will contain $0.05 \times 1000 = 50$ lbs N, 50 lbs P_2O_5, and 50 K_2O. To find the number of pounds required of the constituents, 50/0.20 = 250 lbs of $(NH_4)_2SO_4$, 250 lbs superphosphate and 81 pounds of KCl for a total of 581 lbs. Some 419 lbs of filler must be added to make up the total 1000 lbs. This may be sand, or perhaps calcium sulfate, or some other material. These results show that a lower analysis component could have been used to reduce the amount of filler required, or the analysis could have been higher. A 10-10-10 analysis could not be obtained in 1000 lbs with $(NH_4)_2SO_4$ and superphosphate. The two by themselves would provide a 10-10-0 mixture with no filler. More than one source of N, P_2O_5 or K_2O can be employed.

9. Suppose the recommendations call for 8 ounces of 20-20-20 in 25 gal. How much should be added to a 50 gal concentrate tank which will be fed through a 1:200 injector?

If 8 ounces is required per 25 gal, then $(200/25) \times 8 = 64$ ounces will be required for every gallon of concentrate. Or, $50 \times (64/16) = 200$ pounds of 20-20-20 to be dissolved in 50 gal. It is doubtful that this much could be got into solution. A 1:100 injector would be more suitable as only 100 lbs in 50 gal would have to be dissolved.

10. A soil analysis gives a reading of 50 ppm NO_3. What is this equivalent to in terms of N?

From Table 7-6, the conversion factor is 0.226, or $50 \times 0.226 = 11.3$ ppm N.

11. It may be desirable to convert from ppm to meq/l, or to moles concentration. Assume that one has a 200 ppm nitrogen concentration in the water. What is this equivalent to in meq/l or in mol/l?

In the soil solution, nitrogen may be in one of two forms: NH_4^+ or NO_3^-. The conversion must be to one or the other. If NO_3, then $200 \times 4.425 = 885$ ppm NO_3^-. (Table 7-6).

Keeping in mind that 1 ppm can be thought of as 1 mg in 1 l; then the concentration is the same as 885 mg/l. The equivalent weight of nitrate is 62.0 (Table 7-1).

If 62.0 mg of NO_3^- is dissolved in 1 l, this would give 1 meq/l. Thus, 885/62 = 14.3 meq/l.

It will be difficult to partition the N between NH_4^+ and NO_3^- from NH_4NO_3—if this was the material being used to supply 200 ppm N. Conversion to ammonia (NH_4^+) results in 257 ppm NH_4^+ as contrasted to 885 ppm NO_3^-. Nor can one say that 100 ppm is NH_4^+ and 100 is NO_3^-. The problems arising from use of ppm for fertilizer concentrations are apparent.

Assuming for the moment that our conversion is more nearly correct, how many mol of NO_3^- does 14.3 meq/l represent? A one millimol concentration of NO_3^- is 62 mg in 1 l. So that 14.3 meq/l represents 14.3 mmol. However, if the ion had been divalent, as for example SO_4^{2-} or Ca^{2+}, the above would not be correct. Let us suppose the 200 ppm represents calcium. An equivalent weight of Ca is 20.03, not 40.07 (Table 7-1). So 200/20.03 = 10 meq/l Ca. However, a 1 mmol/l Ca concentration is 40.07 mg of Ca in 1 l. So 200/40.07 = 5 mmol/l of Ca are present.

12. If the recommendation is for 10 meq/l NO_3, what is its equivalent in ppm nitrogen?

Again, the equivalent weight of NO_3 is 62. This amount in mg, dissolved in 1 l will provide one meq/l. So 10×62 = 620 mg NO_3^-, or 620 ppm NO_3^-. From Table 7-6, 0.226×620 = 140 ppm N.

However, if calcium nitrate ($Ca(NO_3)_2 \cdot 4\,H_2O$) is used to supply the 10 meq/l NO_3, the equivalent weight of $Ca(NO_3)_2 \cdot 4\,H_2O$ is 118.1 (Table 7-2). Therefore, 118.1 mg in 1 l will supply 1 meq/l NO_3 and 1 meq/l Ca. In this case, 10×118.1 = 1181 mg $Ca(NO_3)_2 \cdot 4\,H_2O$. The percentage of N in calcium nitrate is 15%. Of the 1181 mg, 0.15×1181 = 177.2 mg N, or 177 ppm N. The 1181 mg of calcium nitrate will also supply 10 meq/l Ca, so 0.17×1181 = 200.8 mg Ca or 201 ppm Ca. Note the difference in ppm where the source of nitrogen is considered.

13. The cost of potassium nitrate has increased sharply. You want to know whether it would be cheaper to shift to KCl and increase ammonium nitrate, at the possible expense of reducing yield, or to remain with KNO_3.

Asume the following costs: NH_4NO_3 = 14 cents per lb
KCl = 6.9 cents per lb
KNO_3 = 39.6 cents per lb

a) NH_4NO_3 contains 33.5% N
Therefore: 0.14/0.335 = 42.8 cents per lb N
b) KCl contains 62% K_2O, but only 51% actual K
Therefore: 0.069/0.51 = 13.5 cents per lb K
c) KNO_3 provides both 13% N and 44% K_2O. The 44% K_2O converts to 36.5% K

Therefore: $\frac{0.396}{0.13+0.365}$ = 80 cents per lb of nutrient (K and N)

One pound of NH_4NO_3 and 1 lb of KCl will cost 56.3 cents as compared to 80 cents of KNO_3 (1975 figures). However, research has indicated (Hartman and Holley, 1968) that KCl will reduce yield. Even if the yield reduction is less than 5%—or 48 flowers per ft^2 as compared to 50—this could amount to a loss in gross return of 20 cents per ft^2, figuring a return of 10 cents per flower.

In all problems, the major mistake is failure to use proper units and to keep the decimal place straight.

7.3.5 Aids to Nutrition Control

There are various tools available to the grower that permit him efficient control of fertilization. They are: (1) plant diagnosis; (2) pH analysis; (3) salinity determination; (4) soil tests, and (5) tissue analyses. Quite often, all of these may be necessary in order to diagnose cultural problems. Seldom, with borderline cases, can a single tool be used to complete satisfaction. In any case, their use requires some knowledge of their limitations, and the variability that one can expect under ordinary conditions. Plant diagnosis is most useful—particularly in assessing validity of the others. It should be emphasized, however, that recommendations can seldom be made unless knowledge is obtained of grower practices prior to, and during, the time the above tools are used. This is often the most difficult information to obtain as too few growers keep regular records of cultural operations. A logbook, into which all practices, applications, kinds of material, concentrations, etc. are entered as soon as they are completed is probably the single, cheapest, most useful record. It removes from the vagaries of memory whether or not the 200 ppm nitrogen application last week was actually 200 not 100 ppm, and whether benches 1 through 10 received it, or benches 10 through 20.

7.3.5.1 Plant Diagnosis

Plants respond to environment of which nutrition is one factor. Unfortunately, knowledge of how, in what manner, how soon, and what the visible changes may be has never been completely written down. To an observant grower, his crop will signal problems without the need of the most sophisticated and expensive equipment. Observational ability is commonly an important distinction between good and poor growers. Again, unfortunately, the ability to observe, to know what to observe, and to be able to draw the proper deductions from the observations, usually requires experience through repeated exposure, intimate familiarity with the crop, and a good picture in the mind's eye of what the crop "ought" to look like. The responses, for example, to an apparent nitrogen deficiency—as described in Table 7-13—could just as well be a problem of excess salinity. Some types of spray injury can simulate chlorosis, resulting from iron deficiency. Trace element deficiencies can mimic herbicide damage, and some types of insect and fungal damage can be confused with nutritional disorder. In such situations, the additional tools mentioned above can be employed to arrive at the right answer. On the other hand, interpretations of soil and tissue analyses are more reliable if the crop is observed. Nitrate levels can be below the minimum recommended in tissue, if the crop appears to be growing vigorously. However, a "hard" crop, with straight, small leaves, short internodes, pale color, etc., combined with a soil and tissue analysis showing low nutrition, certainly reinforces the conclusion that the fertilization rate should be increased.

Table 7-13. Nutrient deficiency symptons for greenhouse flowering plants. (From Collings, 1950)

Symptoms	Deficiency
A. Effects general on whole plant or localized on older, lower leaves	
I. Effects usually general on whole plant, often visible by yellowing and dying of older leaves	
a) Foliage light green, growth stunted, stems slender, few breaks, leaves small, lower ones usually lighter yellow	minus nitrogen
b) Foliage dark green, retarded growth, lower leaves sometimes yellow but more often purplish, particularly on petiole, leaves drop early	minus phosphorous
II. Effects usually local on older, lower leaves	
a) Lower leaves mottled, usually with necrotic areas near top and margins. Yellowing beginning at margin and continuing toward center. Margins later becoming brown and curving under, older leaves dropping	minus potassium
b) Lower leaves chlorotic, usually necrotic in late stages. Chlorosis between veins, veins normal green. Leaf margins curling upward or downward with puckering effect. Necrosis between veins very suddenly	minus magnesium
B. Effects localized on new leaves	
I. Terminal bud remaining alive	
a) Leaves chlorotic, veins remain green	minus iron
1. Necrotic spots usually absent, extreme cases may have necrosis of margins and tips, sometimes extending inward, larger veins only remaining green	
2. Necrotic spots usually present and scattered over leaf surface. Checkered or finely netted effect produced by even smallest veins remaining green. Poor bloom both size and color	minus manganese
b) Leaves light green, veins lighter than adjoining interveinal areas. Some necrotic spots. Little or no dying of older leaves	minus sulfur
II. Terminal bud usually dead	
a) Necrosis at tip and margin of leaves. Young leaves often definitely hooked at ends. Death of roots preceding all the above symptoms	minus calcium
b) Breakdown at base of young leaves. Stems and petioles brittle. Death of meristematic tissue. Tendency to increased branching	minus boron

Examples of obvious nutritional disorders as indicated in Figures 7-11 and 7-12, are situations where the operator has allowed his culture to get out of control. Under ideal conditions, the greenhouse grower uses these tools to "adjust" his fertilization program, and to prevent obvious plant damage. Fast changes, large fertilizer applications, or excessive changes in other environmental factors are undesirable, and, unless utilized with extreme caution, are likely to result in extremes and further damage. The grower establishes "trends" which can be easily arrested when need be by small changes in fertilization. At all times, the plant is under observation as a check on instrumentation reliability. Any tool available to the grower should not be disdained, and, of these, observation is the best.

Fig. 7-11. Chlorosis in rose, brought about by iron deficiency in tissue, may not always be caused by lack of iron in soil (see Table 7-20)

Fig. 7-12. Boron deficiency in carnation. Note that buds on most stems have aborted, accompanied by necrosis of some leaf tissue, and abnormal branching

7.3.5.2 pH Determinations

The acidity of the soil solution is determined by means of a special electrode (Fig. 7-13) which measures the hydrogen ion activity. Once sample extraction is made, actual measurement is precise and quick, and one of these instruments on hand for use by a grower is a valuable adjunct. Variability in the soil sample and

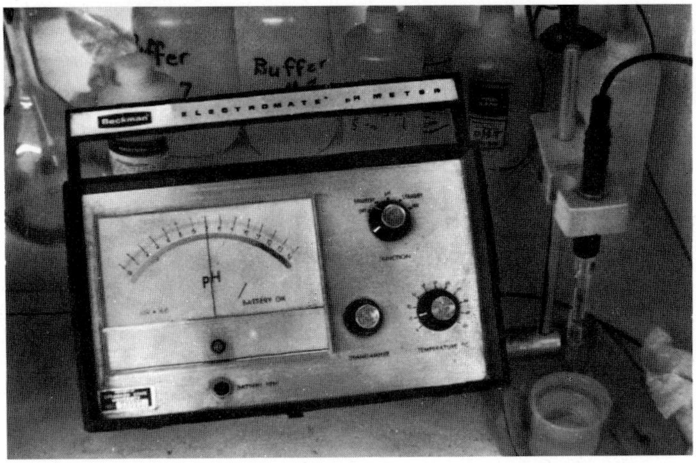

Fig. 7-13. pH meter for determining the acidity or alkalinity of an extracted solution from a soil sample. *Right*: measuring electrode

extraction method, however, reduces accuracy. It is not worth attempting to read the instrument closer than within one-half pH value.

The general range for best growth of most plants is 6.0–7.0. Acid-loving plants such as azaleas and gardenias, however, grow best between 4.5 and 5.5. Blueing in hydrangeas is achieved by reducing the pH to 5.5, whereas the pink color is obtained at pHs above 6.0 where aluminum is not available. As noted earlier, the importance of pH lies in the influence it has on nutrient availability and microbial action of the soil. A series of pH determinations will generally establish a trend which can be corrected in increments so that a suitable pH is maintained.

There are a number of materials that can be used to raise or lower pH (Table 7-14). Aluminum and iron sulfate are quick acting, the former generally being used to rapidly lower pH in such cases as blueing hydrangeas. Sulfur is very slow acting as the result depends upon microbial action and may require 2–3 months at 60° F. However, it is long-lasting. The speed of action of lime to raise pH is determined largely by its fineness. The smaller the particle size, the more rapid the action. One-half to $^{3}/_{4}$-in limestone is often used in inert media to supply a low amount of calcium for long periods, whereas ground limestone action will be rapid.

The reason for the various recommendations in Table 7-14 (amounts to apply for changing pH) arises from the fact that actual change achieved will depend upon the cation exchange capacity of the soil. The higher the capacity, as in a heavy, clay soil, the more that must be applied to achieve a pH change. The amount required also depends upon where on the scale one is starting, and how far one wishes to change the pH. Amounts are also influenced by the climate. Clay loam soils in a warm temperate region may need less than 4 t per acre of finely ground limestone to change the pH from 5.5 to 6.5; but the same soil in a cool temperate region may have a requirement in excess of 4 t. Generally, soils in arid

Table 7-14. Various recommendations for changing pH of soils. These are intended as a guide. The amounts will depend on the desired change and the buffering capacity of the medium (CEC)

Authority	Raise or lower	Material	Amounts	Remarks
Cornell Recommendations (1974)	Lower	Finely ground sulfur	0.5 lb/100 ft^2	Bench soils, 1/2 to 1 pH unit
			0.25 lb/yd^3	Potting soils, 1/2 to 1 pH unit
		Aluminum sulfate	3.0 lb/100 ft^2	Same as above
			1.5 lb/yd^3	Same as above
		Iron sulfate	3.0 lb/100 ft^2	Same as above
			1.5 lb/yd^3	Same as above
	Raise	Limestone, ground	5 lb/100 ft^2	To raise bench soils 1/2–1 unit
			2.5 lb/yd^3	Potting soils, 1/2–1 unit
Furuta (1974)	Raise	Calcium nitrate	6.5 lb/1000 gal	When pH is 5.5 or lower
	Lower	Ammonium sulfate	5.5 lb/1000 gal	When pH is 6.5 or higher
Koths et al. (1968)	Raise	Dolomitic lime	1.0 lb/100 ft^2	To raise pH 1.0 unit
	Lower	Aluminum or iron sulfate	1.0 lb/100 ft^2	To lower pH 0.2 units
Post (1950)	Raise	Limestone or $CaCO_3$	5 to 10 lb/100 ft^2	To raise from 4.0 to 6.0
	Lower	Aluminum or iron sulfate	3 to 10 lbs/100 ft^2	Larger amount to lower from 8.0 to 4.0
Laurie and Kiplinger (1948)	Raise	Limestone	2 to 17 lbs/100 ft^2	Heavier soils and larger changes require higher amounts
	Lower	Finely ground sulfur	2 to 7 lbs/100 ft^2	Silt loam, lesser amount for one pH unit

Fig. 7-14. Electrical conductivity meter for determining total salts in a soil. *Left:* solution extract from soil sample is drawn into pipette; conductivity read from dial in front of instrument

regions are characterized by alkalinity (high pHs), whereas soils in humid regions will generally be below 7.0, or highly acid.

7.3.5.3 Salinity Measurement

Salinity, the total salt concentration of the soil solution, is most easily determined by measuring the electrical conductivity of a solution extract. A total salt reading gives the grower an indication of trends in total salt level, and acts as a check on other analyses. Thus, a high analysis for NO_3 in a soil test, combined with a low salt reading, may suggest an error in the soil analysis if the other nutrients are in reasonable concentrations. Work by White (1957), and later by Schekel (1969, 1971), showed that as salinity increases, growth decreases. Thus, the grower should strive to maintain total salts as low as possible without causing deficiency problems. Sometimes a high salt reading will result from the presence of sodium, whereas other ions may be relatively low. As sodium can replace calcium on the exchange complex, it may be an indication that undesirable soil properties could result if some corrective action were not taken.

The instrument for measuring conductivity is relatively straightforward (Fig. 7-14). Unfortunately, the means of preparing solution samples can vary, so that the grower needs to know how the sample will be prepared if he is to interpret the result for his conditions. Conductivity can be expressed in parts per million (Table 7-15); but is most commonly in mhos $\times 10^{-5}$, millimhos per centimeter or micromhos per centimeter. Table 7-15 shows the relationship between various terms, and the general effect of sample preparation. The problem with 1:2 and 1:5 parts soil to water dilutions is that readings will vary with the root

Table 7-15. Soluble salt interpretations for soils. Safe ranges for majority of greenhouse plants. Comparisons for different extraction methods and units of expression

Extraction method	Units				Ref.
	RD-15 Solubridge (mhos × 10^{-5})	Millimhos per cm (mhos × 10^{-3})	Micromhos per cm (mhos × 10^{-6})	Parts per million	
1 part soil to 2 parts water	50–100	0.5 –1.0	500–1000	700–1400 [a] (325–600) [b]	[1]
	50–175	0.5 –1.75	500–1750	700–2450	[3]
	50–175	0.51–1.75	500–1750	(325–600) [b]	[4]
1 part soil to 5 parts water	25– 80	0.25–0.8	250–800	162–500 [b]	[1]
Saturated paste	<200	<2.0	<2000	<1300 [b]	[5]
Saturated calcium sulfate	261–270	2.61–2.7	2610–2700	—	[2]

[a] Conversion factor to ppm = (mmhos) × (700) × (2) as employed by Waters et al. (1973).
[b] Conversion factor to ppm = (mmhos) (650) according to Richards (1954).
References: [1] Ohio State Florists' Association Bulletin (1961); [2] Ministry of Agriculture, Fisheries and Food (1973); [3] Cornell Recommendations (1974); [4] Waters et al. (1973); [5] Richards (1954).

Table 7-16. Variation in soluble salt readings with type of root medium, using a 1 part soil to two parts water mixture by volume

Medium	Interpretation (mhos × 10^{-5})				Ref.
	Low	Medium	High	Very high	
Sandy soil	Below 25	25– 50	50–100	100–150	Waters et al. (1973)
1:1 peat-sand	Below 33	33– 66	66–130	130–200	
Peat or light weight mixes	Below 50	50–100	100–175	175–275	
Mineral soil	0–50	51–125	126–175	176–200	Cornell Recommendations (1974)
Peat-lite mixes	0–100	100–175	176–225	225–350	

medium (Table 7-16). Artificial mixes can give higher readings without danger. The 1:2 or 1:5 dilution method will also vary with the original moisture content of the sample, with dryer samples giving higher readings. The saturated paste extract method overcomes factors of differing soil types and moisture content by bringing the samples to maximum moisture contents. A saturated paste extract reading of 2 mmhos/cm in Table 7-15 is actually the upper limit under most conditions. Plants will usually do better if EC is maintained between 1.0–

Table 7-17. Soluble salt interpretations according to various authorities

Extraction method	Interpretation (mhos × 10^{-5})					Ref.
	Starvation	Low	Medium	High	Crop failure	
1–2 dilution	less than 15	15– 50	50–100	180–225	Above 225	[1]
	—	0–100	51–175	126–225	176–350 and above	[2]
	—	Below 25–50	25–175	100–275	Above 275	[3]
1–5 dilution	8–25	—	25–100	100–150	Above 150	[1]
Saturated calcium sulfate	—	Below 240	240–270	270–300	Above 300	[4]

References: [1] Ohio State Florists' Association Bulletin (1961); [2] Cornell Recommendations (1974); [3] Waters et al. (1973); [4] Ministry of Agriculture, Fisheries and Food (1973).

Table 7-18. Some recommended soil analyses levels for greenhouse crops, using various modifications of the Spurway system

Plant	Concentration ranges (ppm)				pH	Ref.
	NO_3	P	K	Ca		
Roses	30– 40	3.5–10	10–35	—	5.7–6.8	[2]
	10– 50	5–10	20–40	—	6.0–6.5	[1]
Carnations	30– 40	3.0–8.0	25–50	—	5.8–7.2	[2]
	10– 25	2.0–5.0	20–40	—	6.0–7.5	[1]
	40–100	2+	25–40	150+	5.0–7.8	[3]
Chrysanthemums	30– 40	4–10	25–50	—	6.5–7.2	[2]
	10– 25	2.0–5.0	20	—	5.5–7.5	[1]
Snapdragons	30– 40	4–10	15–35	—	5.8–7.0	[2]
	10– 25	2–5	20	—	6.0–7.5	[1]
Poinsettias	15– 30	5–12	25–45	—	4.8–5.5	[2]
	5– 10	2–5	20	—	6.0–7.0	[1]
Easter Lilies	15– 30	3–5	15–25	—	6.8–7.3	[2]
	10	2–5	20	—	6.0–7.0	[1]
Bedding plants	15– 30	5–12	15–30	—	5.8–7.0	[2]
	2– 10	2–5	20	—	6.5–7.5	[1]
Foliage plants	15– 30	5–12	15–30	—	5.5–6.5	[2]
General	30– 50	4–6	25–35	150+	4.5–6.8	[4]
	20– 90 [a]	5	20–60 [a]	150–200	6.5–6.8 [a]	[5]

[a] Desired level varies with stage of growth (lower with young plants) and location (lower in the ground).
References: [1] Laurie and Kiplinger (1948); [2] Peters (1968); [3] Holley and Baker (1963); [4] Cornell Recommendations (1974); [5] Ohio State Florists' Association Bulletin (1961).

Table 7-19. Concentration of nutrients in the tissue of some greenhouse crops

Plant	Part	Macronutrients (%)							Micronutrients (ppm)					Ref.
		Total N	NO_3	P	K	Ca	Mg	SO_4	Fe	B	Cu	Zn	Mn	
Carnation	5th and 6th leaf pairs from top of laterals	2.0–4.0	0.5–0.7	0.20–0.35	2.9–3.3	1.0–1.5	0.2–0.4	0.1+	50–100	25–100	5–10	25–100	50–150	[1]
Rose	Upper, mature leaves on flowering stem	3.0–3.5	0.1–0.3	0.28–0.34	2.0–2.5	1.0–1.6	0.28–0.32	0.16–0.21	80–120	40– 60	7–15	20– 40	70–120	[2]
Snapdragon	Upper leaves at bloom	5.9	—	0.28–0.33	5.6	1.7	0.69	—	—	—	—	—	—	[3]
Chrysanthemum	Upper leaves at bloom	4.0–5.0	0.8	0.5 –6.5	5.5–1.5	0.8	0.4–	—	50–100	20– 30	5–10	25–100	50–150	[7]
Poinsettia	Mature leaves at bloom	4.0–6.0	—	0.3 –0.7	1.5–3.5	0.7–2.0	0.4–1.0	—	100–300	30–300	2–10	25– 60	45–300	[5]
Azalea	Mature leaves	2.0	—	0.19	0.80	0.22	0.17	—	—	17	—	—	—	[4]
Lettuce	Wrapper leaf	2.5–4.0	—	0.4 –0.6	6.0–8.0	1.4–2.0	0.5–0.7	—	—	25– 45	—	—	—	[6]
Tomato	Young mature leaf	3.0–6.0	—	0.5 –0.8	2.5–4.0	4.0–6.0	0.6–0.9	—	—	40– 80	4– 8	15– 30	60–100	[6]

References: [1] Hanan (1975); [2] Sadasiviah and Holley (1973); [3] Boodley (1962); [4] Twigg and Link (1951); [5] Ecke (1971); [6] Geraldson et al. (1973); [7] Boodley (1964).

Table 7-20. Factors contributing to micronutrient deficiencies in plants. (From Lucas and Knezek, 1972)

1. Boron
   - Soils low in total boron (alluvials, podzols, organics, regosols, low humic greys)
   - Areas of moderate to heavy rainfall
   - Nearly neutral or alkaline soils
   - Dry weather
   - High light intensity
2. Copper
   - Low soil copper (less than 6 ppm in mineral soils; less than 30 ppm in organic soils)
   - High soil phosphorous
   - High soil nitrogen
   - High soil zinc
3. Manganese
   - Slightly acid or alkaline soils
   - Naturally poorly drained soils (organics, ground water podzols, regosols, low humic greys, marly or shelly soils)
   - High soil iron, copper and zinc
   - Dry weather
   - Low light intensity
   - Low soil temperature
4. Molybdenum
   - Low soil molybdenum (acid regosols, acid podzols, acid organics)
   - High free iron (bog iron)
5. Iron
   - Low soil iron
   - Free calcium carbonate ($CaCO_3$)
   - High bicarbonate (HCO_3^-)
   - Moisture extremes
   - High amounts of heavy metals, high soil phosphorous
   - Poor aeration
   - Temperature extremes
   - Heavy manuring (alkaline soils) or low organic matter content (acid soils)
   - Excess soil acidity
   - Genetic differences
   - Root damage
6. Zinc
   - Low soil zinc (low humic greys, alluvials, regosols, organics, high rainfall areas)
   - Calcareous soils
   - Low soil organic matter
   - Cool Temperatures
   - High soil phosphorous
   - Restricted root zones (compacted soils, container-grown plants)
   - Corrals, old orchards, barnyards
   - Liberal applications of nitrogen

1.5 mmhos/cm. A 1:2 sample, with a reading approaching 100×10^{-5} mhos, is usually a signal to begin corrective action.

The higher the salt content of the water supply, the more water that must be applied, and the less the soil can be allowed to dry out. Leaching requirements increase. If fertilizer injection is being used with good water, EC readings and soil analyses can actually run in the deficient range as long as irrigation is frequent. For example, suppose from Table 6-11 the root medium we are using contains at maximum moisture content 8 quarts of water per ft^2. For the common greenhouse soil, about 30% of this will be utilized between irrigations, or about 2.4 quarts. At each irrigation, at least 2.4 quarts per ft^2 must be supplied. Suppose also that fertilizer injection is being used so that the EC of the water supply is 0.75 mmhos/cm. To maintain a 4 mmhos/cm EC in the leachate, 18.8% of the irrigation water must pass through the rootzone (Table 5-1). To calculate the actual amount of water that must be applied:

$1.0 - 0.188 = 0.812$.

Water applied $= 2.4/0.812 = 2.96$ quarts per ft^2.

According to Kofranek and Lunt (1975), for azalea production, 33% of irrigation water having an EC of 0.5 mmhos/cm should be leachate. For the same example above, this means that 3.6 qts/ft^2 should be applied at each watering. If a salt condition already exists, then all water in the soil profile (8 quarts) should be exchanged in order to significantly lower the salinity level.

7.3.5.4 Soil and Tissue Analyses

These analyses offer a means to check on individual elements, and to correct for individual variations. As the result of the higher nutrient level of most greenhouse soils, a separate extraction method was developed in the United States as distinct from methods employed by agronomists for field soils. Almost all greenhouse laboratories in the U.S. use some modification of the Spurway system to extract the sample from the soil. This is considered a weak extraction solution as compared to methods for field soils. The variability, however, means that for reliable interpretation the grower must become familiar with the results from the laboratory he patronizes. The same is also somewhat true for tissue analyses. The differences that can occur in soil analyses are particularly noticeable in Table 7-18.

Probably most important in obtaining reliable soil and tissue analyses is the manner in which the sample is taken. When sampling soils, the $^1/_4$- to $^1/_2$-in top layer should be removed, and then a core sample obtained for the entire soil profile (see Fig. 5-27). Several core samples should be obtained at random. These should be thoroughly mixed, and then the necessary $^1/_4$ lb sample for actual testing removed and forwarded to the laboratory. Inexpensive soil testing kits can be obtained for rough determinations. Nevertheless, care in sample preparation is always required. With tissue samples, the fact that nutrient levels may vary depending upon location and plant age requires strict attention to where the tissue is gathered from the plant (Table 7-19). For most laboratories, at least 5 g of dried tissue is required. Samples are usually obtained infrequently, and mainly employed as a check on other methods.

It should be pointed up that low values of essential elements in soil or tissue analyses are not, in fact, proof that the cause is lack of an essential element in the root medium. EC measurements, pH determination and plant observations will help, but it is important to remember that there are several factors which can cause deficiencies in plants—as outlined in Table 7-20. Problems of this sort are particularly prevalent with iron. Solution to the cause for iron chlorosis (Fig. 7-11) may call into play all the methods just discussed and still require additional work.

References

Baker, D. E., Woodruff, C. M.: Influence of volume of soil per plant upon growth and uptake of phosphorous by corn from soils treated with different amounts of phosphorous. Soil Sci. **94**, 409–412 (1963)

Baker, K. F.: The U.C. system for producing healthy containergrown plants without soil. Calif. Agr. Exptl. Sta. Circ. 347, 1957

Boodley, J.: Fertilization. In: Snapdragons. Langhans, R. W. (ed.). Ithaca: Cornell Univ. 1962, 93 pp

Boodley, J.: Fertilizing. In: Chrysanthemums. Langhans, R. W. (ed.). Ithaca: Cornell Univ. 1964, 185 pp

Brady, N. C.: The Nature and Properties of Soils, 8th ed. New York: Macmillan 1974, 639 pp

Broyer, T. C., Stout, P. R.: The macronutrient elements. Ann. Rev. Plant Physiol. **10**, 277–300 (1959)

Collings, G. H.: Commercial Fertilizers. Philadelphia: Blakiston 1950, 522 pp

Cooper, A. J.: Soil? Who needs it? Am. Vegetable Grower **22**, 18–20 (1974)

Cornell Recomendations: N.Y. State College of Agr.: Cornell Recommendations for commercial floriculture crops. Part I. N.Y. State Coll. and Life Sci. Ithaca: Cornell Univ. 1974, 60 pp

Ecke, P., Jr.: The Poinsettia Manual. Encinitas, CA.: Paul Ecke Poinsettias 1971, 183 pp

Evans, H. J., Sorger, G. J.: Role of mineral elements with emphasis on the univalent cations. Ann. Rev. Plant Physiol. **17**, 47–76 (1966)

Farm Chemicals: Dictionary of Plant Foods. Willoughby, Ohio: Meister 1972, 110 pp

Furuta, T.: Environmental Plant Production and Marketing. Arcadia, CA.: Cox 1974, 232 pp

Geraldson, C. M., Klacan, G. R., Lorenz, O. A.: Plant analysis as an aid in fertilizing vegetable crops. In: Soil Testing and Plant Analysis. Walsh, L. M., Beaton, J. D. (eds.). Soil Sci. Soc. Am., Madison, Wisc., 1973, 491 pp

Gericke, W. F.: The Complete Guide to Soilless Gardening. New York: Prentice-Hall 1940

Gerloff, G. C.: Comparative mineral nutrition of plants. Ann. Rev. Plant Physiol. **14**, 107–124 (1963)

Green, J. L.: Ionic balance and growth of carnations. I. Recent literature. Colo. Flower Growers' Assoc. Bull. **210**, 1–4 (1967a)

Green, J. L.: Ionic balance and growth of carnations. Colo. Flower Growers' Assoc. Bull. **211**, 1–5 (1967b)

Green, J. L.: Ion ratios and competitive uptake of nutrients in carnations. Colo. Flower Growers' Assoc. Bull. **212**, 1–5 (1967c)

Hanan, J. J.: Calcium tissue levels in the carnation. Colo. Flower Growers' Assoc. Bull. **305**, 1–4 (1975)

Hanan, J. J., Holley, W. D.: Hydroponics. Colo. State Univ. Exptl. Sta. General Series 941, 1975, 21 pp

Hartman, L. D.: Ion balance in carnation nutrition. Colo. Flower Growers' Assoc. Bull. **248**, 1–4 (1971a)

Hartman, L. D.: Ion balance in carnation nutrition. II. Colo. Flower Growers' Assoc. Bull. **249**, 1–5 (1971b)

Hartman, L. D.: Effects of irrigation frequency and nutrient solution concentration on yield of carnation. Colo. Flower Growers' Assoc. Bull. **250**, 1–5 (1971c)

Hartman, L. D., Holley, W. D.: Carnation nutrition. Colo. Flower Growers' Assoc. Bull. **221**, 1–8 (1968)

Hoagland, D. R., Arnon, D. I.: The water-culture method for growing plants without soil. Calif. Agr. Exptl. Sta. Circ. 347, 1950

Holley, W. D., Baker, R.: Carnation Production. Dubuque, Iowa: Wm. C. Brown 1963, 142 pp

Ingestad, T.: A new operational definition of nutrient requirements of birch seedlings. In: Proc. Intern. Working Group on Soilless Culture, Gran Canaria, Canary Islands, Nov. 1969, unpublished manuscript

Kofranek, A. M., Lunt, O. R.: Mineral nutrition. In: Growing Azaleas Commercially. Kofranek and Larson (ed.). Univ. Calif., Davis. Publ. 4058, 1975, 108 pp

Koths,J.S., Judd,R., Jaisano,J.: Guide for fertilizing greenhouse crops. Coop. Ext. Ser., Coll. Agr., Univ. of Connecticut, Storrs, 1968, 8 pp
Laurie,A., Kiplinger,D.C.: Commercial Flower Forcing. Philadelphia: Blakiston 1948, 550 pp
Leggett,J.E.: Salt absorption by plants. Ann. Rev. Plant Physiol. **19**, 333–346 (1968)
Loneragan,J.F., Snowball,K.: Calcium requirements of plants. Australian J. Agr. Res. **20**, 465–478 (1969)
Lucas,R.E., Knezek,B.D.: Climatic and soil conditions promoting micronutrient deficiencies in plants. In: Micronutrients in Agriculture. Dinauer,R.C. (ed.). Soil Sci. Soc. Amer. (1972)
Ministry of Agriculture: Fertilizer recommendations. Ministry of Agr., Fisheries and Food Bull. 209, London, 1973, 102 pp
Munson,R.D., Nelson,W.L.: Principles and practices in plant analysis. In: Soil Testing and Plant Analysis. Walsh,L.M., Beaton,J.D. (ed.). Soil. Sci. Am. (1973)
Ohio State Florist: Interpreting soil tests. Ohio State Florists' Assoc. Bull. **377** (1961)
Peters,R.B.: A System of Greenhouse Soil Fertility Control. Allentown, PA.: R. B. Peters 1968, 18 pp
Post,K.: Florist Crop Production and Marketing. New York: Orange Judd 1950, 891 pp
Richards,L.A.: Diagnosis and improvement of saline and alkali soils. USDA Salinity Lab., Agr. Handbook 60, 1954, 160 pp
Sadasiviah,S.P., Holley,W.D.: Ion balance in nutrition of greenhouse roses. Roses Inc. Bull. November Supplement. 1973, 27 pp
Schekel,K.: Influence of nutrient solution concentration. Colo. Flower Growers' Assoc. Bull. **229**, 1–3 (1969)
Schekel,K.A.: Effects of increased soluble salt concentrations on carnation growth. Colo. Flower Growers' Assoc. Bull. **251**, 1–5 (1971)
Stevenson,D.S.: Effective soil volume and its importance to root and top growth of plants. Can. J. Soil Sci. **47**, 163–174 (1967)
Twigg,M.C., Link,C.B.: Nutrient deficiency symptoms and leaf analysis of azaleas grown in sand culture. Proc. Am. Soc. Hort. Sci. **57**, 369–375 (1951)
Wadleigh,C.H., Ayers,A.D.: Growth and biochemical composition of bean plants as conditioned by soil, moisture, tension, and salt concentration. Plant Physiol. **20**, 106–132 (1945)
Waters,W.E., Llewelyn,W., Geraldson,C.H., Woltz,S.S.: The interpretation of soluble salt procedures as influenced by salinity testing procedure and soil media. Proc. Tropical Region, Am. Soc. Hort. Sci. **17**, 397–405 (1973)
White,J.W.: The effects of soluble salts and soil moisture on carnation growth and quality. M.S. Thesis, Colo. State Univ., Ft. Collins 1957, 68 pp

Chapter 8 Carbon Dioxide and Pollution

8.1 Carbon Dioxide

Manufacture of food for plant growth requires CO_2. CO_2 uptake by a green leaf, in sunlight or supplemental illumination, is a common measure of the photosynthetic rate which can ultimately be translated in terms of growth, yield, and quality. Many plants continue to photosynthesize at zero CO_2 levels as liberation of the gas produced in respiration will be utilized. However, growth is minimum, and, as the plant continues to respire in the dark, actual weight loss will occur. As the amount of energy (light) increases, the supply of CO_2 must increase if the process is not to be limited. The general relationships between CO_2 concentrations in air, light intensity, and photosynthesis have been known for many years—for example, the work of Hoover et al. in 1933 (Fig. 8-1). However, it was not until the 1950s that several individuals, Holley, Goldsberry, Wittwer, and Klougart, among others, showed that closed greenhouse systems offered good opportunity to improve production through the elevation of CO_2 levels. In fact, preliminary studies showed that, during the day in a closed structure, CO_2 was usually depressed below that of the outside air (300 ppm) (Fig. 8-2).

8.1.1 Factors Affecting CO_2 Uptake

The amount of CO_2 uptake by leaves depends upon several factors: (1) plant species and variety; (2) radiant intensity; (3) wind velocity; (4) water stress; (5) CO_2 concentration in the air; (6) resistance to CO_2 diffusion through the stomates; (7) previous history of the plant; and (8) leaf age. Temperature also has a direct influence on photosynthesis, with the chemical reaction commonly being doubled for every 10° C rise in temperature over limited ranges. Aikin and Hanan

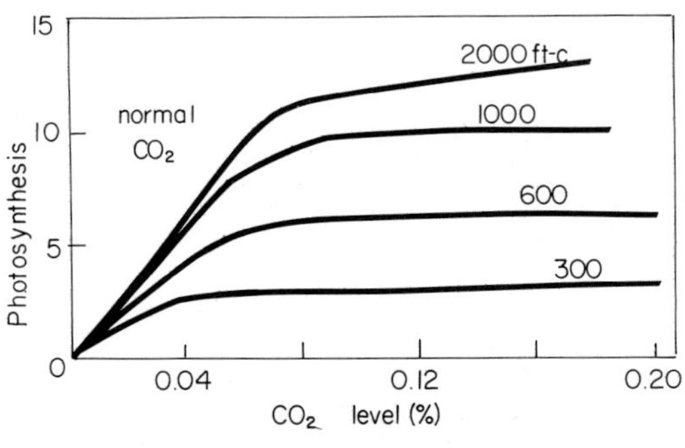

Fig. 8-1. Relationship between CO_2 concentration, light intensity, and photosynthesis in wheat. Note that maximum CO_2 uptake is function of CO_2 concentration and light intensity. Normal CO_2 is 0.030% (300 ppm). (From Hoover et al., 1933)

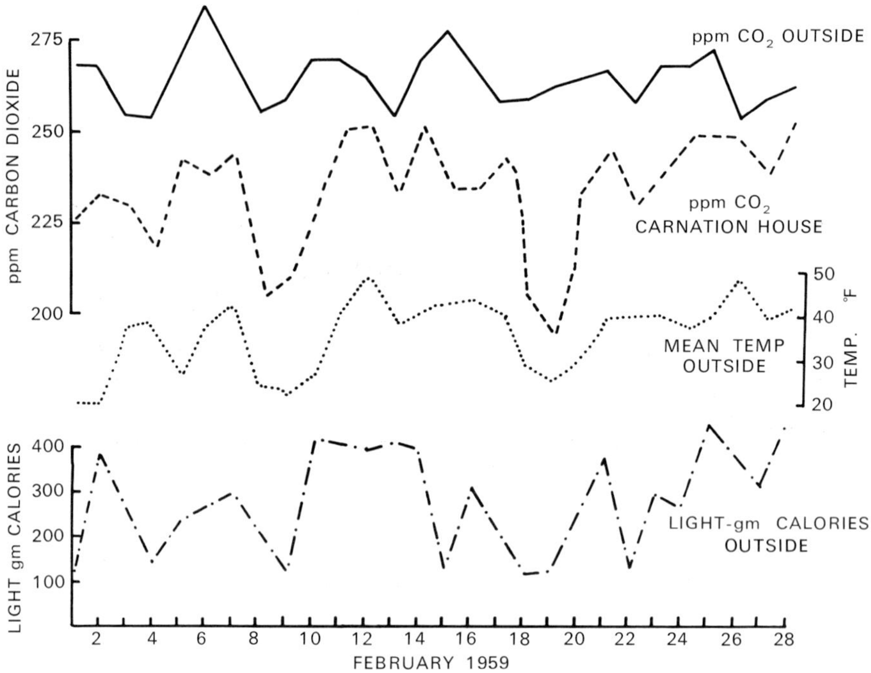

Fig. 8-2. Average CO_2 concentration in unvented greenhouse, during winter compared with outside temperature and total radiant energy per day. No CO_2 injection. Note that because ventilation is reduced CO_2 levels inside greenhouse are lowest when outside temperature is low (Goldsberry and Holley, 1962)

(1975), however, were unable to find a significant variation of CO_2 uptake with temperature under the usual greenhouse temperature fluctuations. These are usually less, during the day, than 10° C.

The effects of species and previous history were discussed in Chapter 2 (Figs. 2-17, 2-18), which showed that at a given CO_2 level, maximum CO_2 uptake is reduced at high light levels if plants are preconditioned (acclimatized) to lower intensities. The interesting varietal deviation in single species has been known among plant breeders for many years in regard to other environmental factors. Studies by Mathis (1972) and Hanan (1973) showed distinct differences between rose varieties as to temperature and CO_2 response. Water stress was covered in Chapter 5 (Fig. 5-7), showing that, as the internal plant water potential decreases (higher stress), photosynthesis begins to decrease rapidly. This effect can be achieved by reducing the water supply to the plant or increasing the environmental demand. In practical terms, high water stress increasingly limits photosynthetic rate as light intensity increases (Fig. 8-3). Under relatively low intensities, stress may vary markedly without affecting CO_2 uptake as light is the limiting factor.

Although increasing CO_2 concentrations will increase photosynthesis under normal conditions, it can be appreciated that the effect will be less the lower the light level (Figs. 8-1 and 8-4). For a time, there was some controversy in European

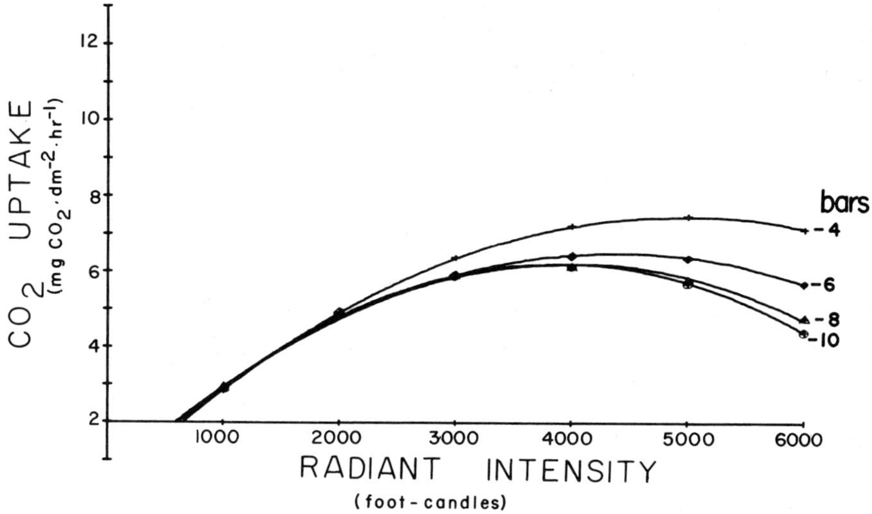

Fig. 8-3. Effect of water stress and light intensity on CO_2 uptake by individual Forever Yours rose leaves. CO_2 concentration at 500 ppm (Aikin and Hanan, 1975)

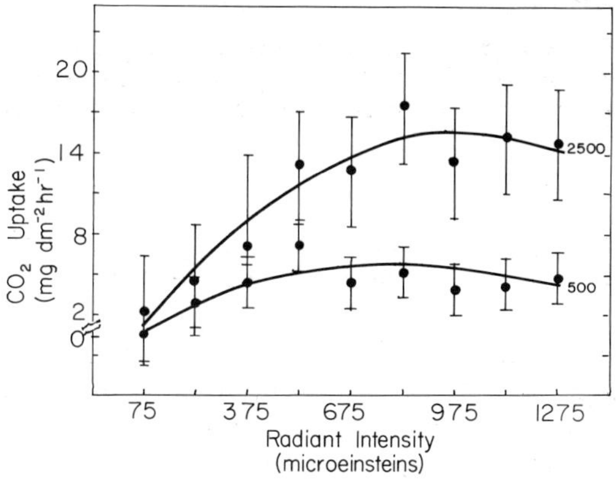

Fig. 8-4. Effect of CO_2 concentration on CO_2 uptake by individual Forever Yours rose leaves. *Vertical lines:* statistical variation which can exist in a single mean. Where they do not overlap, the means are statistically different from each other (Thompson and Hanan, 1975)

literature as to benefits of CO_2 injection. This can be related to the lower radiant intensities in Northern Europe as compared to semiarid regions typified by Colorado. High CO_2 concentrations with supplementary illumination in dark regions offer a synergistic effect, the one aiding the other. At low levels of either, the effect of small additions is much greater. At normal CO_2 levels, maximum photosynthesis for individual leaves of many species has been found at light intensities around 2000 ft-c. As CO_2 increases, the maximum intensity for photosynthesis also increases. For roses, at 500 ppm, the maximum appears to be around 3400 ft-c, while above 1000 ppm, it exceeds 4000 ft-c (Aikin and Hanan, 1975; Thompson and Hanan, 1975). Thus, climatic regions offering the maximum number of clear days, also offer the greatest opportunities for CO_2 injection.

Table 8-1. Influence of wind velocity on maximum rate of photosynthesis of leaves in normal air. (From Gaastra, 1963)

Wind velocity ($cm\ s^{-1}$)	Photosynthesis ($mm^3\ CO_2\ cm^{-2}\ h^{-1}$)
10	79
16	88
42	101
100	109
300	114
1000	118

Wind velocity mainly influences the resistance to CO_2 diffusion from the bulk of air to the leaf. Gaastra, in his definitive study (1959), explored the CO_2 relationships of many greenhouse vegetable plants. As shown in Table 8-1, there is a continuing increase in CO_2 uptake as wind velocity increases. Forty-two cm sec^{-1} corresponds to 83 fpm. This is in the general range for air flow in greenhouses under maximum cooling, using the evaporative fan and pad method. Undoubtedly, one of the benefits from fan and pad cooling is better CO_2 supply. The benefits of forced air circulation in closed greenhouses can be appreciated. Field studies (Lemon, 1973) have shown that, out-of-doors, CO_2 levels in the vegetative canopy can be depressed. Under calm conditions, CO_2 concentration becomes a limiting factor for field production. Similar studies to these just cited have not been carried out for greenhouse bench crops. Gaastra (1959) and others (Pallas, 1965; Lemon, 1973) have shown that stomates partially close as CO_2 concentration increases. This closure affects water loss more than CO_2 diffusion in the reverse direction. Thus, another benefit of elevated CO_2 is reduced plant water stress.

In general, maximum recommended CO_2 concentrations are between 1000 and 1500 ppm. Human toxicity levels are generally considered to be about 50000 ppm (5%). Thompson and Hanan (1975) found slightly higher CO_2 uptake at 2500 ppm as compared to 1500; but, when yield and quality of roses were considered, 2500 ppm was not significantly different. Recommendations for greenhouse CO_2 injection are 1000–1500 ppm on roses. This appears to be the general maximum level for most greenhouse crops. It is apparent, however, that maximum utilization of elevated CO_2 levels is directly influenced by the other cultural procedures. CO_2 injection will offer no benefit if some other factor is limiting photosynthesis and growth.

8.1.2 Effect of CO_2 on Growth

The ultimate effect of CO_2 on growth is a function of concentration and time. Addition of CO_2 in enclosed structures is not always a simple matter. The amount of time CO_2 can be injected will depend upon greenhouse and outside temperatures. As outside temperature drops in the fall, the greenhouse can be maintained for longer periods as a closed system. The effect on available time for CO_2 addition in Colorado was determined for a representative year by Holley et al.

Table 8-2. Percent of daylight hours CO_2 could be added automatically to carnation greenhouses at Ft. Collins, Colo., 1961–1962 (Holley et al., 1962)

Month	Possible h per day	Possible h per month	Lbs. liquid CO_2 per month[a]	Percent of possible time CO_2 added
August	10	310	2.9	8
September	10	300	17.9	52
October	9	279	15.7	46
November	9	270	22.3	72
December	8	248	19.4	68
January	8	248	18.9	66
February	8	224	16.0	62
March	8	248	14.3	50
April	8	240	8.3	30
Total		2367	135.7	

[a] Pounds per 300 ft^2 injected at the rate of 1 ft^3 per h.

Table 8-3. Effect of increased carbon dioxide on yield of Red Gayety carnations (Goldsberry, 1961)

	CO_2 concentration (ppm)		
	200	350	550
Accumulative yield for months ending			
Sept. 30, 1960	13	35	43
Oct. 31, 1960	73	83	83
Nov. 30, 1960	91	94	93
Dec. 31, 1960	105	129	151
Jan. 31, 1961	153	201	236
Feb. 28, 1961	218	272	302
Mar. 31, 1961	274	330	368
Apr. 30, 1961	324	401	436
May 31, 1961	378	490	517

(1962) for carnations (Table 8-2). Even during the coldest period, time for injection was only 72% of the maximum possible. If inside temperature can be elevated, however, periods of CO_2 injection can be 100% of maximum possible. For example, by raising the ventilation temperature on roses to 84° or 86° F, during 1972, Hanan (1973) was able to maintain a completely closed system for three months of the year, with ventilation time less than 6 h per week into April, 1973.

If increased CO_2 is maintained from the time young plants are benched, yield differences will be noted from the start of production (Table 8-3).

If, however, as the result of ventilation requirements, CO_2 cannot be added until outside temperatures are low enough, the effect of injection may not be noted until several weeks later. Thus Holley et al. (1962) concluded that CO_2 had limited effect on the first crop of carnations planted in the summer. On the other hand, yield increases were generated on second year plants in about 13 weeks.

This time lag between application and response resulted in Holley and Goldsberry (1963) concluding that CO_2 injection should not be continued beyond the end of February in Colorado as the flowers produced will be cut in the summer. On roses already in production, Hanan (1973) began to see observable yield differences in 8 to 10 weeks. When rate of accumulative yield was plotted for a period of 32 weeks, the 1500–2000 ppm treatments produced 1.43 flowers per week per plant as contrasted to 1.23 flowers per week per plant for the 500–800 ppm treatments. This is a 16% difference. Combination of high ventilation temperature with 2000 ppm CO_2 produced 33% more flowers than lower ventilation temperature and 500 ppm CO_2.

Some attempts have been made to increase CO_2 even with low ventilation rates. Goldsberry and Holley (1962) stated that ridge ventilators opened for more than 30 in-h (e.g. 3 in for 10 h) resulted in inefficient injection. As many modern greenhouses are force-ventilated, the practice is to automatically shut off CO_2 flow when a ventilation fan comes on. Use of CO_2 will allow day temperatures to be raised (Gaastra, 1959; Holley et al., 1962; Hanan, 1973), so injection can be carried out for longer periods. Fiberglass in Colorado has been shown to decrease ventilation time (Holley, 1964). As a rule-of-thumb, ventilation will usually be required on clear days when the outside temperature is above 55° F, with inside temperatures around 65–70° F. If the temperature differential is 30° F or more, most crops can be grown in closed systems. In fact, recommendations for roses are to raise the ventilation temperature 10° F during the winter months of November through March in Colorado. As elevated CO_2 concentrations cannot be maintained in other months, ventilation must start at lower temperatures in order to avoid undesirable quality.

The effect of CO_2 may not always be an increase in quality and yield. Yield may be increased at the expense of quality. The response can be a shortened period between planting and production, increased size of a potted plant, longer stems, faster seedling germination and growth, or faster maturation of the flowers. Quite often a grower may not employ CO_2 for he must shift his scheduling operations with much higher precision. For example, CO_2 on cut chrysanthemums under regular schedules will cause longer stems which is not always desirable. The photoperiod, pinching and planting dates have to be adjusted to produce the same stem length in a shorter period. Bloom maturation on potted plants may be so fast as to seriously reduce the time the grower has available for selling.

8.1.3 CO_2 Injection

One of the earliest means of raising CO_2 was mulch decomposition. One ton of organic matter will eventually release $1\frac{1}{2}$ tons of CO_2. Results obtained by Caldwell (1963) showed a combination manure and straw mulch produced the most CO_2 during the experiment, followed by straw and then pine shavings. However, decomposition, while perhaps offering a cheap source which can provide side benefits, does not provide uniform levels, and certainly is of no benefit for CO_2 when ventilation is required.

Table 8-4. Estimated CO_2 required per 1000 ft^2 of greenhouse area for carnations (Holley and Goldsberry, 1963)

	ft^3	Pounds CO_2
Sept.	500	58
Oct.	520	61
Nov.	800	94
Dec.	720	85
Jan.	660	78
Feb.	560	66
Total	3760	442

The safest source is liquid CO_2 or dry ice which can be injected from storage tanks. It is also the most expensive; one pound of CO_2 yields approximately 8.5 ft^3 of gas. On the basis of Table 8-2, and assuming 5 cents per pound, the cost between August and April would be 3.4 cents per ft^2 of bench area. Holley and Goldsberry (1963) estimated the amount of CO_2 required per 1000 ft^2 of greenhouse area (Table 8-4). Roses require an estimated $2^1/_2$ times as much as carnations. The cost, from figures in Table 8-4, calculates to about 3.7 cents per ft^2 of bench area (assuming 60% in production), not counting initial investment for storage tanks and flow regulators.

With the relatively cheap natural gas supply in the United States, combustion from natural gas burners is most common. Some kerosene burners have been employed, but not to the same extent as in Europe and Britain. Dutch workers have developed equipment for taking the flue gas directly from boilers and piping it to the greenhouses. A problem in clear-day regions is the fact that boilers are not needed at the time when highest CO_2 levels are required. For the natural gas composition provided in Table 8-5, at 635 mm Hg barometric pressure, each cubic foot of gas will produce 2 ft^3 CO_2, 2.4 ft^3 water, and 880 BTU. At sea level, the energy will be nearer 1000 BTU per ft^3. Assuming a price of 50 cents per 1000 ft^3 gas, the cost for CO_2, based on Table 8-4, calculates to about 0.2 cents per ft^2 of bench area. According to Hand (1972), hydrocarbon fuels such as kerosene or fuel

Table 8-5. Colorado natural gas analysis, 1969

Constituent	% by volume
Methane (CH_4)	85.6
Ethane (C_2H_6)	5.51
Propane (C_3H_8)	1.4
n-Butane (C_4H_{10})	0.09
CO_2	1.0
Isobutane	0.08
Isopentane	0.01
n-Pentane	Trace

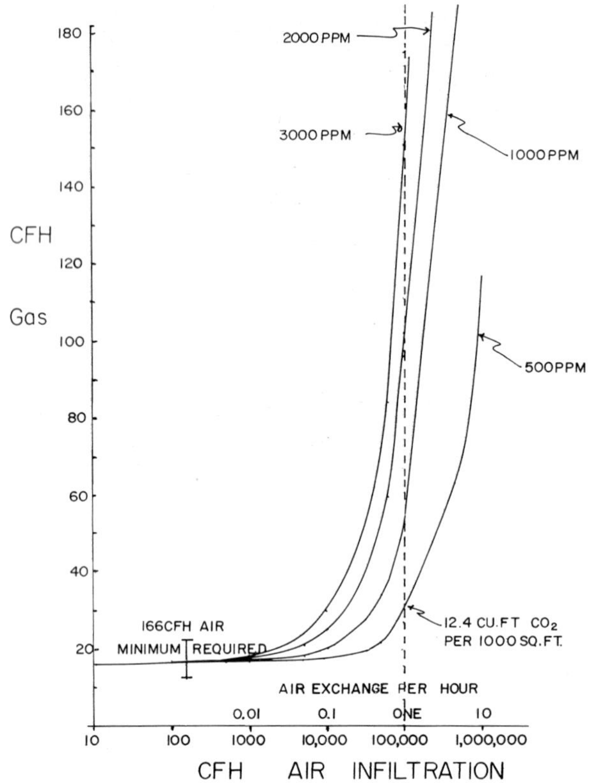

Fig. 8-5. Effect of greenhouse air infiltration rate on $ft^3 h^{-1}$ of gas required to maintain various CO_2 concentrations in a greenhouse with 100000 ft^3 volume. Requirement is based upon 4 ft^3 of CO_2 per 1000 ft^2 of greenhouse area (Hanan, 1972)

oil will produce 3 lb CO_2, $1\frac{1}{2}$ lbs water, and between 18500 and 21000 BTU for every pound of fuel. The cost, while higher, is still much less than for liquid CO_2.

Suggested CO_2 levels are based upon injection rates sufficient to maintain a particular concentration with an actively growing crop, on a clear day, with no ventilation. At concentrations much above 500 ppm, air infiltration of the greenhouse becomes important. At exchange rates of one per hour, even 1000 ppm CO_2 requires nearly three times the natural gas consumption (Fig. 8-5).

Combustion to supply CO_2 carries a penalty. Sufficient combustion air must be provided to prevent any possibility of pollution. Almost all combustion processes produce some ethylene. Hanan (1972) showed the small residue could be diluted if proper air circulation patterns and sufficient makeup are provided. Venting of unit heaters directly into structures is generally frowned upon by code authorities in the United States. But, the majority of growers injecting CO_2 use some type of burner of which Figure 8-6 provides three examples. The outside burner, with an unlimited air supply, has been shown the safest. If inside burners are used, at least 2 in^2 of free-air opening must be provided for every 5000 BTU input rating of the burner. This fresh air must be brought to the combustion chamber. Studies by Hanan (1972) showed that free-blowing unit heaters generally set up individual circulation patterns. A gradual ethylene buildup can be observed as the burner operates, regardless of the amount of infiltration or free-

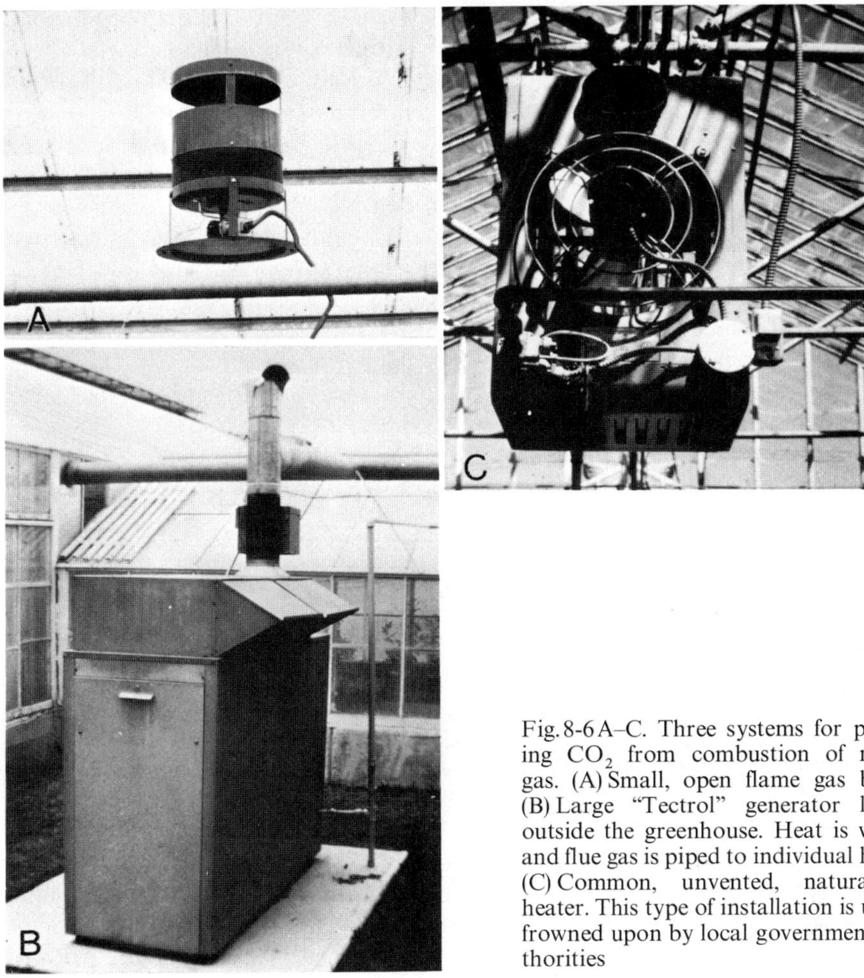

Fig. 8-6 A–C. Three systems for producing CO_2 from combustion of natural gas. (A) Small, open flame gas burner. (B) Large "Tectrol" generator located outside the greenhouse. Heat is wasted, and flue gas is piped to individual houses. (C) Common, unvented, natural gas heater. This type of installation is usually frowned upon by local governmental authorities

air opening. To make use of infiltration, or openings in the side of the greenhouse, the air must be forced to circulate throughout the greenhouse. This usually takes the form of polyethylene ventilation tubes (see Chap. 3) attached to the heater outlet. Wherever this was done, ethylene buildup could not be measured, and associated symptoms of ethylene damage were less likely to be observed. This system does not provide for low temperature situations (0–10° F) where infiltration can be drastically reduced as the cracks in the greenhouse freeze. A special duct to supply air directly to the combustion chamber is a common practice. Other precautions with combustion to supply CO_2 include clean fuels. High sulfur fuels will produce sufficient SO_2 to cause damage. Generally, fuels having less than 0.1% sulfur will be best. It is not recommended for this reason to use open flame burners in structures where sulfur is volatilized for mildew control.

CO_2 injection can begin within an hour after daybreak, and continue to sunset. With fully grown, producing, cut flower crops, we estimate that 4 ft^3 of

CO_2 per hour per 1000 ft^2 will provide any CO_2 level desired with no infiltration in a volume of 100000 ft^3. At an infiltration rate of 1000 ft^3 per hour (0.01 exchanges per h), concentrations up to 3000 ppm can be maintained with a combustion rate of natural gas of 18 ft^3 per hour per 100000 ft^3 greenhouse volume. This is approximately equivalent to 8300 ft^2 of greenhouse area with conventional structures.

One of the products of combustion, of course, is water. As this is vapor, humidity can be raised significantly, with possible undesirable effects unless compensated by dehumidification cycles. In dry climatic regions, with low outside dew points, infiltration will usually be sufficient to compensate for the added water. Data obtained by Cobb (1973, unpublished) suggests this is the reason that mildew on roses in the winter is a minor problem. Shutting off the CO_2 burner and mist system results in a rapid dew point temperature drop so that even though temperature may also drop, relative humidity does not increase significantly. However, in tightly closed, plastic structures the additional water from combustion can result in a severe dripping problem with the possibility of increased disease.

8.2 Pollution

Pollution is contamination of the environment with materials which reduce crop value. Commonly, damage results from airborne substances but, may also be from waterborne materials, or due to the introduction of a phytotoxic compound at the wrong time and place. Far too often, the latter happens from carelessness and inattention to common safety measures. Pollution can be considered the result of our present technological society which manufactures, deliberately or inadvertantly, numerous exotic substances that have potential for damage to plants, humans, and the latters' valued objects.

8.2.1 Air Pollution

Damage resulting from air pollutants may be easily observed (acute) shortly after an episode, or may cause growth reductions of various degrees which can be difficult to discern (chronic). Two or more pollutants can interact so as to cause damage at levels lower than would be caused by either one alone (synergism). The particular effects of pollution will be a function of: (1) species and variety; (2) moisture; (3) light; (4) CO_2 level; (5) temperature; (6) nutritional status; (7) stage of growth; and (8) previous history. As a general rule, the greater the vigor of the plant, the more likely that a given pollution dosage will cause injury. There have been numerous studies indicating that manipulation of the environment to harden the plant, to reduce growth, or in some manner close the stomates, will reduce injury from air pollutants. This results from the fact that to produce damage, most pollutants must enter the leaf through the stomates. Any-

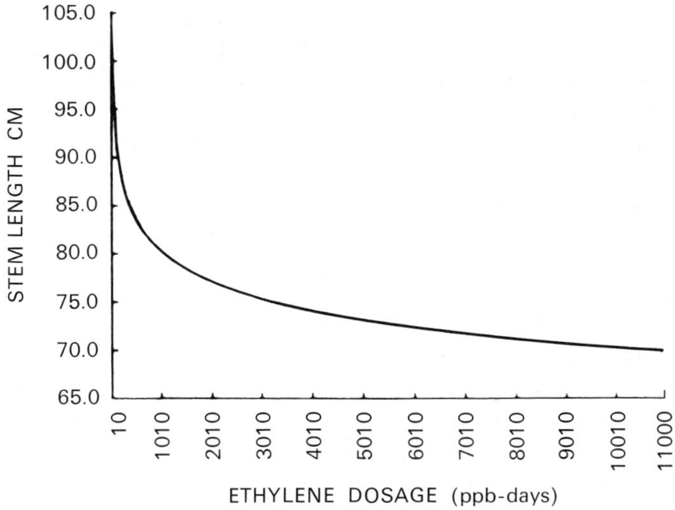

Fig. 8-7. Relationship between ethylene dosage and stem length in carnation (Piersol and Hanan, 1976)

thing that increases resistance to penetration will require higher dosage values to effect injury.

Degree of injury, under any given set of conditions, will usually be directly related to dosage, or the concentration of the material and the time the material is present. This was shown most clearly by Barden and Hanan (1972) and Piersol and Hanan (1976), who showed a direct relationship between keeping life and stem length of carnations and dosage (Fig. 8-7). The keeping life of carnations could be reduced 100% at 70° F, with a dosage of 4400 ppb-h. This dosage can be obtained with any concentration of ethylene and time such as 44 ppb for 100 h.

Problems resulting from air pollution generally arise in urban regions, or close to industrial activities which emit considerable phytotoxic material into the air. It is not unusual to find the degree of pollution directly related to local meteorological and topographical conditions. The prevalence of damaging levels may be downwind of the source, a function of temperature variation with altitude, or may be found to travel up and down a river valley which passes through an urban district. It is understandable that such factors must be taken into account when locating a greenhouse, otherwise, a grower may find himself subject to varying degrees of chronic and acute pollution damage.

For example, temperature usually decreases with height above the ground. Under such conditions, the atmosphere is unstable and pollutant concentration will be reduced by rapid mixing with higher air. Under conditions of rapid heat loss to clear skies, however, the colder ground causes an inversion—or, the temperature increases with altitude. Depending upon the height of the inversion, any pollutants introduced into the air will be confined to a limited air volume. If this persists, pollutant concentration can increase to damaging levels as occurs in the Los Angeles basin or under stagnating high pressure systems. The level of some

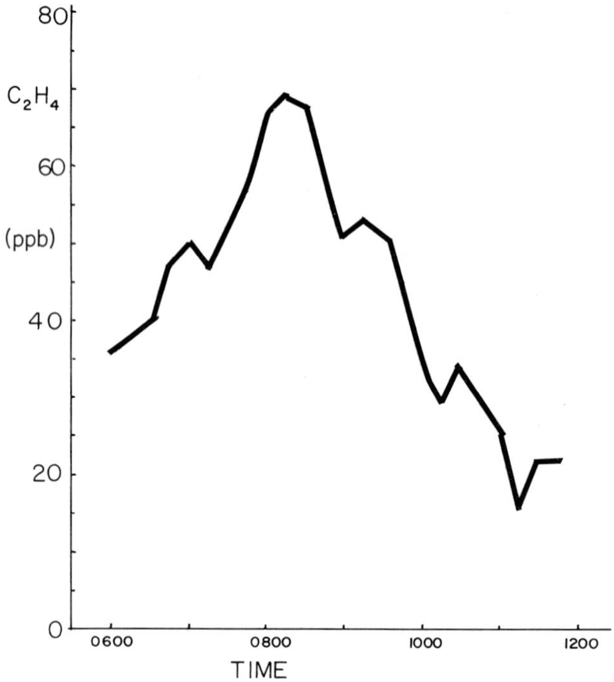

Fig. 8-8. Average variation of ethylene during morning hours close to major highway interchange in Denver, Colorado (Hanan, 1972)

pollutants will show varying concentrations, depending upon human activities such as automobile traffic (Fig. 8-8), or the fact that on weekends most industrial activities are shut down. In the Denver region, it is not unusual to observe movement of pollution down the South Platte river valley in the mornings, and its return upriver during the afternoon and late evening hours.

Almost any material, if in sufficient amount, will cause damage. However, the common air pollutants are generally limited to ozone, peroxyacetylnitrate (PAN) and related oxidants, sulfur dioxide, nitrous oxides such as nitrogen dioxide, hydrogen fluoride, ethylene, and miscellaneous compounds such as ammonia, chlorine, and mercury vapor. There are other substances that can be considered as air pollutants; such as fly ash from coal burning plants which may seriously reduce light transfer of a greenhouse, vapor clouds on cold mornings from generating stations, and dust and dirt from gravel roads, gravel pits, and other mining operations. All of the common pollutants can be detected and measured with suitable equipment. However, this requires sophisticated equipment not available to most growers. The method first employed to detect pollution damage was observation of biological material. This still remains the best test for a grower to employ. That is to grow a few sensitive species in the greenhouse (Tables 8-6 and 8-7). Acute injury symptoms are usually characteristic for a given pollutant. But, they can be easily confused with symptoms resulting from disease or nutritional disorders. Some experience is required, as well as knowledge of general conditions and plant growth, before an individual becomes proficient assessing injury resulting from air pollution.

Table 8-6. List of plant species known to be sensitive to various air pollutants. (After Rogers, 1976)

Ozone	PAN	Sulfur dioxide	Nitrogen dioxide	Hydrogen fluoride	Ethylene
Ageratum	African Violet	Aster	Azalea	Canna	Calceolaria
Endive	Aster	Begonia	Dutch Iris	Cyclamen	Carnation
Aster	Carnation	Broccoli	Lettuce	Gladiolus	Cucumber
Dill	Celery	Brussel Sprout	Petunia	Hemerocallis	Impatiens
Begonia	Dahlia	Canna	Tomato	Hyacinth	Marigold
Fuchsia	Endive	Centaurea	Hibiscus	Iris	Narcissus
Broccoli	Fuchsia	China Aster		Jerusalem Cherry	Nasturtium
Geranium	Impatiens	Chrysanthemum		Sweet Potato	Orchids
Brussels Sprout	Lettuce	(some vars)		Tulip	Philodendron
Gourds	Muskmelon	Coleus		Strawberry	Rose
Carnation	Pepper	Cosmos		Azalea	Snapdragon
Lettuce	Petunia	Violet		Pear	Stock
Celery	Primrose	Endive		Maple	Sunflower
Mallow	Rose	Four O'Clock			Sweet Pea
Chrysanthemum	Tomato	Geranium			Sweet Potato
(some vars.)	Salvia	Lettuce			Tomato
Strawberry	Snapdragon	Mallow			Tulip
Dahlia	Sunflower	Marigold			Marigold
Marigold	Sweet Basil	Morning Glory			
Lilac	Bean	Okra			
Muskmelon	*Poa annua*	Pepper			
Onion	(grass)	Barley			
Palm, Kentia		Oat			
Pansy		Tomato			
Petunia		Zinnia			
Pumpkin		Poppy			
Salvia					
Sweet Basil					
Sweet Potato					
Tobacco					
Spinach					
Radish					

8.2.1.1 Ozone

Ozone is a constituent of the atmosphere at high altitudes. However, nitrogen oxides emitted into the atmosphere at ground level from any combustion at high temperature (automobiles, boilers, etc.) will react in the presence of sunlight to form ozone (O_3). As motor vehicles are a major source of hydrocarbons and nitrogen oxides, relatively high O_3 levels can build up in sunny climates. Ozone is an extremely active material, attacking rubber, wood, plastics, and numerous other substances, and causing rapid degradation. Naturally occurring concentrations range from zero to 500 ppb, with sensitive varieties being injured with dosages of 50 ppb for 8 h. Injury has been reported from most of the larger metropolitan areas of the United States. Ozone probably causes more injury than any other pollutant in the United States, according to Taylor (1973).

Table 8-7. Summary of effects of air pollutants on greenhouse plants[a]. Indicator plants are italics

Pollutant	Sensitive species	Dosage level	General symptoms of acute injury
Ozone	*Petunia*, carnation, chrysanthemum, lilac, *spinach*, radish, *tomato*	1 ppm, 2 h 30 ppb, 4 h	Recently mature leaves most susceptible. Usually develops at leaf tip. Upper surface discoloration with waxy appearance, necrotic lesions, turning white or brown upon drying, "stipple-like" lesions on upper leaf surface, red, purple, black, or brown
Sulfur dioxide	Aster, zinnia, oat, verbena, violet, barley, *tomato*	1 ppm, 1 h 300 ppb, 8 h	Dark, green, water-soaked appearance, lesions dry to a white ivory color, then brown, red, or black, between veins, may have marginal necrosis, chlorosis, and yellowing
Fluoride	*Gladiolus*, tulip, pear, maple	3–4 ppb, 1 h 0.1 ppb, 5 wks	Usually evident at leaf tip first. On narrow leaves may have a dark-colored band separating living and dead tissue, marginal burn, leaf abscission. Necrotic areas turning brown or light tan, may drop out
PAN	Dahlia, petunia, bean, aster, *tomato*, lettuce, *poa annua*	100 ppb, 2 h 10 ppb, 6 h	Bronze or glazed appearance on lower surface of young leaves, abnormal leaf growth, leaves cupping downward and distorted, collapsed tissue near tip of upper leaves. Necrotic areas white to tan-colored
Nitrogen dioxide	Azalea, hibiscus, *petunia*, *tomato*	2.5–6 ppm, 2 h 2.5 ppm, 4 h	Irregular white or brown lesions on middle aged leaves, between veins, leaf abscission. Waxy or glazed appearance on occasion
Ethylene	*Tomato*, carnation orchid, *marigold*	100 ppb, 6 h 5 ppb, 24 h	Hastened senescence of flowers, "sleepiness," petal drop, internode shortening, bud abortion, leaf abscission, epinasty or abnormal growth
Ammonia	*Petunia*, sunflower	17 ppm, 4 h	Whitish discoloration both sides of leaf tissue. Collapse, dark spotting along margins, purple streaking, cooked, green appearance, leaf abscission
Chlorine	Rose, tulip, *zinnia*, maple, begonia, cucumber	100–800 ppb, 4 h 100 ppb, 2 h	Similar to injury from sulfur dioxide, or PAN, white to tan-colored lesions between veins, leaf abscission

[a] Sources: Hindawi (1970); U.S.D.H.E.W. (1970); Hand (1972); Taylor (1973); Rogers (1976).

Recently expanded tissue is most susceptible, whereas very young or old, mature leaves are resistant. The most common symptom is small, light-colored spots on the upper leaf surfaces (Fig. 8-9). Acute injury usually develops at the tip of the injured leaf. The injury might be confused with a massive red spider attack. The upper surface may also appear waxy, with high dosages causing necrotic lesions, the tissue turning white or various shades of brown. On small grains and grasses, white necrotic tissue can develop between the veins, extending across the

Fig. 8-9. Ozone damage on poinsettia; note flecking damage on lower leaves. (Courtesy of O. C. Taylor, Univ. California, Riverside)

larger veins in severe injury, with heaviest injury where the leaf blade bends downward. There may be leaf abscission in some crops.

Long, continued exposure to low dosages can reduce growth, with the development of chlorotic symptoms indistinguishable from senescence or nutritional deficiencies. Ozone may combine with sulfur dioxide to produce injury at levels lower than either alone. Symptoms are similar to ozone injury with upper surface flecking, stipple, necrosis extending entirely through the leaf (bifacial necrosis), and lower surface glazing.

8.2.1.2 PAN

Peroxyacetyl nitrate is one of several compounds associated with smog, and identified initially in California by Middleton et al. (1950). Oxidants such as PAN are secondary compounds from pollutant substances already present, and result from chemical reactions between nitrogen dioxide and simple hydrocarbons in sunlight. Many succulent ornamentals may be injured in a two-hour exposure to 10–20 ppb. Symptoms usually appear on lower surfaces of recently expanded leaves as a silvering, glazing, or bronzing (Fig. 8-10). Very high dosages will cause necrotic lesions on both sides of the leaf, with occasional white to tan-colored banding. The latter results from the fact that maturity in a leaf starts first at its tip and proceeds to the base. So the banding position on the leaf will depend upon stage of maturity at the time of dosage. A pinched or distorted leaf margin may result, with repeated dosages causing hastened senescence, chlorosis, and leaf abscission.

Fig. 8-10. PAN damage on petunia. *Top row:* plants exposed to high light intensity; *middle row:* plants exposed to medium light intensity; *bottom row:* control group. Both upper and middle rows exposed to 50 ppb PAN. (Courtesy of O. C. Taylor, Univ. California, Riverside)

Fig. 8-11. SO_2 injury to poinsettia. (Photo courtesy of H. E. Heggestad, USDA, Beltsville)

8.2.1.3 Sulfur Dioxide

Injury from sulfur dioxide is generally restricted to locations where industrial activity is high, and the sources are burning fuels with high sulfur content, or producing and refining sulfur-containing lead, copper, zinc, and nickel ores. Power generation by coal burning plants is probably the most important single source. Acute necrosis results from absorption of SO_2, reacting with water in the leaf, and producing a sulfite ion which is highly phytotoxic. On broadleaved plants, the major symptom is necrotic spots between the major veins, drying to a white or ivory color (Fig. 8-11). On grasses, injury is usually confined to areas between the veins, giving a streaked effect. The initial injury is usually a grey-green discoloration. Chlorosis can develop with leaf abscission, particularly if the dosage is sublethal. Very low levels of SO_2 can provide a significant portion of the plant's sulfur needs. Generally, sensitive plants will be injured, if they are growing rapidly, after exposure to 480 ppb for 4 h or 280 ppb for 24 h.

8.2.1.4 Nitrogen Dioxide

Injury from this pollutant is not common as atmospheric concentrations seldom reach sufficiently high levels, unless there is an accidental release of NO_2 or NO. Leaf absorption is rapid, with susceptible areas on recently matured or rapidly expanding leaves being injured first. Irregular areas between the veins collapse, die (necrosis), and eventually turn white or light tan-colored. There have been indications of a synergistic effect between NO_2 and SO_2, and considerable growth reduction without acute damage. Susceptibility of leaf tissue can be increased under conditions of low light or total darkness.

8.2.1.5 Ethylene

Ethylene pollution is considered a minor problem by some. But, it is very common for greenhouse operators to pollute their growing area by inattention to combustion within the greenhouse. Ethylene is a growth regulator, produced by any combustion process from tobacco smoking to burning trash. It is found in high concentration in gasoline engine exhausts, and internal combustion engines should not be operated in closed areas where plant material is stored. Ethylene is often produced endogenously by plants, some fungi, and by diseased plants. It is not unusual to find carnations "putting themselves to sleep" when enclosed at

Fig. 8-12. Epinastic response of tomatoes to ethylene. (Photo courtesy of Marlin Rogers, Univ. Mo., Columbia)

Fig. 8-13 A and B. Effect of ethylene on carnations. (A) Exposure of cut flowers to ethylene hastens senescence, reducing keeping life. Control is on the left (Barden, 1972). (B) Effect of continuous dosages on growth of carnations. Note that flowering is hastened in addition to drastic internode shortening (Piersol and Hanan, 1976)

high temperatures in an unventilated container. The common sources in greenhouses are from improperly adjusted, dirty, heating devices burning natural gas or petroleum fuels. Comments on CO_2 production by combustion were given in Section 8.1.

Symptom expression is highly variable. The injury can mimic growth regulator problems, with epinasty (downward bending of petioles) (Fig. 8-12), hastened senescence, leaf and flower abscission, and internode shortening (Fig. 8-13). Snapdragons, calceolaria, and larkspur will drop their florets rapidly. Developing rose buds will abort, leaf abscission will occur, and there will be adventitous sprouts

Fig. 8-14 A–D. Effect of ethylene on roses. (A) Control; (B) leaves initially turn bright yellow between veins, starting at base; (C) and (D) later nearly complete leaf abscission; bud abortion; new breaks beginning near the base (Piersol, 1974)

at the base of the rose plant (Fig. 8-14). Orchids will show sepal injury at 100 ppb for 8 h. Cut flowers should never be stored with fruits or vegetables as some of these will produce sufficient ethylene to cause senescence and leaf abscission. Keeping a bushel of apples in storage was an early practice to defoliate hydrangeas prior to forcing. Flower initiation can be delayed on chrysanthemums. Tomatoes are good indicator plants to show if high levels have occurred. Injury is often associated with severe cold spells when normal infiltration is prevented by the freezing of cracks in the greenhouse structure. Chronic injury in metropolitan regions is common, and the grower may not be aware that it is occurring.

8.2.1.6 Hydrogen Fluoride

Fluoride damage is generally restricted to areas near manufacturing plants such as those refining aluminum ores, phosphate fertilizer plants, brick plants, and steel mills. The problem has been observed in Western Europe for well over a century. Recent problems have resulted from rocket fuel combustion, and some foliage plants are sensitive to fluorine contained in phosphate fertilizers. In leaves with parallel veining, damage first appears at the tip, or at the leaf margin in broadleafed plants. The cells collapse and die, turning a reddish brown or tan color. The line between damaged and healthy tissue is sharp, often set off by a darker brown-red band. On woody plant leaves, the necrotic tissue may drop out, leaving a ragged hole. Sensitive plants, such as gladioli, can be affected at concentrations of 0.1 ppb for five weeks. High concentrations are considered in excess of 3–4 ppb. Young, rapidly growing leaves are most susceptible, with translocation and accumulation causing a buildup to toxic levels.

8.2.1.7 Miscellaneous Pollutants

Damage from ammonia and chlorine usually coincides with an inadvertant, massive release to the atmosphere. Leaves will usually have a cooked appearance, with dark brown, reddish lesions (Fig. 8-15). Trees and shrubs will often be defoliated, grass will turn brown. If the season is early, adventitious buds will allow new leaves to appear. Tan and white colored spots, similar to ozone injury, may appear on vegetables such as cabbage or broccoli.

Fig. 8-15. Chlorine damage on apple leaves. Caused a rapid defoliation of affected trees, followed by regrowth of adventitous buds

Mercury vapor damage is no longer common. The source is usually localized as may occur following mercuric chloride drenches for earthworm control, breakage of mercury containing thermometers, or the use of paint containing mercury to control fungal and algal growth. If such paint is inadvertantly used, the vapor may be trapped by painting the surface with a lime-sulfur mixture (Dimond and Stoddard, 1955). Roses are particularly susceptible, with bleaching of the petals, necrosis of the neck (base or receptacle part of the flower), and dieback of young succulent shoots.

8.2.1.8 Summary

With some exceptions, the grower is limited in preventing pollution to careful site selection. Greenhouses should be located upwind of manufacturing sites or urban regions, and in areas where there is good air drainage. Some time spent with meteorological records can provide information on inversion frequencies and mixing heights, average wind velocities and directions. Local residents may provide information on weather peculiar to the immediate area. Attention should be given to possibility of manufacturing concerns building in the vicinity, possibilities of major highways and heavy traffic, or dust producing operations such as gravel roads and mining. Zoning and code restrictions against dust-producing development, burning, trash disposal areas, etc., are as important to greenhouses as to residences. Much can be done by carrying out regular maintenance on combustion equipment, assuring adequate make-up air to boilers and heaters in all climatic conditions, and that no substance is introduced whose volatile vapors can be toxic to plants intended for economic return.

Varieties and cultivars within a particular species have shown resistance to air pollution. Work by Engle and Gabelman (1966), Cathey and Heggestad (1972, 1973), and others have shown that resistance can be introduced by breeding. Rogers (1976) suggested that increased effort on selection for resistance might be useful. Chemical sprays have provided limited protection, and some chemical growth retardants appear to provide protection from damaging pollutant levels. Filtration of all incoming air is possible, but has not been shown economically feasible for commercial ranges. Activated (brominated) charcoal and potassium permanganate on a suitable support are effective in removing various pollutants such as ethylene.

8.2.2 Inadvertant Pollution

"Inadvertant pollution" refers to the unintended introduction of a material that causes plant damage. This may be via the water supply, in the soil, from spray drift, or in the fertilizer. Damage can be from a variety of chemicals as may occur from improper use of pesticides, soil sterilants, fertilizers, or as shown in Figure 8-16, misjudgement with growth regulators. But, these are usually considered as normally present and necessary in greenhouse culture.

Fig. 8-16. Improper application of the growth regulator B-9 to seedlings

8.2.2.1 Sources and Examples

With present production of numerous esoteric compounds, the greenhouse operator must be on constant guard to ensure his fertilizer, water, and soil supply. The most common injury symptom is referred to as "growth-regulator" injury caused by several types of herbicides. This is a hormone-like distortion, including epinasty, leading to a "fern-leaf" shape and pear shaped fruit in tomato. In lettuce seedlings and other crops such as poinsettias, chrysanthemums, etc., there is a characteristic cupping of the leaves with veins tending to be parallel. Adventitious, above ground, roots may form, there may be increased branching and roughening of the leaves. Usually, only actively growing plant parts are affected. Chemicals causing these symptoms are typified by 2,4-D, 2,4,5-T, MCPB (2-methyl-4-chlorophenobutryic acid) and TBA (2,3,6-trichlorobenzoic acid). Most break down fairly rapidly in soil or plant, but 2,4,5-T, and TBA particularly, may persist for several months.

Circular brown spots on a leaf, with brown scorched areas tending to run down the leaf suggest Paraquat spray drift. Aminotriazole (combination of simazine and amitrole) result in white or pinkish white colors on leaves, followed by collapse of the foliage. Simazine causes yellowing and stunting, followed by leaf scorch and collapse. The uracil and urea herbicide groups cause symptoms similar to simazine and other triazines, but the yellowing tends to follow the veins throughout the whole leaf. Soil analysis is possible for this group as well as most triazines.

The source of any material causing injury may be surprising. Figure 8-17 shows an example of growth retardation from a boiler chemical. In this instance, the grower had used old boiler pipe in his water distribution system. During periods of no flow, sufficient chemical was released into the water to cause the damage shown. There have been several instances of the water supply being contaminated with herbicides or miscellaneous hydrocarbons. This may be as the result of runoff from fields sprayed with herbicides, chemicals used to kill water

Fig. 8-17. Damage to petunias from the use of old boiler pipe in the water supply system

weeds, discharge from manufacturing plants or leakage from storage tanks, the effulent of which somehow gets into the water. It may be necessary to shift water supplies by drilling into a new aquifer, or using domestic water supplies.

Weed killer in lawn fertilizers is a common practice. Occasionally there have been instances where machinery is inadequately cleaned, or some basic material such as phosphoric acid is recycled without adequate processing. The new fertilizer is contaminated with an herbicide with the results shown in Figure 8-18. In the upper example of Figure 8-18, chemical analysis showed 10.9 ppm 2,4-D and traces of Tordon and 2,4,5-T. Figure 8-19 is another example of a persistant herbicide in the soil. A grower should always know the history of his potting and bench soils. Agricultural operations often employ chemicals that are disaster in the greenhouse. Weed spraying by regulatory agencies to control noxious weeds should always make soils close to roads, railway embankments, etc. suspect. Soils for future use, and soil amendments, should never be placed on ground which has been treated with a persistant chemical. Borate-chlorate-bromacil mixtures are examples of persistant herbicides. Spray drift is a very common source of injury, particularly if the applicator used equipment giving a fine spray which is also volatile. Growers should be alert to spraying by homeowners, especially if the spray can be sucked into air intakes. Still another inadvertant source is wood preservative. There is, at present, only one wood preservative that has been shown to be reasonably safe under all greenhouse conditions—copper naphthenate. Materials such as pentachlorophenol and creosote should never be employed as preservatives in greenhouse construction. Their volatile components, depending upon greenhouse tightness, will cause moderate to severe growth retardation and acute injury.

Fig. 8-18. (A) Herbicidal damage to petunia and (B) pepper from traces of various herbicides in fertilizer

Fig. 8-19. Herbicidal damage resulting from persistant chemical in soil

8.2.2.2 Use of Herbicides

Pulling weeds is costly, and a good herbicide sometimes offers a less expensive means for weed control. Preventing weeds inside and outside the greenhouse is a basic sanitation routine for disease and insect control, as well as eliminating competition with the crop and removing unsightly appearances. However, there are a few simple rules that should be followed:

Fig. 8-20. 2,4-D damage to poinsettia. Caused by allowing the chemical to spill on potting bench, from unmarked container

1. Never use an herbicide unless absolutely necessary.
2. Never store herbicides in the same area with fertilizers, pesticides, and other chemicals; or in the same area with potting and bench soils.
3. Never use the same equipment to apply herbicides and pesticides. Apparatus for pesticides should be well marked and kept separate.
4. Never store in an unmarked container (see Fig. 8-20).
5. Read directions on the label. Follow directions with careful regard to mammalian toxicity, proper clothing, and container disposal (see Chap. 9).
6. Certain groups of herbicides should never be used in or near greenhouses. Among these are 2,4-D, 2,4,5-T, and other phenoxy compounds.

Table 8-8 provides a short list of herbicides that have been used successfully in greenhouse culture. The initials *a.c.* or *a.i.* refer to "active chemicals" or "active ingredient" respectively. *WP* is "wettable powder". Recommendations are generally based upon the percent *WP* or *a.c.* to be applied per acre. The actual amount to be utilized for smaller areas can be calculated with the following formula:

$$(\text{pounds a.c. per acre}) \times \frac{\text{ft}^2 \text{ to be treated}}{43{,}560\ \text{ft}^2/\text{acre}} \times \frac{100}{\text{A}}.$$

The pounds a.c. per acre is the application rate, and A is the percentage of the active ingredient in a granular form, pounds of active ingredient per gallon, or percentage in a wettable powder. Where given in percent per gallon, 100 is changed to 1.

Granular materials, or pellets, although usually more expensive, are cheaper to apply. There is no hazard from drift. However, one must nearly always incorporate the pellets into the soil. Herbicides for sprays may be supplied as water soluble concentrates, oil soluble concentrates, emulsifiable concentrates, or as powders or pastes. Wettable powders generally require a surfactant (wetting agent) to keep them in solution. Chapter 9 provides more detail on spraying procedures. When using an herbicide as a spray, every precaution should be taken to prevent spray drift. This most often means low pressure and a coarse spray.

Table 8-8. Herbicides suggested for greenhouse use[a]

Product	Rates	Remarks
TOK WP-50 (nitrogen)	4–12 lb WP/acre	Rates for outdoors on ornamentals. Has been used successfully in greenhouses around base of rose plants. Controls annual bluegrass, crabgrass, nettle, pigweed, purslane. Kills top growth on bind weed (morning glory)
Roundup (glyphosate)	4 qt/100 gal.. 34 ml/gal.	Normally employed on noncropland to control perennial weeds. Reported as used successfully in closed greenhouses on noncrop areas. Best action at temperatures above 70° F, on rapidly growing weeds. Slow acting
Paraquat	1–2 qt/acre 1 oz/3 gal/1000 ft^2	Keep spray off of new foliage, young bark and new breaks. Has no residual action, fumes extremely toxic. Use with 1/10 oz X77 wetting agent. Kills green foliage on contact
Diquat	1.5 lb/acre a.c. 1 oz/3 gal/1000 ft^2	Similar to paraquat, use wetting agent, keep spray off of foliage unless a kill is desired. Control for broadleaf and grass weeds
Karmex 80W (diuron)	0.8–1.6 lb/acre a.c. 2 oz/3 gal/1000 ft^2	Kills annual weeds and grasses, increase to 51 lbs a.c. per acre for noncropland. Used at lower rates as a preemergence spray, but do not use on light soils. In some areas, recommended not to replant within one year after treatment
Princep (simazine)	2–4 lb/acre a.c. 4 oz/3 gal/1000 ft^2	Control annual broadleaf and grass weeds. Do not use in seedbeds or cutting beds. Do not apply to young transplants. Increase to 10 lbs a.c. per acre on noncropland outside the greenhouse
Amizine (amitrole + simazine)	7 lb/100 gal/acre 8 oz/3 gal/1000 ft^2	Controls annual, perennial and grass weeds. Used mostly on woody ornamentals, using a directed spray to avoid contact with stems and foliage. On noncropland will provide a residue for one year
Treflan (trifluralin)	0.5–1 lb/acre a.c.	Controls annual grasses and some broadleafed weeds. Must be incorporated in the soil. Several bedding plants resistant to the material. Will not control existing weeds. Wait a few days before planting. Rate varies with soil type
Telvar 80W (monuron)	4.48 lb/acre a.c.	Noncropland only. Controls most annual and perennial weeds at higher concentrations. Can be used in the greenhouse under benches
Amphyl (disinfectant)	60 ml/gal	Has been sprayed on carnations without injury to kill top growth on oxalis. Will kill top growth on most broadleaf annual weeds. Will reduce breaking on bottom canes of roses if repeatedly sprayed

[a] Sources of information: Bing (1969); Holley (1974); Haramaki (1976); Weed Control Manual (1976).

References

Agri-Fieldman: Weed Control Manual. Willoughby, Ohio: Meister 1976, 190 pp.

Aikin, W.J., Hanan, J.J.: Photosynthesis in the rose; effect of light intensity, water potential, and leaf age. J. Am. Soc. Hort. Sci. **100**, 551–553 (1975)

Barden, L.E.: Effect of ethylene on carnation keeping life. M.S. Thesis. Colo. State Univ., Ft. Collins 1972, 62 pp.

Barden, L.E., Hanan, J.J.: Effect of ethylene on carnation keeping life. J. Am. Soc. Hort. Sci. **77**, 785–788 (1972)

Bing, A.: Weed control. In: Roses. Mastalerz, J., Langhans, R.W. (eds.). Penn Flower Growers, N.Y. State Flower Growers Assoc., Inc. and Roses, Inc. 1969, 331 pp.

Caldwell, M.M.: Carbon dioxide evolution from organic mulches. Colo. Flower Growers' Assoc. Bull. **161**, 1–4 (1963)

Cathey, H.M., Heggestad, H.E.: Reduction of ozone damage to Petunia hybrida Vilm. by use of growth regulating chemicals and tolerant cultivars. J. Am. Soc. Hort. Sci. **97**, 695–700 (1972)

Cathey, H.M., Heggestad, H.E.: Ozone sensitivity and petunia modification by chemicals, Part I. Florists Rev. **152** (3952), 20–21, 39–41 (1973). Part II. Florists Rev. **153**, (3954), 20–21 (1973)

Dimond, A.E., Stoddard, E.M.: Toxicity to greenhouse roses from paints containing mercury fungicides. Conn. Agr. Sta. Bull. **595**, 1–19 (1955)

Dunham, R.S.: Herbicide Manual. ARS, USDA, and U.S. Navy Agr. Handbook 269, Washington, D.C., 1965, 90 pp.

Engle, R.L., Gabelman, W.H.: Inheritance and mechanism of resistance to ozone damage in onion, Allium cepa L. Proc. Am. Soc. Hort. Sci. **89**, 423–430 (1966)

Gaastra, P.: Photosynthesis of crop plants as influenced by light, carbon dioxide, temperature, and stomatal diffusion resistance. Meded. Landbouw. Wageningen. **59**, 1–68 (1959)

Gaastra, P.: Climatic control of photosynthesis and respiration. In: Environmental Control of Plant Growth. Evans, L.T. (ed.). New York: Academic Press 1963

Goldsberry, K.L.: The effects of carbon dioxide on carnation growth. M.S. Thesis, Colo. State Univ., Ft. Collins 1961, 56 pp.

Goldsberry, K.L., Holley, W.D.: Carbon dioxide research on roses at Colorado State University. Colo. Flower Growers' Assoc. Bull. **151**, 1–6 (1962)

Hanan, J.J.: Estimated ethylene levels in the North-Denver-South Platte River valley. Colo. Flower Growers' Assoc. Bull. **264**, 1–4 (1972)

Hanan, J.J.: Ventilation temperatures, CO_2 levels and rose production. Colo. Flower Growers' Assoc. Bull. **279**, 1–5 (1973)

Hanan, J.J.: Use of natural gas for CO_2 production in greenhouses. Colo. Flower Growers' Assoc. Bull. **262**, 1–5 (1972)

Hand, D.W.: Air pollution in glasshouses and the effects of aerial pollutants on crops. Sci. Hort. **24**, 142–152 (1972)

Haramaki, C.: Weeds. In: Bedding Plants. Mastalerz, J.W. (ed.). Penn. State Flower Growers, University Park, PA. 1976, 516 pp.

Hindawi, I.J.: Air pollution injury to vegetation. U.S.H.E.W. U.S. Govt. Printing Off., Washington, D.C. 1970, 44 pp.

Holley, W.D.: Type of greenhouse covering may affect CO_2 utilization by carnations. Colo. Flower Growers' Assoc. Bull. **172**, 1–3 (1964)

Holley, W.D.: Greenhouse herbicides. Colo. Flower Growers' Assoc. Bull. **291**, 4 (1974)

Holley, W.D., Goldsberry, K.L.: CO_2 and temperature recommendations. Colo. Flower Growers' Assoc. Bull. **164**, 3 (1963)

Holley, W.D., Juengling, C.: Effects of additional CO_2 and automated day temperatures on carnations. Colo. Flower Growers' Assoc. Bull. **162**, 1–4 (1963)

Holley, W.D., Korns, C.H., Goldsberry, K.L.: The use of carbon dioxide on carnations. Colo. Flower Growers' Assoc. Bull. **149**, 1–4 (1962)

Hoover, W.H., Johnston, E.S., Brackett, F.S.: Carbon dioxide assimilation in a higher plant. Smithsonian Misc. Coll. **87**, 1–19 (1933)

Lemon, E.R.: Aerial environment and crop yield. Australian Inst. Agr. Sci. **29**, 198 (1973)
Mathis, V.: The effect of high day temperatures on new plantings of Forever Yours, Love Affair, and Cara Mia Roses. Colo. Flower Growers' Assoc. Bull. **270**, 1–3 (1972)
Middleton, J.T., Kendrick, J.B., Schwolm, H.W.: Injury to herbaceous plants by smog or air pollutants. Plant. Disease Reptr. **34**, 245–252 (1950)
Pallas, J.E.: Transpiration and stomatal opening with changes in carbon dioxide content of the air. Science **147**, 171–173 (1965)
Piersol, J.R.: Effect of ethylene on rose growth. Colo. Flower Growers' Assoc. Bull. **286**, 5–6 (1974)
Piersol, J., Hanan, J.J.: Effect of ethylene on carnation growth. J. Am. Soc. Hort. Sci. **100**, 679–681 (1976)
Rogers, M.: Air pollution. In: Bedding Plants. Mastalerz, J.W. (ed.). Penn State Flower Growers, University Park, PA, 1976, 516 pp.
Taylor, O.C.: Acute responses of plants to aerial pollutants. In: Air Pollution Damage to Vegetation. Naegele, J.A. (ed.). Am. Chem. Soc., Washington, D.C. Adv. in Chem. Series 122, 1973, 137 pp.
Thompson, C.J., Hanan, J.J.: Effect of CO_2 concentrations on roses: I. CO_2 uptake by individual leaves. Colo. Flower Growers' Assoc. Bull. **306**, 1–4 (1975)
U.S.D.H.E.W.: Air quality criteria for photochemical oxidants. U.S.D.H.E.W. Nat. Air Pollution Control Adm. Publ. No. AP-63. U.S. Gov. Printing Off. Washington, D.C., 1970
Weed Control Manual. Meister Publishing Co., Willoughby, Ohio 190 pp., 1976

Chapter 9 Insect and Disease Control

9.1 Introduction

All organisms compete with each other for available resources. Under normal conditions, several organisms may occupy the same space in such a way as to afford survival without any one dominating. Generally, the numbers will be several, and the tendency for any one to dominate the biological scene is restricted by its predators or availability of food. The vigor of a population is somewhat less than the genetic potential.

This changes dramatically when man cultivates a species for its usefulness to him. Any organism that tends to reduce that usefulness is in direct competition. Where a minimum return is required before there is profit, competition "skims off the cream". The act of introducing an organism to the exclusion of all others provides conditions where failure to control a competitor can result in serious loss. The cultivated species offers an unlimited food supply and the competitor itself may have no competition.

In this chapter, competitors are those that reduce economic return to the greenhouse grower. Prevention and control assume a significant cost in production. Competitors include diseases caused by bacteria, fungi, and viruses, nematode attack, and damage from several pests such as insects and mites.

9.2 The Greenhouse Climate

To understand problems and limitations to disease and pest control operations, it is important to appreciate conditions under which such procedures are applied if they are to be carried out effectively. In an ecological sense, the environment is quite abnormal as the result of the dense, genetically uniform crops that are grown.

Hussey et al. (1969) provided an excellent discussion of the usual greenhouse environment, distinguishing it from field culture. The environment of greenhouses can be characterized as warm and humid with low air velocities. Temperature variations are slight, relative humidity is seldom less that 60%, and natural enemies, unless introduced, are rarely present. Growth of the crop is rapid and luxuriant. There are long periods in the greenhouse during which the predator can breed and reproduce. Fewer factors, such as extreme temperatures, low humidity, high wind velocities, and natural enemies, restrict predator growth. Control measures such as soil sterilization or pasteurization can further reduce competition in root substrates; control measures may exert powerful selective pressure on the competitive organism. Resistance buildup of mites to acaricides, or of houseflies to DDT, are two examples of selective pressure and the result. Occasionally, the greenhouse may be completely emptied so that neither host or natural enemy can survive.

A complicating factor is the high commercial value of most greenhouse crops, so that the slightest blemish can reduce market value. The consumer is unlikely to return if a purchased plant is damaged, infested with whitefly, or covered with an unsightly spray residue. Healthy plants are usually able to withstand pest attack better, although the rate of predator reproduction is often greater. According to Hussey et al. (1969) there is no evidence that "natural" manures or fertilizers differ in any way from "artificials" as to their effect on plant susceptibility. They can, in fact, introduce pests such as nematodes and weeds.

9.3 Major Diseases and Pests

We do not attempt to describe fully all diseases and pests that can be found in greenhouses, or to describe all possible control measures. The tables are not meant to be complete and we recommend to you several excellent manuals that may be obtained for nominal cost from research stations throughout the world.

In general, we have grouped major predators into broad classifications as indicated in Tables 9-1 through 9-3, and provide some general rules a grower may use to control them. Most manuals describe major diseases and pests for each crop, whereas we have tried to generalize.

Predators show amazing specificity. The fungus *Fusarium* causes several important diseases on greenhouse plants, but *Fusarium roseum f. dianthi* attacks only the carnation, *Dianthus caryophyllus*. Proof that a fungus is the strain *f. dianthi* can only be obtained by inoculating a carnation with the suspected pathogen and observing disease development. Similarly, selection introduced by control measures for mites and aphids may result in strains peculiar to specific greenhouses. Control measures effective in one range may not be effective in another range. Certain viral diseases can only be determined by infecting varieties that will show well defined symptoms.

Any approach to control takes into account the fact that predators respond to environment as well as the host plant. Hussey et al. (1969) point out that at 16° C, in one month, a single female spider mite can produce about 50 offspring. But, at 25° C, the reproductive ability increases to over 100 million in the same period. Stock will not usually be affected by Fusarium wilt unless the soil temperature remains above 21° C, and bacterial stem rot of poinsettias, according to Forsberg (1975), is most prevalent when late plantings are forced under high fertilization rates at temperatures of 24–31° C. Control measures take into account effects on both predator and host, and the cultural procedures employed to produce the crop. Pathogens and pests go through various stages in their development, responding to light, temperature, humidity, etc. The control measure will vary with the growth stage of parasite and host, and the method of attack by the parasite.

Table 9-1. Diseases common to some ornamental greenhouse crops

Type	Causal agent	Common name	Plants attacked
I. Bacteria	*Xanthomonas*	Bacterial leaf spot	Begonia, Geranium, Iris *Dieffenbachia*, Gladiolus, *Anthurium*
		Bacterial stem rot	Geranium
		Pimple	Carnation
		Bud rot	Canna
		Yellow rot	Hyacinth
		Bacterial blight	Stephanotis
		Tipburn	*Philodendron*
		Leaf blight	*Syngonium*
	Erwinia	Bacterial soft rot	Calla lily, Cyclamen, Hyacinth, Iris, Poinsettia
		Bacterial blight	Chrysanthemum
		Slow wilt or stunt	Carnation
		Leaf rot	*Philodendron*
		Stem and leaf rot	Dahlia, *Dieffenbachia*
		Erwinia blight	*Aglaonema, Syngonium*
		Rapid decay	*Pothos*
	Pseudomonas	Bacterial leaf spot	*Dracaena, Scindapsus, Monstera, Philodendron,* Aglaonema, Gardenia, Carnation, Chrysanthemum
		Black leaf spot	Delphinium
		Scab or neck rot	Gladiolus
		Bacterial wilt	Cosmos, Dahlia
		Scale tip rot	Easter lily
		Bud blast	Rose
		Seedling blight	Snapdragon
	Corynebacterium	Bacterial fasciation	Geranium, Sweetpea, Petunia
		Bacterial canker	Poinsettia
	Agrobacterium	Rose crown gall	Rose
	Bacillus	Streak	Sweetpea
II. Fungi			
a) Downy mildew	*Peronospora*	Downy mildew	Snapdragon, Rose
b) Powdery mildew	*Oidium*	Powdery mildew	African violet, Snapdragon, Rose, seedlings and bedding plants
	Microsphaera	Powdery mildew	Azalea, Sweetpea
	Erysiphe	Powdery mildew	Chrysanthemum, Dahlia, Phlox, Delphinium, Hydrangea
	Sphaerotheca	Powdery mildew	
c) Rusts	*Coleosporium*	Rust	Aster
	Uromyces	Rust	Carnation
	Puccinia	Rust	Chrysanthemum, Geranium Iris, Snapdragon
	Phragmidium	Rust	Rose

Table 9-1 (continued)

Type	Causal agent	Common name	Plants attacked
d) Wilts	*Fusarium*	Fusarium wilt	Carnation, Aster, Chrysanthemum, Pansy, Cyclamen, Dahlia, Petunia, Stock
	Verticillium	Verticillium wilt	Chrysanthemum, Cineraria, Dahlia, Rose, Snapdragon
	Cylindrocarpon	Blight and wilt	Azalea
	Sclerotinia	Snapdragon wilt	Snapdragon
	Phytophthora	Phytophthora wilt	Azalea
	Phialophora	Phialophora wilt	Carnation
e) Blights	*Botrytis*	Botrytis blight	Easter lily, Hydrangea, African violet, Cineraria, Geranium, Chrysanthemum, Begonia, Tulip, Orchid, Rose, Snapdragon, Dahlia, Carnation, Cyclamen, Hyacinth, Stephanotis, seedlings and bedding plants
	Briosia	Bud and twig blight	Azalea
	Phyllosticta	Snapdragon blight	Snapdragon
	Sclerotinia	Flower blight	Camellia
		Blight	Stephanotis, seedlings
	Ascochyta	Stem and ray blight	Chrysanthemum
	Ovulinia	Petal blight	Azalea
	Rhizoctonia	Leaf blight	Aster, Azalea, ferns, and seedlings
	Phytophthora	Leaf blight	Calla lily
	Stemphylium	Petal blight	Carnation
	Cercospora	Cercospora blight	Snapdragon
	Alternaria	Alternaria blight	*Schefflera*
f) Rots	*Botrytis*	Stem rot	Begonia, Aster
		Petiole rot	Cyclamen
		Corm rot	Gladiolus
		Bulb rot	Amaryllis
		Bud rot	Gloxinia
		Smoulder	Daffodil
		Gray mold	Poinsettia
		Damping off	Seedlings
	Pythium	Root rot	Easter lily, Poinsettia, Hyacinth, African violet, Calla lily
		Root and stem rot	Chrysanthemum
		Root and crown rot	Begonia
		Blackleg	Geranium
		Black rot	Orchid
		Cutting rot	*Peperomia*, *Aglaonema*
		Damping off	Seedlings
	Fusarium	Basal rot	Easter lily, Daffodil, Iris
		Bud and stem rot	Carnation

Table 9-1 (continued

Type	Causal agent	Common name	Plants attacked
		Corm rot and yellows	Gladiolus
		Fusarium rot	Sweetpea
		Flower and bud rot	Chrysanthemum
	Phytophthora	Root rot	Calla lily
		Foot rot	Aster, Gerbera, Stephanotis
		Stem rot	Easter lily
		Crown rot	Azalea, Kalanchoe, Petunia
		Black rot	Orchid
		Cutting rot	Peperomia
		Root, crown, leaf, and stem rot	Gloxinia
		Stem rot and wilt	Snapdragon
		Damping off	Seedlings
	Rhizoctonia	Root rot	Easter lily, Poinsettia, Sweetpea, Snapdragon, Stock, Aster, Chrysanthemum
		Stem rot	Carnation, Azalea
		Neck and bulb rot	Iris
		Cutting rot	*Peperomia*
		Damping off	Seedlings
	Alternaria	Branch rot	Carnation (sometimes confused with Fusarium branch rot)
g) Leaf spots	*Alternaria*	Leaf spot	Carnation, Stock, Geranium
	Zygophiala	Greasy blotch	Carnation
	Ascochyta	Leaf spot	Aster
	Septoria	Leaf spot	Hydrangea, Gladiolus, Azalea, Chrysanthemum, Carnation, Cosmos, Phlox
	Cercospora	Leaf spot	Azalea, Cosmos, Dahlia, Phlox, *Schefflera*, *Cordyline*, *Ficus*, *Peperomia*, *Crassula*, palms
	Coniothecium	Leaf spot	Calla lily
	Dactylaria	Leaf spot	*Philodendron*
	Glomerella	Leaf spot	Cyclamen
		Anthracnose	Azalea
	Phoma	Leaf spot	Cyclamen
	Phyllosticta	Leaf spot	Cyclamen, Hydrangea, Dahlia
	Ramularia	Leaf spot or stunt	Cyclamen
	Mycosphaerella	Leaf spot	Gardenia, Rose
	Rhizoctonia	Leaf spot	Gardenia
	Botrytis	Leaf and flower spot	Gladiolus

Table 9-1 (continued)

Type	Causal agent	Common name	Plants attacked
h) Cankers and galls	*Phomopsis*	Canker Die-back	Gardenia Azalea
	Exobasidium	Leaf and stem gall	Azalea
	Cryptosphorella	Brown canker	Rose
	Cylindrocladium	Crown canker	Rose
	Leptosphaeria	Stem and graft canker	Rose
	Coniethyrium	Brand canker	Rose
	Phytophthora	Die-back	Azalea
	Botryosphaeria	Die-back and canker	Azalea
	Glomerella	Die-back	Camellia
	Sphaceloma	Scab	Poinsettia
	Myrothecium	Crown and stem canker	Snapdragon
III. Virus		Aster yellows	Aster, Phlox, Canna, Rose, Chrysanthemum, Gladiolus, Geranium, Snapdragon
		Spotted wilt	Calla lily, Geranium, Dahlia
		Mosaic	Easter lily, Carnation, Amaryllis, Gladiolus, Aster, Cineraria, Canna, Coleus, Dahlia, Iris, Petunia, Stephanotis
		Streak	Carnation, Cineraria, Rose, Chrysanthemum, Daffodill
		Mottle	Carnation, Easter lily, Chrysanthemum
		Stunt	Chrysanthemum, Dahlia
		Ringspot	Carnation, Crassula, Peperomia, Hydrangea, Dahlia
		Yellow net vein	Geranium
		Leaf cupping	Geranium
		Chlorotic spot	Geranium
		Bean yellow mosaic	Gladiolus
		Cucumber mosaic	Gladiolus
		Tomato spotted wilt	Chrysanthemum
		Tomato aspermy	Chrysanthemum
		Crinkle or leaf curl	Geranium
		Fleck	Easter lily
		Rosette or yellow flat	Easter lily
		Spring dwarf	Rose
		Breaking	Tulip

Sources: Reinert, 1961; Nichols and Burke, 1965; Jones and Stryder, 1974; Aycock and Daughtry, 1975; Forsberg, 1975; Knauss, 1975a, b; Nichols and Nelson, 1976.

9.3.1 Diseases

Some diseases of ornamental greenhouse crops are presented in Table 9-1. In general, bacterial and fungal diseases are encouraged by warm temperatures and high humidities. There are exceptions such as *Phialophora* wilt of carnations, which develops faster at low temperatures. Most bacteria and fungi are spread by splashing water and require free water for spore germination. However, free water is lethal to the conidia of powdery mildew on roses.

9.3.1.1 Bacteria

Although bacteria are capable of causing rots, leaf spots, blights, and wilts, they are often classified separately as there are relatively few important genera. Predisposing conditions are high humidity, poor air movement and high temperatures—conditions commonly prevalent in propagation beds. Infected plant material is a major means of introduction, but disease can be spread by dirty tools, insects, inoculated soil, splashing water and often gains entry to the host through wounds as may occur in propagation or cutting. Infection can occur through roots, stems, and leaves.

For bacterial leaf spots, the classical symptoms are circular or angular, water-soaked areas on the leaves. These spots may be first dark to light green, gradually turning yellow, then brown or black. In some instances, a halo may be seen around the center when held to the light. Under moist, humid conditions, a

Fig. 9-1. Powdery mildew caused by the fungus *Sphaerotheca pannosa* var. *rosae* on rose. White to gray powdery growth, with leaf distortion and discoloration. Favored by moderate temperatures and high relative humidity approaching 100%. Free water is generally lethal to the conidia

bacterial ooze may be observed, yellow, reddish, clear, or white in color. The spots may coalesce and there can be premature leaf drop. Spots may be sunken, can have reddish brown borders, or purplish concentric rings as on carnations.

Rots and blights caused by bacteria are nearly always characterized as soft rots, leading to complete tissue disintegration and a slimy, often ill-smelling, black mess. The bacteria may gain entry and develop internally as in wilts or stunts. Development may be fast or slow, depending upon environmental conditions and growth stage of the host. In slow wilt of carnation, plants usually show an unthrifty appearance, lighter color, and loss of bloom. Bacterial wilt of carnations, however, may cause sudden wilting without any previous indication. Bacterial blight of chrysanthemums commonly shows terminal wilting with a dull, grayish, water-soaked appearance of the stem at a point one-half to two-thirds of the way up the stem. The vascular tissue of infected plants may be affected on only one side, or all, of the stem, usually black, streaky, slimy, and sticky to the touch. Stickiness and a foul odor of ooze from lesions is a good indication that the causal agent is bacterial.

Species of *Corynebacterium* and *Agrobacterium* cause fasciation and galls on several plants. These are generally abnormal growths such as proliferation of stems and leaves on geranium or galls on rose stems near the soil line. Cankers are localized lesions on a stem with production of an open wound as in the case of poinsettia canker, caused by *Corynebacterium*.

9.3.1.2 Fungi

In many instances, definite diagnosis of disease requires the services of a trained specialist to isolate and identify the pathogen. This is often required for intelligent control and eradication, and is particularly important for fungal diseases. The number of fungal genera that can attack plants is practically without end (Table 9-1). We have classified fungal diseases into broad groupings of blights, rots, leaf spots, wilts, rusts, mildews, and cankers. The easiest to identify are mildews and rusts. Powdery mildew grows a white powdery to cottony mycelium on leaf surfaces (Fig. 9-1), sometimes confined to a single side of the leaf causing leaf deformation and stunting. High humidity is a predisposing condition. Control is directly related to humidity manipulation and outside climatic conditions. Downy mildew of snapdragons is favored by lower temperatures. Rusts are identified by pustules filled with reddish-orange or brown spore masses erupting through the leaf epidermis (Fig. 9-2). The spores require water to germinate, and rust is usually found on plants under drips, or if water is allowed to stand on leaf surfaces for a considerable period.

Wilts, blights, and rots are often difficult to distinguish as, depending upon method of entry and plant part, the same organism may cause symptoms of rotting, blighting, or wilting. Wilts are usually systemic and spread by infected plant material. The organism may invade the root system and grow up one side of the stem so that branches on that side are affected. Or, if young plants are invaded, they may suddenly topple over, wither and die. There is usually a general browning of the vascular tissue, with some black necrosis of the basal stem tissue in the case of aster and chrysanthemum. The plant may appear unthrifty, off-

Fig. 9-2

Fig. 9-3

Fig. 9-2. Carnation rust caused by the fungus *Uromyces caryophyllinus*. Ruptured epidermal areas are filled with reddish-brown, powdery spore masses. Spores require free water for germination

Fig. 9-3. *Botrytis* blight on snapdragons. In this instance the infected leaves, coming into contact with the stem, will infect the stem and eventually result in girdling and death

color, stunted, and produce smaller blooms. The base of the stem can be water soaked. With the exception of *Phialophora* wilt, high temperatures generally hasten growth of the pathogen. No symptoms may be seen until flowering in the case of *Fusarium* wilt of cyclamen. In advanced stages of azalea leaf blight, there may be a sparse to luxuriant brownish mycelium on lesions with powdery white masses of conidia of the fungus *Phytophthora*.

Blighting, which is the rapid withering and decay of tissue, can be caused by several fungi. The genus *Botrytis* is most common, attacking nearly all ornamental crops grown in greenhouses at various stages of growth and development. Predisposing conditions are high humidity, poor air circulation, and high temperatures. The most characteristic signal is the appearance of a brownish or grayish mycelial growth, filled with spore masses which can be readily spread by splashing water or air currents. Leaves may be affected first and the fungus can gain secondary entrance into the stem if the infected leaf comes into contact with the stem (Fig. 9-3). Flowers are very often affected in storage if the spores are present when blooms are stored or transported. First symptoms may be small, water-soaked lesions, enlarging rapidly, and becoming brown to black with age. Petal

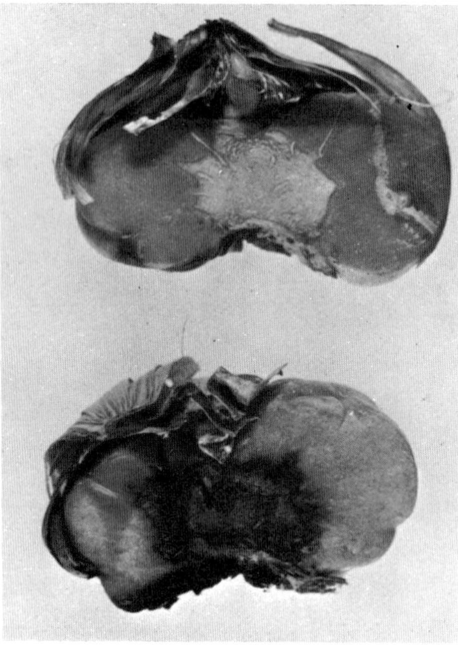

Fig. 9-4. *Fusarium* corm rot or yellows of gladiolus, caused by the fungus *F. oxysporum* f. *gladioli*. It may be of several forms. The upper corm is healthy

blight of azalea, caused by the genus *Ovulinia*, results in eventual rotting and disintegration of the tissue, which turns dark brown to black with petals becoming slimy and falling apart if rubbed gently. Tiny purple spots, soon developing a broad, yellow-green border, are typical for alternaria blight of carnations. Ascochyta ray blight of chrysanthemums is characterized by a one-sided attack of the flowers at various stages of blooming. Flowers turn straw color or brownish, wither, with the discoloration proceeding from base toward tip on each flower.

Rots are generally associated with infested soils and occur on the root system, storage organs, or on the stem at or near the soil line. An example is *Fusarium* corm rot or yellows of gladiolus shown in Figure 9-4. However, there is *Fusarium* bud rot of carnation, spread by the bud mite causing a moist brownish decay of inner floral organs. As with blights, *Botrytis* is a common causal agent. Symptoms may be dark brown to black lesions, separated from healthy tissue by a sharp demarcation line, water-soaked, slimy, odorless, with complete tissue disintegration resulting. Rots caused by *Fusarium* may have pink to red discolorations with brownish or reddish vascular discoloration. If the *Fusarium* fungal mycelium is visible with fruiting structures, it will usually be red. *Pythium* blackleg of geranium (Fig. 9-5) is distinguished from bacterial rot by the shiny, coal black, wet appearance and rapid advance as contrasted to dull black and slow advance of bacterial rot. *Pythium* species are ubiquitous and cause numerous root rots under predisposing conditions of high moisture, poor soil aeration, or root damage. As an example, Figure 9-6 shows effect of *Pythium* root rot on snapdragons where the invasion is not sufficient to kill the plant but causes undesirable wilting on clear days. Feeder roots are often attacked at the tips and rotting may be particularly

Fig. 9-5. *Pythium* blackleg on geraniums

severe if a sterilized soil is inadvertantly reinfested. Crown and root rots generally cause an unthrifty appearance, and the plant may not wilt, although, it can be easily lifted from the soil.

The fungi *Pythium* and *Rhizoctonia* and to a lesser extent, *Fusarium*, *Phytophthora*, *Sclerotium*, *Botrytis*, and *Thielaviopsis* are often instrumental in "damping-off". The complex is the most serious threat to young seedlings and germinating seeds. According to Jones and Stryder (1974), *Pythium* and *Phytophthora* are more likely to cause damping-off in cool, wet soils; whereas, *Rhizoctonia*, *Fusarium*, and *Sclerotium* are more common under warmer, drier conditions. Typical symptoms are rotting stems at or near the soil line and root decay. It may occur in circular areas (Fig. 9-7) with shriveled brown, collapsed, or stunted seedlings (Fig. 9-8). Primary sources of inoculum are infested soil, some seeds, unsterile pots, flats, benches, tools, or hoses. *Rhizoctonia* causes "wire stem" on stock, the tissues turning black and constricting at the soil line, and will attack carnations with a slimy, wet rot of the bark. Under suitable conditions, particularly if the plant is weakened or injured, the common molds *Rhizopus* and *Penicillium* may be found.

Symptomatology for leaf spots varies with the particular pathogen and host. Lesions are generally water-soaked and, in advanced stages, the fruiting structure of the pathogen may be evident in the center of the spots. Predisposing conditions are usually high humidity and poor air circulation, spores being spread by splashing water, sometimes in air currents. For example, there are several fungi causing leaf spots on azalea. Spots caused by *Pestalotia* are silvery gray on the upper surface and light brown below. *Lophodermium* causes silvery-white large spots

Fig. 9-6 A and B. Effect of Pythium root rot on snapdragons. Infection by the *Pythium* fungus in unpasteurized soil does not kill the host, but seriously stunts and prevents adequate water uptake during the day. If pasteurized soil is reinfested with *Pythium*, death and severe wilting of snapdragons at flowering time is likely to occur. In the example shown, (A) shows recovery at night, and (B) is the wilting common during days. The cycle of wilting and night recovery may continue for several days or weeks

with red, raised margins. *Cercospora* produces angular, brown spots on the upper surface with a grayish-brown mycelium in the centers; whereas, *Exobasidium* has distinct yellow spots that are minute and circular. *Septoria* causes irregular, angular brown lesions in late stages with yellow zones around the lesion fairly common. The same pathogen on gladiolus produces small brown or purple brown spots that become finally dark brown with a lighter center. On carnations, the

Fig. 9-7. Damping-off disease on seedlings. Note the several sunken spots where seedlings have rotted or failed to germinate. Strict sanitation is required

Fig. 9-8. Damping-off of petunia seedlings. In this instance, a combination of attack from the fungi *Rhizoctonia* and *Sclerotina*

spots are light brown with purple margins and small black specks, representing the fruiting bodies of the fungus, may be seen in the centers of the spots. On chrysanthemum, the spots are also dark brown with fruiting bodies black. Leaf spot diseases are sometimes referred to as *anthracnose* diseases, which is any fungal disease producing roundish dead spots on leaves or fruit.

Canker producing fungi usually attack stems, producing open wounds in which the mycelium and fruiting bodies of the fungus may be observed in advanced stages. The disease is sometimes referred to as "die-back", particularly when the organism girdles the stem or shoot. The lesions can be slightly sunken with various colors. Red blotch on amaryllis not only produces bright red, vermillion cankers, but can cause the flower stalks to grow at right angles. Longitudinal cracks occur on the gardenia canker caused by *Phomopsis* with the cortical tissue a bright yellow. Poinsettia scab produces conspicuous, raised cankers, frequently uniting to form irregular areas with the centers later becoming somewhat depressed and covered with a grayish to grayish-brown, velvety layer of mycelium. Leaf gall on azalea is fairly spectacular. The thick, fleshy leaves are greatly enlarged and malformed. The mycelium emerges between the epidermal leaf cells with the spore-bearing structure completing its development on the exposed leaf surfaces.

9.3.1.3 Viruses

Plant viruses are generally composed of complex nucleoproteins that are visible under the electron microscope. However, recent research has shown that some diseases formerly thought to be caused by viruses are actually mycoplasms—organisms without cell walls that can exist in two or more forms in one life cycle. More than one virus may infect a plant and sometimes two are required to produce visible symptoms. Symptoms may vary with the season and host development, with some detectable only by grafting, and transmitting the virus to a variety that will produce visible symptoms. Often, the only visible symptoms will be a reduction in vigor which cannot be ascertained without comparison with a virus-free plant. At other times with some viruses, damage will be readily apparent (Fig. 9-9) leading to ultimate death of the host.

Transmission can be by sucking insects, such as thrips, aphids, and leafhoppers, or by juice carried on hands and tools. Control is commonly exclusion of the insect as the case of aster yellows, transmitted by leafhoppers, or careful washing and disinfecting of tools and hands as in the case of tobacco mosaic. Tobacco should never be allowed in tomato greenhouses as the tobacco mosaic readily infects tomatoes.

Viruses are often classified into broad categories as mosaics, yellows, mottles, and streaks. The names roughly characterize the disease symptoms. Mosaic viruses generally produce chlorotic symptoms in a manner described by the term "mosaic"; whereas, "mottling" can be a more or less random yellowing, chlorotic pattern. "Vein clearing", or the loss of green color along the veins, is a common symptom with yellows and mosaic. Tulip "breaking" is probably the oldest known virus disease of plants. Flowers become variegated with leaves striped with light green or even with white. Variegation in flowers consists of bars, stripes, flames, streaks, or feathering of a darker color overlaying a lighter shade of the same color, or on a pure white or yellow. There is reduced size and vigor of the plant with fewer offsets or bulbils. On *Peperomia*, the leaves are distorted and disfigured with concentric zonal markings with each ring beginning as a small translucent spot that grows larger by addition outwardly of narrow bands or lines. The serious stunt virus on chrysanthemums generally causes reduction in size which cannot be diagnosed without a healthy plant for comparison, or by grafting on varieties that produce definite symptoms. Streak virus on carnations generally produces white to purplish-brown flecks in older leaves. With cineraria,

Fig. 9-9. Etched-ring virus in carnations

the symptoms are roughening and curling of the leaves of young plants. Fleck in Easter lily is a combination of two viruses, neither one of which alone will cause symptoms. Geranium leaf cupping is probably the most distinctive and destructive virus disease. The virus changes the shape of the plant to one no longer characteristic of the species with leaves changed to incurved, hairless cups and sinuous veins. Aster yellows may cause a peculiar spindling growth with stems a light, greenish-yellow and abnormal production of sideshoots. Chrysanthemum flowers may be greenish with weak lateral shoots, pale, and upright.

9.3.2 Nematodes

Nematodes, or eelworms, may be found wherever there is moisture. Often less than 1 mm in length, although sometimes up to 8 mm, they are capable of causing serious damage to greenhouse crops. They are a distinct group in the animal kingdom and are best known for their parasitism on animals, including man (roundworms, whipworms, etc.). Isolation and identification of nematodes usually require services of trained personnel. When a plant is removed to test for nematode presence, the root system and soil immediately about the roots should be undisturbed. It may be necessary to incubate the soil and roots for several days and follow various staining procedures. Injury is generally nonspecific, that is, symptoms readily attributable to nematodes are not apparent. Plants may be merely stunted and survive the attack.

Table 9-2. General classification of nematodes attacking greenhouse plants. (From Hussey et al., 1969)

Genus	Common name	Plants attacked	Remarks
Meloidogyne	Root-knot	Crops grown at high temperatures under continuous cropping, most often tomatos, etc.	Unthrifty, wilting plants, producing galls on roots
Heterodera	Cyst nematode Golden nematode Potato-root nematode	Warm temperature, continuous cropped conditions; tomatoes, etc.	Unthrifty, dwarfing, tendency to wilt, minor galling, formation of cysts
Criconemoides	Ring nematodes Spiral nematodes	Roses, carnations	Stubby roots, root necrosis, and proliferation of roots
Pratylenchus	Pin nematode Root lesion nematode	Carnation, amaryllis, gladiolus, begonia, chrysanthemum, lily, narcissus, orchids, etc.	Stunted and yellow top growth, stunted roots and abnormal proliferation
Rodopholus	Burrowing nematode	Several ornamentals	Root lesions and discoloration
Tylenchorhynchus	Stunt or stylet nematode	Azalea	Root growth poor, plants stunted
Xiphinema	Dagger nematode	Roses and other perennial greenhouse crops	Root kinking, particularly near root tips, poor growth, transmits virus
Aphelenchoides	Bud and leaf nematodes, spring crimp or dwarf	Chrysanthemum, fern, strawberry, begonia, gloxinia, Easter lily, orchid, hydrangea, coleus, dahlia, and others	Leaf blotching, apical leaves undersized and misshaped, blind plants, yellowing and necrotic areas on leaves, plants dwarfed and distorted
Ditylenchous	Stem and bulb nematode	Narcissus, hyacinth, tulip, gladiolus, hydrangea, etc.	Pale lesions on leaves, stems thick and distorted, brown rings on bulbs. Flower heads distorted, symmetrical local swelling at point of entry, mainly whitish appearance, entrapped air

Where the root medium is isolated from the surrounding ground, and if proper cultural procedures are followed, nematodes should not be a problem. They can be introduced on infected plants, particularly if soil is adhering to root systems and bulbs. In ground beds, considerable difficulty may be experienced controlling root nematodes as the depth of sterilization is usually limited and the treated area is commonly reinfested from the deeper layers. The pests can be transported in untreated sewage sludge, straw, peat moss, etc. and on dirty equipment. Often, nematodes can be found in association with fungi and plant injury will be greater as lesions produced by feeding allow the fungal pathogen easier entry.

Nematodes require water and air for growth and move to the root system or on the aerial plant parts through water films. Thus, water-logged soils or dry conditions will inhibit activity. Likewise, fine soils will inhibit migration as the spaces between pores may be too small for the nematode to pass through. The same factors also influence control by gaseous or liquid nematicides.

A summary of nematodes attacking greenhouse plants is provided in Table 9-2. Plant parasitic nematodes feed on plant cells by piercing the cell walls and sucking the contents through the hollow mouth spear or stylet. As the nematode population increases, cell death increases with necrosis being the general symptom. If meristematic tissue is attacked, then stunting, distortion, and bud abortion can occur. Some nematodes, in the genus *Meloidogyne* and *Heterodea*, inject fluid into the cell which causes abnormal growth with formation of enlarged cells, galls, or distorted stems, foliage and roots. Damage is usually worse where the crop is grown continuously, allowing a buildup of the nematode population.

The most spectacular damage is caused by the root-knot nematode where attacked roots are stunted, abnormally swollen and galled. There are relatively few fibrous roots and small rootlets may be killed. Giant cell formation close to the root stele interferes with water conduction accounting for frequent wilting under slight drought. *Heterodera* eelworms do not provoke massive root galling, but swollen and stunted roots on seedlings may be mistaken for root-knot. Species of *Heterodera* produce cysts which resist unfavorable conditions for growth. The cyst and rootknot nematodes generally exist as sedentary parasites within the root system as contrasted to the migratory eelworms that "browse" on the roots and enter and leave the roots during parts of their life cycle. *Pratylenchus* is probably the most common and often difficult to isolate and identify. *Xiphinema* species are known to transmit ringspot virus which can cause leaf malformation, chlorosis, and dwarfing in roses.

The genus *Aphelenchoides* has several species which attack hundreds of plant species. They commonly live between cells in the aerial portions of their host, although some shelter between tightly folded leaf or flower petals or in the axils between stem and petiole. The life cycle is generally short and enormous numbers can be produced under favorable conditions. When plant surfaces are dry, the nematodes are restricted to their enclosed habitat. However, when water films persist, the nematodes emerge and swim actively over the plant surfaces.

Foliar nematodes are less likely to be prevalent within greenhouses and are usually introduced on plants previously grown outdoors. On chrysanthemum, symptoms of eelworm damage can be found at all growth stages with interveinal leaf blotching. Damage appears to spread upward as the nematodes migrate upward via water films on the surfaces. Leaf abscission is not common, the leaves hanging down in ragged clusters. If infestation is heavy in the growing points, new leaves will be severely distorted with bud abortion. The stem eelworm *(Ditylenchus)* is able to persist under adverse conditions for several years and increases to astronomical numbers from relatively few nematodes in a good environment. It feeds and breeds only on living plants, being spread in soil contaminated footwear, machinery, plant material, or sometimes in irrigation water. There may be cell enlargement when the plant is penetrated, usually the root-shoot junction of the plant, with proliferation of side and basal shoots, asymmetrical stem growth, twisting, and curling. Growth is stunted with tissue necrosis if the invasion is massive or the plant is sensitive.

9.3.3 Insects and Related Pests

Insects consist of nearly a million species, capable of flying, with a horny external skeleton, an ability to adapt to unusual environments, and with small size conferring advantages of small individual nutritional needs. Their growth is a discontinuous process involving short periods of growth separated by moults in which the old skin is shed. They may undergo simple or complex metamorphosis, the latter when the developmental stages are radically different from the preceeding ones. Table 9-3 groups those causing plant damage according to taxonomic order.

Springtails, crickets, earwigs, and cockroaches are generally less important than many of the others. Where ordinary sanitation and control measures are carried out, they should not be a particular problem.

Thrips, on the other hand, can heavily parasitize many greenhouse crops. Eggs may be laid on or within the tissue, passing through several stages before becoming adults. Adults and larvae feed by piercing plant tissues and sucking the sap. Tissue around the puncture dries out causing a silvery-flecked appearance. On carnation, observable damage such as abnormal twisting of new leaves, necrotic spots, and red streaking of white flowers means that control measures should have been started several days before. Thrips are difficult to observe and they usually enter the greenhouse through the ventilators. Control measures should be initiated early in the spring if damage is to be avoided. Badly infested plant material should be discarded.

There are numerous genera and species in the order Hemiptera that are major pests in greenhouses. Leafhoppers lay eggs in the secondary veins on the underside of leaves. The incubation period may be less than one week in warm weather, the nymphs going through five stages (instars) before reaching adulthood. The adult is usually quite active, particularly when disturbed. The piercing of cells by the feeding stylet causes blanched areas which become confluent and, in severe attacks, the leaves appear variegated or bleached. Whiteflies are serious pests particularly on tomatoes and cucumbers. The adult is noted for its pure white appearance and the fact that large numbers may congregate which leap into flight when disturbed. Eggs are generally laid on the leaf underside. The host is seriously disfigured when black molds grow on the honeydew excreted by the adults and

Fig. 9-10. Nymphs, wingless females and alatae of the aphid *Macrosiphum rosae*. (From Hussey et al., 1969)

nymphs. Affected foliage becomes chlorotic and dies. The incubation period for eggs varies from 30 days at 8° C to 4 days at 30° C, with the complete life occupying three weeks at 21° C.

Scale and mealy bugs are noted for their attacks on tropical foliage plants. The most obvious stage is the female mature scale which becomes an egg laying machine with almost all internal organs sacrificed. Active stages are small and inconspicuous, so that by the time an infestation is obvious, damage has already occurred. Large quantities of honeydew may cause prolific growth of sooty molds. The mealy bug differs in possessing a white, flocculent covering instead of a hard scale. They are generally more active than hard scale types. Hussey et al. (1969) state that there are more than 60 species of hard scale that have been described and they discuss 14 species of mealy bugs that have been found in British greenhouses.

Aphids are commonly found on the stems and lower leaf surfaces (Fig. 9-10). Species may have complex life cycles with alternation between summer and winter hosts and reproduction parthenogenetically or viviparous. The female aphid is capable of producing 50 daughters, each maturing in a week under average conditions. Aphids are important as vectors of plant viruses such as carnation mosaic, chrysanthemum mosaic, lettuce mosaic, tomato aspermy, and tulip breaking, among others. Resistance to organophosphates has become widespread in recent years, though advent of new pesticides has kept abreast of the resistance. With few

Table 9-3. Outline of major insects and related pests of greenhouse plants. (After Hussey et al., 1969)

Group or order	Common name	Plants attacked	Remarks
Collembola	Springtails	Cyclamen, cineraria, lettuce, cucumber, young seedlings	Rarely pests, found mostly in greenhouses infrequently sterilized, attack seedlings at soil line
Orthoptera	Crickets	Seedlings, chrysanthemum, orchids, cineraria	Omnivorous, eating stems of plants at soil line
Dermoptera	Earwigs	Chrysanthemum, cineraria, dahlia, etc.	Damage foliage and flowers, leaves may be skeletonized
Dictyoptera	Cockroaches	Sanseveria, chrysanthemum, orchids, stems and leaves of several species	Foliage and stems, damage at soil line, may be resistant to DDT and BHC
Thysanoptera	Thrips	Lily, arums, azalea, ferns, orchids, fuschia, begonia, *Dracaena*, *Stephanotis*, *Tradescantia*, *Croton*, *Caladium*, cyclamen, carnations, gladiolus, rose, chrysanthemum, etc.	Small, 1–2 mm in length, associated with flowers and foliage, eggs laid on or within tissue, necrosis and malformation, may transmit viruses, silvering of leaves and flower petals on gladiolus
Hemiptera	Plant bugs, Leafhoppers	Chrysanthemum, tomato, cucumber, primula, salvia, *Verbena*, fuschia, and others	Feed nearly always on underside of leaves, leaves may appear variegated or blanched
	Whitefly	Arum, begonia, *Calceolaria*, freesia, fuschia, salvia, *Verbena*, tomato, cucumber, geranium, primula, poinsettia, and others	Adults about 1 mm long, collect in high numbers, laying eggs usually on leaf undersurface, wings pure white, formation of black mold on honeydew
	Scale	Palms, *Asparagus*, *Acacia*, *Dracaena*, *Ficus*, etc., rose, orchids, fern, camellia, amaryllis, etc.	May be small and inconspicuous, causing chlorosis and stunting, ranging from roughly circular to slightly convex, brown, yellowish, reddish scale
	Mealy bugs	Many tropical foliage plants, cacti, gardenia, coleus, chrysanthemum, geranium, tomato, etc.	Related to scale, but possess a covering of white waxy threads instead of hard scale, can cause complete defoliation
	Greenflies Aphids	Most ornamental and vegetable plants	Transmit viruses, very rapid reproductive rate, colonies may be dense, sometimes producing distortion, stunted growth, necrosis
Heteroptera	Plant bugs, capsids, tarnished plant bug	Chrysanthemum, sweetpea, aster, carnations, zinnia	Feed near shoot apices, stems marked with brown calluses, shoots stunted, twisted, blind, foliage twisted, distorted

Lepidoptera	Moths (larvae called cutworms, caterpillars)	Most ornamental plants	May eat through stems at soil surface, skeletonize leaf surfaces, holes in ripening fruit, may eat petals and calyces, damage from adults rare
Coleoptera	Beetles (larvae often called weevils, click beetles, wireworms)	Chrysanthemum, carnations, tomato, lettuce, bulbous crops, cyclamen, vines, etc.	Grubs white, fleshy with curled, wrinkled bodies, may feed on roots, often serious in new ranges, or enter from adjacent infested land, vine weevil causes irregular notches on leaves
Diptera	Flies, crane flies, fungus gnats, midge, leafminer	Chrysanthemum, seedlings, cyclamen, sweetpea, orchids, freesia, rose, azalea, carnation, narcissus	Several types, feeding on roots, formation of galls, tunneling of leaves, shriveled and distorted flowers and buds, distorted growth
Isopoda	Woodlice, sowbugs, millepedes	Cucumber, seedlings, orchids, *Adiantum*, fern, tomato, carnations, chrysanthemum, sweetpea	Generally feed on roots or near soil line, chewing or girdling of stems near ground, moisture critical factor
	Symphilids	Tomato, lettuce, sweetpea, lily, chrysanthemum, African violet	Feed primarily on root system, may initiate root rots from bacteria and fungi, stems may be hollowed out, stunted growth
Acarina	Mites, spider mite.	Carnation, rose, chrysanthemum, cucumber, tomato and others, including cyclamen, bulbs, etc.	Very serious, rapid increase of resistance to chemical miticides, mottling of upper leaf surface, yellowing, fine silken thread or web on heavy infestations
	Slugs and snails	Most ornamentals	Warm temperatures and high humidity favorable, omnivorous feeders, browsing surface vegetation

Fig. 9-11. Damage to cineraria caused by the cabbage looper

exceptions, aphids attacking greenhouse plants are common outdoors and readily enter the greenhouse so that constant attention is required for control.

Plant bugs injure a wide range of greenhouse crops, the adults overwintering in dead leaves and tall grass emerging to take active flight in the spring. Eggs are laid in stems, petioles, or flower buds of young plants. On chrysanthemums, injured shoots from feeding become twisted, stunted and sometimes blind. Generally, the larvae of moths cause most of the damage on the host. The worms may skeletonize leaves and attack stems and flowers (Fig. 9-11) generally feeding at night. Cutworms and caterpillars are common names applied to the larvae of such species as the tomato moth, gothic moth, cabbage moth, carnation tortrix, and South African carnation worm. The larvae of beetles usually cause damage by feeding on the roots of many greenhouse plants. Click beetles and wireworms are sometimes a problem when greenhouses are erected on cultivated land which has become weed infested.

With some exceptions, insects in the orders Diptera and Isopoda are not as much a problem as, for example, whitefly or aphids. The exception to this may be the crysanthemum midge and leafminer. The larvae of the midge burrow into the leaf tissues. Their presence is shown by small, light patches within the leaves. Galls become visible, and in severe cases, the plant may be so heavily galled that leaves are distorted, stems twisted and the flowers misshapen. Infestation is generally caused by bringing infested plant material into the greenhouse. There are midges that attack roses, violets, and cattleyas. Leafminers are widely distributed; the female adults feed upon plant sap, laying eggs below the epidermis; the larvae feed on the palisade tissues gradually burrowing their way through the leaf (Fig. 9-12). The mines are generally white and prompt control is required to prevent foliage from being disfigured. Miners include the tomato leafminer, chrysanthemum blotch miner, chrysanthemum stool miner, and the carnation fly. Fungus

Fig. 9-12. Effect of resistance on attack by leafminer. The undamaged leaf on left was resistant

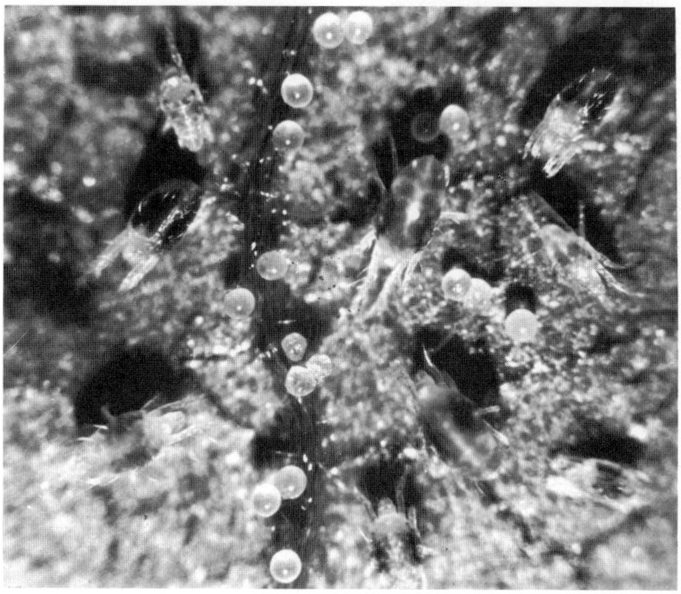

Fig. 9-13. Adults, nymphs, and eggs of the red spider mite. (From Hussey et al., 1969)

gnats are commonly found feeding on dead or decaying tissue. The dark gray or black flies are easily seen near the soil when disturbed. In large infestations, the larvae may feed on live plants, particularly seedlings and rooted cuttings.

Probably the most important pests are Isopods in the order Acarina. This order includes scorpions, ticks, and spiders. While adult insects have three pairs of legs, adult mites have four pairs. A photograph of various stages of the red spider

Fig. 9-14. Typical symptoms of red spider mite attack on roses

Fig. 9-15. Red spider mite webbing on roses. This can be considered as an advance stage of the attack. Control measures should be started before this stage

mite is provided in Figure 9-13. Mites are most active on the leaf undersurfaces, their presence becoming known by a distinctive fine mottling and yellowing (Fig. 9-14). They feed by puncturing the leaf cells and the cells adjacent to the punctures dry out and collapse. In heavy infestations, webs of fine silken strands will be evident and entire plants can be enveloped by sheets of the material (Fig. 9-15). Because of limited movements, resistant mites are often encountered in the same part of the greenhouse year after year. The pest responds to polarized light, availability of food, and photoperiod. Hibernating mites do not resume activity until exposed to several hours below 6° C, so that mites hiding in greenhouse cracks may respond more to outside weather than inside temperatures. Spring emergence is governed by the outside weather. The continued spider problem is largely due to overwintering mites, according to Hussey et al. (1969). The complexity of resistance patterns in different strains of spider mites complicates choice of chemical controls and the ease with which resistant populations have been produced has increased interest in biological, cultural, and hygenic preventive measures. The spider mites belong to the family Tetranychidae. The cyclamen mite, bulb scale mite, broad mite, and fern mite have shining, oval, segmented bodies and are placed in the family Tarsonemidae. The cyclamen mite has been observed on several ornamental species. The females congregate in unopened buds and between folded leaflets, with a life cycle of 10–14 days. Large concentrations of eggs resemble a dust coating. The foliage is usually curled, brittle, twisted, and covered with small lesions. Infested bulbs may be recognized in storage by abnormal dryness, with the outer scales tightly adherent, small and soft. Yellow flecks and streaks, resembling a virus, may occur on affected foliage. Broad mites feed mostly on young and tender foliage, producing misshapen, downward curved leaves, stunted growth, and bud abscission in some host species. Mites in other families are associated with fungi such as *Fusarium*, or bacteria. Infestation is often followed by bud, corm, or bulb rot in freezias, gladioli, hyacinths, tulips, and carnations.

Common pests in humid greenhouses are slugs and snails. They are usually active at night browsing on the vegetation. Some species which tend to be climbers will attack flowers and leaves. Severe damage may occur to roots and bulbs and seedlings may be completely eaten. Most infestations are caused by introducing the pest in soil or compost, on the bottoms of boxes or pots, or if the surroundings of the greenhouse are covered with rough vegetation, they may crawl through side ventilators or cracks in the structure. Symphilid infestation may follow introduction of manure or infested pot plants. The pest feeds on young and tender roots and can initiate root rots by fungi and bacteria. Stems may be hollowed out and damaged host plants wilt readily. Symptoms may be patchy tending to be more frequent along paths, walls, and pipes. Control in ground beds is difficult as the pest migrates to lower depths preventing effective treatment.

9.4 Disease and Pest Control

Methods for controlling or preventing the ravages of diseases and pests are as varied as the organisms themselves. We can, for simplification, divide them into methods which fall under the term "prevention" and procedures which seek to "control". Prevention includes ways by which a competitor can be excluded from the environment, or others which manipulate the environment so an organism cannot gain a foothold. Sanitation, clean stock programs, and environmental control fall in this category. Biological competition, parasitism, chemical programs and resistance are some that would logically fall under control. In certain circumstances, the pest may be present, but can be eradicated by sterilizing the environment or removing the host. It is seldom that any single procedure is satisfactory.

9.4.1 Prevention

Prevention is the cheapest means to avoid undesirable competition. If the pathogen is excluded or prevented in some manner from infesting and parasitizing the host, there will be no need for expensive equipment and chemicals. The possibility that there may be crop injury from phytotoxicity of the applied chemical is avoided. The situation where prevention is adequate in itself is rare, but it is a goal worth pursuing.

9.4.1.1 Sanitation

Good sanitation requires a "state-of-mind" or "awareness" similar to that cultivated by airline pilots and other people whose jobs require that they be "aware". Any training program for employees should include continued emphasis on cleanliness. Where employees have been inculcated and trained with a suitable attitude, the greenhouse operator will usually find that many of his problems have disappeared or never arise. Sanitation includes: (1) washing tools and containers; (2) keeping hose-ends hung up; (3) keeping the environs of the greenhouse weed-free with the outside cleaned and mowed; (4) keeping contaminated tools and feet off of benches and decontaminated areas; (5) closely inspecting all plant material brought into the greenhouse; (6) isolating soils and containers from possible contamination; (7) using disinfecting solutions wherever necessary to prevent recontamination and transfer of diseases and pests. Emphasis on these factors will provide for happier employees and more efficient operation. Cleanliness not only reduces the chances of infesting new areas, but removes possible breeding places for organisms which can infest or inoculate the crop. Weeds and high grass are ideal carry-over areas for such pests as red spider, aphids, and slugs, as well as others. Cautious use of herbicides (Table 8-8) to eradicate plant material immediately around the greenhouse foundation can make a difference in slug and snail infestation. Numerous insects are quite happy if weeds are allowed in or under benches. Chemical control programs become more difficult and costly.

Keeping hose-ends hung up often impresses employees as to the grower's attention to a viable sanitation program. Hose-ends left lying on contaminated ground are probably the most common means of inoculating sterilized soil. Automatic irrigation systems wherever possible go far to reduce the chances of infestation and are one of the advantages over hand watering. Raised benches, unfortunately, are of the height to offer resting places for the feet of lazy people. Individuals should be firmly, but courteously, warned to keep both feet on the ground. This may not always be possible where the crop is grown in the ground or cutting height requires walks on the benches. Some greenhouse operators will provide a footbath filled with disinfectant at the greenhouse entrance (see Table 9-4).

Table 9-4. Disinfectants for use in the greenhouse (Cornell Recommendations, 1975)

Material	Rates	Remarks
LF-10 (65% solution)	1 part LF-10 200 parts water	Drench empty benches, potting benches and shelves. Soak tools and watering systems for 30 min. Flush with clear water. Not a substitute for steam pasteurization. Not effective against *Fusarium*, *Verticillium*, *Thielaviopsis*, or *Cylindrocladium*
	1 part LF-10 100 parts water	Dip pots for one hour after cleaning
Amphyl		(Related product, follow label directions)
Sodium hypochlorite (Clorox) (5.25%)	1 part Clorox 9 parts water	Useful for pots, tools, benches, and watering systems. Soak latter for 30 min. Can be used without flushing. May irritate eyes and skin. Not effective against *Fusarium* and *Verticillium*. Looses potency after standing
Formaldehyde (40%) (Formalin)	1 part Formalin 50 parts water	For sterilizing empty benches, tables, tools, etc. Not effective on nematodes or weeds
Captan + Ferbam + Terraclor	1–2 g each per l water. 50, 75, and 76% WP	Drench for empty benches. Not a substitute for steam pasteurization or for bench soils. Effective against fungi only
Captan + Ferbam	230 g each in sufficient water to cover 100 m^2 50 and 76% WP	Effective on *Fusarium* stem rot of carnations

Employees should be trained to disinfect any tool or object that is likely to come into contact with sterilized soil. A pathogen introduced into a medium where there are no other organisms to provide competition is likely to cause greater crop damage than where competitors are present. Containers that have been set on contaminated soil or used previously, should always be washed and disinfected before being set on potting benches or used again. Transfer of pathogens, particularly viruses, can occur when employees move from an infested area to a clean area without washing their hands. Tobacco in tomato greenhouses should be strictly prohibited. Cutting knives are a good means for transfer. Where there is possibility of transferring a bacterial or systemic disease from one bench to another, buckets of disinfectant should be placed so cutting tools can be dipped after finishing with operations on each bench or in a particular area.

Isolating soils and containers from possible contaminating sources avoids the possibility that a job of disinfecting may not be adequate. An extreme case of isolation can be noted in Figure 9-16, where the high cost of individual plants and the fact that each plant eventually supplies thousands of cuttings, warrants particular caution. Pressurized greenhouses with screens to exclude pests are not economical for production greenhouses, but are common procedures for protecting mother plants in clean stock programs. New plant material should always be closely inspected and observed over several days so that control procedures can be started immediately if there are signs of infestation. Rogueing, or removal of

Fig. 9-16. Method of isolating mother plants in a carnation, cleanstock program. Each plant is in its own container, isolated from the ground, and watered with its own supply. The greenhouse is fan evaporative pad cooled, but exhaust fans are oriented to blow into greenhouse with air taken through fine screens. Pressurization tends to prevent insects from entering. Should a plant show disease symptoms, it can be discarded without danger of contaminating others in house

infected plant material is a common procedure. Any plant showing signs of advanced disease symptoms should be immediately removed and destroyed. Plants showing signs of pest infestation should be isolated and treated before moving them into the main range, or otherwise discarded if eradication is not possible. Some growers will spray the inside greenhouse surface areas during replanting with formalin or amphyl. Where the entire greenhouse is empty, a stronger fumigant can be employed to eradicate pests that may be in cracks and crevices of the house.

9.4.1.2 Clean Stock Programs

Many diseases of economic crops are systemic, capable of being transferred through the propagation stock. Fusarium and Verticillium wilts, bacterial wilts, most viruses, and others can cause serious losses if not eliminated from the plant entirely. Years ago, it was usual to send an employee to the production bed to obtain cuttings. If the plants were infested, then most of the cuttings would also be infested. Furthermore, many mutations have a tendency to increased side shoots. The wild-type mutant of carnation, for example, breaks prolifically and the employee was likely to wind up with most of the cuttings being wild-type. Thus, clean-stock programs are not viable without an ongoing selection program to continually prove the varieties that are to be incorporated in clean-stock production. Novices, and many trained researchers, commonly fail to appreciate the tremendous variation that can exist in so-called, genetically uniform, vegetatively

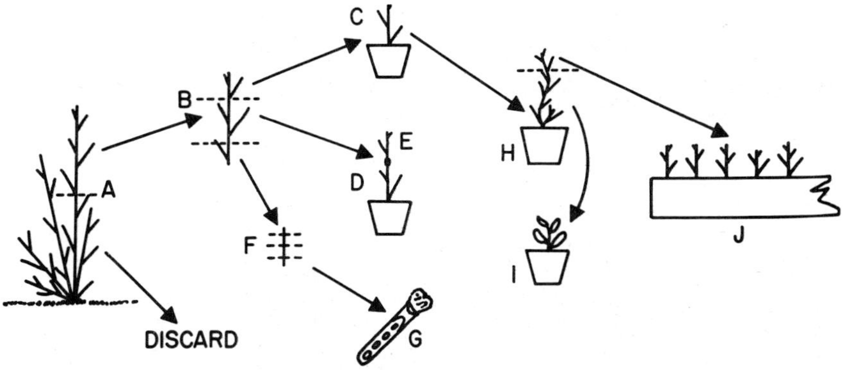

Fig. 9-17. Indexing method for chrysanthemums to eliminate internal bacteria, fungi, and viruses. *A* Original plant selected in a regular reselection or breeding program as desirable for distribution to the trade. *B* The terminal cutting is cut into three pieces. *C* The top part is rooted in sterile media. *E* The middle portion of the original cutting is rooted and a scion of a healthy indicator plant is grafted to detect stunt virus. *F* Basal segments are surface sterilized and placed on sterile agar, *(G)* to detect vascular fungi and bacteria. *H* Juice obtained from rubbings on the lower leaves of the rooted terminal cutting is placed on tobacco plants, *(I)* to detect other viruses. *J* If all tests are negative, cuttings are taken from the now proven mother cutting for increase and eventual distribution. (From Dimock et al., 1964).

propagated clones. This is probably one of the reasons that growers many years ago complained of a variety "running-out". With carnations, individual plants from a clone are selected by trial for the clean-stock program, and after going through the program, must be proved again for possible mutation before cuttings are shipped to the producer. A clean-stock program is a long-term, expensive procedure with the producer often investing $ 75.00 or more in a single plant before he begins to increase it for cutting production.

The simplest procedure is to make a cutting from a selected plant. Several small cross sections are removed from the base, surface-sterilized and placed in sterile liquid media. The terminal part of the original cutting is rooted under sterile conditions. If, after a suitable incubation period, no growth of a contaminating fungus or bacterium is found in the medium containing the basal cross sections, the cutting is assumed free of systemic pathogens and can be grown for cuttings which are then distributed to flower producers. The procedure does not take into account viruses that may be present. Stunt virus threatened the chrysanthemum producers and Dimock (1943a, b, 1947, 1953) and others worked out the system shown in Figure 9-17. As before, a terminal cutting is made and basal cross sections removed to test for internal diseases. The middle part of the cutting is rooted and a scion of a healthy indicator variety is grafted on it. If the stunt virus is present, the scion will show obvious symptoms. The terminal portion is rooted and grown in sterile media. Juice obtained from the lower leaves of this part is tested on tobacco to determine if Aspermy and other viruses are present. If, after about six weeks, test for viruses and internal bacteria and fungi are negative, cuttings are taken from the rooted terminal part for increase and distribution. Any plants shown to be infected are destroyed.

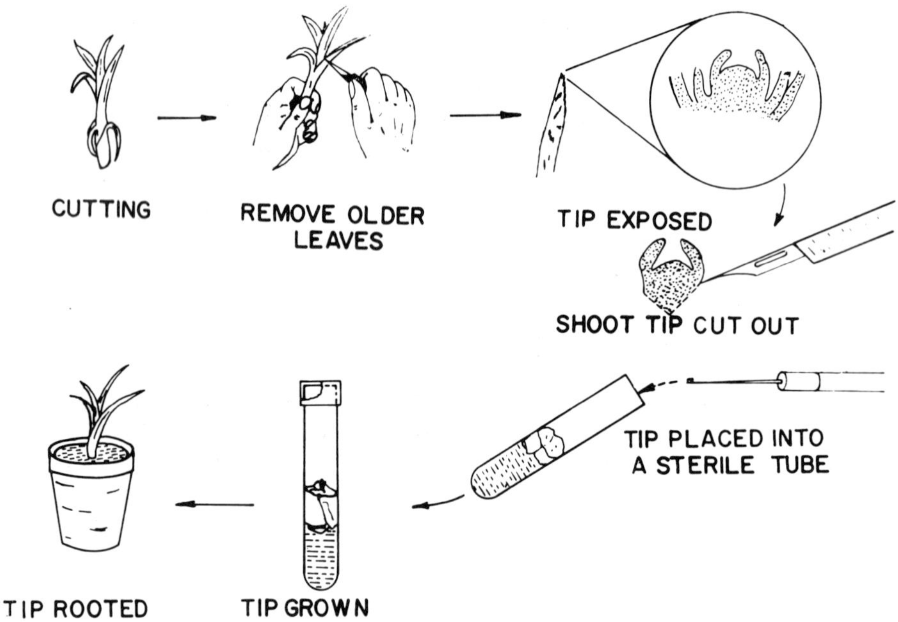

Fig. 9-18. Procedure for growing a carnation shoot-tip in a clean-stock program. A cutting selected in a reselection or breeding program as desirable for production may be first heat treated to reduce the possibility of viruses in the meristem. A 0.2 to 1 mm-long tip is excised and transferred aseptically to a liquid nutrient medium. The tip is supported on a filter paper wick. After 6 to 8 weeks, if there is no evidence of bacterial or fungal growth, the now rooted shoot tip is transfered to a sterile substrate which can be sand or gravel. Cuttings from this mother plant are tested under production conditions to ensure there have been no deleterious mutations. After proving, additional cuttings are placed in a mother block for increase and eventual distribution. (From Baker and Phillips, 1962)

The program outlined above has been shown to be extremely effective for chrysanthemums and is now a standard procedure for propagators. For carnations, Tammen et al. (1956) and Phillips (1962) worked out a different method which depends on the fact that the terminal part of a plant is most likely to be free of internal pathogens. Also, the carnation growth habit means that a sterile shoot-tip can be removed and grown aseptically without running into toxicity complications from surface sterilizing chemicals (Fig. 9-18). Viruses are inhibited by temperatures of 100° F or higher, but carnations can survive if handled properly. The selected cutting can be subjected to a period at high temperature before shoot-tipping. After the tip is rooted, cuttings from the mother plant are tested in a selection program to make sure there have been no deleterious mutations before cuttings from the mother plants are distributed.

It is obvious that clean-stock programs require considerable investment and several years for fruition. The fact that they have been successful is shown in their utilization by commercial propagators and extension of similar programs to other major crops such as geraniums. Propagative material is expensive and sanitation programs to prevent reinfection are a necessity. In recent years, an attempt has been made to reduce costs through so-called rapid asexual multiplication programs. As studied by Ben-Jaacov and Langhans (1972), and others (Davis, 1975), shoot-tips are placed in sterile media and additional shoot-tips are proliferated from the original under aseptic conditions. These are separated, rooted, and grown for production areas. This bypasses the need for greenhouse space and several years to build up cutting stock with attendant possibilities for recontamination. Differences in mutation rates for carnations have not been observed (Hanan and Mueller, unpublished), but there are still difficulties in obtaining rooted plants that can be handled by the producer.

9.4.1.3 Soil Sterilization

The numerous organisms that can survive and multiply saprophytically in soils, but still parasitize economic plants, means that an eradication program for root substrates is an important part of the disease and pest program. *Pythium* and *Rhizoctonia* are ubiquitous soil-borne pathogens. Nematodes can buildup over a period of years to damaging levels, so that programs to treat the soil are a requirement. Prior to the introduction of steam pasteurization, the only practical procedure was to change soils each year, an expensive and time-consuming operation. In recent times, new chemicals have appeared (Table 9-5) that can be utilized in the place of steam. As noted, however, their use requires care as several are toxic to humans and plants, and may have residual effect. Several days may be required for completely aerating the soil before the medium can be replanted. In a tight crop rotation plan, one to three weeks can be too long.

At temperatures of 71° C, almost all pathogenic bacteria, fungi, and most plant viruses are destroyed. Most weed seeds are killed when temperature exceeds 79° C. Although other methods of heating can be used, steam is the most common (Fig. 9-19); Baker (1957) thoroughly discussed methods. Essentially, steam is allowed to flow through the soil to raise the soil temperature to 87° C for 30 min. On ground or raised beds, an impermeable cover is utilized to impound the steam. Difficulties arise with ground beds as it is difficult to force air out of the soil and penetration is limited. In raised beds, the cover is fastened to the bench edges and steam introduced under the cover. The steam is forced through the soil layer and air can be displaced out the bottom. Bulk quantities of soil can be steam pasteurized for potting and it is a standard greenhouse procedure.

Heating to 82° C, or above, destroys all organisms and a considerable danger results from possible recontamination by a pathogen. Without competition from other organisms, the pathogen may grow very rapidly with serious damage. To avoid this, California workers introduced aerated steam. The temperature of steam can be controlled by mixing air and steam. By reducing the temperature to 71° C, several organisms remain alive to colonize the soil and prevent, or reduce, any damage should inoculation by a pathogen occur. However, assuring that all parts of the soil reach an adequate temperature is sometimes difficult, so that aerated steam has not been employed on a large scale.

Steam pasteurization has other advantages in addition to destruction of diseases and pests. The method will usually improve the physical characteristics of the soil structure so that better water and air relationships will commonly result

Table 9-5. Soil treatment materials for control of fungi, nematodes, and weeds (Cornell Recommendations, 1975; Hamlen et al., 1976)

Material	Uses	Min. soil temp (°C)	Remarks
Steam	Weeds, seeds, fungi, bacteria, nematodes, soil insects		See discussion in text. Avoid recontamination, 82° C for 30 min at coolest point. Use an accurate thermometer
Chloropicrin Picfume Niklor Larvacide	Weeds, most fungi, nematodes, soil insects, *Verticillium*	16° C	Applied with hand applicator or self-propelled fumigator. Rate of 2–2$^1/_2$ cc on 25 cm centers, or 5.4 kg/100 m^2. Sprinkle surface immediately or cover with gas-tight cover for 24 h. Aerate for 10 to 21 days. 600 g/m^3 actual. *Toxic to plants*
Vapam, VPM, Sistan, Trimaton	Nematodes, most weeds, and fungi	10° C	1 10^{-2} m with sprinkler or proportioner. Water in thoroughly. Do not plant for two to three weeks after treatment. *Toxic to plants*, although reported as used close to carnation plants
D-D, Vidden D, Nemafene, Dorlone	Nematodes and insects	10° C	Rates vary from 200 to 375 l/ha, applied with hand injector or tractor equipment. Wait 14–28 days before planting. *Toxic to plants*
Vorlex, Di-Trapex	Nematodes, weeds, and fungi	10° C	Applied with hand injector or tractor equipment. Wait 14–28 days before planting. *Toxic to plants*. Cover with plastic
Pestmaster, EDB-85 Bromofume, Soilbrom, Kopifume	Nematodes and insects	10° C	Applied at 30 to 60 l actual per hectare with hand injector or tractor equipment. Wait 14–28 days before planting. Moderately toxic to plants
Formalin	Fungi and bacteria	16° C	160 ml/l of water sprinkled over surface of thin layer of soil and mixed for damping-off control. Soil should be most before treatment. Air until all odor of formaldehyde is gone
Methyl bromide Dowfume MC-2 Brozone	Bacteria, fungi, weeds, insects, nematodes, *Verticillium* not controlled	10° C	Use tight plastic cover, 640 g/m^3 or 180 g/m^2. Air for 3 to 7 days before use. *Do not use* if planting to carnations or salvia. Moderately toxic to plants. Exposure to fumigant for 1 to 2 days. *Hazardous chemical*
Nicofume, Crag fungicide, Mylone, Temik, Vydate L, Mocap, Dasanit mylone	Nematodes, some weeds and fungi	10° C	Read label. See Hamlen et al. (1976) for foliage plants tolerant to nematacides. Make tests with a few plants. Fumes of some listed are toxic

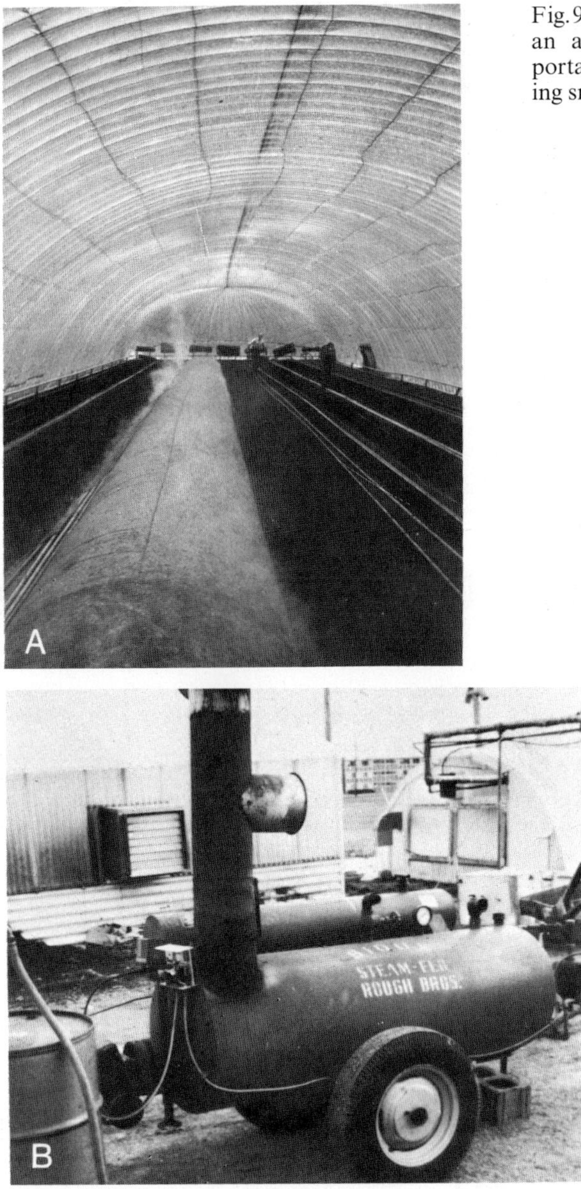

Fig. 9-19. (A) Steaming a ground bed in an air-inflated greenhouse; (B) small, portable steam generator for pasteurizing small soil quantities

in greater growth response. The high temperature will, however, increase the release and solution of chemical elements. If salts are already high prior to steaming, serious injury to the following crop can occur (Fig. 9-20). Steamed soils should be planted as soon as they are cool. Allowing them to stand for several days may cause a gradual buildup of salts to toxic levels. If manure has been added, release of ammonia in toxic quantities can occur. Another advantage of steam, as contrasted to chemical sterilants, is that a crop can be planted almost

Fig. 9-20. Steaming injury to chrysanthemums. Heating causes greater release of salts, and if a bed is allowed to stand for any lengthy period, roots of new plants are likely to be damaged. Ammonia release can also cause similar problems

immediately. The procedure is expensive, fuel costs alone exceeding $ 2500.00 per ha.

Precautions in steaming, also applicable to chemical sterilants, include making sure that the soil is neither too wet or too dry. Bacteria can resist heat if in the dry form, and air-dry soils seldom steam satisfactorily. Too wet soils, on the other hand, require excessive quantities of steam to bring them to the proper temperature and can cause loss of desirable physical properties through "puddling". Soils which are not uniform, or adequately tilled, may show nonuniform crop response later as the steaming was not uniform.

9.4.1.4 Environmental Control

Sanitation, clean stock, and soil sterilization or pasteurization provide three legs of a good preventive program. The fourth leg is manipulation of the greenhouse environment to reduce opportunities for a pathogen or insect to establish itself. Failure to train employees in all four areas, or to neglect meticulous attention to any of them, seriously compromises a prevention program. It is a certainty that a grower will be forced into more expensive efforts.

It will have been noted in Section 9.3, that predisposing conditions for the majority of fungi and bacteria are high temperatures, high humidity, and low air velocities. The best practice a grower can use is to avoid conditions that will cause condensation on the crop. This involves controlling temperatures and relative humidity so that plant and air temperatures do not reach the dew point (see Chap. 5). This quite commonly occurs when day temperatures are turned to the night setting. If the dew point is high, a slight drop in temperature will increase relative humidity to 100%, or plant cooling will cause water condensation on foliage and flowers. Dew point in the greenhouse can be reduced by turning off the evaporative pad system prior to sunset and allowing the exhaust fans to dry out the greenhouse. Plants should never be watered late in the afternoon. High volume, spray irrigation systems are particularly undesirable in this respect as they wet the interior foliage where air movement is nearly nonexistent, and hence the foliage may not dry soon enough to prevent infection.

Some crops should never be wetted, especially when flowering. At this point, improperly trained employees with a hose can cause significant damage. Humidity control is particularly important for powdery mildew on roses as the fungus is most active at relative humidities between 95 and 99%. A common control regime is to ventilate and heat the greenhouse simultaneously in order to drive out moisture-laden air and prevent relative humidity from staying in the critical range. During the winter months in arid climates, the outside dew point is low so that mildew is seldom a problem. As outside air temperatures increase, however, growers tend to turn boilers off to avoid fuel costs, and mildew can nearly always be found during the spring and summer months. Thus, judicious heat and ventilation is doubly critical during those times of the year.

The installation of fan and pad cooling is an effective preventive measure as it permits healthier growth, better ventilation, and reduced temperatures so that disease and pest reproduction cannot reach astronomical levels. Growing conditions that subject the crop to unnecessary stress may hasten disease development so that attention to good cultural procedures and proper environment often contribute markedly to reduced insect and disease problems. However, reproduction rates of some insects such as aphids are influenced by nutrition. Healthy, vigorously growing crops can increase insect fertility. Excessive nitrogen has been known to increase damage from some of the bacterial blights. Prevention by environmental manipulation is, therefore, limited by the crop requirements.

9.4.2 Control

Preventive programs have eliminated some diseases of greenhouse crops such as bacterial and fungal wilts of carnations and chrysanthemums. That they may occur results from reinfection in the cut flower production process. Serious losses from root rots and damping-off can be prevented by soil steaming programs, combined with good sanitation. There are numerous pests and diseases, however, which preventive programs cannot eliminate without some recourse to additional measures. If a pest cannot be eradicated, its ravages must be reduced to a level where reasonable economic return can be achieved. The measures available to the grower are resistance, biological control, and chemical control.

9.4.2.1 Resistance

Breeding for resistance to specific pathogens is an accepted control method for major food crops. Resistance can be provided by one or more genes incorporated in the plant material through appropriate breeding programs, such as rust resistance in wheat, resistance to *Fusarium* in tomatoes, and resistance in potatoes to potato blight. Its importance was recently emphasized in the outbreak of corn blight in the United States, where the genes conferring resistance had not been incorporated in hybrid corn breeding parents. An example of resistance to an insect is given in Figure 9-12.

Difficulty may arise as resistance is often conferred to specific strains of the pathogen, and slight mutations may permit the pathogen to attack a normally resistant host with increased virulence. Programs for rust resistance in wheat must be continuous. Inherited resistance for most ornamental greenhouse crops has not been pursued with great vigor and it is seldom considered as important. Emphasis has been placed on commercially important characteristics such as color or appearance. It is also difficult to combine genes conferring resistance with the many characters required of ornamental plants. Most growers are aware of varietal differences to insect and disease attack, but none of these are great enough, at present, to warrant investment or to be of practical value.

9.4.2.2 Biological Control

The realization of environmental damage that can result from misuse of chemicals and increasing restrictions on pesticide use, has caused greater interest in biological control. This type of control can assume several forms. Aerated steam where some harmless organisms are allowed to survive in order to provide competition is one form. Numerous experiments have studied the effect of introducing a harmless organism to colonize the root medium and prevent unrestricted pathogen growth. Recently, Baker (pers. comm., 1976) has suggested that a nonpathogenic strain of *Fusarium* might be employed to restrict *F. roseum f. dianthi*. Colonization of the ecological niche normally occupied by the pathogenic *F. roseum* could prevent its survival. Natural predators on insects and related pests are provided in Table 9-6, and Hussey et al. (1969) have provided an excellent discussion of the problems and prospects of this type of control. As with predators on economic plants, the greenhouse environment would also offer similar advantages to their predators (Chap. 9.2). For awhile, the British attempted commercial application of the parasite, *E. formosa* as a control for whitefly. The project was discontinued when DDT became available and the rearing technique for *E. formosa* came under criticism. Spider mites and thrips were occasionally distributed inadvertantly on the leaves carrying the parasitized whitefly scales. Other than this example, no attempt has been made to develop commercially acceptable control programs. The use of natural enemies must be integrated with the chemical control program. For example, sulfur vaporization for controlling rose powdery mildew is lethal to the red spider mite predator *Phytoseiulus*.

Other chemicals may be equally lethal to predators as well as the pest and timing of chemical applications would probably have to be very accurate. Even systemic insecticides could affect the predator as it consumed pests killed by the systemic and accumulated a lethal dose.

According to Hussey et al. (1969), the commercial future of biological control depends upon the development of economic and reliable rearing techniques to provide the required number of natural enemies. A practical scheme requires

Table 9-6. Some predators on greenhouse pests. (From Hussey et al., 1959)

Pest	Predator	Remarks
Leafhopper (*Z. pallidifrons*)	*Anagrus atomus* (wasp)	Breeds continuously at proper temperatures, possible if artificially introduced
Whitefly (*T. vaporariorum*)	*C. aphidicola* (fungus) *Encarsia formosa* (wasp)	Requires saturated atmosphere. Interaction dominated by temperature, commercially applied in Britain for a short time
Scale (*P. citri*)	*L. dactylopii* (parasite) *C. montrouzieri* (ladybug)	Both used in combination and introduced artificially
Aphid (*M. persicae*)	A. matricariae (wasp) *C. montrouzieri* (ladybug)	Shown to be relatively successful
Tomato moth (*L. oleracea*)	*P. instigator* (wasp)	Common parasite, but seasonal emergence not well coordinated with the host
Moths (caterpillars)	*B. thuringiensis* (bacillus)	Common control measure (Cornell Recommendations, 1975)
Red spider mite (*Tetranychus* sp.)	*P. persimilis* (Chilean predatory mite)	Control in four weeks, surviving for four weeks after elimination of the host. Sulfur fumes toxic to predator
Rootknot (*Meloidogyne* sp.)	*Crotalaria* and *Tagetes* produce nematicidal substances	Little prospect for successful use

ability to produce several natural enemies in large numbers. This could only be provided by the government, with firm evidence that the industry would contribute its share. In Britain, chemical control of mites on cucumbers costs over $ 400.00 per acre per year, so that any biological control could cost up to $ 200.00 and still be attractive to the grower.

The use of viruses, fungi, or bacteria to attack pests is attractive. The bacterium *Bacillus thuringiensis* to control caterpillars is a successful and common procedure. Other limitations prevent the use of the fungus *Entomophthora coronata* to control aphids. As most fungi and bacteria require high temperatures and humidity for rapid multiplication, the treatment might be worse than the cure.

9.4.2.3 Chemical Control

Chemical control is the last resort. Numbers, formulations, methods of application, and application rates vary from time-to-time, with season, the environment, and with the specific variety. The fact that several governments have seen fit to impose restrictions on chemical usage attests to its misuse by careless and uninformed applicators.

a) Pesticide Safety. Table 9-7 lists common names, some trade names, and their manufacturers, the chemical class, and the acute oral and dermal toxicity of each. Oral toxicity is generally expressed in terms of the "lethal dose", i.e., the milligrams of toxicant required per kg of body weight to kill 50% of a test population. It can be noted in Table 9-8, from the use of "oral", "dermal", and "inhalation" terms, that lethal dosages can be obtained by three major pathways; through the mouth, through the skin, and by breathing. Chemicals with LD_{50}s less than 200 are hazardous, and hence all precautions must be taken in their use. Most instances of known deaths from chemical pesticides have resulted from ignorance and carelessness. Even chemicals in category IV (Table 9-8) become highly hazardous if misused, or if children get into contact with the material. Action often depends on the size of the organism contacting it.

There are some general rules in handling pesticides in which all employees should be trained:

1. Store chemicals in safe, secured areas. This avoids contamination, and if a container breaks, contamination can be localized. Some regulations require that pesticides be kept under lock and key.

2. Never store pesticides in unmarked containers. This is an invitation to disaster. Not only is there the danger that you, as the grower, may use the wrong material causing injury to the crop you are attempting to protect, but there is the possibility that the wrong person may use the chemical for the wrong purpose, such as drinking it.

3. Read the label and understand it before using the chemical. The label provides information on toxicity, the brand name, common and chemical names, the ingredients, uses and directions for use. For example, Benlate® is the brand name used by the manufacturer, benomyl is the common name accepted for the active ingredient, and the chemical name is (Methyl 1-(butylcarbamoyl) 2-benzimidazolecarbamate). The label states the uses for which the pesticide is registered (Table 9-9), and these are the only legal uses for the material. The directions tell

Table 9-7. Toxicity values of common pesticides

Common name	Trade name	Producer	Chemical class	Acute oral LD_{50}	Dermal toxicity
acephate	Orthene	Ortho	Organic phosphate	866–945	—
aldicarb	Temic	Union Carbide	Methyl-carbamoyl	7 [b]	[b]
aldrin	Aldrite and others	Shell Chemical Co.	Chlorinated hydrocarbon	55 [a]	[a]
Bacillus thuringiensis	Biotrol	Abott Labs	Microbial	+15000	—
	Dipel	Thompson-Hayward			
	Thuricide	Sandoz			
benomyl	Benlate	DuPont	Carbamate	+10000	—
	Tersan 1991				
BHC (benzene hexachloride)	Benzahex	Hooker Chemical	Chlorinated hydrocarbon	600–1250	—
bordeaux mixture	—	Chemical Formulators	Inorganic	300 [a]	M
calomel	Calo-gran	Mallinckrodt	Inorganic salt	210 [a]	—
captan	Orthocide	Ortho	Dicarboximide	9000	—
carbaryl	Sevin	Union Carbide	Carbamate	500–850	—
chloranil	Spergon	Bayer	Benzoquinone	4000	M
chlordane	(many)	Velsicol	Chlorinated hydrocarbon	457–590 [a]	[a]
chlorobenzilate	Acaraben	CIBA-GEIGY	Chlorobenzilate	960	—
cycloheximide	Acti-Dione	Tuco (Upjohn)	Antibiotic	2 [b]	S
DCNA (dicloran)	Botran	Tuco (Upjohn)	Nitroaniline	+ 5000	M
DDT	(many)	Montrose Chemical Co.	Chlorinated hydrocarbon	250 [a]	—
DDVP	Vapona	Shell Chemical	Organic phosphate	56–80 [a]	[b]
demeton	Systox	Chemagro	Organic phosphate	2.5–12 [b]	[b]
diazinon	Spectracide	CIBA-GEIGY	Organic phosphate	300–400 [a]	—
diazoben	Dexon	Chemagro	Diazenesulfonate	64 [a]	M
dichlone	Phygon	Uniroyal	Naphthoquinone	1300	M
dicofol	Kelthane	Rohm and Haas	Trichloroethanol	809	—
dieldrin	Dieldrite	Shell Chemical Co.	Chlorinated hydrocarbon	60 [a]	[a]
dimethoate	Cygon	American Cyanamid	Organic phosphate	320–380 [a]	[a]
dinocap	Karathane	Rohm and Haas	Nitrophenol	980	—
disulfoton	Di-Syston	Chemagro	Organic phosphate	2.6–12.5 [b]	[b]
DNOC	Elgetol 30	Bayer	Nitrophenol	20–50 [b]	[b]
endosulfan	Thiodan	FMC Corporation	Dioxathiepinoxide	30–110 [b]	[a]
endrin	—	Shell Chemical	Chlorinated hydrocarbon	5–45 [b]	[b]
		Velsicol			
		duPont			
ferbam	Fermate	Chipman	Carbamate	+17000	M
fixed coppers	(many)	(several)	Inorganic	3000–6000	M

Table 9-7 (continued)

Common name	Trade name	Producer	Chemical class	Acute oral LD_{50}	Dermal toxicity
folpet	Phaltan	Chevron	Phthalimide	+10000	M
		Stauffer Chemical			
fonofos	Dyfonate	Stauffer Chemical	Organic phosphate	8–17.5 [b]	[b]
formaldehyde (formalin)	—	(several)	Methanol	800	M
gamma BHC (Lindane)	(many)	Hooker Chemical	Chlorinated hydrocarbon	88–125 [a]	[a]
heptachlor	Drinox-H34	Velsicol	Chlorinated hydrocarbon	40–188 [b]	[b]
lead arsenate	—	FMC Corporation	Inorganic salt	10–50 [b]	—
lime-sulfur	Orthorix	Chevron	Calcium polysulfides	low	S
malathion	Cythion	American Cyanamid	Organic phosphate	1375	—
mancozeb	Dithane M-45	Rohm and Haas	Carbamate	+ 8000	—
maneb	Dithane M-22	BASF Wyandotte	Carbamate	6750	M
	Manzate D	duPont			—
methomyl	Lannate	duPont	Thioacetimidate	17 [b]	—
methoxychlor	Marlate	duPont	Chlorinated hydrocarbon	6000	—
methyl parathion	(many)	Monsanto	Organic phosphate	9–25 [b]	[a]
		Stauffer			
mevinphos	Phosdrin	Shell Chemical	Organic phosphate	3.7–12 [b]	[b]
mexacarbate	Zectran	Dow Chemical Co.	Carbamate	19 [b]	[a]
mirex	Dechlorane	Allied Chemical	Butapentalene	306 [a]	[a]
nabam	Dithane D-14	Rohm and Haas	Carbamate	395 [a]	M
naled	Dibrom	Chevron Chemical	Organic phosphate	430 [a]	[a]
nicotine	Black Leaf 40	Prentiss Drug and Chem.	Pyridine	50–60 [a]	[a]
oxamyl	Vydate L	duPont	Thioxamimidate	5.4–37 [b]	—
oxycarboxin	Plantvax	Uniroyal	Carboxanilide	2000	—
oxydemeton-methyl	Meta-Systox®	Chemagro	Organic phosphate	56–65 [a]	[a]
oxythioquinox	Morestan	Chemagro	Quinoxalinone	2500–3000	—
parathion	(many)	Bayer	Organic phosphate	13 [b]	[b]
		Monsanto			
parinol	Parnon	Elanco (Lilly)	Pyridine methanol	5000	M
paris green	—	Los Angeles Chem.	Arsenical	22 [b]	—
PCNB	Terraclor	Olin Corporation	Nitrobenzene	200 [a]	M
—	Pentac	Hooker Chemical	Cyclopentadien	3160	—
phorate	Thimet	American Cyanamid	Organic phosphate	2–4 [b]	[b]
piperalin	Pipron	Elanco (Lilly)	Dichlorobenzoate	2500	—

pirimicarb	Pirimor	Plant Protection Ltd.	Carbamate	147[a]	—
	Plictran	Dow Chemical Co.	Hydroxide	540	—
pyrethrum	—	FMC Corporation	Botanical	1500	—
		MGK Company			
rotenone	—	FMC Corporation	Botanical	132[a]	[a]
streptomycin	Agrimycin	Merck Chem. Div.	Antibiotic	9000	M
	Agri-Strep	Charles Pfizer			
sulfotep	Bladafume	Bayer	Organic phosphate	7–10[b]	[b]
TEPP	Vapotone	Miller Chemical	Organic phosphate	1.2–2[b]	[b]
terrazole	Terrazole	Olin Matheson	Thiadiazole	2000	—
	Truban	Mallinckrodt			
tetrachlorvinphos	Gardona	Shell Chemical	Organic phosphate	4000–5000	—
	Rabon				
tetradifon	Tedion V-18	Phillips-Duphar	Sulfone	+14700	—
thiabendazole	Mertect	Merck Chemical	Benzimidazole	3100	—
thiram	Arasan	duPont	Disulfide	780	M
	Tersan				
toxaphene	Strobane T	Hercules Chemicals	Chlorinated hydrocarbon	69[a]	[a]
zineb	Dithane Z-78	Rohm and Haas	Dithiocarbamate	+ 5200	M
	Parzate	FMC Corporation			
ziram	Zerlate	Pennwalt	Dithiocarbamate	1400	M

[a] Moderately toxic.
[b] Highly toxic.
M mild skin irritation, S severe skin irritation. (Condensed from Bohmont, 1976)

Table 9-8. Toxicity tables for grouping chemicals. Oral, dermal, and inhalation ratings of pesticide groups

EPA Category	Pesticide label (signal word)	Oral LD_{50} (mg/kg)	Dermal LD_{50} (mg/kg)	Inhalation LD_{50} (ppm)	Lethal oral dose for a 150 lb (68 kg) man
I. Highly hazardous	Danger Poison	0–50	0–200	0–2000	Few drops to a teaspoon (5 cc)
II. Moderately hazardous	Warning	50–500	200–2000	2000–20000	Teaspoonful to one ounce (30 cc)
III. Slightly hazardous	Caution	500–5000	2000–20000	over 20000	One ounce to 1 pint (0.5 l) or lb (0.5 kg)
IV. Relatively nonhazardous	Caution	over 5000	over 20000	—	Over one pint or pound

Table 9-9. Registered uses of pesticides for greenhouse ornamental crops

Crop	Fungicides	Insecticides
African Violet	Benlate, Captan, Termil, Nemagon, Dexon, Karathane, Terraclor	Dibrom, Nicotine, Sulfotepp, Systox, Vapona, Thiodan
Aster	Benlate, Captan, Termil, Nemagon, Dexon, Karathane, Terraclor	Systox, SBP 1382, Diazinon, Vapona, Kelthane, Thiodan, Malathion, Meta-Systox, Sulfotepp, Dylox
Azalea	Benlate, Captan, Termil, Nemagon, Dexon, Vydate L, Terraclor, Zineb	Systox, Diazinon, Vapona, Thiodan, Malathion, Dibrom, Pentac Sulfotepp
Bedding plants	Banrot (limited), Benlate, Captan, Termil (limited), Nemagon (limited), Truban (limited), Terraclor	Thiodan, Malathion, Dibrom, Nicotine, Meta-Systox (limited), Sulfotepp, SBP-1382
Begonia	Benlate, Captan, Termil (foliage only), Actidione PM (tuberous only), Nemagon, Dexon, Karathane, Terraclor	Systox, Vapona, Kelthane, Thiodan, Malathion, Nicotine, Meta-Systox, Sulfotepp, Dylox
Calceolaria	Benlate, Captan, Termil, Dexon, Terraclor	Systox, Vapona, Meta-Systox, Sulfotepp
Calla	Benlate, Captan, Dexon, Terraclor	Systox, Vapona, Malathion, Meta-Systox, Sulfotepp, Dylox
Carnation	Benlate, Captan, Termil, Dexon, Truban, Ferbam, Phaltan, Maneb, Plantvax, Terraclor, Zineb	Temik, Systox, Diazinon, Vapona, Kelthane, Thiodan, Malathion, Dibrom, Pentac, Plictran, Sulfotepp, Tedion, Dylox
Chrysanthemum	Banrot, Benlate, Captan, Termil, Actidione PM, Dexon, Karathane, Truban, Ferbam, Phaltan, Maneb, Vydate L, Terraclor, Pipron, Agri-strep, Zineb	Temik, *B. thuringinensis*, Systox, Diazinon, Vapona, Kelthane, Thiodan, Malathion, Dibrom, Meta-Systox, Pentac, Pirimor, Plictran, Sulfotepp, Tedion, Dylox
Cineraria	Benlate, Captan, Termil, Dexon, Terraclor	Systox, Vapona, Meta-Systox, Sulfotepp, Dylox
Crocus	Benlate, Captan, Dexon, Terraclor	Systox, Vapona, Meta-Systox, Sulfotepp, Dylox
Cyclamen	Benlate, Captan, Termil, Dexon, Terraclor	Systox, Vapona, Thiodan, Malathion, Dibrom, Nicotine, Meta-Systox, Sulfotepp, Dylox
Daffodils	Benlate, Captan, Dexon, Terraclor	Systox, Vapona, Meta-Systox, Sulfotepp, Dylox
Dahlia	Benlate, Captan, Dexon, Terraclor	Systox, Vapona, Thiodan, Dibrom, Nicotine, Meta-Systox, Sulfotepp, Dylox
Delphiniums	Benlate, Captan, Dexon, Terraclor	Systox, Vapona, Thiodan, Dibrom Nicotine, Meta-Systox, Pentac, Sulfotepp, Dylox
Dutch Iris	Benlate, Captan, Dexon, Terraclor	Systox, Vapona, Meta-Systox, Sulfotepp, Dylox
Foliage plants	Banrot (limited), Benlate, Captan, Termil (limited), Nemagon (limited), Dexon, Truban (limited), Vydate L (limited), Terraclor	Nicotine (limited), Parathion, Sulfotepp

Table 9-9 (continued)

Crop	Fungicides	Insecticides
Fuchsia	Benlate, Captan, Termil, Dexon, Terraclor	Systox, Vapona, Thiodan, Dibrom, Meta-Systox, SBP-1382, Sulfotepp, Dylox
Freesia	Benlate, Captan, Dexon, Terraclor	Sulfotepp
Gardenia	Benlate, Captan, Dexon, Terraclor	Systox, Vapona, Kelthane, Thiodan, Dibrom, Nicotine, Meta-Systox, Pentac, Sulfotepp, Dylox
Geraniums	Banrot, Benlate, Captan, Termil, Dexon, Truban, Terraclor	Systox, Vapona, Thiodan, Malathion, Dibrom, Meta-Systox, SBP-1382, Sulfotepp, Dylox
Gerberas	Benlate, Captan, Dexon, Karathane, Terraclor	Temik, Systox, Vapona, Thiodan, Dibrom, Meta-Systox, Sulfotepp, Dylox
Gloxinia	Benlate, Captan, Termil, Dexon, Terraclor	Systox, Vapona, Thiodan, Dibrom, Nicotine, Meta-Systox, Sulfotepp, Dylox
Gladiolus	Benlate, Captan, Nemagon, Dexon, FORE, Maneb, Vydate L, Terraclor, Zineb	Systox, Vapona, Meta-Systox, Sulfotepp, Dylox
Hyacinth	Benlate, Captan, Dexon, Terraclor	Systox, Vapona, Meta-Systox, Sulfotepp, Dylox
Hydrangea	Benlate, Captan, Termil, Nemagon, Dexon, Karathane, Terraclor	Systox, Vapona, Kelthane, Thiodan, Dibrom, Nicotine, Meta-Systox, Sulfotepp, Dylox
Easter lily	Benlate, Captan, Termil, Dexon, Truban, Terraclor	Temik, Systox, Vapona, Thiodan, Malathion, Dibrom, Nicotine, Meta-Systox, Sulfotepp, Dylox
Orchids	Benlate, Captan, Dexon, Terraclor	Temik, Systox, Vapona, Thiodan, Malathion, Dibrom, Nicotine, Meta-Systox, Sulfotepp, Dylox
Poinsettias	Banrot, Benlate, Captan, Termil, Dexon, Truban, Phaltan, Terraclor	Temik, Systox, Vapona, Thiodan, Dibrom, Nicotine, Meta-Systox, Pentac, Plictran, SBP-1382, Sulfotepp, Dylox
Roses	Benlate, Captan, Termil, Actidione PM, Dexon, Karathane, Ferbam, FORE, Maneb, Vydate L, Parnon, Terraclor, Pipron, Agri-strep, Zineb	Temik, Systox, Vapona, Kelthane, Thiodan, Malathion, Dibrom, Nicotine, Meta-Systox, Pentac, Plictran, Sulfotepp, Tedion, Dylox
Snapdragon	Benlate, Captan, Termil, Dexon, Karathane, Ferbam, Phaltan, FORE, Maneb, Vydate L, Terraclor, Zineb	Temik, Systox, Diazinon, Vapona, Kelthane, Thiodan, Malathion, Dibrom, Nicotine, Meta-Systox, Pentac, Sulfotepp, Dylox
Sweetpeas	Benlate, Captan, Dexon, Terraclor	Systox, Vapona, Meta-Systox, Sulfotepp, Dylox
Tulips	Benlate, Captan, Dexon, Ferbam, FORE, Terraclor	Systox, Vapona, Meta-Systox, Sulfotepp, Dylox

Note that the method of application may limit use of a pesticide; (limited) indicates that the pesticide may not be registered in the plant group; Read the label (Powell and Lindquist, 1975).

Fig. 9-21. Example of protective clothing for pesticide application. Includes rubberized rain gear, boots, and gloves. In tall crops, rain hat may be required in addition to gas mask

the operator when to use the chemical, how to apply the particular formulation, where to apply it, and how much. Safety precautions will be given. Although there are several tables in this chapter providing dosages for pest control, they cannot be used as legal justification for misuse of a chemical, nor do such directions in any publication supersede what is written on the label.

4. Maintain proper clothing and protective gear, depending upon application method and chemical type (Figs. 9-21 and 9-22). Despite discomfort in humid, hot greenhouses, insist as a condition of employment that the gear be used in the proper manner.

5. Insist as a condition of employment that no smoking, drinking, or eating be allowed during application, or until protective equipment is removed, the employee changes clothes, if necessary, and thoroughly washes. Do not allow chemicals into areas where food may be exposed.

6. Be aware of the major symptoms of pesticide poisoning in humans (Table 9-10) with some idea of the acceptable antidotes (Table 9-11). At least one person present in the establishment, should have passed a first-aid course.

7. Keep doctor and emergency phone numbers close to the phone. In the United States, there are emergency poison control centers which can be contacted.

8. Dispose of empty containers in the proper manner. This varies depending upon the container and what was in it.

9. Never work alone. Mix chemicals in a well-ventilated area or outside. Wash protective gloves before removing them and replace frequently. Do not use your mouth to siphon chemicals. When filling tanks, handle hoses to avoid back siphoning.

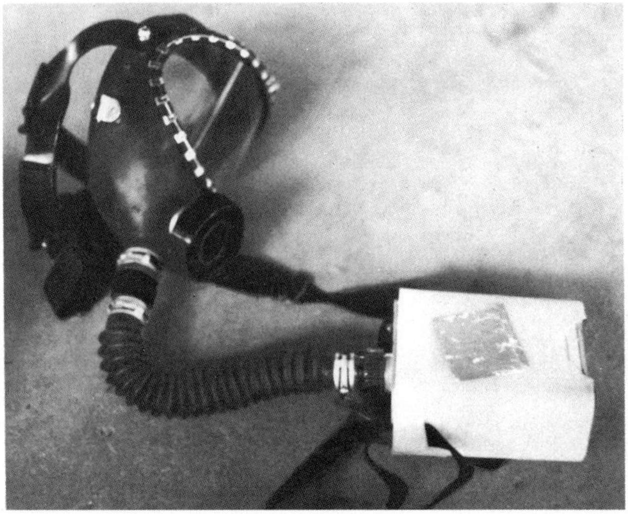

Fig. 9-22. Gas mask with a filter suitable for most organophosphates. Efficacy of gas masks will depend upon their proper maintenance and cartridge replacement

While most chemicals used in greenhouses allow re-entry shortly after application, the grower should be aware of limitations, particularly with food crops. Usually, the label will provide information on how soon workers can handle foliage after an application, or how much time must elapse between an application and sale of the product to the consumer. As noted earlier, a logbook in which all applications, rates, and methods are recorded can serve as a legal record in disputed cases. Where problems in pesticide application arise, such as damage to crops, control failure, etc., a logbook is particularly useful. When a new chemical procedure is attempted, the method should be tested on a few plants to test for phytotoxicity (see Table 9-13).

Chemicals are safe for pest control, if the applicator uses common sense and knowledge. The number of people who have used pesticides for years with complete safety far outweigh the instances of misuse. In several situations, chemical control is the only means for obtaining an economic return in greenhouse production.

b) Application. Chemicals may be applied as smokes, dusts, aerosols, mists, in granular form, as drenches or dips, as baits, or as liquid sprays. Several types of equipment are shown in Figure 9-23. The application method will depend upon the type of material, the pest to be controlled, restrictions on the chemical's use and local conditions. Aerosol bombs, smokes and thermalfoggers require that the volume of the house be known, and are generally easier and quicker to use. Granular materials are often used when applied to soils for systemic insecticides and where the material in others forms is extremely toxic. Granular materials are more expensive, but application costs are less. The highly useful, systemic insecticide, Temik® is an example. The material is extremely toxic, however, requiring protective clothing and cautious use. It can be applied in spreaders as long as the granules are not ground into finer particles.

Table 9-10. General symptoms of pesticide poisoning in humans (U.S. Navy, 1972)

Class	Examples	Toxicity	Symptoms
Chlorinated hydrocarbons	Lindane, DDT, Dieldrin, BHC, Chlordane, Endrin, Dicofol, Mirex, Thiodan, Aldrin, Kelthane	Mild to high	20 min to 4 h. Nausea, vomiting, restlessness, tremor, apprehension, convulsions, coma, respiratory failure, death
Organic phosphates	Malathion, Diazinon, Vapona, Dibrom, Cygon, Systox, Phosdrin, Meta-Systox R, Thimet, TEPP, parathion, Sulfotepp	Moderate to extreme	MILD–Headache, dizziness, weakness, anxiety, tremors, impairment of visual acuity MODERATE–Nausea, salivation, cramps, vomiting, sweating, slow pulse, tremors SEVERE–Diarrhea, pinpoint pupils, pulmonary edema, cyanosis, convulsions, coma, heart block
Carbamates	Sevin, Zectran, Pirimor, Benlate, Fermate, Dithane, Thiram, Vapam, Parzate, Zerlate, Temik	Slight to moderate	Constriction of pupils, salivation, sweating, lassitude, nausea, vomiting, diarrhea, tightness in chest
Arsenicals	Sodium arsenite, paris green	Mild to high	30 min to several h Vomiting, profuse painful, later bloody diarrhea, dehydration, thirst, cyanosis, headache, dizziness, delirium or stupor, skin eruption, convulsions
Halogen fumigants	Methyl bromide	Mild to high	4 to 12 h Dizziness, headache, nausea, vomiting, abdominal pain, lassitude, weakness, slurring speech, mental confusion, mania, tremors, cyanosis
Cyanide fumigants	Hydrocyanic acid, Cyanogas, acrylonitrile, prussic acid	Extreme	Massive dose, death without warning. Bitter, acid burning taste, constriction of throat, salivation, nausea, no vomiting, anxiety, confusion, odor of bitter almonds
Coumarins, Indandiones	Warfarin, Fumarin	Slight to high	After repeated ingestion for several days, bleeding from nose, gums, into urine and stool, pallor, shock
Organic acids	2,4-D, 2,4,5-T, Silvex	Mild	Weakness, lethargy, diarrhea, ventricular fibrillation, cardiac arrest and death
Ureas	Monuron, diuron, Betasan, Bromacil	Mild	Irritation of eyes, nose, throat, and skin. May cause gastroenteritis on ingestion
Miscellaneous	Diquat, Paraquat	Mild to moderate	Lethargy, convulsions, coma

Table 9-11. Medical antidotes for pesticide poisoning

Group	Type	Examples	Antidote
I	Organo-phosphates	Azodrin, Bidrin, Trithion, Vapona, Systox, Diazinon, Cygon, Dyfonate, Meta-Systox, parathion, Phosdrin, TEPP	Atropine sulfate. Protopam chloride (2-PAM) *Do not use* morphine, theophyllin, aminophyllin, barbiturates
II	Carbamates	Zectran, Temik, Lannate, Sevin	Atropine sulphate. *Do not use* protopam chloride
III	Chlorinated hydrocarbons	Endrin, Deldrin, Aldrin, Lindane, Thiodan	Barbiturates Calcium gluconate, intravenously
IV	Inorganic arsenicals	Paris Green, sodium arsenite (Atlas)	BAL (dimercaprol), intramuscularly
V	Cyanides	Hydrogen cyanide, Cyanogas	Amyl nitrite inhalation, Sodium nitrite-intravenously, Sodium thiosulfate-intravenously
VI	Anticoagulants	Warfarin (Prolin), Fumarin, Valone	Vitamin K, Vitamin C useful
VII	Fluoracetates	Sodium fluoroacetate (1080)	Monacetin (glycerol mono-acetate) intramuscularly
VIII	Dinitrophenols	DNOC (Elgetol 30), dinoseb (DNBP, Premerge)	Life supports, Sodium methyl thiouracil to reduce basal metabolism. *Do not use* atropine sulfate
IX	Bromides and carboxides	Methyl bromide (Dowfume), ethylene dibromide, carboxide	BAL (dimercaprol) *before* symptoms, Barbiturates
X	Chlorophenoxy herbicides, Ureas, etc.	2,4-D, 2,4,5-T, monuron, diuron, Diquat, Paraquat, Endothall	Life supports. No antidotes

Note that antidotes given in this table should be prescribed or given only by a qualified physician (Cornell Recommendations, 1975; Bohmont, 1976).

Some materials are injected through the irrigation system. Examples of these are Cygon and Plant-vax. Benlate, Dexon, and Terraclor have also been injected, but these materials require constant agitation to keep them in suspension. If allowed to settle, they will clog fertilizer proportioners and trickle irrigation nozzles. Mist blowers and aerosol bombs carry concentrated pesticides, the former in a high volume, high velocity air stream. The volume of spray is proportionally smaller. Mist spraying is more complicated than hydraulic spraying, requiring skilled and careful operators. Serious plant injury can result from overspraying. Thermalfoggers are fairly common, some requiring special formulations. They are limited to materials that will not be changed readily by the heat. Substances such as sulfur, Dibrom, and TEPP are often volatilized from steam pipes, or sulfur is continuously volatilized in special generators for powdery mildew control. Dusts may leave unsightly residues.

Fig. 9-23(A–F). Several types of chemical applicators. (A) Gasoline-powered mist blower which can be converted to a duster. (B) High pressure hydraulic sprayer for applying wettable powders or emulsifiable concentrates. (C) The "Cornell" nozzle which provides improved coverage for liquid sprays. (D) Several types of smoke generators. (E) Hand-powered duster. (F) Gasoline thermalfogger

Hydraulic sprayers come in various sizes. Large greenhouse ranges will often install a tank and pump in a permanent location and pipe the spray throughout the greenhouse. They range from small $1\frac{1}{2}$ gal (6 l), hand pumped tanks for spot spraying, to several hundred gallon, high pressure (200 to several hundred lb in^{-2}) machines. Hydraulic sprayers can give excellent results if the pressure is maintained, good nozzles are used, they are properly cared for, and the personnel are well trained. Always use clean water, flush sprayers before use, and thoroughly clean them after use. Filtering screens should be kept in place and discs in the nozzles should be replaced frequently. Generally, spraying in greenhouses should continue until runoff, but this should not be excessive.

Table 9-12. Suggestions for choosing the proper adjuvant for pesticide sprays (Bohmont, 1976)

Function	Product
1. Wetting of foliage or spray to give good retention and coverage	Tween 20, Triton X-100, Multi-film, X-77
2. Reduce the rate of spray evaporation and increase drying rate	Bio-film
3. Increase resistance to heavy dews, rainfall, or sprinkler irrigation-weatherability	Nu-film, Plyac, Bio-film, Triton B-1956
4. Enhance penetration and translocation of pesticide	X-77, Tronic, Surfactant WK, Atplus 300, Regulaid, Oils
5. Adjust pH of spray to prevent degredation, prolong life, and reduce re-entry times	Buffer-X, Sorba Spray, Nutrex. Some experimental compounds
6. Improve uniformity of deposit, provide more uniform coverage	Bio-film, Triton B-1956, DuPont Spreader Sticker
7. Improve compatability of pesticide mixes, or of fertilizer-pesticide mixes	Buffer-X, Compex, Sponto 168 D
8. Increase safety to crops, reduce possibility of damage	Some experimental products
9. Reduce spray drift	Fomex, Stull's, Bi-Vert, Dacagin, Norbak

Note that all of the above are approved by the FDA in the United States.
These should not be used to wet soils.

On dense foliage of some crop plants, almost any method of application may provide indifferent coverage. Hussey et al. (1969) state that even where 500 gal per acre is applied (4750 l/ha), 25% of the lower leaf surface may receive no deposit. Unless the person spraying is very skilled, it is rarely possible to spray more than 60 gal (228 l) per h in greenhouses. About 1–1½ gal/100 ft^2 of bed (6 1 10 m^{-2}), will be needed to thoroughly spray insecticides on chrysanthemums or two-year old carnations.

One of the conditions for adequate spraying is thorough coverage. This can be improved by the addition of an adjuvant. Sometimes the pesticide includes a wetting agent, but commonly, judicious use of a wetting agent will greatly improve efficacy of the chemical. Several types are listed in Table 9-12. Adjuvants should be tested on a small number of plants before using on a large scale.

Materials for hydraulic spraying come in a number of formulations such as wettable powders (WP), emulsifiable concentrates (EC), or miscible concentrates. Designation such as 50 WP means that the material is a 50% wettable powder. The letters a.i. refer to "active ingredient".

c) Phytotoxicity. Much damage results because the improper chemical was used, the wrong formulation employed, or conditions at the time of application were improper. There is considerable variation between varieties of the same genus and species, regarding phytotoxicity (Table 9-13). Sometimes, the adjuvant may be the culprit, or the amounts of pesticide must be adjusted when an adjuvant is incorporated. Injury by pesticides is most pronounced on tender, succulent

Table 9-13. Pesticide phytotoxicity

Pesticide	Plant	Remarks
Acti Dione PM (cycloheximide)	Roses	May injure new growth during hot weather
Temik (aldicarb)	Freesias, some pot mums, *Vinca minor*, some Reiger Begonia	Leaf burn on certain varieties of pot mums and begonias, injury on poinsettias at double concentration, on petunias at triple concentration
Banrot	*Crassuala compacta*	Bleached out chlorophyll and some ferns, seedlings
Sevin (carboryl)	Boston Ivy, *Adiantum*, *Pilea*, *Schefflera*, Virginia Creeper	Injury on *Nephrolopis exalta and Peperomia* at higher than normal concentrations
chlorobenzilate	Hydrangeas, some *Prunus*, some roses	
Termil, Bravo (chlorothalonial)	Petunia, Balsam, Fibrous Begonia	Slight injury to blooms
Cygon (dimethoate)	Azaleas, carnations, some foliage	Serious injury to *Acacia notabilis*, *Beloperone*, *Cissus*, *Euphorbia*, *Gloxinia*, seedlings. Causes abnormal growth on carnations if applied to young plants under water stress. Injury noted on numerous other foliage and pot plants
Systox (demoton)	Some Easter Lily, mum and African Violets	Serious injury on *Asparagus*, *Adiantum*, *Anthurium*, *Begonia*, *Euphorbia*, *Fuschia*, *Gardenia*
Diazinon	*Gardenia*, *Hibiscus*, *Pilea*, *Stephanotis*	
Dylox (trichlorfon)	Some carnation, mum and zinnia varieties	
parathion	Numerous plants	Severe injury on *Acalypha*, *Ampelopsis*, *Anthurium*, *Bouvardia*, *Aster*, fern, *Crassula*, *Gladiolus*, *Kalanchoe*, *Gardenia*, *Hydrangea*, *Euphorbia*, some mums, *Monstera*, *Peperomia*, and numerous others. May cause leaf drop on roses if used within 7 days of sulfur, bleaching effect on open blooms of all crops
Pentac	Mums	Some foliage damage on mums, do not apply to open flowers
Pipron (piperalin)	Some roses	Golden Wave roses particularly
Pirimor (pirimicarb)	Some mums	Particularly to Icecap, Goldcap, and related varieties
Plictran	Mums, *Poinsettia*	Injury to mum and poinsettia blossoms, high rates injure roses
Synthrin (resmethrin)	Some mums, calceolaria	Injury to open flowers of Shoesmith varieties, red *Calceolaria*
Tedion (tetradifon)	Some roses and foliage	Especially White Butterfly and Cinderella, causes leaf drop, serious injury on *Cissus*, *Dahlia*, *Ficus*, *Sinningia*, *Primula*, *Kalanchoe*, *Matthiola*, *Cypripedium*
Vapona (DDVP)	Some mums, asters, and snapdragons	Especially Shasta, Pink Champagne, and Nightingale. Serious injury on *Ageratum*, *Aster*

Table 9-13 (continued)

Pesticide	Plant	Remarks
Zectran (mexacarbate)	Maidenhair Fern	High rates cause leaf injury on roses, Phlox, and petunias
Kelthane (dicofol)	Several	Severe injury on *Beloperone*, *Cissus*, *Euphorbia*, *Gloxinia*. Injury on many others if temperature is over 80 °F
Thiodan (endosulfan)	Geranium	Some geranium and mum varieties, injury reported on rose, *Cissus*, *Gardenia*, African Violet, *Scindapsus*
Lindane	Hydrangeas, *Poinsettia*	Severe injury listed for *Cyclamen*, Ferns, *Gloxinia*, *Hedera*, Hydrangea, *Kalanchoe*, Sweetpea, seedlings
malathion	Several	Severe injury on *Asparagus*, Begonia, snap, *Aralia*, *Crassula*, *Fuschia*, Gerbera, *Impatiens*, red carnations, some rose varieties, poinsettia, *Pilea*, Zinnia. Also orchids, sweetpeas, petunias, cucurbits, and *Crassula*
metaldehyde	*Cattleya* and *Phalaenopsis*	
Lannate (methomyl)	Bridal Pink rose	May defoliate Bridal Pink, leaf burn on mums in winter
Phosdrin (mevinphos)	Several	Severe injury on *Pelargonium*, injury on several others such as Begonia, rose, African Violet, *Peperomia*, *Philodendron*, *Lathyrus*, Stock, Tulip, *Primula*, Gerbera, *Campanula*, *Camellia*, *Cissus*, etc.
Dibrom (naled)	Several	Mum, poinsettia, White Butterfly, and Golden Rapture rose, Wandering Jew
Nicotine	Rose, Gloxinia, Hydrangea	Several others such as *Andiantum*, *Asparagus*, snaps, Orchids, *Primula*, *Salvia*, Carnation, Ferns, Gardenia, etc
Meta-Systox (oxymeton-methyl)	Lilies, Rieger Begonias, Geraniums, and some mums	Especially Hurricane, Iceberg, Whitecap and Pennant. *Hedera*, *Peperomia*, *Ficus*
methoxychlor	Mums	Varieties Iceberg, Mefos, and Winter Carnival
TEPP	Several	Such as *Amaranthus*, *Amaryllis*, Anemone, Snap, Aster, Azalea, Carnation, Begonia, Mum, Larkspur, Easter Lily, Petunia, Poppy, Stocks, Sweetpea, Primula, Zinnia, etc.
Karathane (dinocap)	Roses	On roses at high temperatures, severe injury to open mum flowers during hot weather
Truban and Benlate (terrazole and benomyl)	Seedlings	May retard
Morestan	Roses	

Materials listed have been reported to cause injury on the species listed. Note that any pesticide, if misused (e.g. excessive concentrations), is likely to be phytotoxic (Forsythe and Manard, 1969; Anonymous, 1974; Cornell Recommendations, 1975; Rathmell, 1976).

growth, particularly seedlings. Sprays or soil drenches should never be applied when the soil is dry or the plants wilted. Excessively dry atmospheres that increase water stress may cause greater injury. Aerosols, smokes, and fumigants should not be applied when the foliage is wet or the temperature is excessive. Nematicides and soil sterilants can cause injury if the soil is not adequately ventilated before the plants are brought into the greenhouse. Some have toxic residues that damage particular species. Soils with high organic content, or which are wet, will retain chemicals for a longer period.

d) Chemicals for Control. Tables 9-14 through 9-18 provide information on chemicals that can be employed as soil drenches or dips, fungicides as sprays or dusts, and pesticides for insects and related pests. Drenches are laborious to apply unless the liquid can be injected through a hose. Dilution of poisonous substances is not allowed where there is direct connection between the water supply and dilutor. The efficacy of a drench is often limited by its ability to move through the root substrate. Thus, Benlate as a frequent addition to irrigation water where the substrate is gravel works quite well. But, if plants are growing in the soil, the fungicide should be incorporated before planting. Most materials have nearly indefinite shelflife. Some chemicals, however, such as Zectran, cannot be stored indefinitely. Dexon degrades rapidly in light, so it should be used as soon as it is mixed. Other materials such as DDT and Dieldrin have been removed from the market in the United States. The label should always be read to determine if the pesticide is registered for use on the specific crop.

Quite often, two or more chemicals can be combined so that several organisms may be controlled. However, small tests should always be made. Bordeaux, sulfur, lime-sulfur, and oils are seldom compatible with other chemicals. Karathane may undergo decomposition when mixed with others, and emulsifiable concentrates should not be mixed with wettable powders. Most modern pesticides are not compatible with alkaline solutions.

The frequency of application will depend upon residual properties of the chemical, the length of the life cycle of the pest and stage killed, and environmental conditions in the greenhouse. Recommendations vary from less than a week for whitefly and red spider under optimum conditions for their development to several weeks. Sometimes, two or more applications may be made close together, with following applications spaced further apart. Control failure usually results from improper coverage, spray schedules too far apart, or temperature too low. Very often, the grower believes that resistance has developed when actually closer attention to the manner of using the pesticide is required.

Table 9-14. Fungicides for use as soil drenches or dips (Cornell Recommendations, 1975; Nichols and Burke, 1975; Hamlen et al., 1976)

Material	Uses	Formulation	Remarks
Terraclor + ferbam	Dip for Easter Lily bulbs to control *Rhizoctonia*, *Fusarium*	75WP + 76WP	Recommended for field-grown bulbs, 2.4 g + 2.4 g/l
Hot water	Dip for Caladium tubers only	—	Soak for 30 min at 50° C
Thiram	Seed treatment to control damping-off	75WP	2.8–5.5 cc per kg of seed
	Dip for gladioli corms to control Fusarium yellow and corm rot	75WP	Dip for 15 min in 75WP, 15 g/l
Benlate	Drench to control *Cylindrocladium* on Azalea, *Aschochyta* ray blight on mums, *Thielaviopsis* on Poinsettia, *Fusarium* on aster, carnation, Fusarium yellow on gladioli corms	50WP	1.9 g/l for glad corms, 1.8 g/l applied at 4.7 l/m^2 as drench. 3 ppm as constant application through irrigation water on carnations, or incorporate into soil before planting
Dexon	Control of *Pythium*, drench	35WP	For pots, 0.6 g/l, 1 pt/6-in pot; or 18 g/10$^1/_8$ m^2. Degrades in light, cover and do not allow to stand
Terraclor	Drench to control *Rhizoctonia*, *Sclerotinia*, *Sclerotium*	75WP	1.8 g/l, apply 4.7 l/m^2, preplanting drench, do not repeat applications. After planting for Delphinium, 1.2 g/l
Dexon + Terraclor	*Rhizoctonia* and *Pythium*, Drench	35WP + 75WP	12 g + 6 g/10 l, applied to 1 m^2. Cut rates in half for pots, apply $^1/_2$ pt per 6-in pot. Apply immediately. Dexon degrades in light
Captan + ferbam	*Fusarium roseum* on carnations	50WP + 76WP	227 g of each in enough water to cover 100 m^2
Truban	Drench for *Pythium* and *Phytophthora*	25WP or 30 WP	Apply by itself, safety margin for phytotoxicity small. Treat a few plants for trial. For most applications, 3–6 g/10 l 25WP, per 1 sq. m^2. Special instructions for Poinsettias, lower rates on bedding plants
Thiabendazole	*Thielaveopsis* on Poinsettia	60WP	3.6 g/l, applied at 4.7 l/m^2
Dasanit SC, Mocap EC	Dip for nematodes on foliage plants	—	750 ppm a.i., remove soil and dip roots only for 30 min
Vydate L	Dip for nematodes on foliage plants	—	Soak plant material in solution, 3–5 ml/l, 1–30 min
Plantvax	Systemic, specific for rust on carnation	5LC	6–8 ppm in irrigation water
Streptomycin	Bacterial soft rot on *Dieffenbachia*	—	200 ppm, soak canes for 20 min

Table 9-15. Fungicides for use as sprays or dusts (Nichols and Burke, 1965; Ohio State Florist Assoc., 1966; Cornell Recommendations, 1975; Powell and Lindquist, 1975)

Material	Uses	Formulation	Remarks
Benlate	Powdery mildew, *Botrytis*, rose black spot, *Ascochyta* ray blight on mums, *Septoria* leaf spots, *Fusarium* on *Dracaena* and *Pleomele*	50WP	Also *Cercospora* on *Peperomia*, *Cordyline*, *Ficus*, etc.; *Rhizoctonia* on ferns and seedlings, *Dactylaria* on *Philodendron*
Botran	*Botrytis*, leaf spots	50WP	
Captan	Leaf spots, *Botrytis*, *Alternaria* on carnation, rust, black spot on rose, branch rot on Geranium, *Fusarium* on carnation	50WP	
Bravo, Daconil 2787 Termil, Exotherm Thermil	*Botrytis*, *Ascochyta* ray blight	75WP 90% dust Smoke	Use with caution, see label particularly. *Cercospora*, *Fusarium*, and *Rhizoctonia* on foliage plants
Acti-dione PM	Powdery mildew	0.027WP	Antibiotic fungicide
Ferbam (Fermate)	Fungal leaf spots, *Septoria*, *Ascochyta*, Gardenia canker and leaf spot	76WP	Not effective against bacterial diseases or powdery mildew
Phaltan, Folpan	Leaf spots, powdery mildew, rose black spot.	50WP	
Karathane	Powdery mildew	25WP	Be careful at high temperatures, use wetting agent; acts as a miticide
Manzate, Dithane M-22	Fungal leaf spots, *Botrytis*, *Ascochyta*	80WP	Not effective against bacteria or powdery mildew
Dithane D-14 Parzate, Ortho, Nabam	Fungal leaf spots, *Botrytis*, *Ascochyta*		
Parnon	Powdery mildew on roses	4MC	Sometimes used in combination with Pipron
Pipron	Powdery mildew on roses	82MC	
Polyram	Leaf spots, rusts	80WP	
Sulfur	Powdery mildew	—	Wettable powder, volatile, paste on heating pipes
Parzate, Dithane Z-78, Ortho, Zineb Mancozeb (FORE)	Rusts, fungal leaf spots, *Botrytis*	80WP	Adequate for numerous leaf spots on foliage plants, *Alternaria* branch rot of carnations, Anthracnose, Downy mildew of snaps
Streptomycin (Agri-strep)	Bacterial leaf spots, soft rots, bacterial blight	—	
Fixed coppers Bordeaux Lime-sulfur	Fungal leaf spots	—	Various formulations, leave heavy residues, also act as insecticide, used sometimes as dormant sprays

Table 9-16. Pesticides for control of greenhouse pests (Ohio State Florist's Assoc. Bul., 1966; Cutkomp, 1973; Cornell Recommendations, 1975; Powell and Lindquist, 1975; Hamlen et al., 1976)

Pest	Pesticides
Aphids	Systox, Thiodan, Meta-Systox R, Pirimor, Vapona, sulfotepp, Dibrom, Nicotine, Diazinon, Temik, malathion, Cygon, Guthion, Zectran, TEPP
Ants	Chlordane, Diazinon, Dieldrin
Beetles	Diazinon, Sevin, methoxychlor
Caterpillars	*B. thuringiensis*, Vapona, Dibrom, Sevin, methoxychlor, Dylox, Zectran
Centipedes, symphylids	Lindane, Diazinon
Cutworms	Chlordane, DDT, Dieldrin, Sevin
Cyclamen mites	Kelthane, Thiodan, Diazinon, Endrin
Fungus gnats	Chlordane, Dieldrin, malathion, Lindane, Diazinon, Cygon
Grasshoppers	Chlordane, malathion, Dieldrin
Leaf miners	Dylox, Vapona, Nicotine, Diazinon, Dibrom, Temik, Meta-Systox R, Systox
Leafhoppers	Sevin, Diazinon, parathion, Meta-Systox R
Foliar nematodes	Diazinon, malathion, Meta-Systox R, parathion, Systox, Baytex, Temik, Dasanit, Mocap, Vydate L, Fumazone, Nemagon
Leaf rollers	Sevin, Dylox, Temik, Vapona, Dibrom, malathion, parathion
Mealybugs	Diazinon, Lindane, malathion, parathion, Cygon, Sevin, Temik, Vapona, Dithio, Systox, Dibrom, sulfotepp
Plume moth	Sevin, Dylox, DDT, Vapona
Roaches	Chlordane, Diazinon, Dieldrin, parathion
Scale insects	Diazinon, Cygon, malathion, parathion, sulfotepp, Systox, Vapona, Dibrom, Guthion, Zectran, TEPP
Slugs and snails	Metaldehyde (various), Zectran
Sowbugs (pillbugs)	Chlordane, Diazinon, Lindane
Spider mites	Chlorobenzilate, Kelthane, Cygon, Morestan, Meta-Systox R, Pentac, Plictran, Tedion, parathion, sulfotepp, Dibrom, Temik, Vapona, Systox, Diazinon, Di-Syston, malathion, TEPP
Spittle bugs	Thiodan, Lindane, Dieldrin
Springtails	Chlordane, Diazinon, malathion
Tarnished Plant Bug and others	Di-Syston, DDT, parathion
Thrips	Diazinon, Dieldrin, Lindane, Temik, malathion, parathion, Cygon, Di-Syston, Guthion, methoxychlor, Zectran, TEPP, Orthone
Whitefly	Guthion, Diazinon, Thiodan, malathion, parathion, Vapona, sulfotepp, Temik, TEPP, SBP-1382, Zectran, Meta-Systox R, Systox, Cygon, Dibrom

Table 9-17. Pesticide materials for use in the greenhouse (Ohio State Florist Assoc., 1966; Cutkomp, 1973; Cornell Recommendations, 1975; Powell and Lindquist, 1975; Hamlen et al., 1975)

Materials	Formulations	Applied as	Remarks
Temik (aldicarb)	10G	Granular, soil	Systemic, wide spectrum, high mammalian toxicity
Bacillus thuringiensis	Liquid	Foliar spray	Bacterial insecticide (DiPel)
Sevin (carbaryl)	50WP, 80WP	Soil drench, Foliar spray	
Chlordane	50WP, 45EC	Spray, dust Spray, dust, granular	Recently removed from market in U.S. by the EPA
chlorobenzilate	25WP, 25EC	Foliar stray	Not as effective when sulfur used
Cygon (dimethoate)	30.5EC, 23.4EC	Foliar spray, soil drench	Systemic, phytoxic to many plants and under certain conditions
Vapona (DDVP)	81EC	Aerosol, vapor, fog	Injures some chrysanthemums
Systox (demeton)	28.5EC	Foliar spray, soil drench	Systemic, high mammalian toxicity. Apply 5 days before harvesting on ornamentals
Diazinon	50WP, 4E, 50EC, 25EC	Foliar spray, fog, soil drench	Phytotoxic to some plants
Dibrom (naled)	60EC	Aerosol, vapor, fog	
Kelthane (dicofol)	18.5EC, 18.5WP, 35WP	Fog, foliar spray	Incompatible with sulfur
Di-Syston (disulfoton)	15G	Granular, soil	Systemic
Dithio (sulfotepp)	5%, 15% bomb	Aerosol, fog, smoke	
Dylox (trichlorfon)	50SP	Foliar spray	Phytotoxic to some ornamentals
Thiodan (endosulfan)	25WP, 25EC, 50WP	Foliar spray, dust	Phytotoxic to some ornamentals
Baytex (fenthion)	46EC	Foliar spray	Foliar nematodes
Guthion (azinphosmethyl)	22.2EC, 50WP	Foliar spray	
malathion	50EC, 25WP	Foliar, spray, dust, fog	Phytotoxic on several species
metaldehyde	varies	—	For slugs and snails
Meta-Systox R (oxydemetonmethyl)	25EC	Soil drench, foliar spray	Systemic
methoxyclor	50WP	Foliar spray	
Zectran (mexacarbate)	25WP	Foliar spray	Limited shelf life of 2 years
Morestan	25WP	Foliar spray	
nicotine	varies	Foliar spray, smoke	
parathion	15WP, 10% bomb, 25WP, 25EC	Foliar spray, fog, aerosol	High mammalian toxicity
Pentac	50WP	Foliar spray, fog, dust	May be ineffective when temperatures are high, highly effective against resistant spider mites

Table 9-17 (continued)

Materials	Formulations	Applied as	Remarks
Pirimor (pirimicarb)	50WP	Foliar spray	Short residual life, can be used on open flowers
SBP-1382 (Resmethrin)	24.3EC	Foliar spray, fog	Short residual life, best a low temperatures
TEPP	20EC	Foliar spray	High mammalian toxicity
Tedion (tetradifon)	25WP, 12.3EC	Vapor, smoke, dust, foliar spray	Slow acting
Dieldrin	50WP, 18.6EC	Foliar sprays, dust	Not on the market in the U.S.
Lindane	25WP, 20EC	Foliar spray	

WP wettable powder; EC emulsifiable concentrate; SP soluble powder; E emulsion; G granules.

Table 9-18. Suggested application rates for insecticides

Material	Rates	Remarks	Ref.
Temik (aldicarb)	6–11 g/m^2, 10G	Use full safety precautions	
	$^1/_8$ tsp/6-in pot	3–8 weeks between	(1)
	5 g/m^2 or 57 kg/ha	applications	(2)
			(3)
B. thuringiensis (DiPel)	5 ml/l		(1)
Chlordane	3 g/l 40WP, 2 ml/l 45EC		(1)
	3 ml/l 75EC	Soil drench	(2)
	1.2 g/l 40WP, 0.6 ml/l 72EC		(4)
Cygon (dimethoate)	2 ml/l 30.5EC, 2.5 ml/l 23.4EC		(1)
	1.2 ml/l 43.5EC, 1.2 ml/m^2,		(4)
	1.2–1.9 ml/l 2.67 lb.EC		(2)
Diazinon	2.4 g/l 50WP	Soil drench of 4 oz/6-in pot, 3.8 l/m^2	(1)
	1.2 ml/l 50EC, 2% dust		(2)
	2.5 ml/l 25EC, 1.2 g/l 50WP		(3)
Vapona (DDVP)	1 g/m^3 81EC	Vaporize on steam pipes	(1)
	1 g/10 m^3 81EC	Vaporize and close vents 2 h, 10–21° C	(4)
	6 oz smoke generator for 10000 ft^3		
Systox (demeton)	0.6–1 ml/l 28.5EC	Drench, 4 oz/6-in pot, 3.8 l/m^2	(1)
	1.2 ml/l 26EC	Spray	(4)
Dibrom (naled)	1 g/10 m^3 60EC	Vaporize on steam pipes	(1)
Kelthane (dicofol)	1.2 ml/l 18.5EC, 1.8 g/l 18.5WP		(1)
	1.2 g/l 25WP	Not compatible with sulfur	(2)
Dithio (sulfotepp)	5% bomb, 0.3 g/m^3	Repeat every 3–5 days	(1)
	15% bomb, 0.3 g/m^3	Varies with pest controlled	(2)
Di-syston (disulfoton)	6–11 g/m^2 15G	Work into soil and water thoroughly	(1)
	45 g/m^2 10G		(4)
Dylox (trichlorfon)	1.8 g/l 50SP		(1)
	2.4–3.6 g/l		(2)
Thiodan (endosulfan)	1.2 g/l 25WP, 1.2 ml/l 25EC		(1)
	1.2 g/l 50WP, 7.5 ml/l 24EC, 3% dust		(4)
Guthion	2.5 ml/l 22.2EC, 2.4 g/l 50WP		(1)
malathion	1.9 ml/l 50EC, 2.4 g/l 25WP		(1)
	1.2 ml/l 50EC, double for scale		(4)
	1.9 ml/l 57EC		(2)
Meta-Systox R	1.9 ml/l 25EC	Drench, 4 oz./6-in pot, 3.8 l/m^2	(1)
	1.2–1.9 ml/l 25EC	lasts 3–6 weeks	(3)
Zectran (mexacarbate)	2.4–3.6 g/l 25WP		(1)
Morestan	1.2 g/l 25WP		(1)
	0.6 g/l 25WP	No spreader	(4)
parathion	1.8 g/l 15WP		(1)
	1.2 g/l 15WP, 1.2 ml/l 25EC		(4)
	10% aerosol, 0.3 g/m^3	Varies with pest controlled	(2)
Pentac	0.6 g/l 50WP		(1)
	0.5 kg 50WP in water to make a slurry on steam pipes		(4)

Table 9-18 (continued)

Material	Rates	Remarks	Ref.
Pirimor (pirimicarb)	0.3–0.6 g/l 50WP		(1)
SBP-1382 (resmethrin)	1.2–2.5 ml/l 24.3EC	Weekly applications	(1)
Sevin (carbaryl)	1.2 g/l 50WP		(1)
	7.4 g/l 50WP	Can be used as drench	(3)
Tedion (tetradifon)	1.2 g/l 25WP	Slow acting	(1)
	0.6 g/l 50WP, 2.5 ml/l 12.3EC	Can vaporize from steam pipes, 14.2 g/1000 m^2	(4)
TEPP	0.6 ml/l 20EC		(1)
	1 part in 3200 parts water 40EC	Avoid undue soil wetting	(4)
Omite	0.6 g/l 30WP		(2)
	1.8 g/l 30WP		(3)
Metaldehyde	7–10 g/l 50WP	2-week intervals	(3)
Dieldrin	1.2 g/l 50WP, 3.7 ml/l 18.6 EC		(2)
Lindane	0.3 g/m^2, 0.6 g/l 50WP, 0.6 g/l 20EC		(2)

References: (1) Cornell Recommendations, 1975; (2) Cutkomp, 1973; (3) Hamlen et al., 1976; (4) Ohio State Florist Assoc., 1966.
Note: It is the responsibility of the grower to read the label and adjust accordingly.

References

Anonymous: Phytotoxicity of insecticides. North Carolina Flower Growers' Bull. **19**, 2–5 (1974)

Aycock, R., Daughtry, B. I.: Major diseases. In: Growing Azaleas Commercially. Kofranek, A. M., Larson, R. A. (eds.). Univ. Calif., Publ. No. 4058, 1975, 108 pp

Baker, K. F.: The U.C. system for producing healthy container-grown plants. Calif. Agr. Exptl. Sta. Man. **23**, 1957, 332 pp

Baker, R., Phillips, D. J.: Obtaining pathogen-free stock by shoot tip culture. Phytopathology **52**, 1242–1244 (1962)

Ben-Jaacov, J., Langhans, R. W.: Rapid multiplication of chrysanthemum plants by stem-tip proliferation. Hort Science **7**, 289 (1972)

Bohmont, B. L.: Pesticide Users Handbook for Greenhouse Operators. Denver, Colo.: Colo. Flower Growers' Assoc. 1976

Cornell Recommendations for Commercial Floriculture Crops: Part II. Disease, Pest, and Insect Control. Ithaca, NY.: Cornell Univ. 1975, 52 pp

Cutkomp, L. K.: Recommendations for greenhouse and floricultural pest control—1973. Minn. State Florists' Bull. (1973)

Davis, M., Clonal multiplication of carnation by micropropagation. M.S. Thesis, Colo. State Univ., Ft. Collins 1975, 140 pp

Dimock, A. W.: A method of establishing Verticillium-free clones of ornamental plants. Abstr. Phytopathol. **33**, 3 (1943a)

Dimock, A. W.: Coming up-to-date on Verticillium wilt and Septoria leaf spot of chrysanthemum. Chrys. Soc. Am. Bull. **11**, 3–10 (1943b)

Dimock, A. W.: Chrysanthemum stunt. N.Y. State Flower Growers' Assoc. Bull. **26**, 2 (1947)

Dimock, A. W.: Control of three ills of chrysanthemum. U.S. Dept. Agric., Yearbook of Agriculture for 1953, pp. 592–595

Dimock, A. W., Williamson, C. E., Nelson, P. E.: Diseases. In: Chrysanthemums, a Manual of the Culture, Disease and Insects, and Economics of Chrysanthemums. Langhans, R. W. (ed.). N.Y. State Ext. Ser. and N.Y. State Flower Growers' Assoc., Inc. 1964, 185 pp
Forsberg, J. L.: Diseases of ornamental plants. Univ. Ill., Urbana, Publ. No. 3 (Revised), 1975, 220 pp
Forsythe, J., Maynard, J.: The sensitivity of ornamental plants to insecticides and acracides. Hort. Rev. No. 1., Commonwealth Agric. Bur., East Malling, 1969, 66 pp
Hamlen, R. A., Dunn, R. A., Henley, R. W.: Nematode control on commercial foliage crops—1976. Florida Foliage Grower **13**(2), 1–7 (1976)
Hamlen, R. A., Short, D. E., Henley, R. W.: Insect, mite, and related pest control on commercial foliage crops. Florida Foliage Grower **13**, 1–8 (1975)
Hussey, N. W., Read, W. H., Hesling, J. J.: The Pests of Protected Cultivation. New York: Elsevier 1969, 404 pp
Jones, R. K., Stryder, D. L.: Damping off and its control. No. Carolina Flower Growers' Bull. **18(1)**, 1–4 (1974)
Knauss, J. F.: Common diseases of tropical foliage plants: I. Foliar fungal diseases. Florida Foliage Grower **12(10)**, 1–5 (1975a)
Knauss, J. F.: Common diseases of tropical foliage plants: II. Bacterial diseases. Florida Foliage Grower. **12(11)**, 1–7 (1975b)
Meister Publ. Co.: Farm Chemicals Handbook. Willoughby, Ohio: Meister 1976
Nichols, L. P., Burke, O. D.: Diseases of commercial florist crops. Penn. State Univ., Univ. Park, PA, Circ. No. 517, 1965, 31 pp
Nichols, L. P., Nelson, P. E.: Diseases. In: Bedding plants. Mastalerz, J. W. (ed.). Penn. Flower Growers, Univ. Park, PA, 1976, 515 pp
Ohio State Florist Assoc.: Pest control chart. Ohio State Florist Assoc. Bull. **445** (1966)
Phillips, D. J.: The clean stock program and the shoot tip. Colo. Flower Growers' Assoc. Bull. **145**, 1–4 (1962)
Powell, C. C., Lindquist, R. K.: Insect, mite and disease control on commercial floral crops. Ohio State Univ. Bull. 538, Columbus, Ohio 1975, 20 pp
Rathmell, J. K.: Greenhouse notes. Plants and plant phytotoxicity. Penn. State Univ., Univ. Park, PA. 1976, 10, 1–3
Reinert, R. A.: Geranium virus diseases. In: Geraniums, Mastalerz, J. W. (ed.). Penn. State Flower Growers, Inc., Univ. Park, PA, 1961, 100 pp
Tammen, J., Baker, R., Holley, W. D.: Development and production of pathogen free propagative material of ornamental plants. V. Control of carnation diseases through the cultured-cutting technique. Plant Disease Rptr. Suppl. **238**, 72–76 (1956)
U.S. Navy.: Emergency medical treatment for acute pesticide poisoning. U.S. Navy Disease Vector, Ecology and Control Center, Jacksonville, 1972

Chapter 10 Chemical Growth Regulation

10.1 Introduction

Growth regulator chemicals have been defined as organic compounds, other than nutrients, which, in small amounts, promote, inhibit, or otherwise modify any physiological process in plants. Control of growth in horticulture is usually practiced through temperature, water, feeding, photoperiod, and pruning. Chemical growth regulators give us one more set of tools for making adjustments to plant growth and development.

There is always an element of risk involved. Most growth regulators are toxic if applied in supra-optimal doses. The response to regulators depends upon: (1) the cultivar; (2) the part of the plant treated; (3) the stage of development; and (4) the amount and kind of chemical absorbed. The environment at the time a plant is treated modifies the effects and compounds the problem of learning to use regulators effectively with reproducible results.

Regulators are applied in one or more ways: as foliar sprays, aerosols, dusts, as drenches in the growing medium, and injected in plant parts.

10.2 Identified Chemical Groups

10.2.1 Auxin or Auxin-Like Compounds

Auxins and related chemicals promote growth oriented to gravitational forces, stimulate rooting and cambial activity, induce parthenocarpic fruit development, and affect flowering. IAA is the classic compound that occurs naturally in most plants, but it is relatively unstable. Synthetic auxins were the first growth regulators to find widespread use in horticulture. Discoveries of the 30s and 40s are used in common practice today, especially in plant propagation, fruit thinning, and prevention of fruit drop. Indolbutyric acid (IBA) and alpha naphthalene acetic acid (NAA) are commonly used for these purposes.

Triiodobenzoic acid (TIBA) promotes earlier flowering and fruiting on some plants and may also change leaf size, the shape of plants including the orientation of leaves and branches, and has been reported to increase photosynthetic efficiency (Anderson et al., 1965). Two of the most widely used herbicides, 2,4-D and 2,4,5-T are auxins.

For the promotion of rooting cuttings, Shanks (1970) gave broad directions for the formulation of synthetic auxins in liquid dips or in rooting powders.

Liquid preparations for quick dip treatments are popular because of the ease of treatment of large numbers of cuttings. Stock solutions of all the auxin-like compounds are easily made up in ethyl alcohol. Quick dip treatments, in which the final dilutions are made with water, are used in the range of 500–5000 ppm active ingredient (a.i.). Soft cuttings of chrysanthemum, carnation, gardenia, and poinsettia respond to complete immersion in around 500 ppm, although 1000 ppm can be used.

Another common method of treatment of cuttings on a large scale is the incorporation of auxins in talc as a carrier. The basal end of the cutting is dipped in the dust mixture and the excess dust shaken off. Concentrations useful in talc have ranged from 0.1 to 1% a.i. with 0.1–0.3% a.i. being adequate for softwood or green cuttings.

Commercial formulations of both dusts and liquid mixtures are readily available but most large-scale propagators make up their own rooting mixtures to have better control of this important operation.

In treating large numbers of cuttings, the dust or spray may be applied to the cut ends and foliage by a small paint sprayer, or the cuttings may be stuck in the propagating bed and the tops sprayed (Cheever, 1967).

While few new developments in the practical use of auxins have been reported in recent literature, research is continuing on the use of auxins in combination with other growth regulators. Most promising in future research with auxins is their role in increasing certain biochemical growth processes that would increase yields or quality (see metabolic chemicals).

When 2,4-D was applied in doses biologically tolerable to the plants along with certain trace elements, growth and yields of beans and sugar beets were increased (Wort, 1966). Other uses of herbicide auxins in trace quantities have been suggested in recent literature.

10.2.2 Gibberellins

Gibberellins are normal constituents of most plants. They were first discovered by western scientists in the 1950s (Wittwer, 1968), but were known by the Japanese 20 years earlier (Lang, 1957). At least 20 gibberellins have now been identified with some of these available for research in purified form. They are expensive since production is by fermentation processes similar to those used for antibiotics.

While research requires purified forms, commercial usage can be satisfied with relatively impure mixtures of gibberellic acids. Some of the actions of gibberellins are:

reversal of the genotype of genetic dwarfs,
termination of dormancy in various organs,
induction of parthenocarpic fruit,
promotion of flowering in some long-day plants and biennials, and production of development usually regarded as response to photoperiod or to chilling in some plants.

According to Wittwer (1968), gibberellin has revolutionized the production of Thompson seedless grapes for table use. Forty thousand acres (100% of the crop)

Fig. 10-1. Chrysanthemum cv "Dillon Beauregard" sprayed with 100 ppm GA_3 *(right)* at time of disbudding. Pedicles were lengthened to correct "clubby" sprays, stem length was increased 15 to 20%. (Photo by Dr. D. E. Hartley)

are now sprayed at rates of 20 and 40 ppm. Sale of GA_3 reached $ 1 million in 1963 and is increasing substantially each year. Other commercial uses are in wine grapes to loosen the fruit cluster and prevent fruit rots, and on citrus to delay postharvest rind disorders and yellowing associated with aging. The harvest date for lemons is delayed to correspond more to market demand in California. Gibberellins are also being used commercially in the production of F_1 hybrid cucumber seed, the overcoming of dormancy in freshly harvested seed potatoes, to delay bloom in almonds, to increase sugar yield in sugarcane, and to improve the malting properties of barley. Other uses will be found for gibberellins as new gibberellins are available in pure form to researchers.

While few practical uses of gibberellins have been developed for floriculture so far, GA_3 has been used experimentally to terminate dormancy and to promote flowering in azalea and camellia (Stuart, 1966). It increases the size of geranium blossoms (Lindstrom and Wittwer, 1957), and has been used at 250 and 500 ppm to accelerate growth of standard-shaped geranium and fuchsia plants. Starting with rooted cuttings in August and September, it is possible to produce standard plants on 75 cm trunks by the following June. Some cultivars of fuchsia are especially reactive to GA_3. Gibberellins have been used to lengthen pedicels on spray chrysanthemums and the stem length of sprays (Fig. 10-1) thereby avoiding the "clubbiness" of some cultivars at certain times of the year. The chance of applying an overdose has limited the use of gibberellins in floriculture.

The effect of gibberellin on shoot growth is obvious. Lettuce or sunflower respond quickly and dramatically and offer means of bioassay for naturally occurring gibberellins. The effect on root growth is less known. In a few cases, it was found that root growth was retarded in gibberellin treated plants (Halevy, 1969). The decrease in root growth is a consequence of preferential distribution of the available assimilates to the top of the plants rather than the roots (Halevy et al., 1964). Since treated plants are more susceptible to water stress, one should be careful, especially with young or tender plants, or delay treatment until plants have developed good root systems.

Fig. 10-2. PBA sprays release some of the lower buds of carnation from dormancy *(right)*. The buds at the lower six to eight nodes on carnation growths usually remain latent

10.2.3 Cytokinins

Cytokinins promote cell division and enlargement in callus tissues and help maintain chlorophyll in leaves and stem tissue. While reported to delay senescence in lettuce (Werner and Ceponis, 1962) and other green vegetables (Wittwer and DeDolph, 1963), promising results on extending the life of fresh cut flowers have been difficult to reproduce.

Benzyladenine (BA) reduced the respiratory rate of *Anthurium* flowers held at room temperature and reduced chilling injury in shipment (Shirakawa et al., 1964). BA treatment was especially effective on immature anthurium flowers, leading to the possibility of harvesting anthurium at an earlier stage of development. On the other hand, BA accelerated rates of respiration in the preclimacteric phase of apples (Smock et al., 1962).

Cytokinins, especially 6-benzylamino purine (PBA), have stimulated lateral bud growth of such diverse species as *Ilex* (Wright, 1976), *Macadamia* (Boswell and Storey, 1974), apple (Kender and Carpenter, 1972), and *Photinia* and *Rhododendron* (Ryan, 1974). In our laboratories at Colorado State University, PBA induced branching of carnation when used as a wet spray at 200 ppm (Fig. 10-2). Higher concentrations were so effective that quality of resulting flowers was reduced due to overcrowding of stems. A crop of flowers could be started by one PBA spray application.

The number of branches on roses following pruning was increased by expanded foam sprays of BA and PBA (Carpenter, 1975). Foam sprays were more

Fig. 10-3. PBA on Rieger begonia leaf-petiole cuttings. Base of petiole dipped at time of sticking: *(L–R)* control, 50, and 100 ppm. (Photo by Dr. D. E. Hartley)

effective than water sprays at rates of 1000 or 2000 ppm. No shoot growth was promoted from the bud union by these treatments.

Other effects of cytokinins reported are the conversion of genetically staminate *Vitis* plants to functional hermaphrodites (Moore, 1970). Elassar et al. (1974) applied benzyladenine (BA) to whole plants at anthesis and induced parthenocarpy in cucumber. Fruit set in muskmelon was increased (Jones, 1965). Senescence of fruits and formation of abscission zones to cause fruit drop were prevented.

Cytokinins have been used to overcome high temperature dormancy in lettuce seed (Smith et al., 1968) and to counteract certain inhibitors that cause dormancy in apple seed (Badizadegan and Carlson, 1967).

Kinetin with IAA induced aerial crowns in *Asparagus* (Yang and Clore, 1973). Ben Jaacov and Langhans (1972) used kinetin and IAA to modify the type of growth in shoot-tip culture, thereby facilitating rapid multiplication of shoot tips. Heide (1965) reported accelerated regeneration of begonia leaf cuttings due to kinin treatment (Fig. 10-3).

Zeatin, the first naturally occurring cytokinin, was isolated from corn in Dr. Letham's laboratories (Cebalo and Letham, 1967). Ohkawa (1974) isolated zeatin, along with several other cytokinins, from larvae of the insect *Dryocosmos kuriphilus*. Apparently cytokinins have been associated with insect activity in a number of plants.

Cytokinin activity in coleus was reduced by stresses of low pH, low nutrition, low temperature, or the use of growth retardants (Banko and Boe, 1975). Activity was lower in nematode resistant peach rootstocks than in stocks susceptible to nematodes (Kochba and Samish, 1972). Hatch and Powell (1971) found that cytokinins helped to direct transport of hormones in apple seedlings.

10.2.4 Growth Inhibitors

Chemicals which seriously interfere with other response, or stop growth entirely, are often termed "inhibitors". They usually suppress growth of leaf, stem, and flower. The most widely used in this class has been maleic hydrazide. It suppresses apical dominance by completely inhibiting cell division in the apical meristem. Applications of growth inhibitors result in production of plants with short internodes, dark green leaves, several meristems functioning at one time and unexpanded green leaves. Maleic hydrazide is used to limit growth of trees and shrubs along right-of-ways. It has also been used on grass with serious side effects, and in an attempt to control suckers on tobacco. MH controls sprouting on onion and potato (Wittwer, 1968), but is less effective on garlic (El-Oksh et al., 1971). Davidson et al. (1963) used 2500 ppm sprays of MH to stop vegetative growth and promote flowering of *Wigelia.* MH has also been reported to increase cold hardiness in citrus seedlings (Young, 1971), and to increase the proportion of flower buds on a number of apple cultivars (Edgerton et al., 1964). Sachs et al. (1975b) pointed out that the variation in activity from commercial formulations of maleic hydrazide may be due to the surfactant used. TMN-10 was superior to X77 on *Pinus* where inhibition occurred with much less foliar discoloration and needle dieback. The surfactant X77 was better with juniper, pyracantha, and *Callistemon.*

Chloro-IPC (an herbicide) has a similar mode of action. It suppresses sprouting of potato and onion, and causes a side effect on carnation that is similar to that caused by manual pinching. CIPC has shown some promise for inhibition of sprouting of sweet potato roots (Kushman, 1969). There are many identified and unidentified growth inhibitors in plants. The U.S. Borax and Chemical Co. released the first of what will probably be a group of inhibitors called morphactins. Maintain (CF 125) or chlorflurenol has been proposed for plants or weeds to maintain the status quo for a period of time by retarding or otherwise modifying growth (Shanks, 1970). Although it is essentially a dwarfing compound, higher dosages suppress growth and encourage branching. Other effects include promotion of abscission, and prevention of flowering. It is readily absorbed by leaves and roots and translocated within the plant.

Chlorflurenol was used to accelerate propagation of strawberries by spraying single crowns and causing them to multiply (Andrew et al., 1975). Possibly it will prove useful on ornamentals that grow in a rosette such as gerbera and African violet.

Morphactins cause both dicot and monocot seedlings to lose their capacity to respond to gravity or to unilateral light sources (Tognoni et al., 1967). A single full-coverage spray of chlorflurenol at 10 ppm applied at anthesis of the first flowers increased fruit set of tomato under high temperature conditions (Rudich and Rabinowitch, 1974).

Lateral spread of iceplant *(Carpobrotus edule)* was best reduced by sprays of chlorflurenol at 300 ppm (Hield and Hemstreet, 1974). Sachs et al. (1975a) obtained promising results by inhibiting both shoot growth and auxillary bud break of a number of field-grown woody plants with sprays of dikegulac-sodium (Atrinal). This inhibitor may be useful on landscape shrubs since it causes little undesirable injury when applied immediately after pruning. The chemical is sodium 2,3,4,6-di-O-isopropylidene-2-keto-L-gulonate. There will be other responses found for this new group of inhibitors.

10.2.5 Growth Retardants

Growth retardants are chemicals that limit the elongation of stems without malformation of leaf, stem, or flower. There are at least 7 families of chemicals so far that fall into this category (Cathey, 1964, 1965); there will be others. The list to date includes nicotiniums, quartenary ammonium compounds, hydrazines, phosphoniums, substituted cholines, succinamic acids, and ancymidol. Each year, new retardants are being tested for possible use by plant growers. No class of chemicals has been found so far that is effective on all plants, however, ancymidol is widely effective. Growth retardants generally reduce stem elongation, increase color of leaves and root-to-top ratio, increase resistance of plants to stress from drought, air pollution, and salinity. Growth retardants have been used to accelerate flower bud initiation in such woody species as *Rhododendron*, *Ilex*, and *Malus*. No chemical group is more widely used in floriculture except possibly the auxins.

10.2.5.1 Quartenary Ammoniums

Around 1950, a new type of chemical was made available to researchers, Amo1618(2-isopropyl-4-dimethylamino-5-methylphenyl-1-piperdine-carboxylate methyl chloride). This chemical was very active in dwarfing chrysanthemums and some other plants (Cathey and Stuart, 1961). Amo 1618 was effective as a spray (Fig. 10-4), but was never used commercially since other effective and less costly retardants were developed during the 1960s. Amo 1618 represents a family of chemicals that may be exploited in the future.

10.2.5.2 Phosphoniums

The first growth retarding chemical licensed for commercial use on ornamentals was Phosphon (2,4-dichlorobenzyl tributyl phosphonium chloride). It was effective in dwarfing chrysanthemums and lilies, but was injurious when used as a spray. The chemical is persistent in the soil and has a long-lasting effect on plants.

0 100 PPM 500 PPM 1000 PPM

Fig. 10-4. Amo 1618 and most growth retardants affect plants quantitatively. B995 is an exception

Fig. 10-5. Poinsettias are responsive to most growth retardants. (Photo by Dr. D. E. Hartley)

Fig. 10-6. Geranium flower stems may be shortened by a foliar spray of Cycocel, diluted 1-60 and applied three weeks after rooted cuttings are potted. Treated plant on right. (Photo by Dr. D. E. Hartley)

10.2.5.3 Substituted Cholines

Cycocel (2-chloroethyl trimethyl ammonium chloride) was developed by American Cyanamid Co. and distributed to researchers under the designation CCC. The largest projected use for CCC at the time it was released was to increase yields of wheat through increased tillering and the prevention of lodging (Wittwer, 1968). It does cause shorter and thicker stems on wheat and was projected to lead to a major advance in wheat production, especially in Western Europe. Wheat and tomato are among the most sensitive crop plants to CCC. Poinsettia and azalea (Stuart, 1966) are treated with Cycocel commercially. Cycocel is effective both as a drench and spray. Two or three spray applications at 5 to 7-day intervals are required to have the same effect as one soil drench treatment. Comparative effects of growth retardants on poinsettia are shown in Figure 10-5.

A single spray treatment usually is adequate to produce the desired dwarfing effect on vigorous azalea cultivars, induce earlier flower initiation, increase multi-

Fig. 10-7. Cycoced injury on geranium. Yellow leaf margins appeared within a week after spray application but mostly disappeared by five weeks later. Poinsettia may show similar injury. (Photo by Dr. D.E.-Hartley)

ple budding, and prevent the development of new vegetative shoots after flower buds have formed (Shanks, 1970). Treatment may be made 6–9 weeks after the last pinch, allowing an additional 8 weeks for the development of flower buds under good growing conditions. Treated plants may require a few more days to force following storage. Geranium (Fig. 10-6), hibiscus, and gardenia are also dwarfed effectively by Cycocel. Geranium and poinsettia show typical injury symptoms on the older leaves (Fig. 10-7).

10.2.5.4 Succinamic Acids

During the early 1960s, Uniroyal developed the first of this new family of growth retardants and made available to researchers B995 (Succinic acid, 2,2-dimethylhydrazide) in a 5% solution. The designation was soon simplified to B-Nine and registered for ornamental crops. The name Alar (85% WP) was registered for use on some food crops and is the same chemical. Some research publications designate the compound as SADH.

B995 had the widest spectrum of activity for growth retardants up to that time. It has several properties that distinguish it from previous families of growth retardants. The dosage required is the highest of any family and it is relatively safe to apply. Following a foliar spray, it is readily absorbed. Once a response is obtained, any increase in dosage, while not injurious, contributes little to increasing the effects. Sprays of 0.25–0.5% active ingredient have been used in dwarfing most plants affected by it. Higher dosages up to the undiluted 5% have been used without injury but high concentrations are wasteful and costly. Soil drenches with B-Nine or Alar would be too costly, and could be toxic (Cathey, 1969).

An area of partial usefulness is the persistence of its effects. One spray of B995 applied at the proper stage of growth is normally sufficient to produce the desired compact plants (Cathey, 1969). B-Nine is now used commercially on chrysanthemums, azalea, gardenia, hydrangea, and bedding plants. Most poinsettia cultivars are not so susceptible to B-Nine as to Cycocel.

In treating bedding plants (e.g. aster, cosmos, petunia, salvia, zinnia), sprays of up to 0.5% a.i. are made on young plants, as they approach a desirable size for sale, to prevent overgrowing (Shanks, 1970). B-Nine not only extends the salable life of these plants, but helps them to withstand transplanting shock and stress in the garden due to drought or air pollution.

B-Nine (Alar) has also been used to increase hardiness or prevent late growth on woody plants, to reduce storage disorders in apples, and to prolong their storage and shelf life, for the promotion of fruit set, and in extending the life of cut flowers (Larsen and Frolich, 1969).

Because of some disadvantages of B995, such as its relative ineffectiveness under high summer temperatures and the high dosage required, Uniroyal selected a new compound from 32 structures of this family tested on chrysanthemum. UNI-F529 (*N*-pyrrolidino-succinamic acid) was more active during the summer months and equally active with B995 in winter (Cathey, 1969). When compared on an equal weight basis, UNI-F529 was generally more active than B995 when applied to plants of chrysanthemum, coleus, marigold, poinsettia, petunia, and tomato. The visible effects of B995 disappeared sooner than on plants treated with UNI-F529. The two chemicals were nontoxic to the foliage at all levels tested (Cathey, 1969), but did not stimulate growth.

10.2.5.5 Ancymidol

The most recent growth retardant was developed by Eli Lilly and Co. In 1970, EL-531 was distributed to researchers on an experimental basis. By 1971, the product was called Quel, and, later it was registered as A-Rest. This new growth retardant has the common name ancymidol [(α-cyclopropyl-α-4-methoxyphenyl)-5-pyramidinemethanol]. A-Rest is from 80 to 400 times more potent in its growth retarding activity than previous dwarfing compounds (Witte, 1973). The most efficient method of application is by soil drench with enough solution to saturate a moderately dry root ball. It is not mobile in the soil but highly mobile upward in the plant. Ancymidol cannot move laterally in the plant from one vascular strand to another, hence, the possibility of uneven uptake unless the root system has access to an evenly distributed supply. Foliar sprays are effective and can sometimes overcome uneven soil applications (Witte, 1973).

Ancymidol reacts with a wide range of plants; possibly the widest range of species and cultivars of any growth retardant up to this time. Tests to date indicate effective dwarfing of chrysanthemum, Easter lily (Fig. 10-8), poinsettia, hydrangea, azalea, carnation, geranium, many woody plants and vines, a wide variety of annuals, and many foliage plants. Plant injury has been minimal when directions were followed. Spray applications at twice the recommended rate have been safe but wasteful. Soil drench treatments have not caused injury even at several times the recommended rate, according to Eli Lilly and Co.

Ancymidol has shortened internodes on all plants tested. It has also been effective in shortening the petioles of leaves and the flower stalks on plants, such as gerbera, that grow in rosettes. Flower bud formation has been stimulated in bougainvillea and lantana (Witte, 1973).

Fig. 10-8. Ancymidol on Easter lily: *L–R*, control, sprayed, when buds were 2.6 cm long (100 ppm), drenched when plants 10–15 cm tall with 0.5 mg per 15 cm pot. (Photo by Dr. D. E. Hartley)

Howard (1973) disbudded carnations with a soil treatment applied to unpinched cuttings that were in the latter stages of flower bud initiation. Apparently, carnation buds accumulate ancymidol to a toxic level in the initial growth stages. Terminal buds that were already well formed were not affected. Howard described the treatment and type of carnation cutting in detail (1973). Shoot cuttings were taken from stock plants with four or five pairs of leaves. The lower node of the cutting should show some elongation indicating the beginning of flower bud initiation in the shoot apex. After rooting, pot in small containers and apply 15 ml ancymidol at 100 ppm as a soil drench to each pot. Repeat the application after four weeks on all plants that remain vegetative. While this is a new use for growth retardants, it is a most interesting side effect. Obviously, the timing of such a treatment is critical.

Drench treatments with A-Rest have been most effective in research so far. The timing of dwarfing treatments should be just before plants begin elongation. Some cultivars of standard chrysanthemums tend to elongate too rapidly and produce flowers on weak necks. This problem may be avoided by spraying with ancymidol 100 ppm at the disbudding stage (Fig. 10-9). From 0.25 to 0.50 mg a.i. per 15 cm pot has been effective on lilies, hydrangea, and most other pot plants.

Since ancymidol is so recent, many effects of this growth regulant have not appeared in the literature. It is hoped that ancymidol will do everything that other growth retardants do, and on a wider variety of plants.

10.2.6 Abscisin (Dormin)

Until recently, growth inhibitors were known to occur in certain plant tissues that caused dormancy and possibly other related functions. These inhibitors were rather loosely grouped together until Abscisin II was isolated and identified in cotton bolls by Addicott and his students (Ohkuma et al., 1963). This new chemical was identical with Dormin, an inhibitor isolated from maple leaves by Wareing and his associates in Britain (El-Antably et al., 1967). The compound has been

Fig. 10-9. Chrysanthemum cv "Nob Hill" sprayed with 100 ppm ancymidol at disbudding stage to shorten and strengthen stem below the flower. (Photo by Dr. D.E. Hartley)

named abscisic acid (ABA) and has recently been isolated from prune and walnut seed, and several other plant species.

Abscisic acid is a sesquiterpene that functions in a similar manner to short days. It stimulates flowering in several short-day plants and inhibits flowering or stops growth in some long-day plants. It affects tuberization, leaf senescence, and dormancy (El-Antably et al., 1967). From a practical point of view, it is easier to extend day length by lighting than it is to shorten days by black-cloth shading. If abscisic acid is found to substitute for short days, it will be of great practical value (Halevy, 1969). Its use in regulation of dormancy of woody plants may also have practical value. One may visualize future control of woody plants by using long days and gibberellin to stimulate growth followed by abscisic acid to stop growth and cause the onset of dormancy. ABA did not affect shoot growth when applied to woody plants during or after storage (Meyer et al., 1969), so the timing of ABA applications is critical. ABA has not been active in producing leaf abscision.

ABA sprays delayed flowering of petunia, whereas the flowering of short-day plants was not affected (Cathey, 1968b). Daily sprays of ABA for 15 days suppressed the growth of long-day plants growing under long days. The plants formed resting meristems similar to those formed under short days. ABA only partially mimicked the growth characteristics induced by short days. Cathey proposed ABA for possible use to delay carnation crops developing under long days as ABA sprays delayed flowering as much as growing the plants on 8-h days (Cathey, 1968a). When ABA sprays were discontinued, plants resumed normal growth. The delay in flowering of carnation caused by ABA was associated by Cathey with limited elongation and a prevention of the expression of the stimulus in the growing point.

Abscisic acid decreased the photosynthesis rate in apple seedlings while at the same time increasing hardiness (Holubwicz and Boe, 1969). Both N and soluble protein increased as plants passed from pre-dormancy to dormancy.

Abscisic acid occurs naturally in most plants. Three gibberellins and ABA levels were measured bi-weekly through vegetative, storage, and forcing periods of azalea (Sydnor and Larson, 1975). ABA levels in buds were stable until about 24 weeks after the final pinch, when they dropped to undetectable levels in cold treated plants and increased in plants not given cold treatment.

Apices of chrysanthemum plants contained $1\frac{1}{2}$ to 3 times as much abscisic acid as young or old leaves (Sengupta et al., 1974). Plants under long-day treatment contained more ABA than those under short days. Plants sprayed with ethephon and receiving two or three weeks of short days had lower levels of ABA in their apices than plants receiving only short days. Ethephon reduced ABA levels in chrysanthemums under short-day treatment, which may or may not affect their ripeness to flower. Could a combination of growth regulators be devised that would reduce the number of days that black-cloth treatment must be applied during long days?

When apple seedlings were treated with abscisic acid, a water soluble glucoside complex of ABA and glucose was rapidly formed (Powell and Seeley, 1974). Weinbaum and Powell (1975) reported that ABA and its water soluble glucoside diffused from the leaves through the petioles and along the axes of crabapple seedlings. The apparent mobility of the glucoside was greater than that of free ABA. Concentrations of extractable and diffusible ABA were greatest in growing shoot tips. ABA from mature leaves diffused down the petioles more readily than from younger leaves. The enhanced liberation of ABA from older leaves may result in greater quantities of mobile ABA as the season progresses.

Other effects of abscisic acid reported in the literature to this time are:

1. It has hastened the onset of rest and the development of anthocyanin in *Euonymus* (Hemphill and Tukey, 1973).

2. As little as 1 ppm ABA in the vase water decreased water loss from cut roses (Kohl and Rundle, 1972).

3. Spray treatments with ABA inhibited bud break in *Rosa* but not *Syringa* (Cohen and Kelley, 1974). Immersion treatments on several dates during cold storage inhibited bud break in *Rosa* and *Syringa*, and subsequent shoot elongation in *Rosa*. The most effective treatment was 400 ppm. Time of application was most critical in inhibiting bud break on these woody plants.

Since ABA has only recently been available to research workers, many of its uses are still to be found for floriculture.

10.2.7 The Flowering Hormone

Florigen, or the flowering hormone, has been sought by many researchers. James Bonner, one of the pioneers in flowering physiology, has estimated that he has tested over 2000 plant extracts for their ability to induce flowering in vegetative apices (Bonner and Liverman, 1953). Other researchers have spent a good part of their lives seeking this mythical substance. There is no absolute evidence that there is a flowering hormone since it has not been isolated, but there is little doubt that there is a substance which results in the control of gene activity and the redirection of growth at the apical tips. There is adequate evidence that the substance is produced in leaves in response to proper stimuli (Salisbury, 1963). The substance will pass a graft union from one plant to another and the translocation of florigen may be measured. Many plants will change to the flowering condition when only one leaf or segment of the plant receives the proper stimulus. Stephanotis growing in a warm temperature (25° C) and completely vegetative during winter or early spring develops flower buds throughout a vine covering several square meters when only a small segment of the vine is given long days.

10.2.7.1 The Onset of Flowering

Many plants require only age and size to flower. Tomato starts flowering after the production of a given number of leaves. This may be a requirement for a certain minimum metabolite base. Such plants would probably be more difficult to cause to flower than plants that have definite daylength or temperature requirements.

Several of the plant growth regulators cause flowering in some situations. Cytokinins have caused flowering in some short-day plants (Ogawa, 1961), possibly replacing their short-day requirements. Cytokinins and GA have been effective in causing other short-day plants to flower (Pharis, 1972). Abscisic acid has also promoted flowering in short-day plants (El Antably and Wareing, 1966). The gibberellin content has an important effect on flowering of many plants. GA applications can cause some rosette plants to elongate and flower (Lang, 1957). In fact, not long after gibberellins were rediscovered in the West, some researchers were thinking that GA was the flowering hormone.

Many cold-requiring plants can be made to flower with gibberellin treatment. GA can replace the long-day requirement for many, but not all, of the long-day plants such as radish, spinach, and lettuce (Leopold and Kriedemann, 1975).

Gibberellins are probably required at a minimum level before flowering can occur in many species. Evans (1971) suggested that gibberellins would not affect the flowering of plants unless their gibberellin content is limiting.

Growth inhibitors and growth retardants have been reported to slow or stop vegetative growth and trigger the production of flower buds, or cause the production of more flowers (Davidson et al., 1963; Shanks, 1972). TIBA spray applied 15 days after full bloom induced more flowers the following year on triploid apple trees (Dayton, 1966).

10.2.7.2 Suppression of Flowering

Chemicals which suppress or inhibit flowering may function in a number of ways. These may inhibit the biosynthesis of florigen, acting as antimetabolites, or they may prevent translocation of the hormone to the tips, or prevent the development of flower buds. Undesirable side effects such as distortion of the apical meristems, leaves, and stems make antimetabolites useless on floricultural plants. Salisbury (1963) listed a number of chemicals which inhibit flowering regardless of when they are applied. Growth of the flower bud seems to be stopped by these substances. The effective compounds are mostly herbicides, many of which are auxins or auxin-like growth regulators.

The recent emphasis on medical and space research should produce developments that affect the future of florigenic chemicals. Chemicals are needed that determine when and how many flowers will be formed. A chemical that will cause prompt initiation and development of flowers on annuals, biennials and perennials would be highly unlikely. With a few plants and a set of chemicals, this is now possible. The number of flowers per cluster can be regulated on tomato at the two-leaf stage by applying substituted phthlamic or substituted benzoic acids (Cathey, 1965). Bean is also affected by these chemicals in a similar manner. The search for florigenic chemicals may be long and full of disappointments.

10.2.8 Juvenility Cofactors

Hess (1962) and his students isolated at least three lipid root promoting compounds from juvenile *Hedera helix* shoot tissue. These rooting cofactors when added to root promoting auxins may aid rooting of plants that would be difficult, if not impossible, to root. The cofactors are relatively unstable which could cause a problem in commercial preparations.

Stoltz and Hess girdled easy- and difficult-to-root hibiscus cultivars (1966) and found the rooting capacity of the cuttings was increased. Cuttings taken above the girdle rooted better, possibly because of an accumulation of starch and root promoting substances.

Concentration of auxin (IAA) in most tissues decreased 10 days after girdling in both cultivars. Rooting cofactors 1, 2, and 4 were found in the tissues above and below the girdle. Cofactors 1 and 2 increased during the first 10 days but decreased at 20 and 30 days after girdling. Cofactor 4 increased in tissues above the girdle in the easy-to-root cultivar but not in the difficult-to-root cultivar, or below the girdle of either.

There are several levels of juvenility in juniper. Cuttings root best when taken from outdoor stock plants from late November to January. Grafting scions perform about the same way. The rooting cofactor levels are probably optimum under conditions of short days and low temperatures.

Juvenility cofactors have been identified as terpenes, hence are related to gibberellins which are diterpenes.

10.3 Chemicals for Other Specific Uses

10.3.1 Chemicals to Aid Breeding

Male sterility and self-incompatibility are valuable tools of the plant breeder, however, these characters have often made inbreeding difficult. About 3000 species in 250 genera from 70 families have been reported to have some type of self-incompatibility system (Bettancourt, 1972). Almost all species of higher plants produce mutant forms that are male sterile, sometimes coupled with the loss of petals (apetaly). Selected growth chemicals, and other techniques (Eyster, 1941), are now making possible the temporary restoration of fertility. Gibberellins and cytokinins are most promising to date.

Henny and Ascher (1975) removed the self-incompatibility reaction in detached styles of *Lilium longiflorum* with injections of 100 or 200 ppm kinetin. When injections were made before pollination, pollen tubes grew as well from self-incompatible as from self-compatible plants. Lower concentrations of kinetin were either ineffective or gave variable results. Auxins were previously reported to aid in overcoming self-incompatibility in *Lilium* (Henny and Ascher, 1973). Ethephon in low concentrations has stimulated pollen tube growth in peach (Buchanan and Biggs, 1969), hence may help to counteract self-incompatibility. Dayton (1974) applied killed compatible pollen to the stigmas of apple to overcome self-incompatibility.

Male sterile tomato plants grown in solutions containing 300–500 ppm of GA_3 formed stamens and produced viable pollen (Phatak et al., 1966). No report has been found of this treatment being tried on apetalous mutants to restore fertility.

Late flowering and abortion of floral buds is associated with certain male-sterile plasmons in petunia. This phenomenon is so severe that it practically prevents the large scale use of cytoplasmic male sterility for seed production. The late-flowering problem resembles the reaction of fertile petunias to low growing temperatures. Izhar (1972) promoted near normal flowering in cytoplasmic male-sterile petunia by injecting GA_3 into the lower stalk. This treatment had no effect on male sterility.

Cucurbits are especially reactive to growth chemicals. Hybrid cucumber and squash seed are now commonly available due to assistance from chemical treatments. Sex expression in cucumber has long been known to be affected by day length and possibly other external influences.

Gibberellins are especially useful in the regulation of flower sex expression and the production of F_1 hybrid seed. Mitchell and Wittwer (1962) induced staminate flowers to form on gynoecious plants with GA_3, making possible the production of hybrids with parthenocarpic fruits. Sprays of 1000–1500 ppm are required.

The application of TIBA at 10 ppm to monoecious cucumber plants at the first true leaf has produced predominantly staminate flowers, while treatment with CEPA (ethephon) at 100 ppm resulted in predominantly pistillate flowers (Freytag et al., 1970). Granular CEPA applied on the seed at planting time was not as effective as sprays applied twice at 125 or 250 ppm (Cantliffe and Phatak, 1974).

Muskmelon responds to ethephon treatment (Loy, 1971), however, Shimotsuma and Jones (1972) could find no combination of day length and ethophon treatment that would change the sex expression of flowers on watermelon *(Citrullus vulgaris)*.

Benzothiadiazole (MCEB), and possibly similar compounds, act as inhibitors of the ethylene reaction in plants (Augustine et al., 1973). MCEB had no effect on androecious cucumber plants but induced staminate flowers on the gynoecious phenotype when applied at 75 ppm.

We need sterility chemicals that, when applied to plants, are assimilated in the reproductive organs and destroy the pollen mother cells, in some cases the egg cells. Mendok (FW 450), a product from Rohm and Haas, has been used successfully on cotton to enable the production of F_1 hybrid seed (Eaton, 1957). At Colorado State University, FW 450 was effective on zinnia, snapdragon, and other flower crops, but side effects on these plants prevented its commercial use. For example, spray application in the late bud stage was effective on snapdragon and prevented pollen formation by killing the pollen mother cells. The corolla tube failed to develop properly, preventing pollinating insects from fertilizing the treated flowers.

Reduction of the juvenility period of fruit plants and certain woody ornamentals could be of great help to the plant breeder in evaluating seedlings. Zimmerman (1971) reduced the juvenility period for the crabapple by treating the buds with cytokinins and gibberellins even though the young seedlings could not be stimulated to flower by grafting them on older plants. Dayton (1966) induced flowering in triploid apple progenies with TIBA and suggests this method for use in bringing young trees into bearing for earlier evaluation. The juvenility period for woody flowering plants may also be shortened by the use of growth inhibitors (Davidson et al., 1963).

Mutagenic chemicals and other mutagenic agents also offer much to the future of plant breeding. While few, if any, of these agents are growth regulators in the strict sense, Wallace (1964) suggested the following chemicals for use in directed mutation or possible chromosome engineering: alkylating agents, alkaloids, peroxides, formaldehyde, nucleic acid related substances, nitrous acid, and pure oxygen. Not one chemical group is common to all of these. All could cause injury to the gene mechanism of the plant and resulting mutations, if applied in the right way and at a concentration that would not kill the organism. Combination of the chemical with gamma irradiation would offer still greater possibilities.

The production of tetraploids and other phases of ploidy by such chemicals as colchicine and acenaphthene has been practiced by plant breeders since the late 1930s. Theoretically, a sterile triploid or pentaploid can be made fertile by doubling its chromosomes. Haploidy is more recent in plant breeding. Chemicals that are reported to prevent division of the generative nucleus in developing pollen tubes are phenothiazine and acridine derivatives (Gearhart and Rodgers, 1969). The optimum treatment with these kinds of chemicals could possibly enable a breeder to induce haploidy. Haploids can be grown from pollen grains in a few species. Toluidine blue, a vital dye, failed to induce haploidy in tomato and corn (Al-Yasiri and Rodgers, 1971) even though it had induced pseudogamy in mice (Edwards, 1954).

Sterility chemicals and barriers to breeding chemicals may aid the plant breeder and seed producer greatly in the future. In the various systems of incompatibility, researchers are continually investigating ways to remove barriers to breeding. One barrier consists of a factor responsible for heavy protein coats on pollen grains. Trypsin, a protelytic enzyme, may be possibly employed to digest the protein coating (Townsend, 1971).

Chemicals that aid hybridization, counteract sterility and remove genetic blocks in breeding will continue to be developed.

10.3.2 Tailoring Chemicals

Height regulation is now taken for granted. Fruit thinning can be accomplished chemically (Edgerton and Greenhalgh, 1969). We also need chemicals to regulate or alter root structure, for effective uptake of ions and to allow plants to withstand stress. These would allow us to make plants more adaptable to uses other than those now possible. In recent years, work pioneered by Tso (1964) and Cathey et al. (1966) has added chemical pruning and disbudding to the tools of the flower grower.

Fatty acids of 8–12 carbons and fatty alcohols of 8–10 carbon chains were found to kill selectively the terminal meristems in a wide variety of plants. Axillary meristems developed pretty much the same as if the terminals were removed manually. Effective dosages were: herbaceous plants, 0.025–0.05 M, semi-woody plants, 0.05–0.16 M, and woody plants, 0.16–0.27 M. Spray was applied to thorough wetting of the plants. Stuart (1966) pinched several azalea cultivars using C 9 or C 10 methyl esters of fatty acids or a mix of C 8 and C 10s. He used 4–5% stable emulsions on azaleas, finding that lower concentrations either did not kill shoot tips or simply retarded them allowing one or more axillary shoots to develop.

Several researchers, including Cathey et al. (1966), first investigated the possibility of disbudding chrysanthemums by chemical sprays. In the case of disbudding, the lateral buds must be killed or completely inhibited while not injuring the terminal bud; the timing must be exact.

Kofranek and Markiewicz (1967) screened many naphthalene "compounds" for their effects in disbudding chrysanthemums. Some of the "compounds" they tried originally gave highly variable results. After running mass spectrographic analyses, they found 40 chemicals mainly in six chemical groups: acenaphthenes, naphthenes, and 1-methyl naphthene were most effective in disbudding. All of these had a double ring structure (Fig. 10-10) contrasted to the chain structure of the esters and alcohols used for pinching plants.

To date, the number of cultivars of chrysanthemums that can be disbudded are limited to a few derived from English breeding. Most effective has been application of the chemical on the 14th short day, but some effects have been

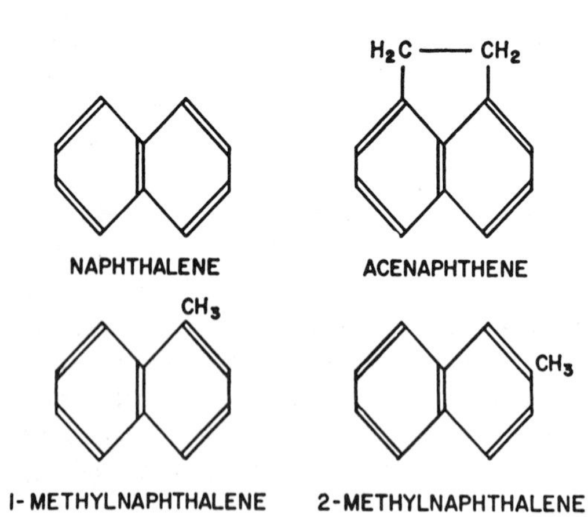

Fig. 10-10. The structural formulas of some of the compounds used by Kofranek and Markiewicz (1967)

obtained on Princess Anne cultivars as late as the 27th short day. None of the disbudding agents have been really effective on cultivars of earlier American origin, suggesting that breeding for ease of chemical disbudding could well be a goal of the future.

Kofranek and Criley (1967) attempted to find reasons why English cultivars are more susceptible to disbudding chemicals than are those of American breeding. They found nothing in the gross anatomy of buds or subtending leaf-cups, or in pubescent coverings to explain this difference. They did find that temperature, time of application, quantity applied and coverage were factors affecting the success of treatment on responsive cultivars. They also found that distortion, chlorosis, and death of the terminal can result from too frequent applications or from high concentrations (Fig. 10-11).

10.3.3 Stress Chemicals

We need chemicals that will limit or prevent water loss without side effects, or that help plants to tolerate drought. We also need chemicals that help plants to tolerate cold, dry air movement, dark homes, air pollutants, and other stresses.

Plaut and Halevy (1966) sprayed wheat plants with Cycocel and B995 before exposing them to various durations of wilting. These growth retardants had little effect on the production from plants watered regularly, but there was a pronounced increase in dry matter and grain production from growth retardants on plants that were subjected to drought. This effect was due to the increased ability of treated plants to generate new shoots and recover after wilting.

Antitranspirants are often used in transplanting certain trees and shrubs. Eight of these formulations were screened by Davies and Kozlowski (1974) for their value in reducing water loss and for possible plant injury. In general, all reduced water loss, at least temporarily. Photosynthesis was also reduced. The effects on transpiration and photosynthesis were greatly influenced by environmental regimes and by the species treated. Some of the eight antitranspirants were phytotoxic.

Zelitch (1965) has tested a number of chemicals for controlling stomate openings. In general, the effective chemicals are in the broad class of respiratory inhibitors. Phenyl mercuric acetate has been one of the most successful. Atraxine, sodium azide, and carbonyl cyanide have been less effective. The assumption is that these chemicals react with membranes of the guard cells to alter membrane permeability, preventing the guard cells from becoming turgid.

Cathey and Heggestad (1972) treated petunia plants with phosphon and Alar before exposing them to ozone at several levels and for various times. The two growth retardants reduced growth and modified ozone sensitivity. The concentration of Alar required to reduce ozone sensitivity was at least twice that used to retard stem elongation. Adding a wax coating and L ascorbic acid to the spray solution increased the protection from ozone afforded by Alar. Chemicals which did not retard petunia growth, such as Cycocel and ancymidol, afforded no protection for ozone treated plants. Some of the cultivars of petunia tested were very resistant to ozone damage.

Fig. 10-11 A and B. Typical injury to the tip of Yellow Shoesmith resulting from two or three applications of 0.5% xylene or 1% Amsco No. 75 (naphtha). Sprays were separated by 3-day intervals. (A) Witch's broom develops when the tip is killed; (B) injury to the floral apex results in an abnormal flower. Note also the greatly thickened stem (Kofranek and Criley, 1967)

Cold resistance, especially of fruit trees and blossoms, is another valuable stress resistance that has been increased by chemical treatment. Abscisic acid increased the hardiness of apple seedlings similar to the cold tolerance gained by gradual exposure to low temperature with short days (Holubwicz and Boe, 1969). Young (1971) increased the cold hardiness of citrus seedlings with MH-30 sprays. Treatment with CCC, B-9 or Amo-1618 failed to harden *Acer negundo* unless it was combined with short-day treatment (Irving, 1969).

As each new growth chemical is released, some researchers test its effects for stress resistance. It is only a matter of time until chemicals will be available in this important plant growth area.

Chemicals that will help acclimate plants to indoor environments are still in the future. We need chemicals that will reduce the time required to acclimate these indoor plants and to reduce the number of failures from poorly acclimated plants. Probably much of the adjustment must be done internally, by increasing the osmotic concentration of the cell sap and by stopping growth.

10.3.4 Ethylene

External applications of ethylene have been used to break dormancy and to effect development or senescence of fruits and other plant parts for many years. Either added to the environment or allowed to accumulate from ripening fruits, ethylene is used commercially to ripen tomatoes and bananas. Ethylene accelerates the senescence of cut flowers. Some species produce sufficient ethylene in closed containers to affect themselves or other species being shipped or stored with them (see Chap. 7).

Unlike other plant hormones, ethylene is a gas at ordinary temperatures, hence its practical use for plant regulation has been limited. Amchem Products, Inc. released a new growth regulator in the late 1960s for testing by research workers who have designated it CEPA, ethephon or Ethrel in their first publications.

Ethrel (chloroethyl phosphonic acid) induces ethylene release by plants, resulting in many responses that could be caused by ethylene gas if the dosages and conditions were known for treatment. Ethrel has greatly advanced ethylene research on the control of plant growth and development, and especially hybrid seed production and mechanical harvesting of tomatoes and other fruits. It may be applied as a foliar spray or as a soil drench.

Halevy (1969) has grouped some of the effects and possible uses of Ethrel on ornamental plants in three categories:

1. Formative Effects. Inhibition of apical bud growth, and stimulation of lateral bud growth, by removing apical dominance and functioning as a pinching agent. Direct application of the chemical to dormant buds at the base of plants may break dormancy and induce bottom breaks. Ethrel has been effective as a growth retardant when used as a soil drench on *Kalanchoe* at 600–1200 ppm a.i., but must be applied either well ahead of, or following, flower bud initiation (Shanks, 1970). Sanderson (1969) reported Ethrel as effective in dwarfing some chrysanthemum cultivars as B-Nine or UNI-F529, although Ethrel may have caused slight delay in flowering. Zieslin et al. (1972) found that a spray treatment on the lower plants promoted renewal canes on roses. To use this treatment most advantageously, it should be timed in relation to pruning and the desired cropping system.

2. *Fruit Drop.* Ethylene stimulates leaf and fruit abscission. A uniform once-over harvest by machine might depend on this as in the case of mechanical harvesting of tomatoes or cherries. In the case of some ornamentals, or fruits grown as ornamentals, it might be desirable to have the flowers but not the mess of dropping fruits over an extended period at maturity. Edgerton and Greenhalgh (1969) have suggested ethephon as a fruit eliminator when applied prebloom or early postbloom, with little or no phytotoxicity.

3. *Promotion of Flowering.* Induction of flowering has been reported for pineapple (Cooke and Randall, 1968), *Plumbago* (Nitsch and Nitsch, 1969), and 3- to 5-year-old apple seedlings (Kender, 1974). Most bromeliads will initiate flowers following a spray of around 1600 ppm a.i. (Shanks, 1970). In case of tank or vase types, the fresh solution may be applied in the vase instead of water.

One of the most interesting and useful applications of ethephon is on cucurbits. Application to seedlings in the early stages of growth causes some species and cultivars to produce all gynoecious flowers, thus facilitating the production of F_1 hybrid seed by insect pollination (Freytag et al., 1970).

Other uses for ethephon reported in recent literature are:

1. Stimulation of latex flow in *Hevea*,
2. Field promotion of ripening in all kinds of fruits,
3. Promotion of abscission layers to facilitate ease of harvesting everything from coconuts to gooseberries,
4. Deflowering of Easter lilies in field bulb-growing (Sanderson et al., 1975),
5. Induction of bulbing in onions under noninductive day length conditions (Levy et al., 1973), and
6. Breaking dormancy of strawberry seeds (Iyer et al., 1970).

Amchem Products Inc. of Ambler, Penna., maintains a list of publications on work done with Ethrel all over the world.

10.3.5 Height Reduction

Plant height inhibition in horticulture may have to be achieved with a combination of chemical agents. Sachs and Hackett (1972) reviewed the literature on this subject and grouped plant height control mechanisms under:

1. Killing the terminal buds of branches or severely inhibiting meristematic activity,
2. Inhibiting internode elongation without disrupting the apical meristematic function, and
3. Causing reduced apical control.

Under 1, they included the use of MH, TIBA, ethephon, and fluorenols. These are chemicals that disturb the apical meristematic function and cannot usually be employed where normal leaf and flower development are necessary. Some of these, as ethephon or TIBA, may increase branching and therefore flowering,

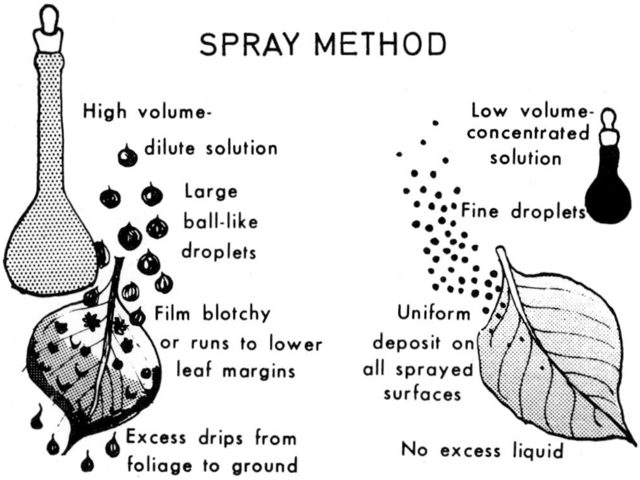

Fig. 10-12. Diagram comparing two methods of applying chemical growth regulators (Cathey, 1969, Part II)

while limiting height, or like ethephon may actually induce flowering on pineapple and plumbago.

Under 2, they grouped the growth retardants already in common use in horticulture such as Cycocel, Alar, and ancymidol. Plants are normal except shorter, with the same number of leaves and internodes. These chemicals may act as antigibberellins. Many possible mechanisms have been postulated for accomplishing this dwarfing.

Under 3, they grouped chemicals that cause the simultaneous growth of many shoot axes. Fatty acid esters cause terminal bud damage while TIBA and the fluorenols induce branching by inhibiting auxin transport. The fluorenols also inhibit elongation of the laterals so a systemic action is indicated.

The timing of application of a chemical for height control is critical. For chrysanthemums it should be applied near the beginning of short days. Poinsettias must be treated early in the cycle, almost at bud initiation to be most effective. On evergreens the application is made shortly after bud break whether in spring or after shearing. Terminal bud inhibitors are more difficult to time. In general, on woody plants, timing is most effective at bud break, even if there are several growths per year.

Most chemicals that reduce shoot growth are also effective in reducing root growth, sometimes with less side effect. Foliar applications of growth retardants are generally 2–3 times as effective in a greenhouse as when applied in the open. This may be in part due to the higher humidities in the greenhouse, possibly even to thinner cuticles on greenhouse grown plants. Cathey (1969) has shown that the effectiveness of chemical dwarfing sprays may be increased up to 12 times by reducing droplet size from 6000 μ to 25 μ in more highly concentrated spray solutions (Fig. 10-12). Considerably less material is used in the smaller droplet spray. Cathey also recommends spraying the lower leaf surfaces which are more absorptive.

Large volumes of spray solution are not required to wet the leaf surfaces thoroughly for safe and rapid application. If droplets are very small, highly efficient surfactants and extremely low volumes of spray solution can do the job by wetting all surfaces of the plant so the chemical forms a continuous film. There is no runoff to collect on the leaf margins and to be lost. All the applied chemical stays on the plant until it is absorbed. A compressed air sprayer, such as that used for spray-painting, or an aerosol fogging machine, delivers the low volume in small droplets needed for efficient application of growth regulants.

The specificity of response of plants to growth regulators is striking. There is no way of predicting how a species, or even a cultivar, will respond, or not respond, due to differences in absorption, transport, or metabolism of the chemical.

10.4 Chemicals Affecting Metabolism

Metabolic chemicals are those that enhance or help conserve carbon fixation in plants. These should enable us to make plants more attractive, could aid in the conversion of sugars to cellulose, and could lead to plants resistant to biological breakdown. One of the greatest possibilities in metabolic chemicals is that of increasing photosynthesis through chemical catalysts.

Dekazos and Worley (1970) reported increased cell wall carbohydrates leading to firmer fruit, and increased anthocyanin pigments when red tart cherries were treated with Alar. Halfacre et al. (1968) increased chlorophyll in young apple trees but decreased net photosynthesis of the leaves with Alar treatments. Alar moved readily from phloem to xylem and induced greater leakage of cell contents indicating that it lowers membrane integrity (Undurraga and Ryugo, 1970).

The herbicide simazine in sublethal doses has given dramatic increases in protein content and yield of beans, peas, and rye (Ries et al., 1968). N-Serve, 2-chloro-6-(trichloromethyl) pyridine, may lessen ammonium or nitrite toxicity, thereby causing much improved growth. The addition of N-Serve to nutrient solutions gave best growth and decreased time to flowering of carnation during suboptimal light conditions of winter (Green et al., 1973).

The possibilities of chemically regulating photosynthesis (Treharne and Stoddart, 1968) and regulating photorespiration, thus increasing net photosynthesis, appear to be attractive, underdeveloped areas of research (Leopold and Kriedemann, 1975).

10.5 Summary

Chemical growth regulation is our newest tool for plant control. Most of the chemicals used have been developed and tested since 1960. Growth retardants make possible the shaping of plants and increasing their tolerance to stresses. Oils and alcohols allow us to prune or disbud some plants. There are growth regulants that may be used to make plants sterile or to restore fertility, to change sex expression in flowers or to induce or promote flowering in some plants. All of this offers exciting possibilities for the future. More needs to be done on interactions between growth regulants and influences of environmental factors on their efficacy. Not nearly enough has been done on metabolic chemicals. The most exciting feature of all is the possibility of increasing photosynthesis or the reduction of respiration in selected plants, and the induction of flowering on all plants.

References

Al-Yasiri, S., Rodgers, O. M.: Attempting chemical induction of haploidy using toluidine blue. J. Am. Soc. Hort. Sci. **96**, 126–127 (1971)

Anderson, I. C., Greer, H. A. L., Tanner, J. W.: Response of soybeans to triiodobenzoic acid. From: Genes to Genus—A Symposium Intern. Mineral and Chem. Corp., Skokie, Ill. Jan. 21–22, pp. 103–115 (1965)

Andrew, L. A., Ormrod, D. P., Evans, W. D.: Chlorflurenol-induced crown division of everbearing strawberries. HortScience **10**, 528–529 (1975)

Augustine, J. J., Baker, L. R., Sell, H. M.: Chemical reversion of sex expression on dioecious cucumber with ethephon and a benzothiadiazole. HortScience **8**, 218–219 (1973)

Badizadegan, M., Carlson, R. F.: Effect of N^6 benzyladenine on seed germination and seedling growth of apple (Malus sylvestris Mill.). Proc. Am. Soc. Hort. Sci. **91**, 1–8 (1967)

Banko, T. J., Boe, A. A.: Effect of pH, temperature, nutrition, ethephon, and chlormequat on endogenous levels of cytokinins in Coleus blumei Benth. J. Am. Soc. Hort. Sci. **100**, 168–172 (1975)

Ben Jaacov, J., Langhans, R. W.: Rapid multiplication of chrysanthemum plants by stem-tip proliferation. HortScience **7**, 289–290 (1972)

Bettancourt, D. de: Self-incompatibility in basic and applied researches with higher plants. Genetica Agraria **26**, 163–216 (1972)

Bonner, J., Liverman, J.: Hormonal control of flower initiation. In: Growth and Differentiation in Plants. Loomis, W. E. (ed.). Ames, Iowa: Iowa St. 1953, pp. 283–303

Boswell, S. B., Storey, W. B.: Cytokinin-induced axillary bud sprouting in Macadamia. HortScience **9**, 115–116 (1974)

Buchanan, D. W., Biggs, R. H.: Peach fruit abscission and pollen germination as influenced by ethylene and 2-chloroethane phosphonic acid. J. Am. Soc. Hort. Sci. **94**, 327–329 (1969)

Cantliffe, D. J., Phatak, S. C.: Response of cucumber to soil and foliar applications of ethephon. HortScience **9**, 465–466 (1974)

Carpenter, W. J.: Foam sprays of plant growth regulating chemicals on rose shoot development at cutback. HortScience **10**, 605–606 (1975)

Cathey, H. M.: Physiology of growth retarding chemicals. Ann. Rev. Plant Physiol. **15**, 271–302 (1964)

Cathey, H. M.: Growth regulators. In: Symposium—A look to the future. Dedication of the Kenneth Post Laboratories. Ithaca, New York: Cornell Univ. 1965, pp. 29–35

Cathey, H. M.: Response of Dianthus caryophyllus to synthetic abscisic acid. Proc. Am. Soc. Hort. Sci. **93**, 560–568 (1968a)

Cathey, H. M.: Response of some ornamental plants to synthetic abscisic acid. Proc. Am. Soc. Hort. Sci. **93**, 693–698 (1968b)

Cathey, H. M.: Enhancing activity of chemical growth retardants. Part 1. Modification of chemical structure of B995. Southern Florist and Nurseryman magazine, January 17, 1–3 (1969a)

Cathey, H. M.: Enhancing activity of chemical growth retardants. Part II. Method of application. Southern Florist and Nurseryman magazine, Jan. 31, 4–5 (1969b)

Cathey, H.M., Heggestad, H.E.: Reduction of ozone damage to Petunia hybrida Vilm. by use of growth regulating chemicals and tolerant cultivars. J. Am. Soc. Hort. Sci. **97**, 695–700 (1972)
Cathey, H.M., Stuart, N.W.: Comparative growth retarding activity of Amo-1618, Phosphon and CCC. Bot. Gaz. **123**, 51–57 (1961)
Cathey, H.M., Steffens, G.L., Stuart, N.W., Zimmerman, R.H.: Chemical pruning of plants. Science **153**, 1382–1383 (1966)
Cebalo, T., Letham, D.S.: Synthesis of zeatin, a factor inducing cell division. Nature (London) **213**, 86 (1967)
Cheever, D.W.: Rooting of carnation cuttings. Colo. Flower Growers' Assn. Bull. **206**, 1–6 (1967)
Cohen, M.A., Kelley, J.D.: Effect of abscisic acid on bud break and shoot elongation in Rosa and Syringa. J. Am. Soc. Hort. Sci. **99**, 185–187 (1974)
Cooke, A.A., Randall, D.I.: 2-Haloethane phosphonic acids as ethylene releasing agents for the induction of flowering in pineapples. Nature (London) **218**, 974–975 (1968)
Davidson, H., Bucovac, M.J., McClean, D.C.: Photoperiodic and chemical control of vegetative growth and flowering of Wigelia. Proc. Am. Soc. Hort. Sci. **82**, 589–595 (1963)
Davies, W.J., Kozlowski, T.T.: Short- and long-term effects of antitranspirants on water relations and photosynthesis of woody plants. J. Am. Soc. Hort. Sci. **99**, 297–304 (1974)
Dayton, D.F.: Chemical induction of flowering in triploid apple progenies. HortScience **1**, 92–93 (1966)
Dayton, D.F.: Overcoming self-incompatibility in apple with killed compatible pollen. J. Am. Soc. Hort. Sci. **99**, 190–192 (1974)
Dekazos, E.D., Worley, J.F.: Effect of succinic acid, 2,2 dimethyl hydrazide on anthocyanin and cell wall carbohydrates in red tart cherries, Prunus cerasus L. cv Montmorency. J. Am. Soc. Hort. Sci. **95**, 703–706 (1970)
Eaton, L.M.: Selective gametocide opens way to hybrid cotton. Science **126**, 1174–1175 (1957)
Edgerton, L.J., Greenhalgh, W.J.: Regulation of growth, flowering, and fruit abscission with 2-chloroethane phosphonic acid. J. Am. Soc. Hort. Sci. **94**, 11–12 (1969)
Edgerton, L.J., Hoffman, M.B., Forshey, C.G.: The effect of some growth regulators on flowering and fruitset of apple trees. Proc. Am. Soc. Hort. Sci. **84**, 1–6 (1964)
Edwards, R.G.: The experimental induction of pseudogamy in early mouse embryo. Experimentia **10**, 499–500 (1954)
El Antably, H.M.M., Wareing, P.F.: Stimulation of flowering in certain short-day plants by abscisin. Nature (London) **210**, 328–329 (1966)
El Antably, H.M.M., Wareing, P.F., Hillman, J.: Some physiological responses to D-L abscisin (dormin). Planta **73**, 74–90 (1967)
Elassar, G., Rudich, J., Palevitch, D., Kedar, N.: Induction of Parthenocarpic development in cucumber by growth regulators. HortScience **9**, 238–239 (1974)
El-Oksh, I., Abdel-Kader, A.S., Wally, Y.A., El-Kholly, A.H.: Comparative effects of gamma irradiation and maleic hydrazide on storage of garlic. J. Am. Soc. Hort. Sci. **96**, 637–640 (1971)
Evans, L.T.: Flower induction and the florigen concept. Ann. Rev. Plant Physiol. **22**, 365–394 (1971)
Eyster, W.H.: The induction of fertility in genetically self-sterile plants. Science **94**, 144–146 (1941)
Freytag, A.H., Lira, E.P., Isleib, D.R.: Cucumber sex expression modified by growth regulators. HortScience **5**, 509 (1970)
Gearhart, J.D., Rodgers, O.M.: Suppression of the generative nucleus division in Tradescantia paludosa by phenothiazine and acridine derivatives. Cytobios. **1**, 17–22 (1969)
Green, J.L., Holley, W.D., Thaden, B.: Effects of the NH_4 and NO_3 ratio, chloride, N-Serve, and simazine on carnation flower production and plant growth. Proc. Florida State Hort. Soc. **86**, 385–388 (1973)
Halevy, A.H.: Recent advances in chemical growth regulation of ornamental plants. Acta Horticulturae **15**, 2–5 (1969)
Halevy, A.H., Monselise, S.P., Plaut, Z.: Effects of gibberellin on translocation and on dry matter and water content of several plant species. Physiol. Plantarum **17**, 49–62 (1964)

Halfacre, R. G., Barden, J. A., Rollins, H. A., Jr.: Effects of Alar on morphology, chlorophyll content, and net CO_2 assimilation rate of young apple trees. Proc. Am. Soc. Hort. Sci. **93**, 40–52 (1968)

Hatch, A. H., Powell, L. E.: Hormone transport of 32 P in Malus sylvestris seedlings. J. Am. Soc. Hort. Sci. **96**, 230–234 (1971)

Heide, O. M.: Interaction of temperature, auxins and kinins in the regeneration ability of Begonia leaf cuttings. Physiol. Plantarum **18**, 891–920 (1965)

Hemphill, D. D., Jr., Tukey, H. B., Jr.: The effects of intermittent mist on abscisic acid content of Euonymus alatus Sieb. "Compactus". J. Am. Soc. Hort. Sci. **98**, 416–420 (1973)

Henny, R. J., Ascher, P. D.: Effect of auxin (3-indolacetic acid) on in vitro compatible and incompatible pollen tube growth in detached styles of Lilium longiflorum Thunb. Incompatibility Newsletter **3**, 14–17 (1973)

Henny, R. J., Ascher, P. D.: Kinetin removes self-incompatibility reaction in detached styles of Lilium longiflorum Thunb. HortScience **10**, 25–26 (1975)

Hess, C. E.: Characterization of the rooting cofactors extracted from Hedera helix L. and Hibiscus rosa-sinensis L. Proc. 16th Intern. Hort. Cong. Gembloux, Belgium: J. Duculot 1963, Vol. **4**, 382–388

Hield, H., Hemstreet, S.: Border growth control of iceplant with chlorflurenol sprays. HortScience **9**, 473–474 (1974)

Holubwicz, T., Boe, A. A.: Development of cold hardiness in apple seedlings treated with gibberellic acid and abscisic acid. J. Am. Soc. Hort. Sci. **94**, 661–664 (1969)

Howard, G. S.: Pot plant carnations now a possibility. Florists' Rev. **152**, (3932), 29–30 (1973)

Irving, R. M.: Influence of growth retardants on development and loss of hardiness of Acer negundo. J. Am. Soc. Hort. Sci. **94**, 419–422 (1969)

Iyer, C. P., Chacko, E. K., Subramantam, M. D.: Ethrel for breaking dormancy of strawberry seeds. Curr. Sci. (India) **39**, 271–272 (1970)

Izhar, S.: Promotion of flowering in cytoplasmic male-sterile petunia by gibberellin. HortScience **7**, 555 (1972)

Jones, C. M.: Effects of benzyladenine on fruit set in muskmelon. Proc. Am. Soc. Hort. Sci. **87**, 335–340 (1965)

Kender, W. J.: Ethephon-induced flowering of apple seedlings. HortScience **9**, 444–445 (1974)

Kender, W. J., Carpenter, S.: Stimulation of lateral bud growth of apple trees by 6-benzylamino purine. J. Am. Soc. Hort. Sci. **97**, 377–380 (1972)

Kochba, J., Samish, R. M.: Level of endogenous cytokinins and auxin in roots of nematode resistant and susceptible peach rootstocks. J. Am. Soc. Hort. Sci. **97**, 115–119 (1972)

Kofranek, A. M., Criley, R. A.: Emulsifiable oils as disbudding agents for chrysanthemums. Florists' Rev. **139** (3611), 24–25, 36, 57–58 (1967)

Kofranek, A. M., Markiewicz, L.: Selected naphthalenes for disbudding agents for chrysanthemums. Florists' Rev. **140** (3617), 20–21, 54–57 (1967)

Kohl, H. C., Rundle, D. L.: Decreasing water loss of cut roses with abscisic acid. HortScience **7**, 249 (1972)

Kushman, L. J.: Inhibition of sprouting in sweet potatoes by treatment with CIPC. HortScience **4**, 61–63 (1969)

Lang, A.: The effect of gibberellin upon flower formation. Proc. Natl. Acad. Sci. U.S. **43**, 709–717 (1957)

Larsen, F. E., Frolich, M.: The influence of 8-hydroxyquinoline citrate, N-dimethylamino succinamic acid, and sucrose on respiration and water flow in "Red Sim" carnations in relation to flower senescence. J. Am. Soc. Hort. Sci. **94**, 289–292 (1969)

Leopold, C., Kriedemann, P. E.: Plant Growth and Development, 2nd ed. New York: McGraw 1975, 545 pp

Levy, D., Kedar, N., Karacinque, R.: Effects of ethephon on bulbing of onion under noninductive photoperiod. HortScience **8**, 228–229 (1973)

Lindstrom, R. S., Wittwer, S. H.: Gibberellin and higher plants. IX. Flowering in geranium (Pelargonium hortorum). Mich. Agr. Exptl. Sta. Bull. **40**, 225 (1957)

Lipton, W. J., Ceponis, M. J.: Retardation of senescence and stimulation of oxygen consumption in head lettuce treated with N^6 benzyladenine. Proc. Am. Soc. Hort. Sci. **81**, 379–384 (1962)

Loy, J. B.: Effects of 2-chloroethyl phosphonic acid and succinic acid-2,2-dimethyl-hydrazide on sex expression in muskmelon. J. Am. Soc. Hort. Sci. **96**, 641–644 (1971)
Meyer, M. M., Jr., Binnie, S. W., Gartner, J. B.: Evaluation of dichlorobenil, abscisic acid, and temperature on shoot growth during and after storage of woody ornamentals. J. Am. Soc. Hort. Sci. **94**, 658–660 (1969)
Mitchell, W. D., Wittwer, S. H.: Chemical regulation of flower sex expression and growth in Cucumis sativus L. Science **136**, 880–881 (1962)
Moore, J. N.: Cytokinin-induced sex conversion in male clones of Vitis species. J. Am. Soc. Hort. Sci. **95**, 387–393 (1970)
Nitsch, C., Nitsch, J. P.: Floral induction in a short-day plant, Plumbago indica L. by 2-chloroethane phosphonic acid. Plant Physiol. **44**, 1747–1748 (1969)
Ogawa, Y.: Über die Wirkung von Kinetin auf die Blutenbildung von Prarbitis nil. Plant Cell Physiol. **2**, 343–359 (1961)
Ohkawa, M.: Isolation of zeatin from larvae of Dryocosmos kuriphilus. HortScience **9**, 458–459 (1974)
Ohkuma, K., Lyon, J. L., Addicott, F. T., Smith, O. E.: Abscisin II, an abscission-accelerating substance from young cotton fruit. Science **142**, 1592–1593 (1963)
Pharis, R. P.: Flowering of chrysanthemums under noninductive long days by gibberellins and benzyladenine. Planta **105**, 205–212 (1972)
Phatak, S. C., Wittwer, S. H., Honma, S., Bukovac, M. J.: Gibberellininduced anther and pollen development in a stamen-less tomato mutant. Nature (London) **209**, 635–636 (1966)
Plaut, Z., Halevy, A. H.: Regeneration after wilting, growth and yield of wheat plants, as affected by two growth-retarding compounds. Physiol. Plantarum **19**, 1064–1072 (1966)
Powell, L. E., Seeley, S. D.: The metabolism of abscisic acid to a water soluble complex in apple. J. Am. Soc. Hort. Sci. **99**, 439–441 (1974)
Ries, S. K., Sweizer, C. J., Chmiel, H.: The increase in protein content and yield of simazine treated crops in Michigan and Costa Rica. BioScience **18**, 205–208 (1968)
Rudich, J., Rabinowitch, H. D.: The effect of chlorflurenol on set and concentrated yield of processing tomatoes. HortScience **9**, 142–143 (1974)
Ryan, G. F.: Chemicals to increase branching of Photinia "Fraseri" and Rhododendron "Exbury Azaleas". HortScience **9**, 534–535 (1974)
Sachs, R. M., DeBie, J., Kretchum, T., Mock, T.: Comparative activity of commercially available maleic hydrazide formulations on several plant species. HortScience **10**, 366–367 (1975 b)
Sachs, R. M., Hackett, W. P.: Chemical inhibition of plant height. HortScience **7**, 440–447 (1972)
Sachs, R. M., Hield, H., DeBie, J.: Dikegulac: a promising new foliar applied growth regulator for woody species. HortScience **10**, 367–369 (1975 a)
Salisbury, F. B.: The Flowering Process. London: Pergaman, 1963, 234 pp
Sanderson, K. C.: Growth retardants for florist crops—A to Z. Florists' Rev. **145**(3757), 29–31, 52–54 (1969)
Sanderson, K. C., Martin, W. C., Jr., Marcus, K., Goslin, W. E.: Effects of plant growth regulators on Lilium longiflorum cv Georgia. HortScience **10**, 611–613 (1975)
Sengupta, S. K., Rogers, M. N., Lorah, E. J.: Effects of photoperiod and ethephon treatment on abscisic acid levels in Chrysanthemum morifolium. Ramat. J. Am. Soc. Hort. Sci. **99**, 416–420 (1974)
Shanks, J. B.: Chemical growth regulation for floricultural crops. Florists' Rev. **146**(3796), 34–35, 50–58 (1970)
Shanks, J. B.: Chemical control of growth and flowering in Hibiscus. HortScience **7**, 574 (1972)
Shimotsuma, M., Jones, C. M.: Effects of ethephon and day length on sex expression of muskmelon and watermelon. HortScience **7**, 73–75 (1972)
Shirakawa, T., DeDolph, R. R., Watson, D. P.: N-6-benzyladenine effects on chilling injury, respiration, and keeping quality of Anthurium andreanum. Proc. Am. Soc. Hort. Sci. **85**, 642–646 (1964)
Smith, O. E., Yen, W. W. L., Lyons, J. M.: The effects of kinetin in overcoming high-temperature dormancy of lettuce seed. Proc. Am. Soc. Hort. Sci. **93**, 444–453 (1968)

Smock, R. M., Martin, D., Padfield, C. A. S.: Effect of N^6 benzyladenine on respiration and keeping quality of apples. Proc. Am. Soc. Hort. Sci. **81**, 51–56 (1962)
Stolz, L. P., Hess, C. E.: The effect of girdling upon root initiation: auxin and rooting cofactors. Proc. Am. Soc. Hort. Sci. **89**, 744–751 (1966)
Stuart, N. W.: Chemical control of growth and flowering in greenhouse azaleas. Proc. Intern. Hort. Cong. East Lansing, Mich.: Mich. St. Univ. 1966, Vol. **1**, p. 221
Sydnor, T. D., Larson, R. A.: Variations in levels of four natural growth regulators during growth and flowering of "Gloria" Azalea. J. Am. Soc. Hort. Sci. **100**, 353–356 (1975)
Tognoni, F., DeHertogh, A. A., Wittwer, S. H.: The independent action of morphactins and gibberellic acid on higher plants. Plant Cell Physiol. **8**, 231–239 (1967)
Townsend, C. E.: Advances in the study of incompatibility. In: Pollen: Development and Physiology. J. Heslop-Harrison (ed.). London: Buttersworths 1971, pp. 281–309
Treharne, K. J., Stoddart, J. L.: Effects of gibberellin on photosynthesis in red clover. Nature (London) **220**, 457–458 (1968)
Tso, T. C.: Plant growth inhibition by some fatty acids and their analogues. Nature (London) **202**, 511–512 (1964)
Undurraga, M. J., Ryugo, K.: The effect of Alar on permeability, a possible explanation for its mode of translocation. J. Am. Soc. Hort. Sci. **95**, 348–354 (1970)
Wallace, A. T.: Mutagenic agents—their use for plant and animal improvement. Agr. Sci. Rev. **2**(2), 1–8 (1964)
Weinbaum, S. A., Powell, L. E.: Diffusible abscisic acid and its relationship to leaf age in tea crabapple. J. Am. Soc. Hort. Sci. **100**, 583–586 (1975)
Witte, W. T.: A-Rest information roundup. Florida Flower Grower **10**, (6) 3–4 (1973)
Wittwer, S. H.: Chemical regulators in horticulture. HortScience **3**, 163–167 (1963)
Wittwer, S. H., DeDolph, R. R.: Some effects of kinins on horticultural crops. Proc. 16th Intern. Hort. Cong. Gembloux, Belgium: J. Duculot 1963, Vol. **2**, p. 428
Wort, D. J.: Effects of 2,4-D, nutrient dusts on growth and yield of beans and sugar beets. Agron. J. **58**, 27–29 (1966)
Wright, R. D.: 6-benzylaminopurine promotes axillary shoots in Ilex crenata Thunb. HortScience **11**, 43–44 (1976)
Yang, H. J., Clore, W. J.: Induction of aerial crowns in Asparagus officionalis L. by kinetin-IAA treatment. HortScience **8**, 490–491 (1973)
Young, R.: Effect of growth regulators on citrus seedling cold hardiness. J. Am. Soc. Hort. Sci. **96**, 708–710 (1971)
Zelitch, I.: Control of stomate openings. Proc. Natl. Acad. Sci. U.S. **47**, 1423–1433 (1961)
Zieslin, N., Halevy, A. H., Mor, Y., Bachrach, A., Sapir, I.: Promotion of renewal canes in roses by ethephon. HortScience **7**, 75–76 (1972)
Zimmerman, R. H.: Flowering in tea crabapple seedlings: Methods of shortening the juvenile phase. J. Am. Soc. Hort. Sci. **96**, 404–411 (1971)

Chapter 11 Business Management

11.1 Introduction

Many large and small flower growing enterprises have excellent technology for production, and mediocre or poor management for the business side of the organization. Students in horticulture may be fascinated by production, but lukewarm or may even subconsciously reject the most rudimentary education in business. Poor management is about the best insurance for business failure. The type of management pretty well determines the degree of success of a flower growing business. If a company has nothing going for it except good management, it will succeed.

Alan Mason (1973), who has spent a lifetime at managing glasshouses, attributes business succees to plantmanship, hard work, and management ability. He asks the question: "How do you express management in a few words?" and answers it: "To think straight; to know what is the eventual aim; to assess the factors; to determine the courses available and to prepare and implement a plan that will achieve the desired result in the most efficient way."

Mason continues: "Another most important asset, and one we fall down on terribly is the ability to stand back and see the whole picture ... Finally, I would say the ability to recognize one's own potential and to develop it to the utmost is paramount. It isn't always the clever chaps that get to the top (very often the reverse), but it's the guy who can maximize his own particular gifts."

What is management? It is the collective name for all those persons responsible for directing business operations. This may be one person in a very small business to a group of people making varying contributions to the running of a large enterprise. Management is also a process and will be so treated in this chapter. It is comprised of the activities of the managers. Decisions are made, group effort must be coordinated, and leadership provided, and some management is provided at all levels of an organization.

The tools of management are planning, organizing, directing, and motivating, and controlling. The neglect of any one of these to the extent that the profitability of an organization is impaired could be considered the beginning of poor management.

11.2 Planning

Flower growers have concentrated too long on production, sometimes on marketing as well, and very seldom on planning. Planning is the thoughtful arrangement of the elements required to make an operation successful. The one word that most characterizes the decade of the seventies would probably be change. In this atmosphere of change, planning is essential to survival. We must be concerned with the future and prepare for facing changes.

Plans are made for the short term (the next crop or the next year's operations) and for the longer term. What are the objectives of the company? These objectives constitute one type of plan. Are they merely related to profit production? What about the desire to have the best operation in the area, in the country? What objectives have been established for quality production? Are there goals related to the community? the state? service to customers? Planning entails the selection of a given path to the future from various alternatives that may be open.

Plans for the longer term must be more general in nature. Here is an example of suggested factors developed for long range planning by each division of one large company (Wrapp, 1957):

The industry
Our position
Competitor's activities
Sales forecasts
Present products
New products
Capital investment requirements
Working capital
Return on investment
Location of new facilities
Manpower requirements
Management controls
Appraisal of strengths and weaknesses
Special problem areas

While this outline was developed for a large industrial company, most of the factors listed could apply to a small company growing flowers, or to a cooperative group of flower growers marketing through a common group of facilities.

Long-term planning should recognize trends that are occurring and try to predict trends for the next decade along with new factors that may influence the industry. A given enterprise or segment of the industry plans their own future with these trends in mind.

11.3 Trends in the Florist Industry

Among the current trends in floriculture that have affected the industry during the past decade and no doubt will be major influences in the next are:

11.3.1 Production

Production is increasing due to better pesticides, better crop planning, relocation in more favorable climates and more intensive use of northern greenhouses. Many obsolete greenhouses have been dismantled. New construction is laid out for better movement of materials and more efficient use of space. There has been expansion of cheaper "temporary" greenhouse structures covered with films, and most greenhouses have better climate control. Other trends have been toward standardized growing media, improved feeding and watering systems and a better utilization of resources. Better varieties are being bred in most crops. The use of meristem culture is out of the laboratory and into commercial greenhouses for propagation of some crops and for disease control of vegetatively propagated crops.

11.3.2 Transportation

Transportation has influenced the growth of floriculture and the location of production facilities. More air transport is available and costs are rising rapidly. Containerization is beginning to be used on a large scale. Truck transport is being used more each year, especially on long hauls. Not only cut flowers but pot plants are being moved thousands of miles. Transport by bus is still important but cost is rising too fast. Air freight rates from underdeveloped countries are relatively low and play a part in the increasing imports from southern climates.

11.3.3 Marketing

The important marketing trends that can be expected to continue are several (Chap. 12). Retail flower shop sales through the flowers-by-wire organizations are increasing rapidly as are retail sales through nonflorist outlets. More producers are selling direct instead of through traditional brokers and commission wholesalers.

Other matters of serious concern in long term planning by flower growers are public policy including taxation, economic conditions, energy, fashion trends, population trends, and shifts, and competition from more favorable climatic areas.

11.4 Directing and Motivating

All persons connected with management must be involved in planning, must accept some responsibility in planning. There is an advantage especially in the execution of plans to involve the key workers in short-term plans.

Managers must be decision makers. Long or short-range plans require that someone in management decides what the path will be. The ability to make good decisions characterizes a good manager. Exact guidelines for decision making are hard to find, but the following six steps can reduce the amount of doubt and agonizing that go along with tough decisions, according to Grimmer, an experienced manager of a large firm (1970a):

1. Do not be afraid to delegate decisions. Decisions reached via a consensus are accepted and carried out more readily.

2. The right timing is important. Allow sufficient discussion but not so much that you get bogged down and are suddenly hit with a crisis.

3. Consider all of your options, but don't let the search become an excuse for vacillation or delay.

4. The practicality of a decision is essential. But remember that a truly practical decision is based on all aspects, including the intangible but essential human ingredients.

5. Plan carefully for the transition period which the results of a decision will require.

6. Be sure that the implementing of the decision allows for review and necessary changes.

Organizing is really the implementation of plans. It includes the provision of physical facilities, capital, and personnel. Creating a good organization structure does not ensure success, for needs change. An organization plan that is satisfactory when a business is started may be obsolete ten years later. Organizing is a continuing function of management. Flexibility is important for the function of organizing consists of the establishment of effective relationships among jobs and people.

We are inclined to favor specialization of jobs in the modern society. Specialization leads to greater efficiency but it also promises boredom in many jobs. The flower grower has a marvelous opportunity to organize along the lines of job enlargement. Many greenhouse jobs are repetitive hence are particularly objectionable to employees. In deciding how far to go in specialization, the manager should consider the matter from the standpoint of both engineering and human relations. Possibly the least desirable and most repetitive jobs can be accomplished by groups. This would then allow the groups to work on more desirable jobs as well.

The job for the organizer then is to analyze each job to be done in flower growing. How much standardization is needed? Can it be simplified? Can it be enlarged and made more interesting by being made a part of a larger assignment? It is remarkable how many jobs around some greenhouses are really not necessary or are done too frequently.

Time is the manager's greatest asset. Often management is spread so thin that planning has to be done off the job. The ability to use time wisely is the mark of a good manager. Grimmer (1970a) suggested that a manager start by organizing his own schedule:

1. The successful grower today is the man who thinks and plans before acting.

2. Hold regular staff meetings, even if it is his own family. Impress on your employees that each is dependent upon the other and success is dependent upon team work.

3. Departmentalizing is the first step to organization.

4. Avoid early morning congestion by cleaning house and doing preparatory work the afternoon before.

5. Make routine preparations in advance—pasteurize soil, arrange for bench space, etc. well in advance to avoid last minute difficulties.

6. Place orders for materials and do all preparatory work for special occasions well in advance to avoid last minute difficulties.

7. Write out in detail all pertinent facts about operation.

8. Place some responsibility with each employee.

9. Good records provide the best way of eliminating repetition of costly mistakes.

10. Proper personnel training pays dividends.

Savings effected by these suggestions would allow the manager more time for planning and control of operations.

As flower growing enterprises become more industrialized, the manager will need the ability to relate people with methods and processes while keeping clear the objectives of the business. The processes will be under increasingly better control due to improved equipment and materials, better scheduling and better trained personnel. The manager will need to understand and accept this systems approach.

It is the manager who sets the job or enterprise in motion; who gives the "green light" to a plan. This includes the issuance of orders and the providing of subordinates with adequate instructions. In most flower growing operations, the manager supplies leadership and instructions to the employees either directly, or delegates these functions to subordinates. In any case, the chain of management should be strongly oriented toward people. When people are the major concern of management, the work objectives are reached on schedule.

In a people oriented organization, it is most important for the owner or manager to understand the needs of employees. These needs fall into three categories. If all can be supplied on the job, management does not have to worry about well-adjusted and productive workers.

Many employers tend to think that satisfying the physiological needs of an employee is sufficient. These needs are food, clothing, shelter, and the things that can be acquired with money. If employees are paid enough, they can cover these needs. If the size of the paycheck were all-important, horticultural industries would have no employees. Other industries pay more. Good planning and superb organization for an efficiently run operation will enable the flower grower to pay better and this must be done.

But, we have a marvelous opportunity to capitalize on other human needs in developing our employees. Social needs are very important to most employees. In fact, most want companionship and shun loneliness. A 15-min coffee break twice each day is not enough to satisfy the social needs of most employees on the job. Management should recognize this fact and group workers so they can talk and enjoy each other's company during working hours. Most employees work best as members of a team. Helping others and being helped by them is rewarding. We can capitalize on this by providing a pleasant working situation.

Egoistic needs are also important to people. These are concerned with the individual's self esteem. Most persons have a deep desire to amount to something. They want to feel that what they are doing is worthwhile. Even though it may be only a part of a process, it can be a very significant part. Time spent by the foreman or group leader in explaining the immediate purpose of the job can pay off in self satisfaction to the worker. Being one's own boss is a part of these egoistic needs. Any arrangement that can allow an employee some autonomy in work will contribute to this feeling.

The nature of work in much of industry makes it difficult for an employee to gain satisfaction of egoistic needs. The flower industry has a unique opportunity in helping even the lowest echelon of workers achieve some measure of self satisfaction. Every employee that has any capacity for learning can be taught specialities of work within the framework of a firm. One can be taught to stick cuttings and propagate and become the *propagator*. There are innumerable opportunities for management to expand on this. Another mixes and pasteurizes soil materials and fills containers for the direct sticking of cuttings. This person should be recognized also by an appropriate title that designates a specialist. This does not mean that all employees work separately. Some may have helpers (new trainees) while all may work in groups at other jobs when needed. In a large cut flower or pot plant range, each employee may be placed in charge of a section and be responsible for as many operations as is feasible. These simple suggestions recognize that people have pride in their job and that this fact should be used by management to help them become better employees. A stable employee feels comfortable about himself, right about other people, and able to meet the demands of life.

One of the real problems and opportunities is that young people often come to the grower for their first job. They have no idea how to work or be productive. The first day on the job and the care in the instructions given these people is extremely important. In fact, our problem in the industry is not just to develop our employees to the point where they can avoid dismissal, but to develop them through leadership and adequate instruction so they can do an outstanding job for us. In this way they earn more for the firm and more for themselves. They also will, more than likely, gain self satisfaction from their job.

Incentive plans are often used in the industry as positive motivators. A grower may be paid 10%, or more, of the net profit from a crop, or a section, or the entire operation. This depends upon the contribution made by the grower and the ease of separation of the segment of operation by adequate accounting procedures. Individual piecework incentive plans are sometimes used. One approach to this is for management to ascertain the lowest possible cost a repetitive job can be contracted for. A more logical way would be to determine by time and motion study what the job is now costing. For example, suppose it now costs 5 cents to disbud a 6-inch pot mum. A manager could offer this, or even 4 cents per pot for the job on a piecework basis. Certain restrictions or penalties would be necessary to safeguard the employer. Employees could be free to set their own working hours as a further incentive. Much work in greenhouses and in producing flowers and vegetables lends itself to payment on incentive plans. The job must be one that is definable and adequate restrictions placed on the employee. The net result is that usually an employee works faster, takes fewer trips to the toilet and makes more money.

As wages increase for greenhouse labor, there is pressure on management to increase productivity and justify this higher labor cost. In Sweden, for example, the world's highest wages are paid. This puts great pressure on the greenhouse operator, especially since there are not enough plants produced in Sweden, and import restrictions are liberal. Greenhouse labor is in short supply as full employment is practically guaranteed by a socialist government. People in the greenhouse industry are constantly studying means of increasing productivity.

They have attempted to establish optimum wages for various jobs in the greenhouse. This includes time and motion studies to establish an acceptable work output and a base rate. This study is often performed by vocational horticulture instructors working with management and the extension adviser. The studies are financed indirectly by the government. However, much of greenhouse work does not lend itself to this categorization.

Mollgaard (1971) gave an example of how incentive pay worked in Sweden at that time. A worker cared for 1820 m^2 of cucumbers from March 15 to September 15. Normal area per man was 1480 m^2. The break even point of a crop was 36 kg per m^2. After this level is reached, there is a sliding scale bonus for each additional kg (differentiated between 1st and 2nd quality in proportion of 9 to 1). The crop harvested for the previous 4 years had been 50 kg per m^2. The rest of the year, the worker is paid the normal wage, even though he may work very little. His income at that time compared favorably to specialists in industry, middle class white collar workers or trade school teachers.

11.5 Controlling

A manager must keep in touch with operations occurring under his direction. Controlling is particularly concerned with locating operational weaknesses and correcting them. Suggested controls that should concern a manager of a flower growing enterprise were suggested by Jerome (1961):

1. Controls used to standardize performance in order to increase efficiency and lower costs. Included could be time and motion studies, improved operations, detailed production schedules, other written procedures.

2. Controls used to safeguard company assets from theft, wastage, or misuse. Important here would be an adequate system of authorization and record keeping as well as division of employee responsibilities.

3. Controls used for planning and programming operations. Such controls would include sales and production forecasts, budgets, various cost standards and comparisons of costs between crops.

4. Controls needed to keep the firm's plans and programs in balance. The major need here would be such controls needed to avoid conflict for capital between current and long-term operations, and still allow reasonable current profit.

5. Controls designed to motivate individuals within a firm to contribute their best efforts. These controls could involve recognizing outstanding performance, awards for suggestions or for finding mutations, some form of profit sharing.

Most flower growers think of the accountant, or the accounting department, as the controlling arm of the organization. While accounting techniques supply most of the records needed for control, accounting is by no means the only control system available to a manager. Dr. John McKeever (1971) studied the accounting procedures of a group of Colorado growers and found that most flower growing firms have a reasonably good accounting system. A large majority of the firms retained the services of public accountants at a reasonable cost. Certain of the procedural steps were performed by the firm manager, or part time bookkeepers, and the reporting function was the responsibility of the accountant.

In addition to preparing financial statements, the public accountant generally prepared all income tax forms and other federal or state report requirements. Generally, the accounting systems seemed to be designed to meet income tax regulations and provided management with something less than was necessary to maintain good management control over business operations. Of the procedural steps in the accounting process, improvement could be made in the classification, reporting, and analytical functions that would result in better cost control of firm operations.

McKeever (1971) suggested a system of classification of accounts which has since been adopted by many Colorado growers. While there is some variation between firms, many use the services of a computer center and a modification of McKeever's classification of accounts to provide them with a readily available cash accounting system at very low cost. Most of the larger commercial banks now provide this computer accounting at current costs as low as $ 5 per month.

Accounts must be coded by numbers as the present generation of computers cannot read letters. A modification of McKeever's suggested classification of accounts:

Three-digit numbers starting with 1, such as 101, 102, to 199 are code numbers for current asset accounts. These include cash in bank, temporary investments in securities, inventory, prepaid expenses, etc. Most growers also include fixed assets in the 100 series along with reserves for depreciation. Land might be coded 150, greenhouses 151, and reserve for greenhouse depreciation 151R. In this case, we add a letter R to the code, but this account does not go through the computer. It is merely a ledger account since it does not involve cash. There is an advantage to having depreciation accounts paired with the asset that is being depreciated. Other assets are also coded in the 100 series.

Most growers code liabilities in the 200 series. Current liabilities such as notes payable, accounts payable, taxes withheld from employees, etc. could be coded 201 to 250 and long term liabilities coded above 250.

Sales income accounts use code numbers 301 to 399 and can be coded by product, even by grade of product, if this is desirable to management. For instance, 301 could be poinsettias in 6-inch pots, 3 plants per pot; 302 could be poinsettias, 6-inch pot, branched plants, etc. Most growers group income sources by crop, or by function such as bedding plants, foliage plants, or Christmas plants.

Growing expenses are coded 401 to 499 and operating expenses 501 to 599. If sales expenses are significant and are not lumped in one category, or if it is important that a thorough breakdown of these is needed by management, they can be coded 601 to 699. McKeever (1971) suggested that growers separate expenses into Growing, Sales, Operating, and Office. There are enough numbers available for classifying expense accounts that all can be included in the 400 and 500 series, or higher series can be used.

Once accounts are classified and coded, the manager can use a computerized accounting service. Every check that is written must be coded with the correct number or numbers of accounts and every deposit slip should be coded crediting the source of funds. Multiple coding is possible on checks or deposits as long as the coding and amounts of money are clearly written. These numbers are fed to the computer by an operator in the computer center.

Account Number 25 00076 **PoudreAccount** DETAIL REPORT

BUSINESS Page 1

TRAEBER GREENHOUSES
P. O. BOX 1622
FORT COLLINS, CO. 80521

JULY, 1974

United Bank of Fort Collins
National Association

CODE	CLASSIFICATION	DOLLAR AMOUNT	MONTH TOTALS	COMMENTS
	..... NON-CODED ITEM	1.00	.00-	NP INCOME, TO 997
290	BORROWED INCOME	3,800.00	3,800.00	
301	SALES - Carnation	326.72		
301	SALES - Carnation	705.97	1,032.69	
304	SALES - Foliage plants	30.99		
304	SALES - Foliage plants	407.26	438.25	
305	SALES - Bedding plants	63.59		
305	SALES - Bedding plants	156.92		
305	SALES - Bedding plants	235.52		
305	SALES - Bedding plants	245.53		
305	SALES - Bedding plants	257.76		
305	SALES - Bedding plants	305.44		
305	SALES - Bedding plants	1,095.49	2,360.25	
401	DIRECT LABOR	919.40-		
401	DIRECT LABOR	1,210.15-	2,129.55-	
402	MANAGEMENT LABOR	250.00-		
402	MANAGEMENT LABOR	250.00-	500.00-	
413	PLANTS	80.00-		
413	PLANTS	94.00-		
413	PLANTS	155.00-		
413	PLANTS	166.00-	495.00-	
414	SEED	26.75-	26.75-	
420	SUPPLIES	203.93-		
420	SUPPLIES	231.00-		
420	SUPPLIES	327.80-		
420	SUPPLIES	1,886.91-	2,649.64-	
422	WATER & SEWER	117.70-	117.70-	
423	GAS	76.28-	76.28-	
424	ELECTRICITY	219.94-	219.94-	
425	REPAIR & MAINTENANCE SUP.	31.90-		
425	REPAIR & MAINTENANCE SUP.	32.87-		
425	REPAIR & MAINTENANCE SUP.	64.43-		
425	REPAIR & MAINTENANCE SUP.	75.44-		
425	REPAIR & MAINTENANCE SUP.	133.90-		
425	REPAIR & MAINTENANCE SUP.	134.83-		

Fig. 11-1. Example of previous month's accounting as provided by a computer service

The computer center supplies management, usually before the 10th of the month, with a print-out sheet containing a complete set of accounts for the month previous (Fig. 11-1) and for the year to date (Fig. 11-2). These accounts include money borrowed, notes paid, money transferred, withdrawals, sales by product or group of products, total income, expenses by all categories, gross profit, and net profit. This accounting is necessarily on a cash basis. To reconcile it to the accrual basis of accounting, the major changes are made in the journal and ledger accounts for preparing balance sheet and income and expense statements. Accounts receivable, inventory, and accounts payable are the major considerations if the accrual system of accounting is used.

As monthly, quarterly, or yearly statements are required by management, relatively few journal and ledger entries are required. The total expenses and sales by classification can be transferred from the summary printout to the journal and posted in ledger accounts. Any miscoded or uncoded items that were paid or received require journal entries. Most of the rest of the accounting necessary is that required by the accrual system of accounting (if used) and balance sheet accounts such as depreciation, inventory, and reconciling liabilities.

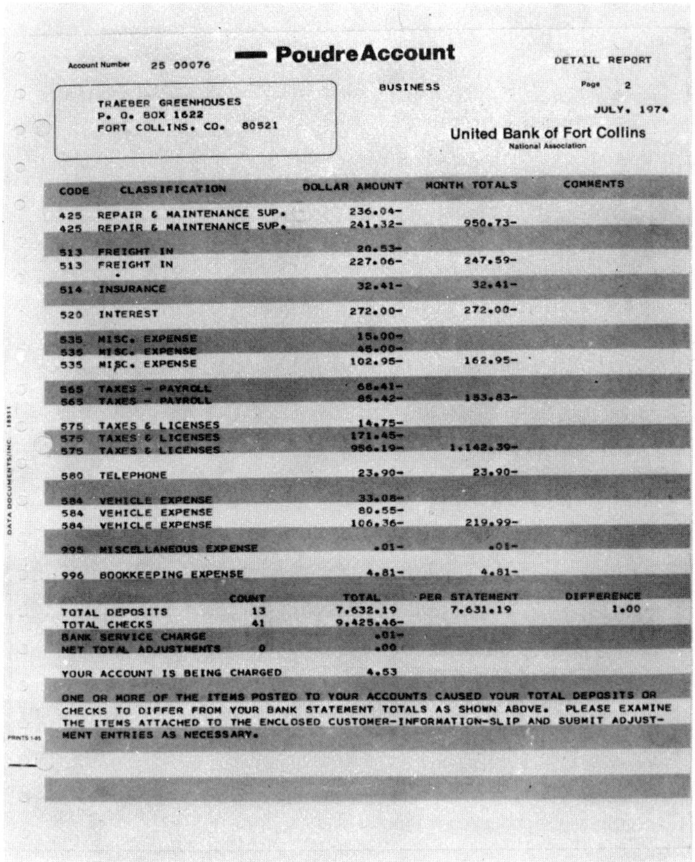

Account Number 25 00076 — PoudreAccount — DETAIL REPORT

BUSINESS — Page 2

TRAEBER GREENHOUSES
P. O. BOX 1622
FORT COLLINS, CO. 80521

JULY, 1974

United Bank of Fort Collins
National Association

CODE	CLASSIFICATION	DOLLAR AMOUNT	MONTH TOTALS	COMMENTS
425	REPAIR & MAINTENANCE SUP.	236.04-		
425	REPAIR & MAINTENANCE SUP.	241.32-	950.73-	
513	FREIGHT IN	20.53-		
513	FREIGHT IN	227.06-	247.59-	
514	INSURANCE	32.41-	32.41-	
520	INTEREST	272.00-	272.00-	
535	MISC. EXPENSE	15.00-		
535	MISC. EXPENSE	45.00-		
535	MISC. EXPENSE	102.95-	162.95-	
565	TAXES - PAYROLL	68.41-		
565	TAXES - PAYROLL	85.42-	153.83-	
575	TAXES & LICENSES	14.75-		
575	TAXES & LICENSES	171.45-		
575	TAXES & LICENSES	956.19-	1.142.39-	
580	TELEPHONE	23.90-	23.90-	
584	VEHICLE EXPENSE	33.08-		
584	VEHICLE EXPENSE	80.55-		
584	VEHICLE EXPENSE	106.36-	219.99-	
995	MISCELLANEOUS EXPENSE	.01-	.01-	
996	BOOKKEEPING EXPENSE	4.81-	4.81-	

	COUNT	TOTAL	PER STATEMENT	DIFFERENCE
TOTAL DEPOSITS	13	7.632.19	7.631.19	1.00
TOTAL CHECKS	41	9,425.46-		
BANK SERVICE CHARGE		.01-		
NET TOTAL ADJUSTMENTS	0	.00		
YOUR ACCOUNT IS BEING CHARGED		4.53		

ONE OR MORE OF THE ITEMS POSTED TO YOUR ACCOUNTS CAUSED YOUR TOTAL DEPOSITS OR CHECKS TO DIFFER FROM YOUR BANK STATEMENT TOTALS AS SHOWN ABOVE. PLEASE EXAMINE THE ITEMS ATTACHED TO THE ENCLOSED CUSTOMER-INFORMATION-SLIP AND SUBMIT ADJUSTMENT ENTRIES AS NECESSARY.

Fig. 11-1 (continued)

11.5.1 The Reporting Function of Accounting

Certain reports are required from the accountant by state and federal governments, investors (present or potential) and management. The two reports prepared routinely are the Income and Expense Statement and the Balance Sheet. These statements may be prepared for any accounting period, but in most flower growing firms are prepared either monthly, quarterly, or annually. Unfortunately, they are not always available from the accountant in time for effective use by management, hence may be afforded a cursory glance and filed away. A statement of income and expenses can be prepared in five minutes of time from the monthly or year-to-date accounts from a computer accounting service.

11.5.2 The Statement of Income and Expenses

This statement matches income for an accounting period with expenses for the same period with the difference between the two being profit or loss for that

Account Number 25 00076 — **Poudre Account** SUMMARY REPORT

BUSINESS Page 1

TRAEBER GREENHOUSES
P. O. BOX 1622
FORT COLLINS, CO. 80521

JULY, 1974

United Bank of Fort Collins
National Association

CODE	CLASSIFICATION	CURRENT MONTH		YEAR-TO-DATE	12
100	RETURNED DEPOSIT ITEMS	.00		705.68–	
	TOTAL DRAWINGS	.00–*		705.68–*	
290	BORROWED INCOME	3,800.00		14,700.00	
	TOTAL BORROWED INCOME	3,800.00 *		14,700.00 *	
295	NOTES PAID	.00		20,440.87–	
	TOTAL NOTES PAID	.00–*		20,440.87–*	
	SALES				
301	SALES – CARNATION	1,032.69	26.95%	9,930.49	19.07%
303	SALES – POT PLANTS	.00		7,383.41	14.18%
304	SALES – FOLIAGE PLANTS	438.25	11.44%	840.28	1.61%
305	SALES – BEDDING PLANTS	2,360.25	61.61%	33,882.84	65.08%
	TOTAL SALES	3,831.19	100.00%	52,037.02 *	99.95%
	GRAND TOTAL SALES	3,831.19	100.00%	52,037.02 *	99.95%
	SALES EXPENSES				
	NET SALES	3,831.19	100.00%	52,037.02 *	99.95%
	GROWING EXPENSES				
401	DIRECT LABOR	2,129.55–	55.58%	9,419.31–	18.09%
402	MANAGEMENT LABOR	500.00–	13.05%	3,131.50–	6.01%
	TOTAL DIRECT LABOR	2,629.55–*	68.64%	12,550.81–*	24.11%
411	FERTILIZER	.00		445.69–	.86%
412	INSECTICIDES	.00		62.87–	.12%
413	PLANTS	495.00–	12.92%	1,664.08–	3.20%
414	SEED	26.75–	.70%	1,338.62–	2.57%
420	SUPPLIES	2,649.64–	69.16%	6,542.94–	12.57%
422	WATER & SEWER	117.70–	3.07%	549.95–	1.06%
423	GAS	76.28–	1.99%	3,552.33–	6.82%
424	ELECTRICITY	219.94–	5.74%	1,504.62–	2.89%
425	REPAIR & MAINTENANCE SUP.	950.73–	24.82%	1,777.42–	3.41%
	TOTAL GROWING EXPENSE	4,536.04–*	118.40%	17,438.52–*	33.49%
	GROSS PROFIT	3,334.40–*	87.03%	22,047.69 *	42.35%
	OPERATING EXPENSES				
501	ACCOUNTING	.00		350.00–	.67%
512	DUES & SUBSCRIPTIONS	.00		44.00–	.08%
513	FREIGHT IN	247.59–	6.46%	436.89–	.84%
514	INSURANCE	32.41–	.85%	771.97–	1.48%
520	INTEREST	272.00–	7.10%	6,166.94–	11.85%
525	LEGAL	.00		15.00–	.03%
535	MISC. EXPENSE	162.95–	4.25%	556.17–	1.07%
545	OFFICE EXPENSE	.00		30.74–	.06%
565	TAXES – PAYROLL	153.83–	4.02%	756.44–	1.45%
575	TAXES & LICENSES	1,142.39–	29.82%	1,939.58–	3.73%
580	TELEPHONE	23.90–	.62%	150.04–	.29%
584	VEHICLE EXPENSE	219.99–	5.74%	1,422.07–	2.73%
	TOTAL OPERATING EXPENSES	2,255.06–*	58.86%	12,639.84–*	24.28%
	TOTAL EXPENSES	9,420.65–*	245.89%	42,629.17–*	81.88%
	OPERATING PROFIT	5,589.46–*	145.89%	9,407.85 *	18.07%
600	OTHER INCOME	.00		26.50	.05%
601	OTHER EXPENSE	.00		689.90–	1.33%
995	MISCELLANEOUS EXPENSE	.01–	.00%	9.42–	.02%
996	BOOKKEEPING EXPENSE	4.81–	.13%	28.63–	.05%
	NET INCOME	5,594.28–*	146.02%	8.706.40 *	16.72%
997	NON-POSTED INCOME	1.00		2,512.22	
998	NON-POSTED EXPENSE	.00		4,246.30–	

Fig. 11-2. Example of an account summary provided by a computer service

period. This statement is required by federal and state tax agencies with an accepted form being used by accountants for all types of businesses. The general form of the statement follows:

1. Income in the several product categories used by the firm followed by total income from sales.
2. Less cost of goods sold.
3. Equals gross profit on sales.
4. Less expenses in the several major categories.
5. Gives net operating profit (loss).
6. Other income not directly related to the flower growing business would then be added and other expenses subtracted to give the Net Income before taxes for the accounting period.

This form of reporting applies particularly to retailers who buy and sell. In order to make it more applicable to flower growing, McKeever (1971) has recommended that accounts be separated into Growing expenses, Sales expenses, Operating expenses, and Office expenses. All expenses directly attributable to the growing of flowers become the "Cost of Goods Sold" section in reporting. Operating expenses are those overhead expenses too difficult to allocate to growing, sales, or office.

A simple way that the manager can look at income and expenses each month is by means of the accounting print-out sheet from the computer center. The total for every expense classification is before him and the income from every product category. If he wants more detailed information on sales or expenses, he may need to use more detailed coding of the accounts.

11.5.3 The Balance Sheet

Another standard form of reporting, used by business to show the financial position of the firm at the time it is prepared, is the balance sheet. It is a statement of the company's assets and liabilities at the end of a given business day. All assets are listed in the first section and all indebtedness and financial obligations in the second section. The differences between indebtedness and assets required to balance the two is stockholders' or owners' equity. In this time of rapid inflation, it is necessary for a manager to keep at least two balance sheets—one for tax purposes and another for his own or potential investor's use. The value of land, greenhouses, and other investments are often not realistic on the balance sheet prepared for income purposes so it is necessary for the manager to have more nearly the real values in his thinking and planning for the long term. He may find that he cannot afford to continue operating when real values are considered. Possibly the assets should be sold and the money invested in other enterprises or securities.

As a firm develops a history of financial operations over a period of years, reporting becomes more meaningful. It is customary to prepare income and expense statements and balance sheets for the three or four years previous and the current year. A budget for the next year may also be included.

11.5.4 The Analytical Function of Accounting

The real guts of accounting procedure, and the opportunity for best control and planning by management, can come from the analysis of accounting data. Of accounting procedures studied by McKeever (1971), the analytical function was overlooked and not used by flower growers at that time. The analytical function is the responsibility of the manager and not that of the accountant. The financial reports are a source of facts, but unless these are analyzed by the manager in deciding alternate courses of action with the limited resources of the company, the entire accounting procedure becomes an exercise. Not that the accounting process does not have other objectives. It provides figures from which costs can be calculated and it is required for tax purposes. But management of a business enterprise is primarily a decision making process, based upon all the facts a manager can get.

11.5.4.1 Profitability Ratios

Profitability ratios provide management with information on past performance that should facilitate good planning and control. There are several profitability ratios that can be calculated from the Income and Expense Statement and from the Balance sheet:

1. The Net Operating profit to Gross Sales ratio reflects the performance of management in the use of resources to produce profits. Also, it shows the percent profit on gross sales.

2. Net Income before Taxes to Gross Sales may be the same as the former ratio, but becomes important if the firm has income or expenses from other sources.

3. The ratio of Net Income before Taxes to Owners' Equity reflects the return on invested capital. This ratio can be improved in a number of ways. Obviously an increase in profits relative to investment will do it. An increase in sales by changing to more intense crops, or expanding facilities through the use of long-term borrowed funds will increase the ratio. Most firms find it beneficial to work with borrowed funds as long as the additional income generated exceeds the cost of borrowing.

4. Gross Sales to Owners' Equity is computed by dividing Gross Sales by owners' equity. It reflects the times of turnover rather than a percent. This ratio indicates the number of times owners' equity is being used. This ratio can be influenced by the equity structure of the firm. Undercapitalization increases the ratio while overcapitalization reduces it. There is a balance between invested capital and long-term borrowed funds for every industry. Many flower grower firms are undercapitalized, while some older firms with no appreciable debt are the reverse. Lending institutions consider invested capital as well as quality of management, long-run growth potential, current economic conditions, and several other factors in arranging loans to flower growers.

5. The Gross Sales to Net Fixed Assets ratio is computed with fixed asset costs less accumulated depreciation. It indicates to management how well fixed assets are being utilized in generating sales. This is also expressed in turnover

times. Management should usually strive to increase this ratio as much as possible, since most flower firms operate on high priced land and have extensive fixed assets. Ratios of Gross Sales to:

6. Average Work Force.
7. Square meters of Bench Space and
8. Square meters of Greenhouses, are other statistics that may be helpful to management in making year to year comparisons. Ratios computed on the relationship of Square meters of Bench Space to:
9. Net Operating Profit and
10. Total Expenses are other useful statistics. Most all growing firms compute these, and many keep current on them since the expenses per square meter have been rising steadily within the memory of most growers. There are several other profitability ratios that can be calculated by management if they are needed. All of these profitability ratios over a period of years become trends for a company. Occasionally they may be compared with financial statements from other growers. From these trends a manager can separate the controllable from the non-controllable aspects of the total operations.

11.5.4.2 Liquidity Ratios

Liquidity ratios measure the debt paying ability of a company. Liquidity is important to the manager in establishing a line of credit with his banker. Current Assets to Current Liabilities is the most universally used liquidity ratio. A desirable ratio is 2:1 for most industries, but this is seldom achieved by growers. Managers should compute this ratio for every accounting period and keep it as high as possible.

The Liquid Assets to Current Liabilities ratio is also known in business as the "Acid Test Ratio". In computing this ratio, only those assets that are cash or convertible to cash immediately are used. A ratio of 1:1 would probably be considered good. In the industry much of the assets at a given time might be tied up in growing crops, therefore, a ratio of 0.40 to 1 for this ratio would probably be common. Having the necessary cash or near cash to meet current obligations is an indication of good management control over resources, but having cash lying idle in a checking account is wasteful. Extra cash should be invested in short-term securities of which there are a variety available.

11.5.4.3 Leverage Ratios

Leverage ratios are used to measure the relative risks being borne by the owners versus creditors. If debt is relatively high compared to owners' equity, the risks are with the creditors and the owners would be operating on high leverage. On the other hand, a company with little or no debt would have low leverage. Working on the equity of others is common in real estate and some other industries. This practice may result in enlarging the profit return on owners' equity if the return on borrowed capital exceeds the cost of borrowing. It has its pitfalls, however. Because of fear of indebtedness, some growers hesitate to borrow funds. Extremes of non borrowing or too high indebtedness can severely hamper the health of a business.

Other liquidity ratios that are calculated by management, that are often useful, compare Owners' Equity with: (1) Current Liabilities; (2) Longterm Liabilities; (3) Total Liabilities; and (4) Fixed Assets.

One miscellaneous ratio not involving financial data that was recommended by McKeever (1971) is Labor Turnover. As an outsider to the industry, he was impressed with the high turnover of labor in many flower growing concerns. He suggests that the total separations during the year be divided by the average work force. In McKeever's opinion, the cost of finding, hiring, training, and retaining a qualified employee in the greenhouse may range from $ 500 to $ 1500. In most industrial companies, the normal attrition rate for employees is estimated at 5–10%. In recent years, the turnover for flower growers has been many times this. McKeever concludes his labor recommendations with the following: "Management can reduce its labor turnover by using better selection, training, and retention policies and procedures, and creating a work environment that is conducive to the achievement of the personal goals of employees. Labor turnover is costly and deserves the attention and consideration of management."

11.6 Return on Investment

Dr. David Bowen of the Colorado University School of Business uses the chart in Figure 11-3 for teaching florists and nurserymen how to analyze their businesses. By filling in the blanks from business records, the performance of the enterprise

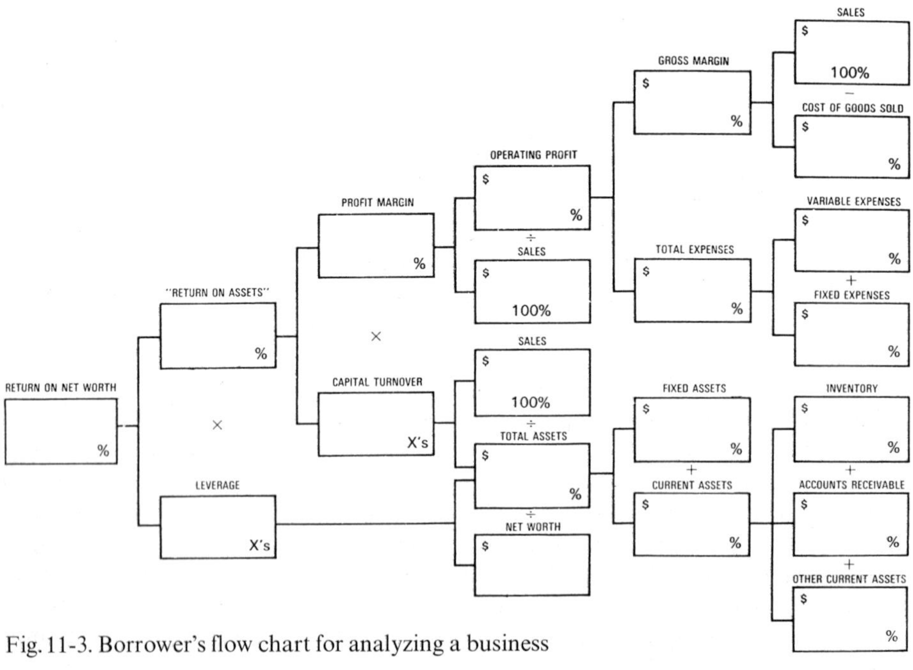

Fig. 11-3. Borrower's flow chart for analyzing a business

can be determined. Or, by starting at the *left side* with desired goals, one can estimate what must be done in sales, expenses, etc. to reach these goals. The chart may be simplified if "Return on Assets" is the end result desired. But, the important figure desired by most managers is that of return on net worth. How does the business return on invested capital compare to the 8 or 10% that money could possibly earn from a "safe" investment?

The northern greenhouse grower must continually battle for survival. Among the things that he has to do are (Grimmer, 1970b):

1. Grow the right crops—those that are in demand and that can be adapted to the climate he can supply—that fit his goals.
2. Time crops for peak demand, whether it be for holidays or the outdoor planting season.
3. Sell his crops well. Know costs and price accordingly. Sell as nearly 100% of production as possible. There is even a market for design grades, but it requires an effort to find it.
4. Control expenses and the direction of his company.
5. Purchase wisely.
6. Use all materials and resources wisely.

11.6.1 Production Costs

There are several ways to calculate production costs. All may be used in a complex growing operation. If one crop is produced, say carnations or roses, the total cost of running the operation is attributable to the crop and calculations are relatively simple. If we assume that these total costs are $40 per m^2 of bench area and the crop occupies the total space for the full year, we must produce and sell more than $40 worth of flowers per square meter from the area in order to make a profit. Assume that we produce 300 roses per square meter and sell these at an average net selling price (after selling costs) of 12 cents each. We have a gross return of $36.00 and a cost of $40.00 or a net operating loss of $4.00. If our production is 300, we must sell the roses at an average net price of 13.3 cents in order to break even. To make this crop profitable, we must raise production per unit of area, improve the grade and quality to command a higher price, or sell the flowers at a higher average price. One other alternative open to management is the reduction of the $40.00 cost of production through a carefully planned system of cost control. In fact, alert management will explore and possibly employ all of these alternatives.

Cost of production becomes only slightly more complex as other crops are added. Assume that another firm grows roses in 60% of the area on a year around basis. The remaining 40% is used for poinsettias and bedding plants. A simple system of calculation would be to charge roses with 60% of the total expense and the other two crops with 40%. But, this could be misleading and entirely unfair to the rose crop. More accuracy requires a breakdown of the expenses into specific costs (attributable to the crop) and overhead costs that are allocated according to space occupied "X" time. Specific costs include plants, royalties paid, containers, soil mix, and any supplies or costs used for that crop. To complete the example,

Table 11–1. Approximate pots per square foot of bench area at several spacings; multiply by 10 for pots per m^2

Pot size		Spacing		
Inches	cms	Pot tight	1/2[a]	1/4[b]
3	7 1/2	11.80	5.90	2.95
4	10	7.10	3.55	1.78
5	12 1/2	4.75	2.38	1.19
6	15	3.40	1.70	0.85
7	17 1/2	2.56	1.28	0.64

[a] Alternate rows removed.
[b] Alternate rows and alternate pots in remaining rows removed.

we need to allocate space to the crops. We will assume that 3000 square meters of greenhouse is used for roses in which 1800 m^2 of bed area are planted. Poinsettias and bedding plants are produced in the other 1000 m^2 of greenhouse with a bench area of 800 m^2 due to better utilization of space. Poinsettias are charged for the space from August to December and bedding plants are produced from February to July. The house is either closed in January, or bedding plants pay for the space.

Cost of production of the rose crop becomes $^3/_4$ (overhead costs) + specific costs of the rose crop. Flower grading may be charged as a cost of marketing rather than production, but should be charged to roses. Flower cutting is usually charged to production.

Cost of production of poinsettias would become $^1/_4$ (overhead) × 5/12 + specific costs such as pots, soil mix, plants, sleeves, etc. Only 5/12 of the overhead for the area is charged to poinsettias as the crop occupies the area 4 months but pays for August as well.

Cost of production for the bedding plants becomes $^1/_4$ (overhead) × 7/12 + specific costs for containers, soil mixes, seed, plants, and any supplies used by bedding plants. If one or more of the crops requires more than its proportionate share of direct labor, electricity or any other category of overhead, this should be adjusted.

To apply these costs per plant or per flat, calculate the cost per square meter of bench area for the crop and apportion it on the basis of time the space is occupied. Table 11-1 presents a simplified way of estimating bench space required for several sizes of pots at three most commonly used spacings (Appendix B).

A 15-cm, pinched poinsettia may be produced in 0.1 m^2, and a 15-cm unpinched, 3 plants-per-pot poinsettia, occupies 0.15 m^2. The overhead charged to the latter would be 50% more than for the smaller plant. Three plants instead of one would also add to the specific cost to produce the larger plant.

Most growers continually update their overhead costs since they are increasing each year. The total of budgeted expenses less supplies, plants, and seed for specific crops is divided by the total bench area to give a budgeted overhead expense. This total, divided by 52, gives the overhead per week charged to a crop. The time a bench is empty must be charged to the preceding crop or the crop that follows. A handiflat module of bedding plants could serve to illustrate this simplified method of cost calculation. The standard handiflat occupies 0.19 m^2. A flat of tomato plants grown from April 1 to May 19 in jiffy 7s would cost 7 weeks of overhead + cost of flat, seedlings, transplanting and 50 jiffy 7s.

In calculating costs, an allowance should always be made for loss, either through disease or from produce being unsold. In an efficiently run operation, 5% is usually adequate for loss. This loss is added to overhead costs and specific costs to give the projected cost of production of the item ready for sale. The cost of a poinsettia becomes:

Specific costs
\+ *overhead costs*
Subtotal
\+ *5% allowance for loss*
Cost of production

The cost of marketing also must be considered, along with a reasonable profit when pricing the item for sale. The selling price may have to be based on competition, supply and demand as well as product cost. It is up to management to choose the profitable crops, grow them correctly, price them profitably and market them successfully so there will be profit produced for the firm. Without adequate cost information, the manager may base his prices entirely on competition and what the traffic will bear.

11.6.2 Alternate Crops

One of the short-run decisions that continually faces management is choice of crops. Are there alternate crops that can be produced to improve the profitability of the operation? The alternative may be the same crop in a different form, or size, to reach a different market. Or the manager may consider a different crop that has not been produced by the firm before. A list of questions and considerations that should apply to all new crops is included in Table 11-2.

Table 11-3 shows one way of projecting production and profitability for snapdragons. Assume the snapdragons are to be grown single stemmed and 4 crops per year can be finished on the same bed area. This gives a potential of 480 stems

Table 11–2. Questions and considerations by management on any alternate crop

1. Present status and trends in the crop. History and development and how it became a commercial crop
2. The geography of its distribution in present marketing of flowers and plants
3. Commercial yields, planting distances, other economics of production
4. Opportunities for innovation in its culture
5. Production timing and how it is done; when available in markets?
6. Principal problems encountered in producing the crop
7. Special points in marketing
8. A list of up-to-date references on the crop

Table 11–3. Projection on single stemmed snapdragon (Antirrhinum) with percent marketed of planted and net selling price above cost of marketing as the variables. This assumes 10 cm by 7 1/2 cm spacing and 4 crops finished per year

Percent marketed of planted	Flowers sold per m^2	Net selling price in cents					
		10	11	12	13	14	15
66	320	32.00	35.20	38.40	41.60	44.80	48.00
75	360	36.00	39.60	41.20	46.80	50.40	54.00
83	400	40.00	44.00	48.00	52.00	56.00	60.00
90	430	43.00	47.30	51.60	55.90	60.20	64.50

per year per m^2. From general experience, it is clear that 100% of the plants planted are not cut as flowers. What is the usual loss rate of this? We can make a table that includes % marketed of plants planted, flowers sold per m^2, and multiply these by a range of possible selling prices. We see immediately from the table that % marketed becomes very important along with net selling price (after cost of marketing is deducted). To complete the projection, a manager asks and must answer to his satisfaction—"At what price can I sell snapdragons? Are there methods that I can develop that will enable our firm to market 85% or 90% of the plants that we transplant?" For obviously the profitability of single stemmed snapdragons goes up with higher % survival and higher prices.

11.6.3 Summary

Management is the collective name for all those persons responsible for directing a greenhouse operation. Management is also a process. The tools of management are planning, organizing, directing, and motivating, and controlling. Neglect of any of these functions can seriously limit profits.

Planning must be for both short and long term. Flower growers often get so bogged down with near-term operations that they do not allocate sufficient time to studying trends and planning for the long haul.

Organizing is the carrying out of plans on schedule. The manager must provide the capital, facilities, and the personnel when they are needed.

The manager sets the job or business in motion and keeps it moving in the right direction. The flower growing enterprise must be people oriented to reach the objectives of management on schedule. It is most important for a manager to understand the needs of people. Floriculture can fulfill these needs equally well or better than any other industry.

Controlling is the function of management concerned with recognizing operational weaknesses and correcting them. Performance must be standardized, production schedules should be as detailed as necessary, and some procedures should be written down in simple terms.

Accounting provides the figures to management but unless the manager studies and acts on them, accounting may become just an exercise for tax purposes rather than the best tool the manager has for cost control. In order to have accounts timely, computer services are suggested at nominal cost. Analysis of accounting data should be done promptly and regularly, comparing them with prior years and with the firm's objectives.

Finally, the manager must know costs of production in order to market profitably and to compare with alternatives.

11.7 Suggestions for Further Reading

Drucker, Peter F.: 1964. Managing for Results. Harper and Row, Publishers. New York: 240 pp.

Haynes, W. W.: 1963. Managerial Economics: Analysis and Cases. The Dorsey Press, Inc., Homewood, Ill. 618 pp.

Longnecker, Justin G.: 1964. Principles of Management and Organizational Behavior. Chas. E. Merrill Books, Inc., Columbus, Ohio. 676 pp.

McGregor, Douglas: 1960. The Human Side of Enterprise. McGraw-Hill Book Co., Inc., New York: 246 pp.

References

Grimmer, W. W.: Greenhouse management—suggestions for improving your management skills. Talk presented at the Penn. Florists Convention, Penn. State Univ., 1970a

Grimmer, W. W.: Can midwest greenhouses continue to exist? Florists' Rev. **147**(3798), 26, 68–71 (1970b)

Jerome, W. T., II.: Executive Control—The Catalyst. New York: Wiley 1961, pp. 32–33

Mason, A.: Business efficiency. The Grower **79**, 1432 (1973)

McKeever, J. L.: A recommended system for accounting and cost control in a flower growing firm. Research project sponsored by the Colorado Flower Growers' Ass'n., Denver, Colo. (unpublished) 1971, 210 pp

Mollgaard, H. B.: A report from Sweden: optimum wage hunt. Florists' Rev. **148**(3837), 26, 62–64 (1971)

Wrapp, H. E.: Organization for long range planning. Harvard Business Rev. **35**(1), 44 (1957)

Chapter 12 Marketing

Marketing channels for floriculture in the United States were worked out for the first time by Fossum (1950) (Fig. 12-1). Almost 40% of all flowers produced were sold to the consumer by conventional retail florists and 21% were sold by retail growers in 1949. This emphasizes the extreme importance of retail florists to flower and plant producers. When an industry has this sort of established market, a thorough understanding of this market and its problems is absolutely essential. The trends in retailing must be known by the producer and he should offer every means of help to the retailer in expanding sales. This is done to some extent through Allied Florist Associations and the American Florist Marketing Council, but more can be done for the benefit of the industry.

The producer has the greatest stake in the floral industry, but he should not be required to do all of the promotion. In fact, he is not doing his share in the present market. Industry-wide advertising has made a beginning through the efforts of the American Florists' Marketing Council. Unfortunately, the degree of participation is not nearly enough to do an adequate job of advertising for a $2^1/_2$ billion dollar industry. The several flowers-by-wire services spend much more in advertising flowers and services in consumer magazines. Some advertising is also done by special groups such as Colorado Flower Growers' Assn., Inc. and Roses, Incorporated. Flowers and plants are promoted in local areas through Allied Florist Associations with the producers, wholesalers, and retailers serving that area cooperating.

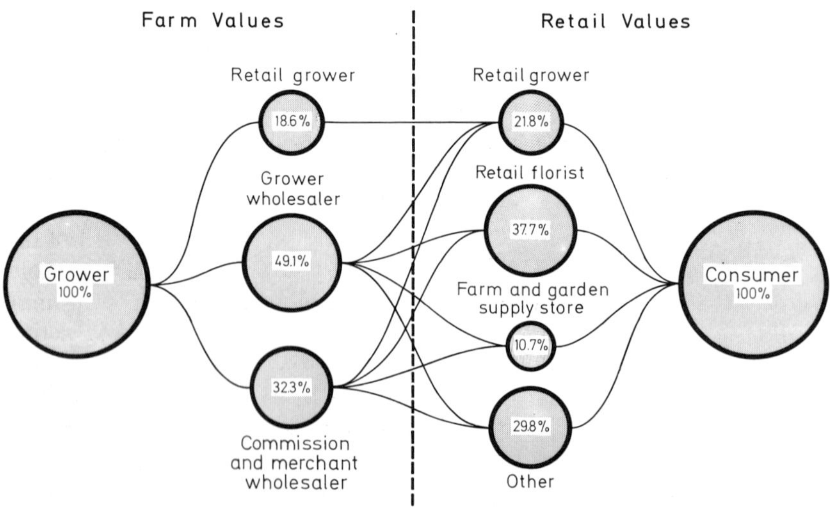

Fig. 12-1. Marketing channels for ornamental greenhouse crops in U.S. (Fossum, 1950)

We have a needed product and we can assume people want it. What do we really have to sell? Flowers have color, freshness, shapes or forms, many have desirable scents, and they are alive! Flowers and plants can satisfy the customer's needs to express moods, to congratulate friends and neighbors, to express love or sympathy. No other product or group of products can do these things for a consumer in such a wonderful way. We need to tell the buying public this story and keep on telling it.

"Allieds" are associations of retail florists, growers, wholesalers, and allied tradesmen serving a specific city or area. They meet regularly to discuss common problems and to know one another. Each contributes a small percent of gross sales to public relations and local advertising of flowers and plants.

The American Florists' Marketing Council is the public relations arm of the Society of American Florists. One of the oldest florists trade organizations in the world, the Society of American Florists, serves retail florists, growers, wholesalers, and allied tradesmen. Each member contributes an allocated percentage of gross sales to AFMC funds. These monies are used in promoting the sale of floriculture products. Among the many services performed by AFMC are national ad campaigns tied in with outdoor billboards, full color mailers to members, and display ads to be used in shop windows and on delivery vehicles. Sponsorship of national television programs and sports spectaculars have been included in recent years. SAF through AFMC also combats practices that discriminate against the use of flowers and provides a continuing program of education in schools on the use of plants and flowers.

12.1 Trends in Retail Production

12.1.1 The Retail Florist

The number of retail florists (not associated with crop production) increased 30% from 1949 to 1959 (Fossum, 1961) and the value of their sales increased 70%. In a later economic analysis of the U.S. Florist Industry (Fossum, 1967), the total value of flower and plant production for 1966 was 450 million dollars. This amounted to $2.25 per capita or $7.50 per household, and is pitifully small compared to such flower consuming countries as West Germany, Sweden, or Finland. In 1967, 25000 retail florists sold about one billion dollars worth of the products of floriculture plus services. This compared to $500 million sold through nonflorist outlets [1]. The percent of flowers sold in retail florist stores may vary from country to country, but it is a very significant segment of the marketing organization for the flower and plant industry and should be examined in depth.

The retail florist shop began around 1900 when a need arose for the supply of fresh flowers and services not then provided. Before that time flowers were sold

[1] Economic Res. Ser., U.S. Dept. of Agric., Statistical Bulletin No. 529, Washington, D.C., May, 1974 (Raleigh, 1974).

from retail greenhouses or direct from larger producers to hotels, restaurants, and morticians. Pfahl (1968) estimated that on the average there is a need for a new retail florist for every 10000 people. The retail florist business is characterized as an easy entry business as requirement for investment capital is low and long time training is minimal. There are disadvantages to the retail florist business such as long working hours at times and low pay, but the large majority of florists are dedicated people who love working with flowers and plants and they treasure being their own boss.

Retail florists have been criticized by producers for the "high" markup they place on the flowers and plants that they sell. Another criticism is that retailers do not "push" flowers that are in heavy supply in order to relieve glutted markets.

It must be understood that retail florists are customers of producers and must be appreciated as such. The retail florist does not care what it costs to produce a crop. As a business manager, the florist must make a profit on his enterprise and this is done by sharp buying, good selling, efficient operation, and giving good service to customers.

Many retail florists do most of their business over the telephone and are not located for high traffic sales. Thus, it is not feasible for them to "push" flowers that are in heavy supply. On the other hand, they depend on "occasion" sales that are relatively inelastic. However, several researchers have found that special sales can increase business for the retail florist.

Von Oppenfeld et al. (1957) found that small flowering plants and cut flowers were readily moved as cash-and-carry "specials". Special sales were more effective on weekends when combined with advertising such as window banners and adequate displays. New customers were attracted by the specials. Many old customers bought specials in addition to their usual purchases. The number of special sales made increased from week to week as the practice was continued. More specials were bought for home use than for gifts or special occasions.

In a later study, an attempt was made to develop new and improved promotional appeals and to test their effectiveness in increasing sales and profits in retail flower shops (Pfahl et al., 1959). Consumer surveys were made in two comparable communities before and after a promotional program of weekend cash-and-carry specials. The special sales in all shops of the community were combined with newspaper and radio advertising, open houses, and special newspaper publicity. Lectures and demonstrations on the use and care of flowers in the home were presented to civic and service organizations, and to high school classes. Sales and profits increased significantly in the city where the promotion was done and sales continued to increase after the promotional program. Sales of weekend specials did not replace service sales. These researchers concluded that specials can be sold successfully on a continuing basis as volume of cash-and-carry items remained relatively constant throughout a test period of more than a year. The survey data indicated little or no change in consumer attitudes toward flowers.

Pease, one of the researchers in the previous study, reported on another study in 1962. He kept records of dollar volumes of floricultural sales made in an isolated county serviced by one florist-greenhouse from 1949 to 1959. The florist embarked on a program for increasing taste and preference for flowers by winning the public's confidence, by familiarizing the public with the business and by devising merchandising techniques. The physical plant was enlarged progressively from year to year. Landscaping service was initiated throughout the county. Special merchandising services were offered to nonprofit organizations. An unusually beautiful wreath was designed that added much to the life of funeral flowers. An informality was maintained in the flower shop that had not been present before.

Pease concluded that innovations must be adopted progressively and constantly to maintain or increase a high level of taste preference and sales volume in a retail flower shop. In other words, continuous sales promotion, by whatever means, is necessary to maintain an increasing demand for flowers.

Markup on flowers sold varies from store to store and is based on the cost of doing business. As volume of sales increase, markup usually is reduced. (For example, a nonperishable item such as nylon stockings, costing 50 cents might sell for 89 cents in a high-traffic supermarket and $ 2.00 or more in a jewelry store or retail florist shop.) The markups in these stores are legitimate and are those required by the nature of the operation. Delivery, credit, and other services besides low traffic volume, influence the retail florist's markup.

Flower and plant producers should appreciate the tremendous built-in market created by the retail florist industry and do what can be done to help this industry increase sales. The retail florist is a customer of producers and wholesalers, not an agent for them.

12.1.2 Wire Services

One of the greatest flower marketing ideas and much copied in other industries are the flowers-by-wire services. The Florist Transworld Delivery Association was first and started around 1892 (Pfahl, 1968), followed by TELEFLORA, Interflora, and Florafax.

A local florist member of a wire service either sends or receives orders for flowers to or from any other member of the association. Most wire services have international connections including arrangements with major credit card companies such as Carte Blanche and Diners Club.

To send an order for a customer, it is either telephoned or telegraphed to a member florist at the destination. The customer pays for the flowers plus minimal charges for clearance and transmission of the order. The sending florist collects for the order and sends this to the clearing house. Regular statements from the clearing house reconcile incoming and outgoing orders for all members. Slightly different arrangements on orders characterize the different wire services.

The wire services have increased flower sales by spending more advertising dollars than any other industry group. Their advertising on television and in consumer magazines has done much to popularize flowers as business gifts, and to remind consumers of occasions when a gift of flowers or plants is appropriate and in good taste.

Pfahl (1968) estimates FTDA membership at 11000 florists in the United States and Canada plus 24000 overseas florists in 90 different countries. FTDA, together with Interflora British Unit and Fleurop, are affiliated in an international floral delivery network, Interflora, Inc., with some 33000 members.

TELEFLORA includes some 9000 member subscribers in the U.S. and Canada, with international affiliations in Britain, Europe, Australia, and New Zealand, Latin America, and the Orient. Florafax Delivery, Inc. had some 6000 subscribers in 1968, and through an agreement with the United Flowers-By-Wire Service, Ltd., in Canada, offered marketing facilities to 7000 florists of the United States and Canada and increasing international connections. While many florists belong or subscribe to more than one service, it is readily apparent that wire services greatly enhance the network for selling the products of floriculture.

12.1.3 The Retail Grower

The simplest yet possibly the most complex operation in floriculture is the family-operated retail greenhouse. Earlier in the history of floriculture, this type of business was the backbone of the industry. The "country florist" weathered the depression years about the best of any segment of agriculture. Today there is a desire among many environmentally inclined young people to develop businesses around a small greenhouse operation. The opportunity is there but the business can be extremely complex.

Most small greenhouses that have a retail shop find that the shop often makes a profit at the expense of the greenhouse. If there are two operations, they should be separated, and each one made profitable.

The retail grower has a marvelous opportunity to survey his local market and produce what is needed. The location of such an operation is less important than one would think. People love to come to greenhouses, especially to buy their garden plants and supplies. Why would they not want to buy their house plants and small bunches of cut flowers, or a flowering plant, on regular visits? Because of the product freshness, service, price, or some other reason, the retail grower can be extremely successful. Often the most loyal customers come for information as well as plants.

Marketing can be a developing function for the retail grower. Advertising by newspaper along with word-of-mouth, direct mail, and a developing reputation for quality plants can put the retail grower in an enviable position. Alternate means for selling some of the product may be needed at times. Excess production might be moved through garden centers, other retail businesses for special promotions, or through neighboring retail florists and plant shops.

The point is that the retail grower knows his customers. He can control a rewarding life for himself and his family, and could, in many respects, be considered the apex of the floral industry.

As growers enlarge and specialize, they often expand their marketing by selling direct to retail outlets. The product may be delivered by the producer on a regular basis or may be shipped by bus, truck, or air. The producer may assume one or all phases of marketing in this manner, but accurate costs should be kept for comparison with alternate methods of marketing.

12.2 Commission Wholesalers

The major wholesale arm of the U.S. floral industry is the large group of wholesale commission florists. Producers consign their flowers to these merchants located in every city. These wholesalers often perform many functions. They receive, unpack, condition, sell, repack, and ship, collect for, and remit to producers with sales statements at regular intervals. There are many fine producer-wholesaler-retailer relationships of long standing. Some producers find it necessary to sell through a number of commission wholesalers, especially if they are specialists in one or a few crops. If one wholesaler is oversupplied, others in this case may be able to sell any surplus production.

When marketing through commission wholesalers, the producer should work as closely with them as he would with a sales department within his own firm. The wholesaler will help him plan what to grow, the cultivars and grades most in demand and when demand is best. The producer needs to be in daily communication with his wholesaler.

The wholesale commission florist is also a buyer of products not available from his producers. He locates sources for field grown flowers or greens, and obtains regular shipments in order to have a complete line for his customers. He may also stock supplies for the retail florist and the grower.

Sales of large quantities of flowers are often made by wholesalers to other wholesalers (jobbers) on a direct sale basis at the going price less a discount. This practice results in a producer paying what amounts to two commissions, as his regular wholesaler may charge 20% and the discount to a jobber may be 15 or 20%. When flowers are in heavy supply, and the demand is low, holding to a set price may be less important than cleanup.

12.3 Grower-Owned and Public Markets

Boston, San Francisco, and other cities have flower markets owned by the flower growers. Each grower can purchase a space or a stall in the market entitling him to display and sell his stock. Rules for operating the market are decided by an elected governing board. Some markets operating in this manner allow card holding buyers to participate early in the marketing day then throw sales open to everyone after a certain hour. This usually assures a cleanup of the market each day.

A corporation made up of nearby growers of plants and flowers controls the Boston Flower Exchange, Inc. (Jarvesoo, 1966). The building is owned by the corporation which leases stall space to growers, who may use the space for selling their flowers in person, through employees, or through commission men. The rules of the Exchange state that only flowers and plants grown in New England and shipped directly to the market by stall holders shall be offered for sale on the floor of the Exchange.

A stall holder may assign leased space to a commission wholesaler, but these wholesalers must be approved and admitted by the Directors of the Exchange. In the early 60s, there were about equal numbers of commission men and growers selling direct.

Stall space and stall rent are assigned on the basis of greenhouse area operated, with a grower required to rent space at the rate of one stall per 2000 m^2.

At Copenhagen, the wholesale market is used for selling fruit, vegetables, and flowers (Rathmell, 1971). Chartered by the city in 1954, the market structure is a cooperative corporation. Each member pays a minimal annual fee plus annual rent for a stand. Center of the market is a sales hall of 28 700 m^2 with about 20 ha total space for parking and adjacent buildings. Most of the members live within 30 km of the market, so produce is brought in trucks or cars. Buyers may rent parking space on an annual basis.

Each grower parks his truck or car inside at his stall space, with spaces for some 850 vehicles in the building. Total sales for 1969 were almost $ 39 million with flowers and foliage plants accounting for about $ 21 million.

Buyers push carts around the building picking up their needs which they move to their outdoor parking space. The market is open every weekday and is self-supporting.

Mexico City and Sao Paulo, Brazil, have extensive public flower markets subsidized by the government. For a small fee a flower grower can park his truck or van under cover, and sell to all comers. Since flower marketing days are designated, buyers, especially those who plan to ship to other areas of the country, are in much evidence.

12.4 Cooperatives

While there are few pure cooperatives in flower marketing, some element of cooperation is evident in most marketing groups. Farmer cooperatives have been notably successful in marketing other agricultural products. In economic need such as too wide handling margins, too narrow outlet for products, too many growers competing for the same customers, or like reasons; farmers or flower growers may find it advantageous to form a cooperative. They must assure themselves of capable management and they must have adequate volume to hold unit costs down. They must meet the usual problems of business efficiency. Failures of most cooperatives have been due to the same ills that plague any business concern. They can be lumped together in the term poor management.

An ever-present problem is one involving membership relations. It is involved in the organizational pattern, the financial structure, and the day-to-day operations. It is no simple matter to maintain a cooperative as a member-controlled association, one in which the members know enough about it to give it intelligent direction.

A cooperative can perform all of the functions of a wholesale commission florist and at the same time do more for its grower-members. Its functions are in two groups:

1. The primary function is to market the flowers for grower-members at the best possible price and at the lowest cost. To do this job well the wholesale house has all the problems and trials of any wholesale selling establishment, whether cooperative or independent. Management must maintain good trade relations by:

a) Advertising.

b) Having good salesmen on the road calling on and servicing customers; salesmen who know the products; some of the difficulties involved in producing them; why they are not always of equal quality, etc.

c) Helping to coordinate the production of grower-members to insure the desired flow of flowers into the pool at different times of the year. This necessitates crop planning and enforcement of planting schedules.

d) Constant research and improvement in shipping and handling methods to make their operations more efficient and to cut shipping costs.

2. A second function of the cooperative includes additional services it can perform for grower-members.

a) A grower service department that tests soil, supplies disease free stock, assists in pest control, crop planning, makes feeding recommendations, etc.

b) Cooperative buying of supplies to save money for the members.

c) Centralized grading of flowers for grower-members may be provided.

12.4.1 How the Pool System of Marketing Works

Each grower brings in his daily cut of flowers to the wholesale house with his grading list. Or, the ungraded flowers are brought to a centralized grading facility. Trained receivers check the accuracy of the grower's grading and counting. They do not hesitate to downgrade any bunches that have been graded too optimistically. After checking the grower's grade list and making any changes necessary, the flowers enter the pool with those from other grower-members. Flowers coming through centralized grading are automatically received in the pool. Pool periods may be one or more weeks.

If all flowers are sold, the grower receives a check for all flowers brought in less the cost of selling them and other deductions. If only 90% are sold of a given grade, the grower receives 90% for the number he delivered in that grade and so on. Even though a stable selling price is maintained, the return to the grower fluctuates according to the grade of flowers he brings in and the percent of cleanup on each pool.

Naturally, there are many problems inherent in such a system of marketing. Grading should be centralized and as objective as possible, and grade standards should be adhered to year around. The management of the wholesale house should be given adequate authority.

12.4.2 Auction Selling

Another form of cooperative is the clock auction perfected first by the Dutch growers. Every grower from the United States who has visited Aalsmeer, Holland, has been impressed by their fast, efficient, and inexpensive method of selling flowers and plants. The new auction at Aalsmeer sold $ 180 million worth of flowers and plants in 1975[2]. There are some 500 buyers in attendance at the auction, many of them living there or nearby. The new auction at Aalsmeer is the world's largest and most modern. There are almost 10 ha or floor space that combined two older auctions (Rathmell, 1971). Six clocks may operate at one time selling cut flowers exclusively.

Efficiency is featured in the handling of flowers. An endless chain moves the flowers from the receiving area to the sales area then to the wholesaler's area for packing and dispatch. Most of the flowers sold at Aalsmeer go to the international trade, especially to West Germany, Switzerland, and France. Since the auction is located near Schiphol Airport and Amsterdam, transport facilities are excellent. Another large auction is located in the Westland section of Holland and many smaller auctions are in other towns.

How does an auction market work? Buyers occupy permanent seats in the market (Fig. 12-2). Flowers or plants are arranged on movable tables in a standing hall prior to the auction and may be inspected by the buyers. When the auction begins, the flowers are moved in front of the buyers and described briefly as to number, type, and grower. The clock is started at a high price and the hand moves down slowly. As the price a buyer is willing to pay is approached, he may punch a button on the arm of his seat and stop the clock. His buyer number flashes on the face of the clock and he is free to buy all or part of the lot of flowers. If he takes only part of the lot, this is recorded and the hand is started again for selling the balance of the lot. As all of the lot is sold, the table moves back to the standing hall and another table takes it place.

After flowers are sold they move to coolers and may be packed for shipment to many sections of Europe. The buyers may be exporters, wholesalers, or they may be agents for wholesalers, retailers, or street peddlers. All sales are in cash and the cost of selling to the grower may be no more than 5–8% of the value of the sale. Holland and Denmark have had successful auction markets for around 50 years. Germany, Sweden, and Canada have started auctions in recent years. The new flower auction at Nice, France, is also quite successful.

The grower's auction is a cooperative association founded and owned by the flower growers of a specific area. It is simple, ingenious, and effective, but this system has not been tried in the United States. The traditional wholesale commission florist says it will not work here. As it has developed in Holland, it is a workable, proven system. Long years of experience have shown that grower, commission man, and retailer alike benefit by it. No one willing to cooperate is put out of business.

There are certain essential rules that must be followed. All participating growers must agree to sell 100% of their produce through the auction. Violators of this agreement are fined 1% of their previous year's turnover for the first infraction, 2% for the second, and are expelled the third time. Expulsion of a grower-member has never happened at Aalsmeer.

---

[2] Annual Report of V.B.A. (Dutch Marketing System), Aalsmeer, Holland, 1975.

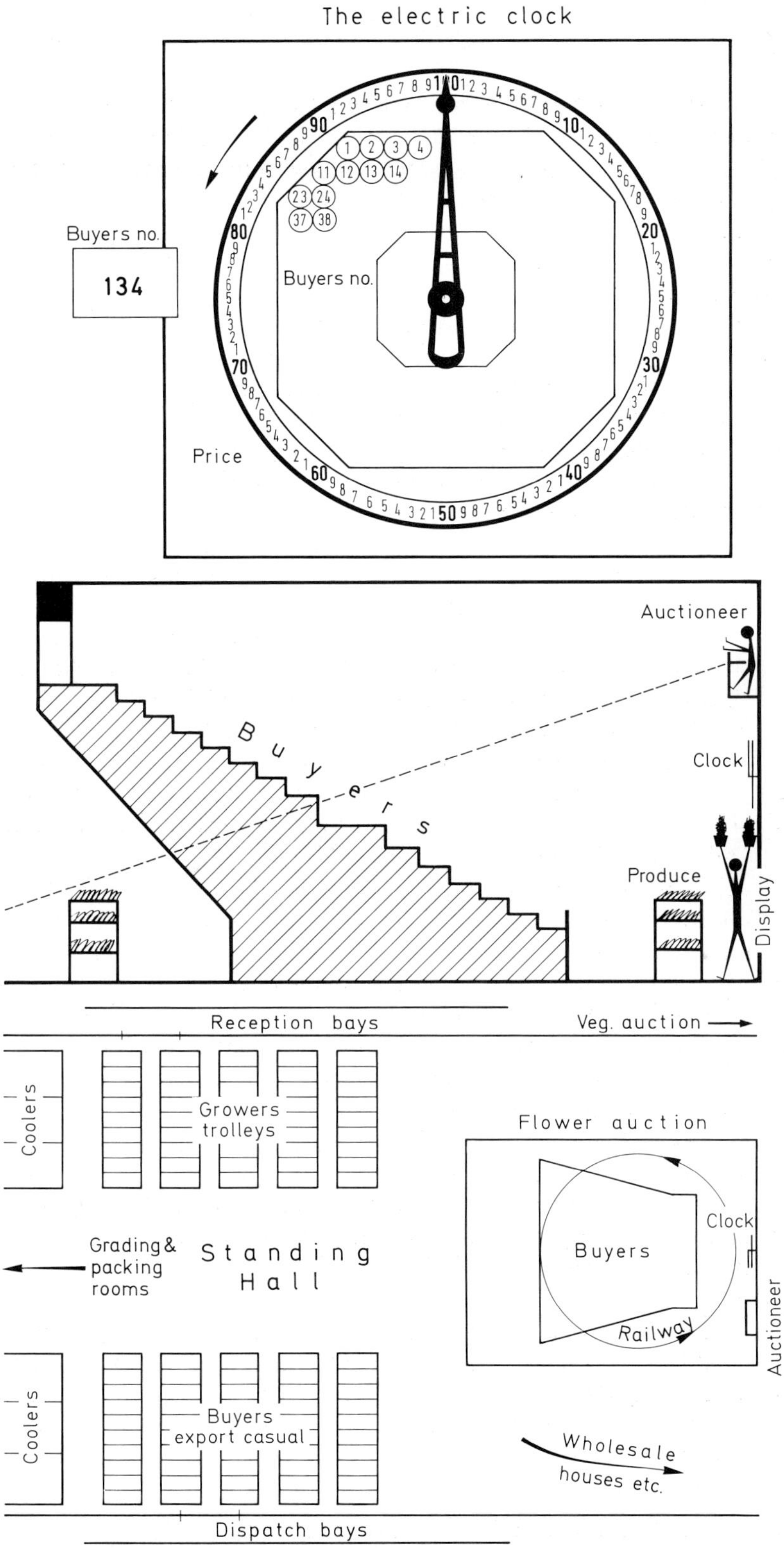

Fig. 12-2. Schematic showing how an auction clock works

As in other cooperatives, the grower members elect the management of auctions and sound business principles must be followed.

Five experiences with auction markets will be cited here. All areas but one were using commission wholesalers or direct selling before an auction was started.

By 1953, the region around Neuss, Germany, had increased flower cultivation to the point that sales were lagging behind production. A quicker and more efficient method of marketing was needed so the growers cooperated in founding an auction market patterned after the one at Aalsmeer.

In Straelen, Germany, no flowers were grown prior to 1952. Competition from southern grown vegetables, principally from Italy, forced the conversion of glass houses to flower production. The auction market had already been used many years to sell their vegetables so they sold their flowers the same way.

Vancouver, British Columbia, started a flower and plant auction in 1963. The auction had only six growers at the start but increased to 40 members the first three years. They had many problems the first few years, but the auction market will probably survive. Two major problems were that growers sold only a part of their crops through the auction and the auction system was against the tradition of marketing in America.

On the other hand, Toronto, Ontario, has most recently established an auction that has been successful from the beginning. By the end of 1974 there were 110 members and a \$ 20000 dividend was distributed to members (Anonymous, 1975). The successful Toronto auction may guide the way to formation of auction markets in the United States.

Approximately 200 growers move their flowers through an auction market south of Malmo, Sweden (Rathmell, 1971). Flowers are displayed in shipping boxes by grouping in rows on the floor of part of the building. Thus, roses can be inspected in one group, carnations in another line, etc. Growers display their flowers from 7:30 a.m. to 8 a.m. The buyers may inspect them until 8:30 a.m. at which time the auction starts. A clock is used for selling, but flowers are not brought into the auction room. Each buyer has a list of the items to be sold with the grower's number. As lots are sold, buyers stop the clock and buyer number, price, and lot number are recorded. The buyer repacks his purchases, often to fill orders that he has from retailers.

The Malmo auction has been credited with improving the quality of cut flowers grown in this area that produces more than half the cut flowers grown in Sweden.

In summary, growers who market their flowers and plants through auctions do not want to return to former methods of marketing.

Among the advantages they claim for auction selling are:

1. It is fast.
2. Buyers cannot take advantage of producers because of perishability of the product.
3. Competition is keen among the buyers.
4. Sales are for cash.
5. Growers are partly reimbursed if flowers are not sold for some unforeseen reason.

6. All flowers are sold the same day they are brought to the auction. There are no leftovers to depress the price of future production.

7. Auctions have increased trade and production of flowers in every area where they have been established.

8. The buyers and producers find easy cooperation in public relations, advertising, and promotion of sales.

9. Auctions are efficient and economical. Savings in overhead and selling costs are phenomenal.

12.4.3 Marketing Orders and Agreements

Another type of cooperation in Agriculture can be accomplished through Federal Marketing Orders. The Agricultural Marketing Agreement Act of 1937 provides general authority for the use of marketing orders in the U.S. Since that time, marketing orders have been used on such diverse commodities as milk, cotton, fruits, and vegetables. An order can be set up for roses, carnations, or other floral crops. A marketing order can do many things to solve major marketing problems. There are also some things an order cannot do.

Marketing orders can regulate the quality that can be shipped, helping to keep inferior grades off the market. Quality can be regulated by allotting market shares among producers based upon sales during a base period. Short-term rate of flow can be controlled to avoid market gluts. Orders can also provide market information and provide funds for research, advertising, product promotion, and prohibit unfair trade practices.

Marketing orders cannot control production or fix prices at any level. In developing a marketing order, the producers and handlers of a commodity should draft a preliminary proposal after analyzing the marketing problems and deciding that a marketing order would help them. Solid support among a group of producers is essential, and there must be general agreement on the types of regulation that will be proposed. The USDA and the Secretary of Agriculture offer technical assistance in this development. Other countries may or may not have legislation to permit marketing orders.

As a marketing order develops toward approval, hearings must be held among producer groups, handlers, and even consumers. The Department of Agriculture issues a recommendation based on the evidence of need presented at these hearings. Time is allowed for those interested to study the proposed order and submit exceptions. A final order is then put to a vote of the producers of the commodity with two-thirds majority usually required to activate an order. An order can be amended by the same procedure. An order can be terminated by a simple majority. Up to this time there have been no marketing orders in floriculture, but orders have been considered by grower groups.

A marketing agreement, signed by handlers of a commodity, may be issued along with a marketing order and carries the same provisions. The law permits handlers of any agricultural commodity to enter into marketing agreements with the Secretary of Agriculture. An agreement may be used without an order to serve the same purposes, but the agreement is binding only on those who sign it.

12.5 Advertisement

We are fortunate in floriculture to have products that are universally desirable. But, successful marketing of these products requires much more than a need by customers. We have to let consumers know more about our products, what our products can do for them, where the products can be bought and at what price. This requires product promotion. Flowers and plants are competing with other products in the consumer's mind. Advertising should be stepped up to improve our status in this competition. If you have a good product, put your name on it. Few brand names or trade marks have been used in floriculture, yet most of us are familiar with Diamond walnuts, Calavo avocados, and Jiffy 7s. In the relatively few instances where trade marks have been used in floriculture the effects have been good. A product must have an identity to benefit from promotion.

12.6 Mass Marketing of Floral Products

The demand for flowers fluctuates seasonally and with certain holidays. Through the regular channels from producer to wholesaler to retailer this demand is inelastic. To add a degree of elasticity to the demand for flowers requires marketing through other than conventional channels and this has been developing over the past decade. There is more elasticity to the market for flowering pot plants and even more for foliage plants because of the wider types of retail outlets in which these are sold.

For the first time, the U.S. Census of Business attempted to enumerate sales of floricultural products in nonflorist outlets in the 1972 Census (Fossum, 1975). Of the approximately 2.5 billion dollars spent on floriculture at retail, 62% went to retail florists and 38% to nonflorist retail establishments. Almost 25000 retail florist shops sold $ 1.6 billion worth of flowers, plants, and services in 1972. Despite the trend toward diversification by retail florists, 93% of their sales were of cut flower and plant oriented items. Since census data from the nonflorist trade lacked some detail on floriculture, Fossum estimated flower and plant sales in nonflorist retail outlets at just under $ 1 billion in 1972.

While retail florist sales are increasing each year, it is generally recognized that nonflorist sales are increasing at a faster rate. Nonflorist or mass market sales of flowers represent a future growth area in flower marketing so should be examined in depth.

12.6.1 Flowers and Plants in Mass Markets

As early as the 1940s, Prof. Alex Laurie and his students at Ohio State University were investigating prepackaging of cut flowers with the objective of merchandising our product in stores with high traffic (Laurie and Kiplinger, 1950). Dr. Max Brunk, Professor of Marketing at Cornell University, worked on a similar project

(Brunk and Hampton, 1953). Work with plant sales in supermarkets was done at Texas A & M University (DeWerth and Sorenson, 1955) and at Ohio State (Kiplinger and Sherman, 1962). While investigation of these potential sales outlets for flowers began in the 40s and 50s, the supermarket industry did not begin to accept flowers as potential profit makers until around 1965 or later. One of the first surveys of mass marketer's attitudes toward floral products was done by the Agricultural Marketing Service (Moore, 1959).

Information was collected from eight local, regional, and national food chains (2300 stores) and eight variety chains (2200 stores) in 1955 and again in 1958. Pot plants represented the majority of floral product sales in 1958. Cut flower sales decreased in mass market outlets from 1955 to 1958. Seven of the eight firms preferred Wednesday, Thursday, and Friday for delivery of floral products. Highest sales of cut flowers usually occurred on Saturday.

Most of the firms surveyed sold cut flowers seasonally. Spring, fall, winter, and summer were ranked in order of importance on the basis of cut flower sales. Pot plants were sold throughout the year by all firms. The retail margin averaged 45% for cut flowers and 43% for plants, although a flexible pricing policy was used by most firms for pricing cut flowers. Spoilage losses averaged 10% for cut flowers and 5% for plants.

Mass market outlets obtained cut flowers from wholesale florists, or large grower-shippers or assemblers. Pot plants were obtained mostly from wholesale growers. Agreements were used by five of the eight firms.

Problems faced by mass market outlets in establishing a merchandizing program for cut flowers included: (1) not enough consumer demand to make the sale of cut flowers profitable, (2) physical characteristics which make these products difficult to handle, (3) almost complete lack of packaging materials and display equipment to maintain quality, appeal to the consumer, and provide ease of product handling, (4) difficulty of obtaining good quality products during peak demand periods, and (5) inefficient and inconvenient delivery arrangements.

Moore concluded this report with: "The home use of flowers is indicated as a potential area for market expansion of floral products if these products are conveniently available, easily accessible, and reasonably priced ... mass market outlets that have tried to satisfy this home-use demand ... have had adequate profits on sales."

Since 1970, Dr. George Kress, Professor of Marketing at Colorado State University, has developed a special rapport with the supermarket industry and has completed a number of surveys by telephone and by mail to determine their attitudes on plants and flowers as profit makers, and the problems they face in supply, display, and perishability of product. The first study was carried out in the summer of 1971 among 79 supermarket chains. The following year, in a second telephone survey on the sale of cut flowers, 69 chains (almost 8000 stores) were contacted to determine the level of activity in cut flowers, to compare their present involvement with that a year earlier, and to try to ascertain the chain's plans for future involvement with floricultural products (Campbell and Kress, 1972).

Of the 69 chains contacted, about 13% handled cut flowers on a regular basis, 27% handled cut flowers on a seasonal basis, and 60% did not handle cut flowers.

The same chains were much more deeply involved with plants, as 42% handled pot plants regularly, 46% handled them on a seasonal basis, and only about 12% did not handle pot plants.

These figures can be misleading since there was considerable variation among chains and their involvement with cuts and pots. Approximately 39% of the chains contacted did not handle cut flowers. Over 27% of them were involved with cut flowers on a regular basis, but not necessarily in every store. Some chains (17%) handled cut flowers regularly in some stores and seasonally in others. In some chains this arrangement results from the fact that an increasing number of them are including formal flower sections in their new stores, enabling them to handle cut flowers on a regular basis. Their older stores would handle cut flowers on a seasonal basis. Store size and customer volume also influenced the nature of involvement with floral products in individual stores.

Pot plant involvement was much greater (90%) than cut flowers (60%). It was obvious that supermarket representatives favored plants since these require less care, are more durable, last longer, and have a larger per unit profit.

Due to their proximity to a wide variety of flower growers, stores in California were much more involved with cut flowers than were stores in other sections of the country. By removing the California stores from the results of the survey, 10% of the stores handled cut flowers regularly and 28% on a seasonal basis, while almost 62% of the stores did not handle cut flowers at all. This indicates that reliability and variety of supply is especially important to the development of cut flower sales in supermarkets.

12.6.2 Future Involvement

Of the chains interviewed, 44% planned to increase their involvement with cut flowers, whereas 37% planned to increase activity in plants. This small difference was probably due to the fact that present involvement with pot plants was much greater. Only one of the 69 chains planned to decrease their current involvement with cut flowers. Six chains of the 27, not handling cut flowers at the time of the interview, planned to start handling them in the near future.

Since each of the 69 chains was interviewed a year earlier, it was possible to compare their involvement with flower marketing for the two years. Almost 35% of the chains increased on pot plants from a higher level the previous year. Approximately 7% decreased involvement with both cut flowers and pot plants in the year.

When Campbell and Kress questioned the chains on purchase arrangements, they found that about half bought cut flowers outright with the remainder using some sort of consignment arrangement or leased space within their stores to outside firms. The vast majority of chains (82%) bought pot plants outright. Of those chains leasing space to outside firms, they usually charged 10–20% of retail sales.

The major suppliers of cut flowers to the supermarket chains were growers and wholesale florists. There was generally one buyer in the chain's organization who purchased flowers for all stores. In only five out of 42 chains handling cut

Fig. 12-3. In the most recent decade, flowers have been made more and more available to people, wherever they are

flowers did individual stores have the authority to make purchases. In those chains where a store was allowed to buy flowers, these purchases were usually limited to a supplier approved by the chain headquarters.

The chains did not want to get involved with written contracts with their suppliers but preferred to buy at the current market price.

12.6.3 Practices

12.6.3.1 Placement of Flowers Within a Store

The amount of space allocated floral products varied greatly from store to store; from as little as 2 m^2 to as much as 20. The front of their produce section was the favorite location for flowers by 44% of the chains and an additional 34% kept flowers as near the front entrance as possible. It was clear that supermarket managers felt that it was critical to display flowers in high traffic areas. The produce people within each store were usually responsible for the upkeep of the flower and plant display.

Almost half of the chains used buckets for displaying cut flowers. Open or closed coolers were used by around 15% and combinations of these were used by others. Several chains used different types of displays in different stores. The stores that had formal flower sections tended to use display coolers most.

12.6.3.2 Sales

The majority of chains contacted used a 40–50% mark-up of selling price for cut flowers. In his third and most comprehensive survey of supermarket involvement with floral products, Kress and his students received replies from 231 firms representing 23704 stores (Kress, 1975a). Of that number 11670 stores were selling floral products on a regular or seasonal basis, and another 882 planned to add them in the future.

Kress' first study was to determine involvement with fresh flowers. His second included plants as well as flowers. This third survey included questions on all three major categories: cut flowers, pot plants, and bedding plants. Every firm that handled floral products sold pot plants in 1975, while 83% of these also handled bedding plants in season and 59% of those handling floral products sold cut flowers in some or all of their stores.

12.6.3.3 Source of Supply

Local growers were the principal supply sources for pot plants and bedding plants. Cut flowers were purchased from local growers, distant growers, and wholesale florists (Kress, 1975a). About half of the firms had some sort of contractual arrangement with their suppliers of bedding and pot plants.

12.6.3.4 Methods Used to Display and Store Floral Products

Cut flowers were displayed in open buckets by 87% of the stores while 12% of the stores selling cut flowers used open or closed coolers. A third of all stores with flowers had permanent displays for pot plants. Most stores handling bedding plants in the spring used a large outdoor display area.

In most firms selling flowers, the average space devoted to these products was less than 7.5 m^2 but in those stores with formal flower sections, space used for flowers averaged 12.8 m^2. A fifth of all the stores selling flowers had progressed to identifiable, permanent flower sections.

Less than 3% of the stores with flowers employed fulltime flower attendants. Only 22 firms had special central warehouse facilities for flowers, and most of these were chains with 76 or more stores.

While the vast majority of firms realized the importance of providing their employees with special training on handling and display of flowers, less than half did so (Kress, 1975a).

12.6.3.5 Pricing Procedures and Sales Results

Kress summarized the information obtained from the 231 firms as follows:

The majority of firms marked up their floral products between 35 and 50% of selling price, with higher markups usually on cut flowers (50%). Slightly lower markups (40%) were used on bedding and pot plants.

The weekly sales in stores with seasonal flower sections averaged $205 per store. Per square meter sales averaged $40.10 for regular displays to $48.90 for stores with seasonal displays. Almost $^2/_3$ of the 2479 stores providing sales data had sales of less than $400 per week, but 122 did have sales in excess of $1250 per week. Pot plants accounted for 69% of the flower sales in those stores with regular flower sections. Cut flowers and bedding plants accounted for 20 and 11% of sales in these stores. In the majority of firms (all stores, regular and seasonal displays), pot plants accounted for 75% of flower sales and cut flowers for 10% or less.

Flower sales in supermarkets were adversely affected by summer months. About 30% of the firms discontinued handling flowers from June to August. In a few firms (22–28%) flower sales maintained the same level as they were during the rest of the year.

When analyzing the results of his survey of 231 firms (Kress, 1975a), additional questions arose. He enclosed these questions with the report mailed to each responding firm and compiled the answers to give us some excellent socio-economic information on supermarket sales of flowers. Sixty-two firms representing 5000 U.S. supermarkets responded to these questions (Kress, 1975b).

The firms were equally divided on whether income of the average customer had an influence on its flower sales. The firms who felt income was a factor stated that flowers are an "impulse" purchase.

The majority of those responding to the question about age of customer felt there was little or no relationship between age of customer and flower sales. Of the 25% that did feel age was a factor, most stated that younger people (30 and under) bought more flowers than older ones.

Three out of five respondents felt the ethnic background of a store's customers had little or no influence on its flower sales.

Plant and flower sales averaged around 7% of produce sales and 1% of total store sales. Six firms indicated that flower sales were more than 15% of all produce sales. The methods of these firms should be studied by the floral industry.

Kress tried to relate floral sales to other products sold in supermarkets. Various firms indicated that flower sales were similar to sales of candy, soup, or juice, and drinks. All firms stated that flowers carried a much higher margin of profit than other lines with similar sales volume.

Flowers seemed to be "recession proof" since 45% of the respondents stated that their sales did not decrease during the economic downturn of 1974–1975; in fact, 42% stated their sales actually increased.

12.6.4 The Track Record

After two or more decades, just how much have mass markets influenced producers of plants and flowers? retail florists? wholesale florists? The leading thinkers and marketers of the florist industry had predicted in the forties and fifties that development of mass market sales would revolutionize the industry. Worldwide growth of floriculture has been equal or greater than the rest of the economy, and no doubt much of this growth has been due to innovative marketing.

By early 1972, nonflorist outlets were accounting for half the pot plants sold in the U.S. (Berninger, 1972b). It was expected that it would be only a matter of time before these outlets would effectively penetrate the market for cut flowers. Berninger cited the interest from supermarkets in floral products and consumer response as reasons for their expansion of floral departments. The demand for flowers was cutting across all socio-economic lines. Few growers at this time were geared to direct mass-volume selling. According to Berninger, one California flower producer launched a firm in 1970 specifically to supply supermarkets and delivered the production from 500 acres directly to chain warehouses.

Kress reassessed the impact of mass marketing on the cut flower industry in late 1972 and stated that it was clear that effects on cut flowers up to that time were far from revolutionary. Why? Kress felt that some explanations were provided by looking at those outlets which supposedly had the most potential for mass marketing of flowers—the supermarkets. Many supermarkets had not seen fit to move into cut flowers because of their questionable performance for some markets that did handle them. Little in the way of facts and figures existed on how well fresh flowers had done for the reputedly successful chains such as Alpha Beta, Stop 'N Shop, and Handy Andy. These firms and others deeply involved with fresh flowers were satisfied with sales results, but "satisfied" is not easily translated to sales or profits. Kress added that much of the information on fresh flowers moving through the supermarket industry's grapevine at the time was negative since it came from those firms that had bad experiences with flowers. Kress also listed mistakes that had been made by supermarkets with fresh flowers. These apply largely to situations where the flowers are sold in self-service displays.

12.6.4.1 Poor Store Location

Even with prices at \$ 1.49–\$ 1.98 a bunch, fresh flowers cannot be successfully sold in all supermarkets. Flowers are quasi-luxury goods and thus are dependent upon the income of the store's customers. Flower sales have usually been poor in those stores primarily serving low-income clientele. Kress continued: "At this time neither the flower suppliers nor the supermarkets really have strong evidence as to what makes a 'good' flower store." The flower suppliers were too concerned with getting their products into supermarkets and were not carefully analyzing the sales potential of each store.

12.6.4.2 Handling

A second finding of Kress was that placement of flower displays within the store was critical to successful sales. The flowers have to be visible to all shoppers moving through the store and in a position where the customers will encounter them early in the shopping trip. The ideal situation is a total flower section where cut flowers are displayed with plants, floral accessories, and even artificials. A total flower atmosphere is created.

Kress suggested that the supplier has to provide store managers with evidence as to why his flowers should receive a favored location. An outstanding sales per square foot record is the kind of evidence that causes a store manager to listen. A problem facing the supplier is that sales of flowers have to build over a period of time. Many store managers want instant results so they may move flowers to less desirable locations. Kress suggests that a slightly higher return be allowed a store during the first month in order to get flowers in the most favorable location.

Lack of interest by supermarket personnel is another reason for disappointing sales by some stores, cites Kress. Flowers sold primarily on a self-service basis out of coolers or buckets are generally serviced by a supplier daily or every other day. Shoppers are constantly rooting through the flowers, checking various stems and combinations. Many do not put the stems back in water. Unless frequent attention is given the display, disarray and losses will result. Kress felt that the attitudes of store personnel toward fresh flowers was neutral.

Kress listed a fourth problem that had arisen in the sale of flowers in supermarkets as that of questionable displays. He criticized the supplier for this, saying that they do not pull dying flowers soon enough, or do not restock the display often enough. The potential sales must be accurately predicted so the display is adequately stocked but not overstocked. Kress also suggested that some sort of display "gimmick" be developed to give display cases an appearance of being full even though half empty, as store personnel are trained to take pride in full shelves. Kress also admonished suppliers to be prompt in resupply as the obligation to regularly service a store is taken on when that store becomes a customer.

12.6.4.3 Potential

In 1973, Walter Weir, partner in the international marketing consultant firm of La Borie-Weir, S.A., stated that they were convinced the potential mass market for floral products was around $ 2 billion. They were also convinced that this market could not be exploited by retail florists. Weir projected a total U.S. floral industry of $ 8 billion by 1980 and stated that most critical planning is needed to develop the mass market for flowers else it is likely to blow up in everyone's face. At the time, he felt that there were too many, too diverse and unorganized elements developing and exploiting mass market sales (Weir, 1973).

Production of floral products cannot be turned on and off. Extra greenhouse space and months of care and attention are required to produce extra volume. No grower is going to invest the capital this calls for without some assurance of return. One large customer can do this if he has a plan. A dozen small customers cannot do this. If volume is not increased to meet increased demand, price will go up to defeat the whole scheme.

Computers will help in predicting weekly needs. Assuming that 32000 major supermarkets in the U.S. would average $ 60000 in floral sales annually, the potential is $ 2 billion.

Weir continued that not all supermarkets have had good experience with flowers. That they have not may be due chiefly to the fact that they treat mass floral marketing as a part of their produce departments. *Floral marketing and produce marketing are about as dissimilar as they can be. The produce mentality finds it difficult to conceive flowers as a product*, Weir added.

Weir's concern was the threat of having the market destroyed before the promise of having it built can be realized. He suggested that some major marketer must take over at least 25% of the market and bring order to it. The threat of destruction comes from too many minor marketers in the act.

12.6.5 Overview—Flowers in Supermarkets

The purpose of this final section is to answer the question, "What are the implications of these data to supermarkets and the firms that grow and supply floral products?"

While almost one out of every two supermarkets handles some type of floral product, adoption has not been as rapid as had been predicted. Many people predicted that flower sections would become commonplace in the majority of supermarkets. Stores associated with voluntaries or cooperatives have been fairly slow in taking on these products.

Part of the reason for limited involvement of affiliates has been the lack of flower sales information available to them. They have left such involvement up to the individual stores.

In previous studies (Kress, 1972, Campbell and Kress, 1973), the participating firms emphasized the need to educate their produce people on the care and display of floral products. The comments received in the 1975 study repeat these same sentiments. Many firms indicate that a major hurdle to the successful sale of flowers is the inability of their store personnel to maintain these products in a reasonably attractive state.

It would seem the firms themselves would have initiated programs to provide such training, but only 43% of them have done so. It will continue to be a critical problem until the firms themselves take a more active role in educating their own employees on the proper care of floral products.

Potted plants account for almost 70% of the dollar sales of flowers in supermarkets. The overall impression from the 1975 survey is that the supermarkets are well satisfied in their experiences with potted plants and plan to become even more deeply involved. They are enthusiastic about such holiday-oriented plants as lilies and poinsettias. Potted mums and foliage plants are also very successful, but on a year-round basis.

Saturation of the plant market may occur in the next two or three years. This possibility should be more of a warning to growers than to the supermarkets.

The potential of cut flowers for supermarkets is still a mystery. Certain firms have had outstanding success, others have been disappointed. It would seem that those firms who are willing to properly display and care for cut flowers have good success with them. These products require daily attention and many firms are not willing to provide the necessary attention.

This lack of solid evidence pertaining to the supermarkets success with cut flowers is the prime reason only 10–12% of the stores selling cut flowers use a permanent cooler to display them. Before more firms make the heavy dollar commitment needed to purchase such units, they will have to see greater evidence of their potential. The purchase of such coolers generally commits the store to year-round sale of cut flowers and many are not willing to make such a commitment at this time.

At present, the majority of the bedding and pot plants sold in supermarkets are obtained from local growers. But, when big chains are involved, regional offices often have the opportunity to buy a large number of plants at a "good" price from a large distant grower. They make such a purchase and then distribute them to their stores, who now must either cancel or cut back on their order with local growers. So, while many firms state they have "contracts" with growers, the growers view these as somewhat shaky agreements since many have had such "contracts" ignored by the chains.

Word of this has spread among flower growers and many are apprehensive about getting too heavily involved with supermarkets. There is a "rule of thumb" followed by many growers—no more than 15–20% of their output to any one firm and no more than a total of 30% of their output to chains.

The sales return per square meter from flowers in stores with *seasonal* displays is $ 48.90. This figure is $ 12.00 more than the average sales per square meter of $ 36.80 realized for produce items [3]. Another attractive feature of flowers is that their markups (40–50%) exceed those for most produce items (25–35%). The sales per square meter of *regular* flower displays is $ 40.10, a figure which also exceeds the average sales per square meter of produce items in general. Thus, the sales and net margin aspects of flowers make them potentially attractive additions to produce operations.

Managers should be aware that buying quality products at a reasonable price is a safer route in the long run, than trying to get the lowest priced merchandise and foregoing some quality. It only takes one batch of low quality or questionable merchandise to do irreparable harm. Having a supplier who continually provides good quality floral products is especially critical to those supermarkets whose personnel have only limited knowledge of proper plant care.

12.6.6 Outstanding Examples

12.6.6.1 The Migros Stores of Switzerland

Migros has over 350 outlets that sell flowers, plants, and supplies (Rathmell, 1974). Migros founded in 1925 to sell food items, has gradually moved into sales of almost everything. It is a cooperative with a million or more members. Every part of Switzerland has access to a Migros store, some being served by "stores on wheels" that travel the back roads. Flowers and plants accounted for $2^1/_2$% of the total sales in 1972, or over $ 750000 in a country of $7^1/_2$ million people.

Migros first handled flowers and foliage plants in 1947. Carnations are the top sellers. Migros owns flower farms in Spain and imports from Holland and all reliable producers if quality and price are right.

[3] "Produce 74", *Progressive Grocer*, Vol. 53, Number 3 (March, 1975), pp. 47+.

Migros is a pioneer in flower packaging. Most flowers are sold in small bunches. A sealed polyethylene pack was developed for Migros by an outside research firm in the early 60s. Flowers in a special gas atmosphere are protected from handling damage, yet are clearly visible.

Migros had 110 permanent flower sections in as many stores in 1974. The remaining 290 branch stores sold flowers on a seasonal basis. All displays are self-service. Flower counters are located away from the food section but in high traffic areas. Counters are attended by a sales staff trained in flower care by Migros in special courses. The cooperative also operated five flower shops that are separate from supermarkets.

The regional cooperatives have local suppliers with some growing for Migros on a contract basis. The majority of pot plants are produced by Swiss growers. Contrary to U.S. supermarket sales, 45% of flower sales by Migros are cut flowers with pot plants accounting for 35% and house plant and gardening supplies the balance.

Migros has a main flower distribution center in Berne and imports about 80% of its cut flowers, primarily from the Aalsmeer auction market. The local cooperative managers buy plants for their districts. Contract growing for Migros by the Swiss grower can be obtained by either an oral or written agreement.

Flower promotions are done on a regular basis. A national promotion involves all stores featuring the same flower. At other times a promotion may be done by one of the 12 regional cooperatives.

Every new Migros store must sell flowers. High quality at reasonable prices has made flowers a part of the Swiss way of life, all due to the Migros approach.

12.6.6.2 Buketten Stores

In the early 70s, an innovation in flower selling was started on the outskirts of Malmo, Sweden's third largest city located in the area of flower production in that country. By 1974, A.B. Buketten had 85 bouquet shops in Sweden, as well as franchised stores in England, Norway, and Finland (Rathmell, 1975). Starting from a base of 7 ha of greenhouses, the marketing idea was conceived to arrange flowers for the consumer. These were packaged and offered in company controlled outlets at reasonable prices. In addition to cut flowers, *Adiantum* and pot plants are grown and sold through the Buketten stores.

Flowers and greens come into the assembly line of the bouquet factory of about 1000 m^2 work area and an equal amount of refrigerated rooms. Standard bouquets are made in specially-built machines. Nine workers per machine make 2500–3000 bouquets per h. Bouquets are packaged in polyethylene sleeves with a color code on the plastic to enable shops to know the age of the flowers. A different color is used each day with a solid line for one week and a dotted line the next. Fern is added to each bunch of carnations or roses. Mixed bouquets are done manually by workers alongside an endless belt and these are sleeved at the end of the line (Rathmell, 1972).

All bouquets are placed in standard boxes that hold 25 bunches and are covered by a foam plastic liner. A code number is used on each box to identify the flowers inside and date of packing. Packed boxes are stacked on dollies, wheeled to a large cold store to await loading on vans for delivery.

Buketten produces almost half the flowers and buys the rest at auctions or from contract growers. Buketten has contract growers in Sweden, purchases flowers and plants direct from growers in other countries, and has its own company for buying in Holland and Denmark.

Buketten shops advertise in the daily newspapers and include prices. Prices for mixed bouquets tend to remain the same throughout the year, although numbers of flowers in a bouquet may be varied with availability and cost. In 1970, the firm packed around 10 million bouquets. Sales growth of the company has been excellent.

The Buketten shops have reasonable prices on plants as well as bouquets. A self-service basket for the customer is available near the door. Other gardening supplies such as potting soil, containers and plant food can be picked up along with flowers and plants.

The bouquets are displayed in water on special flower carts. A packet of flower preservative is given with each flower purchase, amounting to 6 million packets in 1971. The interior of the self-service shops uses special lighting to accent the reds in flower colors.

Rathmell concludes with: "Buketten has changed the flower buying habits of Sweden." Actually, what Buketten has done is make flowers accessible at a reasonable price and let the customer know about it.

12.7 Floriculture and the Economy

Max Brunk, Professor of Marketing at Cornell University, has worked longer with the Floral Industry than any other economist. In a thought-provoking speech to the industry in 1969, he took a critical look at the industry and the socio-economic forces affecting it. The U.S. economy was expanding over and above inflation at a long time, constant rate. Brunk felt that market opportunities for flowers are best indicated by miltiplying per capita income by number of people. At this time, the total market had grown to a half trillion dollars. By January of 1976, the market had grown further to 1.3 trillion according to the U.S. Commerce Department. The major problem with the economic environment, he cautioned, is that it encourages lethargy. A flower grower or seller can continue to do the same total business with less effort.

Real income in constant dollars has been increasing for a full generation of 30 years. As real income and affluence increase, people demand different services and seek better products from the market. What were once regarded as luxuries become necessities. Social customs and values change rapidly.

A distinct change in the U.S. economy around 1940 suddenly shook us out of a long period of economic lethargy according to Brunk. Disposable incomes remained almost constant through the early part of the century and the great depression of the 30s. Not one force but several happening at the same time set us off on a new course of creating wealth at a constant rate of almost 3% compounded annually for 30 years. He listed preparation for World War II, inflation of wartime and women joining the work force as major contributors to this socioeconomic change. Many families found life more rewarding with more than one wage earner. Not all families stopped at two wage earners. There were not enough workers during these years and as wages rose, a backlog of technology was injected into the economic stream, increasing productivity of each person. "Moonlighting" became common. Consumer credit expanded rapidly.

We became a nation of home owners with mortgages rather than a nation of tenants. We learned to spend before we earned, getting on an economic treadmill that we have not been willing or able to abandon. Even our attitudes toward retirement have changed. For those whose health will stand it, retirement payroll supplements income from a second career begun after retirement.

Dr. Brunk feels that the floral industry has so long been in lethargy that it needs the type of shock treatment experienced by the total economy in 1940. In his speech, he predicted that some day, some technology, some innovator, or some new social custom will provide that shock. Unless it comes, Brunk felt from experience with flower marketers in all categories, that the industry would continue to drift in an affluent market or eventually fade away or be replaced.

Dr. Brunk presented seven positive guidelines for action by the floral industry which will always apply to any floral group anywhere.

Guideline 1. Learn to work together as an industry. Learn to have confidence in, listen to and react to the ideas and opinions of growers, wholesalers, suppliers, and retailers. Recognition that you have far more in common interests than in conflict might well enable you to capitalize on some new opportunities.

Guideline 2. Identify your competitor not in terms of product handled or service rendered, but in terms of consumer values offered. You must become more sensitive to the changing social and economic environment in which you do business. Such evaluations call for the cold, calculated look of an unemotional, disinterested party and the power to see ourselves as others see us.

Guideline 3. Within your own industry, seek out ways to become distinctive, to become identified with a specific value, to serve a definite and distinguishable market segment. It must be recognized that all effort in market development should be directed at differentiation of product and service designed to fulfill identifiable use values. A common product or service is marketable only at a price approaching cost of production. A different product or service leaves room for profit as well.

Guideline 4. A reasonable amount of company growth and industry growth should be planned for and realized from the efforts of the firm or industry itself. Every firm or industry should have a long-range plan for growth with appropriate goals and defined methods for achieving them. People who work with living plants should recognize that vigor comes from the struggle for survival.

Guideline 5. Emulation of a competitor is a poor substitute for thinking and initiative. Every emulative action reduces the capacity of a firm or industry to perform.

Guideline 6. Pricing is to marketing what labor efficiency is to production. Both the greatest errors and the greatest ignorance in marketing lie in pricing.

Most of us assume that prices are the sole product of supply and demand. Many of us price by a system of default. Prices create both supply and demand. The greatest opportunities in flower marketing lie in the area of understanding and using efficient systems of pricing.

Guideline 7. Recognize, analyze, and capitalize on the nature of demand for your product and service. Two extreme forms of demand exist simultaneously for the floral industry. The necessity market of occasions has a highly inelastic demand but it coexists right along with the market for flowers and plants for the home, a luxury market with extreme opportunity for growth. The luxury market is far more responsive to economic growth and it requires the best of merchandising skill. No other industry has such a dichotomous market, or such a marketing challenge.

12.7.1 Trends in Marketing

12.7.1.1 The Consumer

An attempt was made to identify some of the social environment changes that will affect the florist business of the future (Brunk, 1969).

Inflation will continue in the future, perhaps at an accelerated rate in many countries. This will speed up technology and maintain fuller employment. Real incomes will continue to rise. More luxury products will become necessities.

Family life will be materially altered. People will be far more mobile. They will engage in more social activities outside the home.

Attitudes toward the female are undergoing enormous change. Not only does she carry more of her weight as a family wage earner, but she is more independent than ever before. The male-female relationship is changing so that the female receives a different kind of attention from the male. She is no longer wooed and courted as in the past. She receives fewer overt demonstrations of affection from the male. Conspicuous displays of affection are less popular. If the female is to have the pleasure of flowers, in many cases, she will provide them. The female will do even more of the buying than she has in the past.

In a congested society, man has always tried to set himself apart from his fellow man. As society becomes more affluent, the methods of differentiation, of gaining social prestige and attracting attention to one's self change. How can the floral industry capitalize on this trait? How can it contribute to new fads and customs in floral uses?

In an affluent society, quality displaces quantity. The choice of variety becomes more important. Appreciation of the arts becomes more widespread. Products and services must be upgraded to meet this demand.

Increasingly, we are being married and buried on the run. Long established social customs are being overturned. We look at these two major events of life with increasing realism and less emotion. What will it mean to the floral business?

If the consumer has become anything in the recent era of the 1970s, he has become an economic "animal" (Kokus, 1975). That is, with wildly gyrating price fluctuations, combined with shortages and substitute buying, his instincts became honed to bargain—sale purchasing and seeking out top quality for dollars spent. In addition, the era has begun to teach him self-discipline in either doing without the product or service desired, or deferring purchase, or cutting back on the quantity bought.

The consumer has educated himself in borrowing, the cost of credit, and above all, alternate ways of investment. The more sophisticated consumers have become knowledgeable in small differences in interest rates received or paid out and have learned to move savings from savings and loan to treasury bills to corporate bonds, or to whatever investment source returns the most, commensurate with the risk involved.

The consumer has reached an all-time high in credit consciousness. Consumers have steadily reduced personal debts during late 1974 and 1975. Consumers rarely pay off more debt than they take on for any extended period (Kokus, 1975).

More than ever, we have become a credit card society. This has mixed blessings for the floral industry but most of them are good. Consumers use the credit card for better debt management. From the standpoint of the retail florist, the credit card allows easier sales and facilitates customer convenience. Larger and even multiple sales are encouraged. Cash flow and collection problems are minimized for the retailer, never known as particularly good in this area. These credit conveniences acknowledge the life styles of the upwardly mobile young professionals and the well-to-do mature consumers of our present society, but most of all they increase sales.

One final note on the consumer and influences on consumer buying habits. This involves consumer confidence and the overall effects of asset value and a feeling of well being this has on confidence or willingness to buy. This confidence derives from a feeling of affluence which is the total of all sources of income plus savings and the value of investments.

Housing, referred to earlier, is one major contributor to this consumer affluence. The mortgage debt on housing is at a manageable level. Equity value of homes has been rising rapidly due to inflation. Consumer savings rates are well above normal.

The stock market has contributed more than its share to consumer confidence. From a low Dow Jones Average in December, 1974 of about 570, a relatively steady increase capped by a sharp rise in January–February, 1976, came back to challenge the 1000 level, previous record high. Whether personal income levels, savings, or value of investments, the consumer's feeling of affluence has never been greater. This promises well for the near term market for floral products. It is up to the floral industry to make products available in the places convenient for the consumer and in the quality and variety wanted. Above all, these products must be priced at a profit to the industry but at a level considered reasonable by the consumer. If these stipulations are met, the floral industry in the U.S. will have met the most interesting challenge in its entire history.

12.7.1.2 Trends in Volume Expansion

Voigt (1972) identified the innovative leaders of the floral industry and interviewed them in depth. His purpose was to explore the current new developments in marketing of flowers and to assess the future possible influences of these developments in the business environment. Voigt had remarkable cooperation with the business leaders of floriculture because this type of person wants to help the industry's future. The firms interviewed were those characterized by vigorous past growth whose continued expansion virtually assured their growth in the future.

Observations on sales growth made by Voigt from these interviews were: (1) retail florists operate at lower average sales and experience less absolute sales growth; (2) growers having wholesale outlets were the largest enterprises with average sales just under four million dollars. They had vertically integrated functions but had not, as yet, completely integrated by performing the retailing function; (3) combination or fully integrated florists had experienced the most growth in sales volume in the previous five-year period. Is this a trend? Any combination of functions seemed to be causing firms to grow faster; (4) wholesale growers had experienced steady sales growth throughout the previous 5- and 10-year periods. The same was true of those firms considering themselves solely as wholesalers. In other words, these successful wholesalers had achieved remarkable sales growth without formal integration or market structuring with other firms with whom they did business.

Voigt asked specific questions about trends. At the time, mass marketing was on the minds of most industry leaders. Expansion via a mass marketing method of some type seemed to be a dominant trend. Growers, wholesalers, and retailers were either pariticipating in mass marketing or were considering this participation.

Combining functions (integration) such as growers becoming shippers or wholesalers or retailers; wholesalers becoming retailers; retailers investing in wholesaling and/or production was an extremely significant trend among these floral leaders.

Imported flowers were another trend almost equally mentioned with integration. Many wholesalers and retailers were handling imports. Some, including growers, were participating in the production of flowers outside the U.S. for import. Other trends mentioned by many of these industry leaders for volume expansion included:

1. The entry of outsiders into the floral industry (such as United Brands in foliage plants).
2. The Bachman-Pillsbury European Flower Markets (franchising).
3. Product Promotion to even greater levels.
4. Going public.
5. Multistores.
6. Mergers and acquisitions.
7. Delivery coordination.
8. Contract production or purchasing.
9. Mechanization and automation.
10. Packaging or prepackaging.
11. Direct sales, and either
12. Diversification or
13. Specialization.

Voigt concluded his assessment of trends in methods of expansion in the floral industry with: "The study involved a selected sample of rapidly growing flower businesses which could represent at least one-twentieth of the industry and which might well point the way for our industry's future business environment and the methods employed in that achievement.

"Mass marketing and volume expansion dominate all trend activity. A broadening and deepening of the market is inevitable. The size of market will mushroom if future industry growth even closely resembles our sample's last five years' growth.

"Expansion will bring about changes in marketing methods and market structure. Volume, integration of activities, and market influence will become increasingly evident.

"A greater emphasis on markets, consumers, and distributional methods will occur. Attention to business management methods will be mandatory. Volume will be a big key for success.

"The flower industry and markets will become more scrambled as traditional service retailers provide more attention to high traffic locations. Mass marketers will provide less leadership or allegiance to the flower industry.

"Current industry members will vertically integrate more activities and thereby 'wear many hats'. Trade associations may have severe adjustments as members become large, integrated firms allocating their resources primarily from their individual, influential standpoints.

"This study requires serious contemplation and objectivity for the individualized appropriate adjustments by anyone. However, adjustments are those of a dynamic, growing, rewarding industry—with an almost completely uncharted area of mass marketing and demand potential. Many florists of all sizes and types in varying markets will not only survive but flourish, providing an orientation to customers and business management methods are employed."

12.7.1.3 Trends in Production and Supply

There have been two great expansion periods in U.S. floriculture. The first following World War I and during the 1920s resulted from greatly expanded new greenhouse construction plus the wholesale conversion of northern greenhouses from the production of vegetables to flowers. This latter happening was due to the development of refrigerated rail transportation for winter vegetables produced in the South and West.

The second expansion period followed World War II and is still going on. While new greenhouse construction played a part initially, expanded technology such as new pest controls, better environment controls, and new varieties enabled the grower to increase greatly production in a given area. Relocation to better climates in the South and West, the widespread construction with saran, plastic film and rigid plastic greenhouses, and air transportation contributed to this expansion period. A new chrysanthemum industry based on technology and climate was started under saran houses in Florida. San Diego County in California became a factor in carnation production during the 1950s and in pot plants for shipment by truck around 1970.

To add to this increased supply of flowers, imports from Latin America became a significant factor in the U.S. market for the first time in the late 1960s. American capital and technology was blended with Latin American business acumen to develop almost too much supply in too short a time. This Latin American production was based in part on cheap labor and an export subsidy from several of the governments. The supply overgrew its market in the U.S. Some marketing chaos resulted that hurt U.S. growers temporarily. They in turn acted through the U.S. Governmental agencies to get subsidies removed from foreign competitors. Labor costs began rising in Latin America due to labor unions and to fringe benefits legislated by the various governments. At this time imports have stopped their unmanageable rise of the late 1960s and early 1970s and will probably not again become the problem to U.S. floriculture that they were in those years. For a time, imports of carnations and chrysanthemums doubled each year. At the same time, there was unprecedented greenhouse expansion in central California. Cut roses never became an import problem nor did pot plants. The Latin Americans could not get sufficient planting stock of roses while pot plants with soil cannot be exported to the U.S.

West Germany and Switzerland have experienced an import problem because of their continued affluence. Outstanding marketing through Migros stores turned imports to an advantage for Switzerland. The German flower industry has survived the flood of imported flowers by placing them in every outlet accessible to consumers and by greatly expanding public relations on a cooperative basis (Hunt and Kirk, 1971).

12.7.1.4 Other Trends

Fewer flower producers will grow larger volumes (Kokus, 1975). This trend toward larger and larger operations has been developing for at least two decades and will probably continue.

Due to rising fuel prices and possible shortages of fuel, flower and foliage plant production will continue to expand in the South and West. Additional greenhouse construction in northern areas for the production of cut flowers will be limited.

Labor costs will continue to rise. This means that production efficiency must be kept at a maximum level even if it means redesigning operations for layout to achieve mechanization.

We cannot continue to rely so heavily on air transport. Almost every major holiday causes a crisis in transportation. Long distance truck transport will probably be used to an even greater extent than at present.

While the retail florists will continue as relatively small operators, except for a few with multistores, the retailer will be in an even more enviable position to provide consumer information service. We are in a consumerism movement as a trend in the 1970s with the operators of flower shops, plant stores, and garden centers major contact points for consumers. This can be a real opportunity and a challenge. The gap in floricultural knowledge between the floral expert and the supermarket will continue to widen making the service area an ever increasing opportunity for the retailer.

Consumer groups will be identified, catered to, and more information on the use of floral products will be made available to them.

New production greenhouses will be designed for maximum use of solar heat and to reduce heat loss by increasing the use of plastic films.

Floricultural markets have never been more variable nor have they had more promise for growth. There is a place for everyone in this growing industry.

Berninger summarized our situation in 1969: "Change is all about us and great things will happen to our industry in the 1970s and beyond. You can try to resist change, but generally you won't succeed. Just as we have many marginal operations today, so too will we have them in 1975 and 1980. Perhaps you will be content for yours to be one of them. If not, and I certainly hope you don't decide in favor of the rocking chair, then chart a positive course of action. Identify your role as you'd like to see it in future years. To do so, however, you must identify the changes we will face as an industry. When looking ahead, you are usually on the conservative side.

"Many new parties will enter the flower business in the immediate future. I personally believe their presence will make it truly an exciting period. They will bring new concepts that should revolutionize our industry's techniques and thus rapidly expand the consumption of flowers. You can occupy a most productive position in the flower industry, if you approach the tasks ahead with enthusiasm and a real desire to face reality and change."

References

Anonymous: "Produce 74", Progressive Grocer Vol. 53, (3), March, 1975, 47 p (1975a)

Anonymous: Toronto clock reports second growth year. N.Y. State Flower Industries Bull. **54**, 4 (1975b)

Baker, M., Goodrich, D. Jr.: Flower retailing by mass outlets. N.J. Agr. Exptl. Sta. Bull. **817**, 31 p (1968)

Berninger, L. M.: A new era for the florists' industry: What's in store for the 70's? Florists' Rev. **145**, (3758) 23, 52–54 (1969)

Berninger, L. M.: Patterns are changing in wholesaling flowers. Florists' Rev. **149**, (3870) 23–25 (1972a)

Berninger, L. M.: The supermarket business looks at prospects in flowers. Florists' Rev. **150**, (3885) 36–37 (1972b)

Brunk, M. E.: A look at the 1980's: Will florists still be around? Florists' Rev. **145**, (3757) 27–28, 46–51 (1969)

Brunk, M. E., Hampton, R. N.: Problems in selling prepackaged roses. Roses Incorporated Bull. **189**, 11–12 (1953)

Campbell, J., Kress, G.: The sale of cut flowers in supermarkets. Colo. St. Univ. Exptl. Sta. Sp. Rept. 1972, 17 p

Courtenay, H.: Knowing the product and the consumer helps create happy customers at a profit. Florists' Rev. **157**, (4078) 39, 83–85 (1976)

DeWerth, A. F., Sorenson, H. B.: Marketing cut flowers and pot plants for home use. Tex. Agr. Exptl. Sta. Misc. Pub. **138**, (1955)

Fossum, M. T.: Floricultural and Ornamental Horticultural Economics. Alexandria, Va.: Soc. Am. Florists 1950, 150 p

Fossum, M. T.: Trends in floriculture. Florists' Rev. **127**, (3296) 23–25, 34 (1961)

Fossum, M.T.: Trade in Horticultural Specialties. 1890–1950. U.S.D.A. Bur. of Agr. Econ. Marketing Rept. No. 33, 1963
Fossum, M.T.: Economic analysis of the Florists' Transworld Delivery Association, 1955–1966. Detroit: FTD Headquarters 1967
Fossum, M.T.: Census of business reveals expenditures for floricultural products in all retail outlets. Florist **9**, (7) 48–50 (1975)
Gatty, R., Wolf, D.: Merchandising prepackaged cut flowers in supermarkets. N.J. Agr. Exptl. Sta. Bull. **799**, 20 p (1961)
Havas, N.: A graphic view of the retail florist industry. U.S.D.A. Econ. Res. Ser. Marketing Res. Rept. No. **788**, 48 p (1967)
Hunt, A.R., Kirk, J.H.: The marketing of flowers in the United Kingdom. Wye Coll. Marketing Ser. No. **7**, Univ. London, 93 p (1971)
Jarvesoo, E.: Wholesale florist industry in Massachusetts. U. Mass. Agr. Exptl. Sta. Bull. **556**, 51 p (1966)
Kiplinger, D.C., Sherman, R.W.: Florist crops for mass market outlets. Ohio Agr. Exptl. Sta. Res. Bull. **928**, 37 p (1962)
Kokus, J.: Special report on the state of the economy, the consumer and changes in life-style, and the U.S. floricultural industry in the 1970's. Alexandria: Soc. Am. Florists 1975, 40 p
Kress, G.: Mass marketing and the wholesale florist. Colo. St. Univ. Agr. Exptl. Sta. Sp. Rept. May, 1972, 17 p
Kress, G.: Mass marketing—a reassessment. Colo. St. Univ. Agr. Exptl. Sta. Sp. Rept. 1973, 7 p
Kress, G.: Floral products in supermarkets. Colo. St. Univ. Agr. Exptl. Sta. Sp. Rept. April, 1975a, 32 p
Kress, G.: Floral products in supermarkets. Colo. St. Univ. Agr. Exptl. Sta. Sp. Rept. July, 1975b, 6 p
Laurie, A., Kiplinger, D.C.: Commercial Flower Forcing, 5th ed. Philadelphia: Blakiston 1950, 550 p
Moore, E.J.: Expanding the retail market for floral products. U.S.D.A. Agr. Marketing Ser. AMS **286**, 1959, 11 p
Oppenfeld, H. von, Schwartz, J.F., Krone, P.R.: Merchandising in retail flower shops. Mich. St. Univ. Agr. Exptl. Sta. Bull. **412**, 28 p (1957)
Pease, R.W.: Expanding the retail demand for floricultural products sold by florists. W. Va. Univ. Agr. Exptl. Sta. Bull. **477**, 22 p (1962)
Pfahl, P.B.: The Retail Florist Business. Danville, Ill.: Interstate Printers 1968, 435 p
Pfahl, P.B., Trotter, C.E., Pease, R.W.: Merchandising for profit in retail flower shops. Penn. St. Univ. Agr. Exptl. Sta. Bull. **659**, 26 p (1959)
Raleigh, S.M.: Ornamentals production and marketing trends, 1948–1972. Econ. Res. Ser. U.S.D.A. Statistical Bull. **529**, 100 p (1974)
Rathmell, J.K., Jr.: Flower markets get big. Florists' Rev. **149**, (3865) 12, 54–55 (1971)
Rathmell, J.K., Jr.: Rathmell reports from Sweden. Florists' Rev. **149**, (3873) 25, 72–73 (1972)
Rathmell, J.K., Jr.: Marketing is a science and big business in Switzerland. Florists' Rev. **155**, (4015) 74 (1974)
Rathmell, J.K., Jr.: Oil crisis crunches thoughts of expanding Buketten operation. Florists' Rev. **155**, (4023) 39, 78–79 (1975)
V.B.A. (Dutch Marketing System): Annual report for 1975. Aalsmeer, Holland
Voigt, A.O.: Exploration of new developments in the flower industry. Penn. St. Univ. Dept. Agr. Econ. Sp. Rept. 1972, 15 p
Voigt, A.O.: Marketing maneuvers, Florists' Rev. **157**, (4068) 11–12 (1975)
Weir, W.: Opportunity in flower marketing. The Packer, Flower Marketing Supplement, August 25, 1973

Appendix A Conversion Tables

Not all are familiar with the exponential method of writing figures employed in these conversion tables to save space and avoid cumbersome zeros. It is only necessary to write the significant digits, and then multiply by ten, raised to the power shown, in order to obtain the correct decimal point. For example: $1=10^0$, $10=10^1$, $100=10^2$, and $1000=10^3$. If 2568 is written exponentially, it could be written as 2.568×10^3, 0.2568×10^4, 25.68×10^2, etc. Or, the decimal point is moved to the right the number of places indicated by the exponent. Sometimes the decimal point must be moved to the left. A negative exponent indicates this: $0.1=10^{-1}$, $0.01=10^{-2}$, and $0.001=10^{-3}$. Suppose we have the number 0.0000002568. This can be written as 2.568×10^{-7}, 25.68×10^{-8} or 0.2568×10^{-6}. In multiplying, the exponents are added, if divided, they are subtracted. Only the significant digits need be multiplied or divided. For example: $(2.34 \times 10^6) \times (0.56 \times 10^3) = (1.3104) \times 10^{6+3}$ or 1.3104×10^9. Written out the number is 1 310 400 000. If the same numbers are divided; the answer is $(2.34/0.56) \times 10^{6-3}$ or 4.1786×10^3 which is 4 178.6.

To indicate square or cubic units, 2 or 3 as an exponent is used. For example: 10 cubic yards would be yd^3, or 10 square feet would be ft^2. If one has 5 ounces per cubic yard, the symbolism uses a negative exponent for the "per", or 5 oz yd^{-3}. Sometimes we wish to show so many grams per liter, which could be written as $g\,l^{-1}$, the negative exponent one being used to express the relationship.

Table 1. Conversion for length

	mm	cm	m	km	in	ft	yd	mi
1 millimeter (mm)	1	0.1	10^{-3}	10^{-6}	0.04	0.003	0.001	6.2×10^{-7}
1 centimeter (cm)	10	1	10^{-2}	10^{-5}	0.39	0.03	0.01	6.2×10^{-6}
1 meter (m)	10^3	100	1	10^{-3}	39.4	3.28	1.1	6.2×10^{-4}
1 kilometer (km)	10^6	10^5	10^3	1	3.9×10^4	3.3×10^3	1.1×10^3	0.62
1 inch (in)	25.4	2.54	0.03	2.5×10^{-5}	1	0.08	0.03	1.6×10^{-5}
1 foot (ft)	304.8	30.5	0.3	3.0×10^{-4}	12	1	0.33	1.9×10^{-4}
1 yard (yd)	914.4	91.4	0.91	9.1×10^{-4}	36	3	1	5.7×10^{-4}
1 mile (mi)	1.6×10^6	1.6×10^5	1.6×10^3	1.60	6.3×10^4	5280	1760	1

To convert from units in left column, multiply by the factor under the appropriate column heading.
Note: 1 Angstrom (Å) $=10^{-7}$ mm $=10^{-8}$ cm $=10^{-10}$ m $=3.9\times10^{-9}$ in; 1 micron (μ) $=10^{-5}$ mm $=10^{-4}$ cm $=10^{-6}$ m $=3.9\times10^{-5}$ in; 1 millimicron (mμ) = 1 nanometer (nm) $=10^{-8}$ mm $=10^{-7}$ cm $=10^{-9}$ m $=3.9\times10^{-8}$ in; 1 mil $=10^{-3}$ in $=0.0254$ mm

Table 2. Conversion for volume

	m^3	cm^3	l	ft^3	yd^3	in^3	bu
1 cubic meter (m^3)	1	10^6	10^3	35.3	1.3	6.1×10^4	28.4
1 cubic centimeter (cm^3)	10^{-6}	1	10^{-3}	3.5×10^{-5}	1.3×10^{-6}	0.061	2.8×10^{-5}
1 liter	10^{-3}	10^3	1	0.035	1.3×10^{-3}	61.02	0.028
1 cubic foot (ft^3)	0.028	2.8×10^4	28.3	1	0.037	1728.0	0.804
1 cubic yard (yd^3)	0.756	7.6×10^5	764.1	27	1	4.7×10^4	21.71
1 cubic inch (in^3)	1.6×10^{-5}	16.4	0.016	5.8×10^{-4}	2.1×10^{-5}	1	4.7×10^{-4}
1 bushel (bu)	0.035	3.5×10^4	35.24	1.24	0.046	2.2×10^3	1
1 quart (qt) (fluid)	9.5×10^{-4}	947.0	0.947	0.033	1.2×10^{-3}	57.7	0.031
1 gallon (gal)	3.8×10^{-3}	3785.0	3.8	0.134	4.9×10^{-3}	231	0.107
1 ounce (oz) (fluid)	2.9×10^{-5}	29.57	0.029	1.04×10^{-3}	3.9×10^{-5}	1.8	8.4×10^{-4}
1 cup	2.3×10^{-4}	236.5	0.237	8.4×10^{-3}	3.1×10^{-4}	14.5	6.7×10^{-3}
1 pint (pt) (fluid)	4.8×10^{-4}	473.0	0.473	0.017	6.2×10^{-4}	28.9	0.013
1 peck	3.2×10^{-4}	9.1×10^3	9.1	0.311	0.012	537.6	0.25
1 teaspoon (tsp)	5.0×10^{-6}	5	5×10^{-3}	1.8×10^{-3}	6.6×10^{-5}	0.308	1.4×10^{-4}

Table 2 (continued)

	qt	gal	oz	cup	pt	peck	tsp
1 cubic meter (m^3)	1.1×10^3	262.6	3.4×10^4	4.2×10^3	2.1×10^3	3.1×10^3	2×10^5
1 cubic centimeter (cm^3)	1.1×10^{-3}	2.6×10^{-4}	0.034	4.2×10^{-3}	2.1×10^{-3}	1.1×10^{-4}	0.2
1 liter	1.06	0.263	33.8	4.2	2.1	0.110	200
1 cubic foot (ft^3)	29.95	7.46	957.3	119.7	59.8	3.21	560
1 cubic yard (yd^3)	808.0	202	2.6×10^4	3.2×10^3	1.6×10^3	87.4	1.5×10^4
1 cubic inch (in^3)	0.017	4.3×10^{-3}	0.55	0.069	0.034	1.9×10^{-3}	3.3
1 bushel (bu)	32.0	9.3	1.2×10^3	148.9	74.4	4.0	7.1×10^3
1 quart (qt) (fluid)	1	0.25	32.0	4.0	2.0	0.107	192.0
1 gallon (gal)	4.0	1	128.0	16.0	8.0	0.428	768.0
1 ounce (oz) (fluid)	0.031	7.8×10^{-3}	1.0	0.125	0.063	3.4×10^{-3}	6.0
1 cup	0.25	0.063	8.0	1	0.5	0.027	48.0
1 pint (pt) (fluid)	0.5	0.125	16.0	2.0	1	0.054	32.0
1 peck	9.3	2.34	294.1	37.0	18.5	1	592.0
1 teaspoon (tsp)	5.2×10^{-3}	1.3×10^{-3}	0.17	0.02	0.031	1.7×10^{-3}	1

To convert from units in the left column, multiply by the factor under the appropriate column heading.

Table 3. Conversion for area

	mm^2	cm^2	m^2	ha	km^2	in^2	ft^2	yd^2	Ac	mi^2
1 square millimeter (mm^2)	1	0.01	10^{-6}	10^{-10}	10^{-12}	1.6×10^{-3}	1.1×10^{-5}	1.2×10^{-6}	2.5×10^{-10}	3.9×10^{-13}
1 square centimeter (cm^2)	10^2	1	10^{-4}	10^{-8}	10^{-10}	0.16	1.1×10^{-3}	1.2×10^{-4}	2.5×10^{-8}	3.9×10^{-11}
1 square meter (m^2)	10^6	10^4	1	10^{-4}	10^{-6}	1.6×10^3	10.8	1.2	2.5×10^{-4}	3.9×10^{-7}
1 hectare (ha)	10^{10}	10^8	10^4	1	10^{-2}	1.6×10^7	1.1×10^5	1.2×10^4	2.5	3.9×10^{-3}
1 square kilometer (km^2)	10^{12}	10^{10}	10^6	10^2	1	1.6×10^9	1.1×10^7	1.2×10^6	247.1	0.39
1 square inch (in^2)	645.2	6.45	6.5×10^{-4}	6.5×10^{-8}	6.5×10^{-10}	1	6.9×10^{-3}	7.7×10^{-4}	1.6×10^{-7}	2.5×10^{-10}
1 square foot (ft^2)	9.3×10^4	929.0	0.09	9.3×10^{-6}	9.3×10^{-8}	144	1	0.11	2.3×10^{-5}	2.3×10^{-5}
1 square yd (yd^2)	8.4×10^5	8.4×10^3	0.84	8.4×10^{-5}	8.4×10^{-7}	1296	9	1	2.1×10^{-4}	3.2×10^{-7}
1 acre (Ac)	4.0×10^9	4.0×10^7	4.0×10^3	0.4	4.0×10^{-3}	6.3×10^6	4.4×10^4	4.8×10^3	1	1.6×10^{-3}
1 square mile (mi^2)	2.6×10^{12}	2.6×10^{10}	2.6×10^6	259.0	2.6	4.0×10^9	4.4×10^4	3.1×10^6	640.0	1

To convert from units in the left column, multiply by the factor under the appropriate column heading.

Table 4. Conversion for weight

	kg	g	t	ton	lb	oz	gr
1 kilogram (kg)	1	10^3	10^3	1.1×10^{-3}	2.2	35.28	1.5×10^4
1 gram (g)	10^{-3}	1	10^{-6}	1.1×10^{-6}	2.2×10^{-3}	0.035	15.43
1 metric ton (t)	10^{-3}	10^6	1	1.1	2.2×10^3	3.5×10^4	1.5×10^7
1 short ton (ton)	907.2	9.1×10^5	0.907	1	2000	3.2×10^4	1.4×10^7
1 pound (lb)	0.45	453.6	4.5×10^{-4}	5×10^{-4}	1	16	7000
1 ounce (oz) (avoirdupois)	0.028	28.35	2.8×10^{-5}	3.1×10^{-5}	0.063	1	437.5
1 grain (gr)	6.5×10^{-5}	0.065	6.5×10^{-8}	7.1×10^{-8}	1.4×10^{-4}	2.3×10^{-3}	1

To convert from units on the left, multiply by the factor under the appropriate column heading.

Table 5. Conversions for velocity

	cm sec^{-1}	m sec^{-1}	km h^{-1}	ft sec^{-1}	ft min^{-1}	mi h^{-1}
1 centimeter per second (cm sec^{-1})	1	10^{-2}	0.036	0.033	1.97	0.022
1 meter per second (m sec^{-1})	10^2	1	3.6	3.28	196.9	2.24
1 kilometer per hour (km h^{-1})	27.78	0.28	1	0.91	54.6	0.62
1 foot per second (ft sec^{-1})	30.48	0.30	1.1	1	60.0	0.68
1 foot per minute (ft min^{-1}, fpm)	0.51	5.0×10^{-3}	0.02	0.017	1	0.011
1 mile per hour (mi h^{-1}, mph)	44.7	0.45	1.6	1.47	88.0	1

To convert from units in the left column, multiply by the factor under the appropriate column heading.

Table 6. Conversion for weight per unit volume

	$kg\ m^{-3}$	$g\ cm^{-3}$	$g\ liter^{-1}$	$lb\ ft^{-3}$	$lb\ yd^{-3}$	$lb\ bu^{-1}$	$oz\ in^{-3}$
1 kilogram per cubic meter ($kg\ m^{-3}$)	1	10^{-3}	1	0.062	1.69	0.077	5.8×10^{-4}
1 gram per cubic centimeter ($g\ cm^{-3}$)	10^3	1	10^3	62.43	1.7×10^3	78.75	0.527
1 gram per liter ($g\ liter^{-1}$)	1	10^{-3}	1	0.062	1.696	0.079	5.8×10^{-4}
1 pound per cubic foot ($lb\ ft^{-3}$)	16.02	0.016	16.02	1	27	1.24	9.3×10^{-3}
1 pound per cubic yard ($lb\ yd^{-3}$)	0.6	6×10^{-4}	0.593	0.037	1	0.046	3.4×10^{-4}
1 pound per bushel ($lb\ bu^{-1}$)	12.96	0.013	12.87	0.806	21.74	1	7.3×10^{-3}
1 ounce per cubic inch ($oz\ in^{-3}$)	1.7×10^3	1.898	1.8×10^3	108	3×10^3	132.98	1

To convert from units on the left, multiply by the factor under the appropriate column heading.

Table 7. Units and conversion factors for pressure (mass per unit area)

	$dyne\ cm^{-2}$	(b) bar	(mb) millibar	(A) atmosphere	mm Hg	in Hg	psi	$kg\ cm^{-2}$	cm water
1 $dyne\ cm^{-2}$	1	1.0×10^{-6}	1.0×10^{-3}	1.01×10^{-6}	0.75×10^{-3}	0.03×10^{-3}	1.5×10^{-5}	1.0×10^{-6}	1.0×10^{-3}
1 bar	1.0×10^6	1	1.0×10^3	0.987	750.2	29.53	14.51	1.02	1017
1 millibar	1.0×10^3	1.0×10^{-3}	1	1.01×10^{-3}	0.750	0.0295	0.015	0.001	1.017
1 atmosphere	0.99×10^6	1.013	1013.3	1	760	29.92	14.7	1.003	1030
1 mm Hg	1.33×10^3	1.33×10^{-3}	1.333	1.31×10^{-3}	1	0.039	0.019	1.4×10^{-3}	1.36
1 in Hg	3.3×10^4	0.034	33.9	0.0334	25.4	1	0.49	0.035	34.42
1 psi	6.7×10^4	0.069	68.95	0.068	51.7	2.04	1	0.07	70.07
1 $kg\ cm^{-2}$	0.98×10^6	0.981	980.7	0.968	735.6	28.96	14.22	1	999.1
1 cm water	1030	9.8×10^{-4}	0.983	9.71×10^{-4}	0.738	0.029	0.014	1.0×10^{-3}	1

To convert units in the left column, find the multiplying factor in the appropriate column heading.

Table 8. Conversion factors for flow (volume per unit time)

	gpm	CFM	CFS	1pm	CMS
1 gallon per minute (gpm, gal min^{-1})	1	0.134	2.23×10^{-3}	3.79	6.31×10^{-5}
1 cubic foot per minute (CFM, ft^3 min^{-1})	7.48	1	0.017	28.32	4.72×10^{-4}
1 cubic foot per second (CFS, ft^3 sec^{-1})	448.8	60.0	1	1699.2	0.028
1 liter per minute (1pm, 1 min^{-1})	0.264	0.035	5.89×10^{-4}	1	1.67×10^{-5}
1 cubic meter per second (CMS, m^3 sec^{-1})	1.59×10^4	2118.9	35.31	6.0×10^4	1

To convert units in the left column, find the multiplying factor under the appropriate column heading to the right.
Note: Water loss from vegetation is often described in equivalent depth of water, either in inches or millimeters. Thus, one inch of water is equivalent to 0.6 gal or 2.4 l covering a square foot area. One millimeter of water is equivalent to one liter or 0.26 gal covering one square meter. One acre-foot of water = 325828.8 gal = 5837.0 m^3 = 2363.16 m^3 covering one hectare, or 23.6 cm water covering one hectare.

Table 9. Units of illumination

Units	Lux	Phot	Milliphot	Foot-candle
1 lux	1	0.0001	0.1	0.09
1 phot	10000	1	1000	929
1 milliphot	10	0.001	1	0.929
1 foot-candle	10.76	0.001	1.07	1
1 lumen cm^{-2}	10^4	1	1000	929

Table 10. Conversions, energy per unit area

	Joule cm^{-2}	BTU ft^{-2}	watt-h m^{-2}	g-cal cm^{-2}
1 Joule cm^{-2}	1	0.881	2.78	0.239
1 BTU ft^{-2}	1.136	1	3.15	0.271
1 watt-h m^{-2}	0.3597	0.317	1	0.086
1 g-cal cm^{-2}	4.19	3.69	11.624	1

Table 11. Units of power per unit area (intensity)

	erg sec^{-1} cm^{-2}	Langley min^{-1}	g-cal min^{-1} cm^{-2}	BTU h^{-1} ft^{-2}	watt cm^{-2}
1 erg sec^{-1} cm^{-2}	1	1.43×10^{-6}	1.43×10^{-6}	6.47×10^{-9}	10
1 Langley min^{-1}	6.99×10^5	1	1	221.13	0.0698
1 g-cal min^{-1} cm^{-2}	6.99×10^5	1	1	221.13	0.0698
1 BTU h^{-1} ft^{-2}	1.54×10^8	4.52×10^{-3}	4.52×10^{-3}	1	3.16×10^{-4}
1 watt cm^{-2}	0.1	14.32	14.32	3.16×10^3	1
1 watt m^{-2}	1000	1.43×10^{-3}	1.43×10^{-3}	3.17×10^2	10^{-4}

Table 12. Fundamental units of energy and power

	Energy (work)						
	erg	Joule	g-cal.	kilo g-cal	BTU	watt-h	kilowatt-h
1 erg	1	10^{-7}	2.39×10^{-8}	2.39×10^{-11}	6.02×10^{-6}	2.78×10^{-11}	2.78×10^{-14}
1 Joule	10^{7}	1	0.239	2.39×10^{-4}	9.48×10^{-4}	2.78×10^{-4}	2.78×10^{-7}
1 g-cal	4.19×10^{7}	4.19	1	0.001	3.97×10^{-3}	1.16×10^{-3}	1.16×10^{-6}
1 kilo g-cal	4.19×10^{10}	4.19×10^{3}	1000	1	3.97	1.16	1.16×10^{-3}
1 BTU	1.66×10^{5}	1.06×10^{3}	2.52×10^{2}	0.252	1	0.293	2.93×10^{-4}
1 watt-h	3.59×10^{10}	3.60×10^{3}	8.60×10^{2}	0.860	3.41	1	0.001
1 kilowatt-h	3.59×10^{13}	3.60×10^{6}	8.60×10^{5}	8.60×10^{2}	3.41×10^{3}	1000	1

	Power						
	erg sec^{-1}	Joule sec^{-1}	g-cal min^{-1}	BTU min^{-1}	watt	microwatt	kilowatt
1 erg sec^{-1}	1	10^{-7}	1.43×10^{-6}	5.69×10^{-9}	10^{-7}	0.1	1.0×10^{-10}
1 Joule sec^{-1}	10^{7}	1	14.34	0.0569	1	10^{7}	10^{-3}
1 g-cal min^{-1}	6.98×10^{5}	6.98×10^{-2}	1	3.96×10^{-3}	6.98×10^{-2}	6.99×10^{4}	6.98×10^{-5}
1 BTU min^{-1}	1.76×10^{8}	17.57	252.52	1	17.57	1.76×10^{7}	1.76×10^{-2}
1 watt	10^{7}	1	14.34	0.0569	1	10^{6}	0.001
1 microwatt	10	10^{-7}	1.43×10^{-5}	5.69×10^{-8}	10^{-6}	1	10^{-9}
1 kilowatt	10^{10}	10^{3}	1.43×10^{4}	56.9	10^{3}	10^{9}	1

1 watt $cm^{-2} = 14.34$ cal cm^{-2} min^{-1}.
1 watt-h = 3600 Joules.
1 watt = 1 Joule sec^{-1}.

Table 13. Conversion: Centigrade to Fahrenheit—Fahrenheit to Centigrade

Centigrade to Fahrenheit			Fahrenheit to Centigrade		
°C	0.0	0.5	°F	0.0	0.5
+100	+212.0	+212.9	+130	+54.4	+54.7
99	210.2	211.1	129	53.9	54.2
98	208.4	209.1	128	53.3	53.6
97	206.6	207.5	127	52.8	53.1
96	204.8	205.7	126	52.2	52.5
95	203.0	203.9	125	51.7	51.9
94	201.2	202.1	124	51.1	51.4
93	199.4	200.3	123	50.6	50.8
92	197.6	198.5	122	50.0	50.3
91	195.8	196.7	121	49.4	49.7
90	194.0	194.9	120	48.9	49.2
89	192.2	193.1	119	48.3	48.6
88	190.4	191.3	118	47.8	48.1
87	188.6	189.5	117	47.2	47.5
86	186.8	187.7	116	46.7	46.9
85	185.0	185.9	115	46.1	46.4
84	183.2	184.1	114	45.6	45.8
83	181.4	182.3	113	45.0	45.2
82	179.6	180.5	112	44.4	44.7
81	177.8	178.7	111	43.9	44.2
80	176.0	176.9	110	43.3	43.6
79	174.2	175.1	109	42.8	43.1
78	172.4	173.3	108	42.2	42.5
77	170.6	171.5	107	41.7	41.9
76	168.8	169.7	106	41.1	41.4
75	167.0	167.9	105	40.6	40.8
74	165.2	166.1	104	40.0	40.3
73	163.4	164.3	103	39.4	39.7
72	161.6	162.5	102	38.9	39.2
71	159.8	160.7	101	38.3	38.6
70	158.0	158.9	100	37.8	38.1
69	156.2	157.1	99	37.2	37.5
68	154.4	155.3	98	36.7	36.9
67	152.6	153.5	97	36.1	36.4
66	150.8	151.7	96	35.6	35.8
65	149.0	149.9	95	35.0	35.3
64	147.2	148.1	94	34.4	34.7
63	145.4	146.3	93	33.9	34.2
62	143.6	144.5	92	33.3	33.6
61	141.8	142.7	91	32.8	33.1
60	140.0	140.9	90	32.2	32.5
59	138.2	139.1	89	31.7	31.9
58	136.4	137.3	88	31.1	31.4
57	134.6	135.5	87	30.6	30.8
56	132.8	133.7	86	30.0	30.3
55	131.0	131.9	85	29.4	29.7
54	129.2	130.1	84	28.9	29.2
53	127.4	128.3	83	28.3	28.7
52	125.6	126.5	82	27.8	28.1
51	123.8	124.7	81	27.2	27.5

Table 13 (continued)

Centigrade to Fahrenheit			Fahrenheit to Centigrade		
°C	0.0	0.5	°F	0.0	0.5
50	122.0	122.9	80	26.7	26.9
49	120.2	121.1	79	26.1	26.4
48	188.4	199.3	78	25.6	25.8
47	116.6	177.5	77	25.0	25.3
46	144.8	115.7	76	24.4	24.7
45	113.0	113.9	75	23.9	24.2
44	111.2	112.1	74	23.3	23.6
43	109.4	110.3	73	22.8	23.1
42	107.6	108.5	72	22.2	22.5
41	105.8	106.7	71	21.7	21.9
40	104.0	104.9	70	21.1	21.4
39	102.2	103.1	69	20.6	20.8
38	100.4	101.3	68	20.0	20.3
37	98.6	99.5	67	19.4	19.7
36	96.8	97.7	66	18.9	19.2
35	95.0	95.9	65	18.3	18.6
34	93.2	94.1	64	17.8	18.1
33	91.4	92.3	63	17.2	17.5
32	89.6	88.7	62	16.7	16.9
31	87.8	88.7	61	16.1	16.4
30	86.0	86.9	60	15.6	15.8
29	84.2	85.1	59	15.0	15.3
28	82.4	83.3	58	14.4	14.7
27	80.6	81.5	57	13.9	14.2
26	78.8	79.7	56	13.3	13.6
25	77.0	77.9	55	12.8	13.1
24	75.2	76.1	54	12.2	12.5
23	73.4	74.3	53	11.7	11.9
22	71.6	72.5	52	11.1	11.4
21	69.8	70.7	51	10.6	10.8
20	68.0	68.9	50	10.0	10.3
19	66.2	67.1	49	9.4	9.7
18	64.4	65.3	48	8.9	9.2
17	62.6	63.5	47	8.3	8.6
16	60.8	61.7	46	7.8	8.1
15	59.0	59.9	45	7.2	7.5
14	57.2	58.1	44	6.7	6.9
13	55.4	56.3	43	6.1	6.4
12	53.6	54.5	42	5.6	5.8
11	51.8	52.7	41	5.0	5.3
10	50.0	50.9	40	4.4	4.7
9	48.2	49.1	39	3.9	4.2
8	46.4	47.3	38	3.3	3.6
7	44.6	45.5	37	2.8	3.1
6	42.8	43.7	36	2.2	2.5
5	41.0	41.9	35	1.7	1.9
4	39.2	40.1	34	1.1	1.4
3	37.4	38.3	33	0.6	0.8
2	35.6	36.5	32	0.0	0.3
1	33.8	34.7	31	−0.6	−0.3

Table 13 (continued)

Centigrade to Fahrenheit

°C	0.0	0.5
0	32.0	32.9
– 1	30.2	29.3
2	28.4	27.5
3	26.6	25.7
4	24.8	23.9
5	23.0	22.1
6	21.2	20.3
7	19.4	18.5
8	17.6	16.7
9	15.8	14.9
10	14.0	13.1
11	12.2	11.3
12	10.4	9.5
13	8.6	7.7
14	6.8	5.9
15	5.0	4.1
16	3.2	2.3
17	1.4	0.5
18	– 0.4	– 1.3
19	– 2.2	– 3.1
– 20	– 4.0	– 4.9
21	5.8	6.7
22	7.6	8.5
23	9.4	10.3
24	11.2	12.1
25	13.0	13.9
26	14.8	15.7
27	16.6	17.5
28	18.4	19.3
29	20.2	21.1
30	22.0	22.9

Fahrenheit to Centigrade

°F	0.0	0.5
30	– 1.1	– 0.8
29	1.7	1.4
28	2.2	1.9
27	2.8	2.5
26	3.3	3.1
25	3.9	3.6
24	4.4	4.2
23	5.0	4.7
22	5.6	5.3
21	6.1	5.8
20	6.7	6.4
19	7.2	6.9
18	7.8	7.5
17	8.3	8.1
16	8.9	8.6
15	9.4	9.2
14	10.0	9.7
13	10.6	10.3
12	11.1	10.8
11	11.7	11.4
+ 10	–12.2	–11.9
9	12.8	12.5
8	13.3	13.1
7	13.9	13.6
6	14.4	14.2
5	15.0	14.7
4	15.6	15.3
3	16.1	15.8
2	16.7	16.4
1	17.2	16.9
0	17.8	18.1
– 1	18.3	18.6
2	18.9	19.2
3	19.4	19.7
4	20.0	20.3
5	20.6	20.8
6	21.1	21.4
7	21.7	21.9
8	22.2	22.5
9	22.8	23.1
10	23.3	23.6
11	23.9	24.2
12	24.4	24.7
13	25.0	25.3
14	25.6	25.8
15	26.1	26.4
16	26.7	26.9
17	27.2	27.5
18	27.8	28.1
19	28.3	28.6
20	28.9	29.2

Table 14. Miscellaneous conversions and constants

1. Conductivity to milliequivalents per liter: meq/l $= 10 \times EC \times 10^3$
2. Conductivity to parts per million: ppm $= 0.36 \times EC \times 10^6$
3. Grains per gallon to parts per million: ppm $= 17.1 \times$ grains per gallon
4. Milliequivalents per liter to parts per million: multiply meq/l for each ion by its equivalent weight
5. Conversion of phosphorous to equivalent P_2O_5: $P \times 2.29 = P_2O_5$
6. Conversion of potassium to equivalent K_2O: $K \times 1.20 = K_2O$
7. 1 gallon of water = 8.34 pounds avoirdupois
8. 1 cubic foot of water = 7.48 gallons = 62.42 pounds avoirdupois
9. 1 inch of water over one square foot = 5.2 pounds
10. 1 acre-inch of water = 226 512 pounds
11. 1 acre-foot of soil = 2000000 pounds
12. 1 cubic foot of soil = 70 to 105 pounds
13. Parts per million to percent % = ppm $\times$ 0.0001
14. 1 cubic yard of an amendment or mulch will cover:
    81 ft^2 if 4 inches thick
    108 ft^2 if 3 inches thick
    162 ft^2 if 2 inches thick
    324 ft^2 if 1 inch thick
    648 ft^2 if 1/2 inch thick
    1296 ft^2 if 1/4 inch thick

Table 15. Number of pots and flats per bushel

Pot size	No. per bushel	Pot size	No. per bushel
$2^1/_4$″ standard	370	2″ azalea	80
$2^1/_2$″ standard	220	4″ azalea	40
3″ standard	150	6″ azalea	22
4″ standard	55	5″ low pan	50
5″ standard	30	6″ low pan	39
6″ standard	18	7″ low pan	17
7″ standard	11	11 × 22″ flat (8 pak)	5.9
8″ standard	7	11 × 22″ (8 pak, 9-cell)	7.8
12″ standard	2		

Table 16. Number of pots per area of bench

Pot Size	Pattern	Number per 100 ft^2	
		Clay	Plastic
3″	Square	1175	1498
	Hexagonal	1357	1730
4″	Square	711	857
	Hexagonal	821	989
5″	Square	476	554
	Hexagonal	550	639
6″	Square	341	387
	Hexagonal	393	447
7″	Square	256	285
	Hexagonal	295	330
8″	Square	199	219
	Hexagonal	230	253

Appendix B Symbolism

Å	Ångström (unit of length)
A, Ac	Acre (English unit of area, Area)
atm (A)	atmosphere (unit of pressure)
B	boron (micronutrient)
b	bar (metric unit of pressure)
BTU	British thermal unit (unit of energy)
bu	bushel (English unit of volume)
C	Centigrade or Celsius (metric unit of temperature, Carbon)
Ca	calcium (macronutrient)
cal	calorie (unit of energy)
cb	centibar (metric unit of pressure)
cm, cm^2, cm^3, cc	centimeter [metric unit of length, area or volume (square or cubic centimeter)]
cm sec^{-1}	centimeter per second (metric unit of velocity)
CEC	cation exchange capacity (capacity of a soil to adsorb ions with a positive charge)
CFM	cubic feet per minute (English unit of flow)
CFS	cubic feet per second (English unit of flow)
Cl	chlorine
CO_2	carbon dioxide
Cu	copper
dy	day (unit of time)
EC	electrical conductivity
F	Fahrenheit (English unit of temperature)
Fe	iron
ft, ft^2, ft^3	foot [English unit of length, area or volume (square or cubic foot)]
ft-c	foot-candle (English unit of illumination intensity)
fpm, ft min^{-1}	feet per minute (English unit of velocity)
fps, ft sec^{-1}	feet per second (English unit of velocity)
g	gram [metric unit of mass (weight)]
g-cal	gram-calorie (unit of energy, work)
g-cal cm^{-2} min^{-1}	gram-calorie per square centimeter per minute (unit of intensity, power per unit area)
g cm^{-3}	grams per cubic centimeter (metric unit of density)
gal	gallon (English unit of volume)
gpm	gallons per minute (English unit of flow)
gr	grain [English unit of mass (weight)]
H	hydrogen
ha	hectare (metric unit of area)
HCO_3	bicarbonate
hp	horsepower (English unit of energy or work)
h	hour (unit of time)
Hz	Hertz (unit of frequency, cycles per second)
in, in^2, in^3	inch [English unit of length, area or volume (square or cubic inch)]
in Hg	inches mercury (English unit of pressure, referring to pressure at the bottom of a column of mercury so many inches high)
K	potassium

kg	kilogram [metric unit of mass (weight)]
kg cm^{-2}	kilogram per square centimeter (metric unit of pressure)
km, km^2	kilometer (metric unit of length or area)
km h^{-1}	kilometer per hour (metric unit of velocity)
kw, kw-h	kilowatt (metric unit of power, kilowatt-hour)
l	liter (metric unit of volume)
lb	pound [English unit of mass (weight)]
lb ft^{-3}, lb yd^{-3}	pounds per cubic foot or yard (English unit of density)
lx	lux (metric unit of illumination intensity)
ly	langley (power per unit area, radient intensity)
M	molar (unit of concentration, solutions)
m, m^2, m^3, CM	meter [metric unit of length, area or volume (square or cubic meter)]
m, mi, m^2	mile (English unit of length or area)
m sec^{-1}	meter per second (metric unit of velocity)
mb	millibar (metric unit of pressure)
Mg	magnesium
μ	micron (unit of length)
ml	milliliter (metric unit of volume)
min	minute (unit of time)
Mn	manganese
meq l^{-1}, meq/l	milliequivalents per liter (concentration of ions in solution expressed in equivalents per unit volume)
mm, mm^2, mm^3	millimeter [metric unit of length, area or volume (square or cubic millimeter)]
mm Hg	millimeter mercury (metric unit of pressure, referring to pressure represented at the bottom of a column of mercury so many mm high)
mmhos	millimhos (unit of electrical conductivity)
mph	mile per hour (English unit of velocity)
mo	month (unit of time)
mμ	millimicron (unit of length)
N	nitrogen
N	normal (unit of concentration, solutions)
Na	sodium
nm	nanometer (metric unit of length, used in wavelength)
NO_3	nitrate
O	oxygen
oz	ounce [(fluid or avoirdupois), English unit of volume or weight]
P	phosphorous
pH	acidity or alkalinity of the root medium, the negative logarithm of the hydrogen ion concentration
ppb	parts per billion (unit of concentration, by volume or by weight)
ppm	parts per million (unit of concentration, by volume or by weight)
psi, lb in^{-2}	pounds per square inch (English unit of pressure)
qt	quart (English unit of volume)
RH	relative humidity (the ratio of actual concentration of water vapor in air to what could be present at saturation, both conditions at the same temperature, expressed as a percentage)
S	sulfur
SAR	sodium adsorption ratio (refers to capacity of soil to adsorb sodium)
SO_4	sulfate
sec	second (unit of time)
t	ton (metric or English units of weight, are not the same, Time, Temperature)
T	Temperature (Absolute)
tbs	tablespoon (English unit of volume)
tsp	teaspoon (English unit of volume)

w, w-h	watt (metric unit of power, applied mostly to electrical power, watt-hour, energy)
wk	week (unit of time)
yd, yd^2, yd^3	yard [English unit of length, area or volume (square or cubic yard)]
yr	year (unit of time)
Zn	zinc
ψ	potential (refers to energy or pressure in reference to water relationships of plants and soils)
μmhos	micromhos (unit of electrical conductivity)

Appendix C Definitions

Abscission	Generally refers to leaf drop brought about by the rupture of cells in the petiole, allowing the leaf to separate from the plant.
Absolute humidity	The mass of water vapor per unit volume of moist air (grains per cubic foot or milligrams per cubic meter).
Acaracide	A chemical used to control mites and ticks.
Acclimatization	Refers to the adaptation of a plant to a new climate or environment.
Acid soil	A soil with more hydrogen and aluminum ions than hydroxide ions; pH below 7.0.
Acrylic	A glassy thermoplastic that is incorporated in rigid plastics to aid in weatherability.
Active ingredient (a.i.)	The active agent in a chemical or mixture expressed as a percentage.
Adjuvant	A chemical or agent added to a pesticide mixture which helps the active ingredient do a better job.
Adsorption	The attraction of ions or compounds to the surface of a solid. It is not the same as absorption.
Aeration	The process by which air in soil is replaced by air in the atmosphere.
Aggregate	Soil particles held in a single mass or cluster such as a clod, crumb or block.
Air infiltration	The air exhange between the outside and inside of the greenhouse, usually expressed as the number of complete air exchanges per h.
Air-inflated greenhouse	Normally refers to a double plastic film cover firmly attached on all edges to a superstructure. The cover is kept rigid by forcing air between the layers at pressures up to 1.25 lbs/ft^2.
Air porosity	That part of the soil volume filled with air at a given moisture content. In a soil layer 15 cm deep (6 in), those pores drained at a suction of 15 cm water.
Air-supported greenhouse	Usually refers to a type of greenhouse supported wholly by air pressure from 3–5 lbs/ft^2. No internal superstructure required to support the roof and they are normally cylindrical or quonset-shaped.
Alkali soil	A soil with a pH of 8.5 or higher, with sodium comprising 15% or more of the total exchange capacity.
Alkaline soil	Any soil with a pH greater than 7.0, with more hydroxide ions than hydrogen ions.
Amendment	Any substance added to a soil to alter its physical properties, generally to make it more productive.
Ammonification	Process by which ammonium nitrogen (NH_3) is released from nitrogen-containing organic compounds.

Anion	A negatively charged ion (e.g., NO_3, SO_4^{2-}, Cl^-).
Anion exchange capacity	The total of exhangeable anions that a soil can adsorb.
Anionic surfactant	A surface-active additive to a pesticide, having a negative surface charge. Most wetting agents are of this type, performing better in cold, soft water.
Anthocyanin	A chemical pigment in plants, usually the red color in fruits and flowers. There are also blue anthocyanins.
Anthracnose	Any of various fungal diseases of plants in which roundish, dead spots appear chiefly on leaves or fruit.
Antidote	An immediate treatment to counteract the effects of a poison.
Aspirate	To force air over or through a device—such as an aspirated thermostat shelter.
Asset	Any resource that has value, including cash, inventory, notes receivable, equipment, real estate, etc.
Assimilate	As a noun: A substance taken up by a plant. As a verb: To absorb or incorporate.
Auxin	A class of substances, occurring naturally in plants, and regulating growth of the plant. Effective in minute amounts. Examples are indolebutryic acid (IBA), indoleacetic acid (IAA), and napthalenacetic acid (NAA).
Available water	The part of the water in the soil which can be readily absorbed by plant roots. In most greenhouse soils, generally that amount retained between suctions at bench or pot capacity and approximately 300–500 cm water.
Bactericide	A chemical used to control bacteria.
Base-saturation percentage	Percentage of the total cation exchange capacity of the soil which is saturated with cations other than hydrogen or aluminum.
Beam	Any support that resists loads that try to bend it. Examples are joists, purlins, rungs in a ladder.
Bench capacity (BC)	The amount of water retained against gravity by a shallow, freely draining soil in a bench. Soil moisture suction in the upper soil surface is equal to the depth of the soil expressed as cm water, and suction in the bottom layer equals zero.
Bioassay	The determination of the biological activity of a substance by testing its effect on the growth of an organism.
Biological control	Control of pests by means of predators, parasites and disease producing organisms.
Blight	Any disease that causes plants to wither and decay. Can be applied to any condition that kills, withers or checks growth of plants.
Box coil	Two or more heating lines stacked vertically, and into which the heating medium (steam or hot water) is introduced simultaneously to all lines in the stack.
Breaking	Discolored streaks or blotches in a normal, solid color of a flower, usually caused by a virus.
Buffering	The resistance of a soil or solution to changes in acidity or nutrient composition.
Bulk density	The weight of a dry soil per cubic centimeter volume. May be expressed also as pounds per cubic yard.

Calcareous soil	A soil containing sufficient calcium carbonate to visibly effervesce when treated with hydrochloric acid.
Cambium	The layer of meristematic tissue between the phloem and the xylem, responsible for annual growth rings in trees and for the increase in diameter of trees.
Candela (candle)	A unit of luminous intensity.
Canker	A localized lesion on a stem which usually results in the corrosion and sloughing away of tissue with the final production of an open wound.
Capillary conductivity	A measure of the readiness with which the soil transmits water.
Capital	Wealth (money or property) owned or used in a business by a person or corporation. Wealth in whatever form used to generate more wealth.
Capital assets	Any resource of value usually held for long-term gain.
Capitalization	The total invested in a business by the owner(s). Or, the total assets available to the business. To "undercapitalize" is to operate more on borrowed capital, or to have insufficient funds to operate the business properly.
Carbon-nitrogen ratio	The ratio of organic carbon to nitrogen in the soil.
Cation exchange	The interchange between a cation in solution and another cation on the surface of a soil particle such as clay or humus.
Cation exchange capacity	The total of exchangeable cations a soil can adsorb.
Chelate	A type of chemical compound in which a metallic ion (e.g. iron) is firmly combined with a molecule by means of several chemical bonds.
Chlorosis	Failure of the chlorophyll (green) to develop. Chlorotic plants range from light green to almost white. Generally occurs between the veins on leaves.
Clay	A soil particle smaller than 0.002 mm diameter.
Clone	A group of organisms (plants) obtained from a single individual by various means of asexual reproduction.
Colloid	Soil particles having a small diameter in proportion to surface area (e.g. clay or humus).
Column	Any support that resists loads that try to shorten it. Examples are studs or posts holding up a roof, chair legs, and side rails of a ladder.
Commercial fertilizer	A fertilizer containing impure salts of the major macronutrients.
Compatible	Two or more compounds capable of being mixed together without affecting the other's properties.
Complete fertilizer	A fertilizer containing all three of the major fertilizer elements N, P, and K, the concentration expressed as percent N, P_2O_5, and K_2O.
Condensate	Condensed steam. Sometimes refers to the water condensed from the air on a cold greenhouse roof.
Conductivity	The ability of a soil to conduct water as the result of a pressure (potential) gradient.

Constant feeding	The term applied to the application of dissolved fertilizers in the irrigation water at each watering.
Corrective maintenance	Maintenance performed as the need is discovered.
Cortex	The tissues between the epidermis and vascular tissue of a stem or root.
Cultivar	A variety of a plant (species) that has been produced only under cultivation.
Cytokinin	A substance, active in minute amounts, which affects various plant physiological processes.
Cytoplasmic	Pertaining to the protoplasm of a cell, exclusive of the nucleus.
Deficiency symptoms	The appearance of plants indicating that one or more essential elements are lacking.
Deflocculate	To separate the individual particles of a soil—to break down the soil structure, or to "puddle" it. Loss of aggregation.
Dentrification	The chemical reduction of nitrate or nitrite to gaseous nitrogen.
Dermal toxicity	The passage of pesticides through the skin with toxic effects on the organism.
Desiccation	Drying up.
Detergent	A chemical (not soap) having the ability to remove soil and grime.
Dew point (T_d)	The temperature at which any further cooling will cause water to condense from the air.
Discrete	Separate or distinct.
Dormancy	Interrupted growth. A resting period during which no growth will occur even if the environment is suitable. May require special treatment (e.g. low temperatures, short days) before growth will resume. A method of overcoming adverse environmental conditions.
Double poly-air inflated	Refers to a type of greenhouse covering where two polyethylene layers are laid over the superstructure, fastened at the edges, and air forced between them to separate the cover.
Dry bulb temperature (T_a)	Actual air temperature as measured by an unwetted device or thermometer bulb.
Dry feeding	The application of fertilizers in a dry form to the soil surface, or mixed into the soil prior to planting.
Eave	The connection between the sidewall and roof.
Electrical conductivity	The reciprocal of electrical resistance, expressed as "mhos". A method for expressing salt concentration in a soil or water.
Emulsifiable concentrate (EC)	A material produced by dissolving the pesticide and an emulsifying agent in an organic solvent.
Equity	Value of a property beyond the total amount owed on it in terms of mortgages, liens, etc. The amount of capital invested by the owner, or what would be due him after all debts were paid.
Erg	Basic unit of energy in the centimeter-gram-second system.
Etiolated	Blanched or yellow as the result of insufficient light. Lack of chlorophyll which requires light to develop.

Evaporative-pad	Major part of a greenhouse cooling system, through which air is drawn by the exhaust fans. Heat is extracted from the air to evaporate water in the pad, thereby lowering the air temperature.
Evapotranspiration	The combined water loss due to transpiration from plants and evaporation from soil surfaces.
Exchangeable sodium percentage	The extent to which the exchange capacity of a soil is occupied by sodium.
F_1 hybrid	Plants resulting from a cross of two pure inbred lines. The first generation from such a cross.
Fan-jet	Special tube pressurizing fan, used to force air through a polyethylene tube in the greenhouse, distributing air throughout its length.
Fasciated	Abnormal growth. Growth of several buds that fuse laterally in their formative stage.
Fertilizer	Any organic or inorganic material of natural or synthetic origin which is added to the soil to supply essential elements for plant growth.
Fiber bloom	The exposure of glass fibers on the surface of fiberglass reinforced panels due to the deterioration of the polyester resin in which they are imbedded.
Fiber-reinforced plastic (FRP)	A greenhouse covering composed of glass fibers embedded in a plastic resin.
Field capacity (FC)	The amount of water retained in a field soil against the force of gravity about 24–36 h after flooding. Suction for deep soils will usually range between 300–500 cm water at field capacity.
Fire rated panel	A fiberglass reinforced plastic panel that is manufactured with specific additives, resulting in a flame spread index (I_s) of less than 35. Standard panels have ratings approaching 100 or more. For comparison, the flame spread ratings (I_s) for gypsum-board is 5–20 and exterior grade Douglas fir plywood 120–200.
Fixed asset	A resource not readily convertible to cash, such as real estate, structures, etc.
Flaccid	Flabby. Lacking firmness.
Fleck	Translucent lesion visible through the leaf.
Flue	A tube, pipe or shaft for the passage of smoke, hot air or gas as in a chimney or boiler.
Foliar feeding	The application of dissolved fertilizers to the leaves of plants, the fertilizer being absorbed through the epidermis.
Foot-candle (ft-c)	Illumination produced on a surface, all points of which are at a distance of one foot from a uniform point source of one candela—or illumination per square foot.
Frequency	The number of waves of electromagnetic radiation passing through a given point in an interval of time (waves per second).
Fungicide	A chemical used to control fungi.
Gable	The triangular end of the greenhouse, bounded by the roof on two sides and the end wall on the bottom.
Gall	A tumor on plant tissue caused by the stimulation of fungi, bacteria, or insects.

Genotype	The genetic constitution of an organism or group of organisms.
Gibberellin	A naturally occurring growth regulator. Promotes cell elongation and often used to enhance flower initiation in some plants.
Gram-equivalent weight	The weight in grams of the equivalent weight of the material, or molecular weight divided by the valence (e.g. K = 39.1, Ca = 20).
Gram-formula weight	The weight in grams equal to the formula weight of the material (e.g. KNO_3 = 101.1).
Gram-molecular weight	The weight in grams, equal to the molecular weight of the material, can be considered synonymous with gram-formula weight.
Green manure	Plant material incorporated into the soil while green.
Gross sales	All income from selling the products of the greenhouse before any deductions are made for costs in producing the product.
Gutter post	Main, vertical, supporting member located under the gutter, usually in a ridge and furrow greenhouse range, supporting gutter and roof.
Gynoecious	A plant having only the female parts of the flower. Lacking male sex organs.
Gypsum block	An electrical device employed to measure the matric potential of soil water.
Hard water	Water containing the ions calcium and magnesium, usually in conjunction with bicarbonate or sulfate.
Herbicide	A chemical used to erradicate unwanted plants or weeds.
Hermaphrodite	An organism in which the reproductive organs of both sexes are present.
Humus	The most stable portion of soil organic matter, remaining after decomposition of plant and animal residues.
Hydrophobic	A term applied to materials which resist wetting.
Hyphae	The thread-like elements of the fungus (mycelium).
Illumination	The process by which a surface is illuminated or lighted.
Infect	Penetration of a host plant by the pathogen, eventually resulting in an expression of a disease in the host.
Infest	The act of placing a pathogen in a location where it may induce disease in the host plant.
Infiltration	The downward entry of water into the soil.
Inhibitor	A growth regulator which prohibits growth.
Injector	(see proportioner)
Inoculum	The quantity of bacteria or fungal spores capable of causing disease on coming into contact with a susceptible host.
Insecticide	A chemical used to control insects.
Insolation	Solar radiation received on a given body or over a given area.
Instar	Any of various stages in the life cycle of an insect. Between molts.
Ions	Atoms, groups of atoms or compounds which are electrically charged (e.g. anions or cations).
Irradiance	The process by which a surface is radiated—may be any radiation including visible.

In this book, a chemical substance applied to plants to change their growth characteristics or habits—affects various physiological processes of plants.

lity The ratio between actual vapor content of air and the amount of vapor the air could hold at saturation at the same temperature—expressed as a percentage.

A growth regulator which slows down or reduces growth such as elongation or leaf expansion. Growth does not cease.

The highest part of the roof of a greenhouse, usually forming a major structural component of the greenhouse.

ow A type of greenhouse construction where individual houses are combined at the gutters, usually to form one open area under the entire roof.

The degree of slant or angle of divergence from the horizontal. A standard pitch of a greenhouse roof is considered 26.50°. This is a rise of six feet for every 12 feet across the width of a house.

A soil containing sufficient soluble salts to reduce productivity.

The amount of soluble salts in the soil, expressed in terms of ppm, electrical conductivity (EC) or equivalent terms.

Chemicals which are found in soils and dissolve in water. Generally found as cations (calcium, sodium, potassium, etc.) and anions (sulfates, nitrates, bicarbonates, etc.).

One or more panes of glass supported to form a cover such as a cold frame or panel in a greenhouse roof. Generally capable of being moved such as a ventilator sash.

A structural member supporting a sash, particularly applicable to the main supporting members of a glass-covered greenhouse, holding the glass panes.

y The inability of a plant to fertilize itself even though both sexes are present. Fertilization may occur, but the genetic result may be lethal. Or, fertilization is inhibited at some point in the process.

Generally refers to the loss of keeping life of cut flowers. The gradual progression toward maturity and ultimate death.

Generally taken to mean those wavelengths shorter than 5000 nm. Sometimes referred to as "visible" radiation.

A louvered device to keep out wind, rain, and backdrafts. May be operated by air pressure or mechanically.

A soil particle 0.05–0.002 mm in diameter.

n A greenhouse cover consisting of a single layer of plastic film.

tion The ratio between sodium and calcium plus magnesium.

Water containing high amounts of sodium.

(see soil test).

The liquid in the soil and the materials dissolved in it.

The chemical analysis of a soil to determine acidity, total salt concentration and concentration of major elements.

A curve depicting the relationship between soil water potential and the soil's water content.

Juvenility	In reference to plants, a period in early stages of growth during which flowering is difficult if not impossible to induce.
Latent	Refers to buds or growth which are dormant or undeveloped, but have the potential of starting to grow.
Leaching	The process of removing soluble material from a soil by passage of water through the soil.
Lesion	A diseased or injured region.
Lethal dose (LD)	Method for expressing toxicity of a compound, usually as the milligrams of chemical per kilogram of body weight of the test animal required to kill a percentage of the test population (e.g. LD_{50}, LD_{25}, etc.).
Liability	Debt of an individual or business. Liabilities may include notes payable, mortgages, leins, etc.
Liming	The addition of calcium, sometimes with magnesium, to the soil in the form of calcium carbonate, hydrated lime or limestone.
Liquid asset	A resource readily convertible to cash.
Loads Dead load	The weights of the permanent structural material.
Loads Live load	Additional fixtures, such as steam and water pipe, unit heaters, and in many cases, hanging baskets. Does not include snow, wind, earthquake, or dead loads.
Loads Snow load	Vertical loads applied to the horizontal projection of the building roof.
Loads Wind load	Loads caused by wind blowing from any horizontal direction. Tornado winds are excluded.
Loam	Textural class name for a soil having 7 to 25% clay, 28 to 50% silt and less than 52% sand.
Long-wave radiation	Sometimes called "thermal radiation". Generally taken to be radiation at wavelengths longer than 5000 nm.
Lumen	The luminous power emitted by a point source of one candela.
Luminous energy	Visible radiation.
Lux	Illumination on a surface of one square meter at a distance of one meter from a uniform source of one candela.
Macronutrient	A chemical element necessary in large amounts for plant growth, usually supplied artificially in a fertilizer.
Major elements	(see macronutrients).
Male sterility	Refers to plants in which the flowers lack male parts, or there are no stamens present. Important in breeding as the plant cannot be self fertilized.
Matric potential	Attraction of water by colloidal material in the plant or colloids in the soil.
Metamorphosis	A change in form, structure, shape, or substance. More specifically, the physical transformation of an insect while going from one stage to another.
Mhos	A unit of electrical conductivity.
Micronutrient	A chemical element necessary in very small amounts for plant growth.

Milky FRP panels — White blotches that occur in panels that have been improperly cured in the manufacturing process or stored outside where moisture can get between the sheets and solar energy causes increased temperatures.

Mineral soil — A soil consisting largely of mineral matter, with organic matter usually less that 20%.

Miscible — Capable of being mixed.

Miticide — A chemical used to control mites.

Molal solution — A one molal solution contains one molecular weight of the material in 1000 g of water (e.g. 101.1 g KNO_3 in 1000 g water).

Molar solution — One liter of a molar solution contains one molecular weight of the material dissolved in enough water to make one liter.

Molecular weight — The sum of the atomic weights of all the atoms in the molecule.

Monoecious — A plant having both male and female organs in the same individual—hermaphroditic.

Mulch — Any material spread upon the soil surface to protect soil and plant roots from effects of freezing, crusting, evaporation, etc.

Mutant — The result of a sudden departure from the parent type, generally caused by a change in a gene or chromosome.

Necrosis — Localized death of living tissue.

Net profit — Return to the greenhouse owner after all expenses involved in producing a product have been deducted. Usually determined before taxes.

Nitrification — The conversion of ammonia to nitrate.

Normal solution — One liter of solution containing one equivalent weight of a material. Or, the molecular weight of the material divided by its valence and dissolved in enough water to make one liter.

Organic soil — A soil usually containing 20% or more organic matter.

Osmotic potential (Ψ_0) — Applicable to both soil or plant. The pressure that would have to be applied to a solution to prevent water from moving from a less concentrated solution, when the two solutions are separated by a semipermeable membrane. In soil, Ψ_0 refers to the soil solution. In plants, the term is in reference to the concentration of the cell sap.

Overhead — Term applied to the continuing expenses involved in running a business such as rent, maintenance, utilities, taxes, etc. Not directly related to expenses in growing and marketing a crop.

Parthenocarpic — Refers to an organism which is capable of developing an egg without fertilization.

Partial pressure — The pressure exerted by an individual gas in a mixture of gases (e.g. oxygen in air).

Pathogen — An organism capable of inducing disease in a host plant. The casual agent.

Pedicel or peduncle — A small stalk bearing the flower or flowers of a plant.

Penetrant — Wetting agents wh the surface.

Perimeter heat — Heating system lo hung on the sidew

pH — Expression of the alkalinity of the sc

Phenotype — The observable cc sulting from the in

Phloem — The food conducti

Photo degradation — Radiation in the fo decomposition of be present for long

Photometer — An instrument for visible wavelength

Physiological — Physiology deals w physiological pert

Plant analysis — (see tissue analysis

Plant pressure potential — Pressure in a plan (turgor pressure).

Plant water potential — The pressure of w matric, and pressu in sign, to diffusi "stress", usually ne

Porosity — The volume of a sc

Pot capacity (PC) — The amount of wa draining soil in a p

Preventative maintenance — A planned progra and facilities.

Proportioner — A device or system cally injected into

Psychrometry — A method of deter the temperature d thermometers.

Purlin — A horizontal part the roof, located b

Pyranometer — An instrument for from the whole a reflected radiation

Pyrheliometer — An instrument for (sunlight) perpendi

Radiant intensity — The rate of flow, p radiant energy.

Radiation — The transfer of en photons—may be

Radiometer — An instrument for

Reentry interval — The length of time the field to conduc

Reflectance — The ratio of radia to that incident up

Regulator

Relative humi (RH)

Retardent

Ridge

Ridge and furr

Roof pitch

Saline soil

Salinity

Salts

Sash

Sash bar

Self-incompatabilit

Senescence

Short-wave radiation

Shutter

Silt

Single layer fil

Sodium adsor ratio

Soft water

Soil analysis

Soil solution

Soil test

Soil water characteristic curve

Soil water potential (soil moisture stress)	The equivalent negative pressure or suction of water in the soil. It is the sum of osmotic, matric and gravitational potentials. The largest term is usually matric potential. May be expressed in any convenient pressure units.
Solar radiation	Radiation emitted by the sun—sunlight.
Solu-bridge	An instrument for determining salinity in a soil extract by measuring electrical conductivity.
Spreader	A substance which increases the area that a given volume of liquid will cover.
Staminate	A male flower—having stamens but no pistils or female parts.
Starring	A crazing in fiberglass reinforced plastic panels created by hail.
Steam trap	A device that allows condensate water to return to the boiler, but prevents passage of steam from the heating coil into the condensate return.
Sticker	A material added to a pesticide to increase the tenacity with which it remains on the surface.
Sump	A pit or tank, located at the lowest point in a drainage system to collect water. May be a tank for return condensate, which is pumped back into the boiler.
Superphosphate	A nonsoluble fertilizer containing phosphorous.
Surfactant	A material which reduces the surface tension between two unlike materials such as oil and water. A spreader or wetting agent.
Systemic	A term used to describe pesticides or diseases that enter and are distributed, or act, within the plant.
Tensiometer	A pressure device used to determine the soil moisture suction or matric potential of the soil water.
Terrestrial radiation	The thermal radiation of the earth and atmosphere. Generally long-wave radiation.
Thermodegradation	High temperatures contributing to the breakdown or depolymerization of plastic material.
Threshold proportion	The ratio between a soil and its amendment when air porosity is minimum.
Tissue analysis	The chemical analysis of plant tissue to determine concentrations of essential elements in the plant.
Tolerance	The maximum amount of pesticide that may remain on the raw agricultural product.
Total salts	(see salinity).
Trace element	(see micronutrient).
Translucent	So clear as to permit passage of light.
Transmittance	The ratio of radiant or luminous energy transmitted by a body to that incident upon the body.
Transpiration	The process of water loss from a plant, generally occurring through the stomates, and a direct function of available radiant energy and internal plant water potential.
Trombone coil	A type of heating method, employing pipe, in which the medium (steam or hot water) circulates in a series of loops, generally twice the length of the greenhouse, before returning to the boiler.

Truss	Main supporting structure of the greenhouse roof. May have any number of designs, commonly supported at the ends by the gutter or sidewall posts. Can be of steel, wood, pipe or aluminum.
Unit heater	A heating device with its own fan. Heat may be supplied from natural gas, oil, steam or hot water.
Valence	The combining capacity of atoms or groups of atoms (e.g. sodium, potassium, nitrate, etc. are monovalent—one electrical charge; sulfate, calcium, and magnesium are divalent—2 electrical charges).
Vapor pressure (e_s)	The partial pressure of water, as a gas, in a gaseous mixture such as air.
Wavelength	The linear distance between two similar points on adjacent waves (centimeters per wave).
Wave number	The number of waves per unit distance (per centimeter).
Wet bulb temperature (T_w)	Temperature of an aspirated thermometer bulb encased by a wick saturated with water. The equilibrium temperature reached by an evaporating liquid surface.
Wettable powder (WP)	The type of chemical that can be mixed with water.
Wetting agent	A compound which reduces surface tension and causes a liquid to contact plant surfaces more thoroughly.
Xylem	The main water conducting tissue of plants, on the inner side of the cambium in woody plants, eventually becomes the wood.

Subject Index

Advanced Series in Agricultural Sciences

Co-ordinating Editor: B. Yaron
Editors: G. W. Thomas, B. R. Sabey, Y. Vaadia, L. D. Van Vleck

Volume 1
A.P.A. Vink

Land Use in Advancing Agriculture

94 figures, 115 tables. X, 394 pages. 1975
ISBN 3-540-07091-5
Distribution rights for India:
Allied Publishers Private Ltd., New Delhi

Contents: Land Use Surveys. – Land Utilization Types. – Land Resources. – Landscape Ecology and Land Conditions. – Land Evaluation. – Development of Land Use in Advancing Agriculture.

Volume 2
H. Wheeler

Plant Pathogenesis

19 figures, 5 tables. X, 106 pages. 1975
ISBN 3-540-07358-2
Distribution rights for India:
Allied Publishers Private Ltd., New Delhi

Contents: Concepts and Definitions. – Mechanisms of Pathogenesis. – Responses of Plants to Pathogens. – Disease-Resistance Mechanisms. – Genetics of Pathogenesis. – Nature of the Physiological Syndrome.

Volume 3
R.A. Robinson

Plant Pathosystems

15 figures, 2 tables. X, 184 pages. 1976
ISBN 3-540-07712-X
Distribution rights for India:
Allied Publishers Private Ltd., New Delhi

Contents: Systems. – Plant Pathosystems. – Vertical Pathosystem Analysis. – Vertical Pathosystem Management. – Horizontal Pathosystem Analysis. – Horizontal Pathosystem Management. – Polyphyletic Pathosystems. – Crop Vulnerability. – Conclusions. – Terminology.

Volume 4
H.C. Coppel, J.W. Mertins

Biological Insect Pest Suppression

46 figures, 1 table, 314 pages. 1977
ISBN 3-540-07931-9
Distribution rights for India:
Allied Publishers Private Ltd., New Delhi

Contents: Glossary. – Historical, Theoretical, and Philosophical Bases of Biological Insect Suppression. – Organisms Used in Classical Biological Insect Pest Suppression. – Manipulation of the Biological Environment for Insect Pest Suppression. – A Fusion of Ideas.

Springer-Verlag
Berlin
Heidelberg
New York

Ecological Studies

Analysis and Synthesis
Editors: W.D. Billings, F. Golley, O.L. Lange, J.S. Olson

Volume 4

Physical Aspects of Soil Water and Salts in Ecosystems

Editors: A. Hadas, D. Swartzendruber, P.E. Rijtema, M. Fuchs, B. Yaron
221 figures, 61 tables. XVI, 460 pages. 1973
ISBN 3-540-06109-6

Contents: Water Status and Flow in Soils: Water Movement in Soils. Energy of Soil Water and Soil-Water Interactions. – Evapotranspiration and Crop-Water Requirements: Evaporation from Soils and Plants. Crop-Water Requirements. – Salinity Control.

Volume 6
K. STERN, L. ROCHE

Genetics of Forest Ecosystems

70 figures. X, 330 pages. 1974
ISBN 3-540-06095-2

Contents: The Ecological Niche. – Adaptations. – Genetic Systems. – Adaptive Strategies. – Forest Ecosystems. – How Man Affects Forest Ecosystems.

Volume 8

Phenology and Seasonality Modeling

Editor: H. Lieth
120 figures. XV, 444 pages. 1974
ISBN 3-540-06524-5

Contents: Introduction to Phenology and the Modeling of Seasonality. – Methods for Phenological Studies. – Seasonality in Trophic Levels. – Representative Biome Studies. – Modeling Phenology and Seasonality. – Applications of Phenology.

Volume 9
B. SLAVÍK

Methods of Studying Plant Water Relations

With contributions by B. Slavík, J. Catský, J. Solárová, H.R. Oppenheimer, J. Hrbáček, J. Slavíková, V. Kozinka, J. Úlehla, P.G. Jarvis, M.S. Jarvis
181 figures. XVIII, 449 pages. 1974
ISBN 3-540-06686-1
Distribution rights for the Socialist countries: Academia Publishing House of the Czechoslovak Academy of Sciences, Prague

Contents: Water in Cells and Tissues. – Water Content. – Water Exchange between Plant Roots and Soil. – Liquid Water Movement in Plants. – Water Exchange between Plant and Atmosphere. – Table Appendix.

Volume 13

Epidemics of Plant Diseases

Mathematical Analysis and Modeling

Editor: J. Kranz
46 figures. X, 170 pages. 1974
ISBN 3-540-06896-1

Contents: The Role and Scope of Mathematical Analysis and Modeling in Epidemiology. – Automatic Data Processing in Analysis of Epidemics. – Multiple Regression Analysis in the Epidemiology of Plant Diseases. – Nonlinear Disease Progress Curves. – Simulation of Epidemics.

Volume 19

Water and Plant Life

Problems and Modern Approaches
Editors: O.L. Lange, L. Kappen, E.-D. Schulze
178 figures, 66 tables. XX, 536 pages. 1976
ISBN 3-540-07838-X

Contents: Fundamentals of Plant Water Relations. – Water Uptake and Soil Water Relations. – Transpiration and its Regulation. – Direct and Indirect Water Stress. – Water Relations and CO_2 Fixation Types. – Water Relations and Productivity. – Water and Vegetation Patterns.

Springer-Verlag Berlin Heidelberg New York